Taxonomy

Taxonomy

A TEXT AND REFERENCE BOOK

RICHARD E. BLACKWELDER

Department of Zoology
Southern Illinois University
Carbondale, Illinois

JOHN WILEY & SONS, INC.

New York · London · Sydney

TO

Waldo L. Schmitt

Who has labored for sixty years in the
scientific service of his government

Who has served taxonomy actively for
at least fifty years

Who has supported at critical periods
most of the organizations of taxonomy,
whether in his specialty or not

Who has shown a keen understanding of
the implications of taxonomy and its
interactions with other sciences

Who has inspired taxonomists by his
friendship, enthusiasm, and support

Who founded and has largely sustained
the Society of Systematic Zoology

Preface

The number of books designed to aid in the teaching of taxonomy is very small, and books of any sort on taxonomy are by no means abundant. In fact, in the last quarter century there have been no more than a dozen that attempted to deal with any considerable part of the subject.

The taxonomic competence of biologists confronted with taxonomic problems of one kind or another is extremely varied. I have therefore attempted to include the entire range of taxonomic activities in an arrangement that will allow the student to go as far into any aspect as his needs require and his background permits. This does not mean that all aspects of taxonomy are fully covered. Where these are dealt with in other books in detail, it is believed to be better to refer the student to these books than to duplicate a large amount of discussion.

In classes at all levels, whether in general zoology or in an advanced specialty, I make it a practice to reiterate to students that in all books they read and in all lectures they hear there is constant reflection of the ideas, beliefs, and background of each writer or lecturer. They must recognize that his choice of words and the meaning intended will almost certainly be different from those of other writers and speakers. The implications of facts, as well as their selection, the interpretation of correlations and of theories, the organization of thoughts, the selection of supporting arguments, even the entire approach to the field— all these will be diverse.

The same philosophy applies to the present book. There is an attempt to present much factual material from as many viewpoints as practicable; differing interpretations are often discussed in detail; and my conclusions as to the most satisfactory solutions are presented. However, it is not intended that these be taken as final answers to the innumerable questions that can be raised in systematics. Almost every teacher will have different conclusions, which he will attempt to justify to his class; every teacher will see exceptions not noted here; every

vii

teacher should go into some aspects in greater detail. In short, each user should accept this book as a statement of the views of *one* taxonomist, bolstered by what he believes are the views of others, as quoted or referred to.

The relevance of any given part of the text will depend in large measure on the nature of the course. In vertebrate-oriented courses it will seem that there are sections about subjects that are no longer serious problems in vertebrate taxonomy, such as the description of new genera and species, types of both genera and species, and classification of groups. In courses heavily slanted toward evolutionary aspects of systematics, this approach will be found inadequately covered, and some argumentation may be rejected. In paleontological courses the stratigraphic and temporal aspects will doubtless require supplementing. And in courses primarily used to serve ecological interests, the instructor will want to emphasize and illustrate the taxonomic implications in this field. All this is good. A textbook should never be used as a source of all the final answers but rather as a point of reference for developing or at least understanding other possible views and interpreting other ideas, even contradictory ones.

This book is arranged in six Parts: (I) introduction, (II) the elements of taxonomic practice, (III) the animal diversity that is the basis and the major problem of classification, (IV) the advanced practice of classification, (V) the theoretical aspects of systematics, and (VI) the special requirements of zoological nomenclature. It is specifically planned to be used as a textbook in two courses: a beginning course on the nature and practice of taxonomy and an advanced one on the theory and technicalities of taxonomy. In the first of these the student would use about half of the book, leaving the remainder for reference and extra reading. In the advanced course the first half would be used for review. The student would begin with a study of classification and proceed to consideration of the theoretical aspects and the endless problems of nomenclature, which reflect almost as many problems of taxonomic nature. The study of the diversity serves as a background for both courses, inasmuch as it is the basis for all comparative studies, upon which all taxonomy or systematics rests. To discuss all aspects of animals that are relevant to taxonomy and classification would be to discuss all that is known of all animals, so these sections are no more than illustrative.

The book includes some discussion of the philosophical nature and basis of taxonomy and the controversies about these, but in the main it is concerned with the practical tasks of the taxonomist and will be relevant regardless of the views of the user as to the purpose, basis, or implications of taxonomy and its results. It is a book *about* taxonomy *for* taxonomists.

R. E. *Blackwelder*

CARBONDALE, ILL.
NOVEMBER, 1966

Acknowledgments

Aside from the many who have contributed ideas and stimulation for this book, often without being aware of it, including those whose books are often referred to herein, there are three persons who have taken an active part in its preparation and without whose help it would certainly have had many more faults than now. I take pleasure in thanking them individually here.

Waldo L. Schmitt, of the U.S. National Museum, retired, has done more to encourage present-day taxonomists than any other person. He has inspired taxonomists to take pride in their field; he has contributed largely to the understanding of taxonomy in other fields of science; and he has aided others in the search for support of research, in the protection of valuable collections, and in obtaining study facilities for taxonomists. He has been an enthusiastic supporter of this book, and a friendly critic of the manuscript. Although he was not directly responsible for its inception, his interest over the years in academic, professional, and practical taxonomy was one of the prime reasons for my undertaking the task. His detailed review of the five non-nomenclatural parts has resulted in numerous improvements.

John C. Downey, of Southern Illinois University, has been the indispensable critic of the book. Nothing aids an author so much as constructive criticism, and few things are so difficult to obtain in acceptable form. With an outlook on systematics more receptive to the evolutionary influences, with research interest in the evolutionary trends and processes related to taxonomic problems, and with experience in population studies, numerical methods, and computer techniques, Dr. Downey has tempered my statements and strengthened the arguments designed to separate these things from taxonomy. Because of him the book has fewer faults. Those that remain are doubtless in paragraphs which through some circumstance he did not see.

William I. Follett, of the California Academy of Sciences, is one of the most respected taxonomists among those who have undertaken the difficult study of zoological nomenclature. He is one of the best examples of the modern nonprofessional taxonomist, as his formal training was in law rather than biology. That this "deficiency" has been corrected is evident in the high regard placed on his opinion, both in the taxonomy of fishes and in problems of nomenclature in general.

The present book contains a strong critique of the latest code of nomenclature. So many points had to be criticized that there was real fear that the result would look like rejection of the rules *in toto*. Mr. Follett's agreement on all the major points makes me confident that these apparently harsh judgments are appropriate and necessary. His review was limited to Part VI, for which many useful suggestions were offered.

Grateful acknowledgments are made to Dr. R. C. Moore, the University of Kansas Press, the Geological Society of America, and the McGraw-Hill Book Company for permission to use the figures of conodonts; also to Longmans, Green and Company, Ltd., London, for use of a quotation from Roget's *Thesaurus*. A long quotation on parataxa is from an unpublished document by Drs. R. C. Moore and P. C. Sylvester-Bradley, whose permissions are gratefully acknowledged. The article on "Applied Systematics" by Dr. W. L. Schmitt is from the Smithsonian Annual Report for 1953.

R. E. B.

Contents

PART I   **Introduction**   **1**

1   Definitions and perspectives   3
2   The place and importance of taxonomy   17
3   Academic and professional taxonomy   38

PART II   **Practical Taxonomy**   **49**

4   The use of classification   51
5   Identification   60
6   Curating   76
7   Recording the data   84

PART III   **The Diversity to be Classified**   **103**

8   The diversity of individuals   105
9   The diversity of kinds and groups   121

PART IV   **Classification, Naming, Description**   **133**

10   Comparative data   137
11   Species and subspecies   162
12   The practice of classification   182
13   The use of names   210
14   The use of literature   239
15   Descriptive taxonomy   279
16   The publication of data   308

PART V   **Theoretical Taxonomy**   **329**

17   Taxonomy as a science   331
18   The nature of classification and of species   351

PART VI   **Zoological Nomenclature**   **373**

19   Rules of nomenclature   379
20   The nature of scientific names   407
21   Names of taxa in the higher categories   435
22   Names of the family-group taxa   449
23   The names of genera   474
24   The names of subgenera   497
25   Genotypy and the types of genera   502
26   The names of species   535
27   The names of subspecies   568
28   Epithets   577
29   The types of species and subspecies   589

      Bibliography   **605**

      Index   **663**

Taxonomy

PART I

Introduction

"Taxonomy is at the same time the most elementary and the
most inclusive part of zoology, most elementary because animals cannot be discussed or treated in a scientific way until
some taxonomy has been achieved, and the most inclusive
because taxonomy in its various guises and branches eventually gathers together, utilizes, summarizes, and implements
everything that is known about animals, whether morphological, physiological, psychological, or ecological." (Simpson,
1945)

Chapter 1. Definitions and Perspectives   3
    Definitions   3
        Taxonomy   3
        Classification   3
        Systematics   3
    The goals of taxonomy   5
    History of taxonomy   6
        Chronology   6
        Taxonomic thinking   8
        Linnaeus   10
        Taxonomic history of groups   11
            Birds   11
            Insects   12
            Nematodes   13
            Pogonophorans   14
            Comment   14
        Taxonomic reference works   14
    Summary   16

Chapter 2. The Place and Importance of Taxonomy   17
    What taxonomists do   18
    The employment of taxonomy   19
    "Applied Systematics"   24
    Summary   37

Chapter 3. Academic and Professional Taxonomy   38
    Taxonomists   38
    Training of taxonomists   39
    Academic taxonomy   40
        Teachers   42
    Professional taxonomy   42
    The organizations of taxonomy   43
    Summary   48

CHAPTER 1

Definitions and Perspectives

Taxonomy has been defined as the study of the principles and practices of classification. If we read into these words a broad meaning, the statement represents well the subject matter of this book. Although the theory of classification is discussed in the advanced sections, the practice of taxonomy is the main subject matter of many of the chapters. What is done, how it is done, and why it is done—these are the things to be studied if one is to obtain a working knowledge of taxonomy or even an understanding of its contributions and potentialities.

Definitions

In order to establish the perspective of this book and give a basis for further discussion, there are here discussed the three principal words sometimes used to signify the field of the organizing of information about kinds of animals, with definitions to show their intended meanings herein.

Taxonomy, as the term is employed in this book, refers to the day-to-day practice of dealing with the *kinds* of organisms. This includes the handling and identification of specimens, the publication of the data, the study of the literature, and the analysis of the variation shown by the specimens.

Classification is the arrangement of the individuals into groups and the groups into a system (also called a classification) in which the data about the kinds determine their position in the system and thereafter are reflected by that position.

Systematics. Both taxonomy and classification, and all the other aspects of dealing with kinds of organisms and the data accumulated about them, are included in systematics, which is the general term that covers all aspects of the study of kinds. Therefore, systematics is the study of the kinds and diversity of organisms, their distinction, classification, and evolution.

3

These words are not universally used in these meanings. The dictionaries define them all as the science and practice of classifying. Inasmuch as there are several aspects to the field of classifying organisms, it is foolish to use all three words with the same general meaning and be left with no terms for the other concepts that we wish to discuss. The distinction listed above is that proposed by Blackwelder and Boyden in 1952,[1] who were simply recognizing a usage that was common among taxonomists of their acquaintance. Most writers have accepted some distinction between these words, but there is substantial disagreement on how to apply the terms. Ferris in 1928 used "systematics" for all the general aspects but defined taxonomy as the arrangement of the groups, which is classification as defined above. Simpson in 1961 used "taxonomy" for the study of the theoretical aspects of classification, the philosophy of systematics—the metataxonomy or methodological taxonomy. These differences reflect the fact that no standard usage has yet been attained.

Nevertheless, it is not intended to imply here that these words represent completely distinct activities. They are all part of the effort to organize and record the diversity of organisms. At best the definitions and discussions overlap. It will be the intent in this book to use the term taxonomy when dealing principally with the practical aspects of segregating, describing, naming, and recording the kinds of organisms (the work that occupies most of the time of the people who call themselves taxonomists); the term classification when referring chiefly to the grouping of like things into a system; and the term systematics when referring to all the activities of a biologist who is studying and recording the diversity of organisms, the origin and maintenance of the diversity, and the methods by which these studies are pursued.

During the past several decades some discussions of taxonomy have used terms that may be either derogatory in tone, ostentatious in implication, or pedantic in application. Among these may be cited typological, neo-Adansonian, non-dimensional, The New Systematics, biological species concept, numerical taxonomy, alpha-, beta-, and gamma-taxonomy, morphotype, archetype, holomorph, biospecies, biosystematics, hypodigm, and onomatophore. These terms are sometimes helpful for clarification of ideas, but in taxonomy they usually are not really needed. All of them will be defined or used somewhere in this book. Some of these terms reflect the fact that there still are controversies in taxonomy over its goals, its basis, and its

[1] The nature of systematics. Syst. Zool., vol. 1, pp. 26–33.

methods. These differences of opinion are mostly at the theoretical level and scarcely affect the practicing taxonomist in his day-to-day labors. Most of them are discussed in chapters 11, 17, and 18.

In summary: as used in the following discussions, taxonomy is what taxonomists do. A wide variety of activities is involved, including techniques of many sorts, analysis of data, synthesis of organization schemes, bibliography, publication, interpretation of nomenclature rules, and the more obvious identification, description, and naming of species.

The goals of taxonomy

In Linnaeus' time, the few taxonomists that were working on animals faced the problem of distinguishing the few thousand kinds that were available to them. As the work progressed, more and more areas of the world were explored, and more and yet more animals were discovered in a rapidly increasing number of very distinct groups. No matter how many species were described, there were always more awaiting treatment, and it is doubtful if new species have ever been described as fast as they were collected. All large museums undoubtedly contain thousands of undescribed ones.

Some groups of animals were more available to, or more popular with, taxonomists than other groups, with the result that the work of finding and describing the kinds from all over the world progressed more rapidly in those groups. It is believed that there are now few kinds of birds still to be discovered, whereas new kinds of nematodes and mites are being described in far greater numbers now than ever before. Popularity has no fixed relation to biological importance, and it must be recognized that the job of making known the animals of the earth is so far from finished that even the general pattern of relative abundance of all the different major groups is not yet known. The pronouncement by Mainx in 1955 that "this work has been completed" is so far from true that even his qualification "at least for certain groups of the animal and plant kingdoms," is inadequate to justify such a statement. Probably the only animal group of which this statement is true is the small class Aves.

It would be a mistake to leave the impression that recognition and description of new species is the only, or even the primary, purpose of taxonomy. This is only a prelude to the major job, which is classification. Classifying kinds is generally recognized as part of the job of taxonomy, but it is often overlooked that the more important and vastly more difficult part of taxonomy is that of keeping track of all the information discovered. Because there are at least a million kinds of animals, it is a colossal job to keep track of hundreds or thousands of facts about each kind; keep them available in a way that will permit extrapolation from them; and arrange them in a system that will permit unlimited additions of new data at any time.

The first requirement is a system of classification by means of which the kinds can be kept in order. An alphabetical file would admirably serve this purpose, but taxonomists have found that a great deal of additional advantage can be obtained if animals with like features are grouped together into classes at various levels. This was one of the ways known to the ancients for handling and increasing knowledge, and it is a normal part of the subject matter of Major Logic, which deals with definition, division, grouping, and argumentation as instruments of knowledge.

Taxonomy has been under constant pressure from other branches of biology to provide classifications and identifications to organize the knowledge of all kinds of animals. It has not been able to keep up with these demands. This is not because of inadequacy of its classifications or its concepts, but largely because the labor involved is far beyond the limitations of the few professional taxonomists, even with the assistance of the more numerous non-professionals.

Specifically, the goal of taxonomy is to build and maintain a classification and naming system capable of identifying and grouping all the kinds of animals that exist; to keep it consistent with all the facts discovered about animals; to use it to organize all the facts available; and to elicit new facts about correlations and groupings. Its basic purpose is thus to systematize data for the use of other disciplines and to uncover new facts in these data themselves.

History of taxonomy

The development of many biological ideas which had an effect on the direction of science can be traced in a chronology of important dates. In taxonomy, however, it appears that even the most determined chronologist would have difficulty in citing many specific dates that have marked turning points or major breakthroughs. Most of those that can be cited are of less pertinence than might appear, and others represent occurrences whose supposed effect on taxonomy has turned out to be largely illusory.

The following dates would usually be cited in a chronology of taxonomic events:

1758 Linnaeus' 10th ed. of *Systema Naturae*
1842 *Strickland Code* of British Association for the Advancement of
        Science

1859 Darwin's *Origin of Species*
1889 1st International Zoological Congress adopted the *Blanchard Code*
1900 Rediscovery of the work of Mendel
1901 *Règles Internationale de la Nomenclature Zoologique* adopted by 5th Congress
1904 International Commission on Zoological Nomenclature formed
1913 Plenary Powers granted to International Commission on Zoological Nomenclature
1930 Setting of a new nomenclature deadline
1940 *The New Systematics* published
1942 *Systematics and the Origin of Species* by E. Mayr; first book to say that taxonomists must study the processes of speciation rather than just its products
1953 Publication of the *Copenhagen Decisions* . . . and the W. I. Follett summary (1955)
1961 Publication of new *International Code of Zoological Nomenclature*

In this list there are three dates—1859, 1900, and 1940—which were expected to have a great effect on taxonomy. Such an effect has been claimed, but there is little evidence that evolution, genetics, or The New Systematics had any obvious effect on the *actual classifications* produced by the taxonomists.

For example, some writers have cited three periods in the history of taxonomy: (1) the study of local faunas, (2) the acceptance of evolution, and (3) the study of populations. Although these writers have repeatedly stated that Darwinian ideas revolutionized taxonomy, it is doubtful if anyone can tell by examination of the *taxonomic work* of any nineteenth-century taxonomist whether he accepted the evolution theories or not, unless he specifically stated that he did so. Certainly the bulk of taxonomists continued unchanged their classifications based on the visible features of their specimens. Only a very small number of taxonomists even drew family trees or speculated on the phylogeny of the animals.

Again, the repeated assertion by some that the study of populations is part of taxonomy and actually the only really modern aspect falls on deaf ears for the most part. All taxonomists study individual organisms and species. When they choose to study populations as such, they are at that moment studying ecology or evolution, which are not directly part of taxonomy.

Eight dates: 1842, 1889, 1901, 1904, 1913, 1930, 1953, and 1961 refer to standardizing efforts in the sub-field of zoological nomenclature. They are turning points in the attempt to obtain universality and stability in the names of taxa. These dates mark only the publication of the various sets of rules or other events; they are not the dates of universal adoption of them in practice. Because of this, these dates were not of immediate importance. (Some additional comments on the history of nomenclature will be found in Part VI of this book.)

The date of 1758 is often listed as one of very great importance, because the system of naming therein adopted came to be universally accepted within about forty years. This was the date of publication of the first zoological treatise to use the binominal or two-name system for naming animals and to use it consistently. The circumstances of this event are discussed in a later paragraph.

But this supposedly new development was neither new nor a major aspect of taxonomy. There had been similar two-name systems used before 1758, and the system did not affect the taxa or groups of organisms, merely the names for them. The date 1758 is of more direct importance in zoological taxonomy because it has been adopted as the starting point in nomenclature. No older names are nowadays cited in synonymies. This has also resulted in nearly universal abandonment of all pre-1758 literature for taxonomic or scientific purposes.

The remaining date in this list, 1942, marks the publication of what is generally taken as the first American book on The New Systematics, and the first book to promulgate the so-called "Biological Species Concept." This refers to the population aspect mentioned above. The "biological species" is the species of the evolutionist. It may theoretically be also the species of the taxonomist, but there are no instances in which we can prove that they are the same.

Some ecologists and evolutionists find the taxonomic concept of species inadequate for their work on populations and phylogeny. It is thus reasonable for them to say that the classical species of the taxonomist is not a satisfactory concept for speculations in theoretical biology. It does not follow that any other sort of species need be or can be used *in taxonomy*. Eventually, it may prove possible to reconcile the two concepts.

Taxonomic thinking. The origin of taxonomy as a way of thinking is lost in antiquity. Even before man evolved on the earth, animals recognized kinds among themselves. They recognized their own kind, the kinds which might be dangerous to themselves, and the kinds that would provide them with food. When man began to transmit ideas by means of speech, he used certain words to refer to the kinds of organisms about him, being often quite discriminating in his distinctions.

The ancient Greeks knew enough kinds of organisms to need to group them for easy reference; so they applied their principles of logic and developed classification systems, involving comparison, grouping, hierarchies, genera, and species. Even then it was evident that several aspects of classifying result from the nature of the human brain, because the very tendency to classify is fundamental to intelligence and to communication as man practices it.

The systems of classification based on these ancient ideas have recently been labeled as inadequate, inappropriate, impossible, and completely inacceptable. The recent arguments are sound, but they are not relevant. The later systems are not based on the part of the ancient logic which is challenged. They are based on simple ideas of comparing, dividing, grouping, categorizing, and naming.

Linnaeus chose features from the systems of his predecessors and established them in a more workable combination. This involved the comparison of species and groups of all animals, and the comparison of their features as then known. Before many decades had passed, the system was so extensive, so uniform in its application, and so well integrated that it is hard to see how it could have been changed.

Many writers have thought that the Darwinian revolution in biology must have affected taxonomy. Some have even claimed that it did, but no one has cited examples. What happened is that many zoologists began to study comparative anatomy in the hope that it would prove and explain evolution. They were not interested in species but in classification at the highest levels. It is seldom mentioned that their efforts ended, around the turn of the century, with a vast knowledge of animal anatomy but with a complete failure to resolve the basic evolutionary problems. In the meantime, taxonomists had been continuing their work largely uninfluenced by these new ideas.

A second revolution was also expected by some zoologists, after the publication of New Systematics books in 1940 and 1942. It was assumed that taxonomy would be changed by the neo-evolutionary ideas and speciation theories. The supposed change of taxonomy was proclaimed many times by a few voices, to which scarcely any opposition was heard. Although some taxonomists have worked with the evolutionary concepts, and many have made use of the new perspectives and data available, few or none have changed their taxonomy, except for gradual improvement in method and product (as envisioned in Huxley's 1940 book *The New Systematics*).

It has not been so clear *why* there was no change after Darwin and again after Huxley. It was not because taxonomists rejected any particular part of the evolution or speciation theories or directly rejected genetic knowledge and population ideas. The simple fact seems to be that the sort of change expected is not possible in the taxonomic system. A second likely reason is that the system worked so effectively for taxonomists that they had no wish to alter it. Whichever reason has been dominant, there has been very little real change in the system in two hundred years. There is no reason to think that it will be changed in the future, although it is quite conceivable that it might be replaced with some other system.

There have been some attempts in recent years to deny that modern classification is still largely practiced on the principles and methods of Aristotle. These attempts have been unconvincing. Whatever we may now think of the philosophical aspects of the variety and evolution of organisms, there has never been any practical alternative to grouping animals and kinds according to their shared features or attributes. This is still the only way available to us, and all taxonomists employ it, either consciously or unconsciously.

Linnaeus. It is sometimes implied that the Swedish naturalist Carolus Linnaeus (1707–1778) invented our present systems of classification and nomenclature. As has been pointed out, neither one was new in Linnaeus' time—he accepted the logical methods of his predecessors. His contribution was the uniform application of one particular naming system to all organisms.

The idea of genus and species had been used for centuries and had consistently been applied to animals. Linnaeus' idea was simply to cite the names of both the genus and the species together in a two-part form. Even this was not a new idea, but the use of it in a uniform manner in a major monographic work brought its advantages to the fore so that it rapidly displaced all other systems. His students enthusiastically developed his system and spread it throughout the educated world.

The success of Linnaeus' form of the classical methods was partly due to another factor, often overlooked. This was the use of a precise terminology, which gave definiteness to his descriptions and made it possible for others to recognize his species and groups with assurance. This terminology was in Latin, a language known to all educated persons of that time.

It is not intended here to detract from the importance of Linnaeus as the founder of modern zoological taxonomy: his influence can hardly be overstated. His system has proved to be infinitely extendable to cover millions of organisms he never dreamed of. The changes in his classifications made by later workers possessing new information are no reflection on his work. With the animals known to him, and the attributes which could be detected in them at that time, his work was of a quality seldom equaled.

Taxonomic history of groups. When one considers a particular group of animals, and scans the development of its taxonomy over the span of 200 years, it is often possible to see a gradual change of viewpoint at several points in the history. Four examples are given briefly below. These were of course chosen for the diversity of their history. They probably represent the extremes of the varying situation found among the groups of living organisms, but further factors could doubtless be illustrated among extinct groups.

Birds have been well known to man far longer than many other groups. When the modern taxonomic study of birds began in the eighteenth century, the rather small number of kinds were commonly recognized and could be dealt with in a general work like the *Systema Naturae* of Linnaeus along with other animals. Such general works were the rule during the rest of that century, with broad coverage but little illustration.

After the end of the eighteenth century, general works on birds among other animals were being replaced by works on the birds of individual countries or regions or monographs on particular groups of birds. There grew up many learned societies, which published journals often devoted entirely to birds. Large ornithological books containing colored plates, often in multi-volume sets, became almost commonplace. Oftentimes, the hand-colored plates were the real reason for the book, there being a minimum of text. Exploration of the far corners of the world served as a basis for elaborate regional monographs.

Starting in the first decade of the nineteenth century, there was a long series of studies of the classification of birds. Successive schemes were not always built on the previous ones, but they generally reflected current knowledge of bird anatomy and natural history. By the end of the century, the basic arrangement had been fairly generally agreed upon, and only minor changes and subdivisions were made thereafter. Also by the end of the nineteenth century most species of birds had been described and named. In recent decades it has become rare to discover new species or even new subspecies anywhere in the world. In a sense the taxonomy can be said to be finished. Interest has turned largely to other aspects of ornithology such as speciation; but it is a little strange that there is still a great deal of work to be done on the comparative anatomy of birds; this may lead to revisions of the classification at some future time.

Insects were numerous among the animals included in the *Systema Naturae*. Although some 2,100 species were therein included, mostly from Europe, this substantial number forms no more than 25% of those now known from Europe and no more than 3% of those now known from the entire world.

By the end of the eighteenth century several prolific scientists had begun to specialize on individual groups of insects, classifying and monographing the known species and greatly increasing the knowledge of these kinds. At the same time, studies were begun on the structure and life history of these forms, increasing the naturalness of the classifications and forming a base for many later aspects of entomology.

In the first half of the nineteenth century, these monographic works continued to be produced in moderate numbers; but there appeared also a number of large sets of faunal studies, frequently with numerous elaborate illustrations. As with the birds, the illustrations were sometimes the main excuse for the set, being likely to deal with the showy insects, especially the butterflies. It was during this period that the regional or national nature of many books led to the describing of widespread species under several names. These synonyms began to accumulate in lists in the monographic studies, and these lists increased in length through the century.

The number of known species of insects, even in a region like Britain, was now so large that it was impractical to deal with them taxonomically as a single group, and all taxonomists became specialists on an order, a family, or even a genus. This led to a very uneven development of the groups: The butterflies and beetles were much studied, but groups such as scale insects and parasitic wasps were neglected.

By 1850, the number of kinds of insects described had grown so large that it was recognized that centralized indexing was going to be essential. The literature had also become enormous and required aids for recovery of data. In England W. F. Kirby and W. S. Dallas, in Germany C. E. A. Gerstaecker and W. F. Erichson, and in America S. H. Scudder were founding bibliographic and indexing services such as the insect portions of the *Zoological Record* and the *Archiv für Naturgeschichte*. Several large bibliographic catalogs were started at about this time, including the twelve-volume *Catalogus Coleopterorum*, by Gemminger and Harold (1868–1876) and the numerous catalogs of the collections of the British Museum (Natural History) from 1844 to 1870 and later. The Insecta portion of the *Zoological Record* was often nearly as large as all other parts combined. Several national checklists of particular orders passed through many editions, scarcely keeping up with the large numbers of new species and new records.

In the late nineteenth century a series of multivolume faunal monographs were begun, such as the *Biologia-Centrali-Americana* and the *Fauna of British India*. The parts on insects in these sets were so large that they frequently threatened to bankrupt the entire plan. Financing has become so difficult in recent years that these are seldom produced.

In the twentieth century, the influence of new ideas and techniques in zoology was felt in taxonomy in the form of new types of data available for classification. High-quality revisionary and monographic work was increasingly broad in the data employed and the methods used to obtain these data. The rising costs of publication made monographing discouraging work, but nothing less than this most detailed and complete type of study would serve the modern demands. The number of taxonomists working on insects was often less than the number working on vertebrates, although there were at least ten times as many species of insects to be studied.

The economic importance of insects put increasing pressure on the taxonomists for identification and classification. Governments spent considerable amounts on identification but too often considered classification as something to be desired but not really provided for.

There have been estimates of the total number of species of insects in the world as high as ten million. It seems likely, therefore, that much less than half the species of the world are now known, in the sense of having been segregated and named. Most of those described so far are known from a few collections and only as adults. Only a small fraction of the families or genera of the world have been monographed, and only another small fraction have been treated in any other sort of comprehensive taxonomic work. Regional studies are common in Europe and North America, but the bulk of insects of the world are still poorly known or unknown.

Nematodes were known to Linnaeus, but only two genera and three species were listed in the *Systema Naturae*. By the beginning of the nineteenth century, these worms were only beginning to be distinguished from other types of worm-like animals. The parasitic nematodes were studied throughout the century, especially their anatomy and peculiar life cycles. As late as 1866, in the first important monograph of the group, the free-living forms were still much neglected.

Because of the medical, veterinary, and horticultural importance of nematodes, the literature on them is enormous. This literature has been cataloged in a unique set of many volumes issued from 1902 onwards, in which thousands of papers are listed by author. The subject index consists of hundreds of thousands of cards in the Bureau of Animal Industry, U.S. Department of Agriculture, far too large a mass of data to be published.

Monographic studies have been few. One reason may be that the animals present relatively few characters of usable sort for such work. They are in this sense one of the so-called difficult groups. Most of the major works are influenced by the parasitological interest in these animals, and the parasitic, free-living, and plant-parasitic forms are usually treated separately.

It has been estimated that the twenty thousand or so known kinds of nematodes may represent as little as a hundredth part of the ones which actually exist. In either case they are a major group of animals of which the present knowledge is substantial but no doubt still fragmentary.

Pogonophorans, or beard-worms, form the most recently discovered phylum of animals. No one knows exactly when the first one was seen by man, because specimens have been in museums for years without being recognized as anything unusual. The first known collection was in 1899–1900, but the specimens could not be identified with any known phylum until 1951. In the meantime other specimens from the depths of the sea were collected in 1933, described as a new class in 1935, and raised to phylum in 1944.

The first monograph of this group was published in Russian in 1960 and translated into English in 1963. The phylum at that time consisted of one class, two orders, five families, 14 genera, and 67 species. New species are being discovered every year, and the phylum is now in a period of rapid taxonomic growth.

Comment. Even these brief sketches show that many factors have influenced the development of taxonomy, and that the history of the various groups has been not only not synchronous but not even closely parallel.

It is sometimes stated that the taxonomy of any group passes through three stages, descriptive, systematizing, and evolutionary. These three, called alpha-, beta-, and gamma-taxonomy, do have some relevance to the broader field here called systematics, but the evolutionary studies (gamma) are in reality not part of taxonomy. The use of alpha and beta for the descriptive taxonomy and classification, respectively, seems merely to substitute symbols for self-explanatory terms.

There is little evidence in the taxonomy of the animal kingdom as a whole that taxonomists are actively trying to pass from the alpha level through the beta level to the gamma level. Most taxonomists work primarily at descriptive taxonomy and do a varying amount of classification. Only a few attempt work that is really evolutionary. Recent attempts to combine taxonomy with evolutionary studies have not been highly successful. The evolutionary work must be done largely outside of the range of taxonomic concepts and methods.

Taxonomic reference works. The tremendous growth in knowledge of animals in the century after Linnaeus led to two developments: one was the idea of evolution in the minds of Darwin and Wallace, and the other was the need for centralized recording of the growing literature and cataloging of known species.

A group of taxonomists in England, including A. C. L. G. Günther, A. Newton, E. C. Rye, W. S. Dallas, and others, started, in 1864, *The Record of Zoological Literature* . . . , which for five years was a private venture. In 1870, with the name changed to *The Zoological Record* . . . , support was provided by the British Association for the Advancement of Science, and in the following year an independent Zoological Record Association was formed. It continued its sponsorship until 1886, when the Zoological Society of London assumed the task.

Few zoologists today realize that *The Zoological Record* suspended publication for a period of years from 1906 through 1914, during which time the work was continued under the *International Catalogue of Scientific Literature*. The Zoological Society of London continued to collaborate in its preparation, and in 1915 resumed full control. Most sets do not show this variation in sponsorship. Other organizations now cooperate in the preparation and distribution of the annual volumes.[2]

Although the *Zoological Record* has probably been of greater use to English-speaking taxonomists, it was not the first annual index of animals and their literature. In 1835, A. F. A. Wiegmann in Berlin had founded the *Archiv für Naturgeschichte*, which has been continued ever since and has been indexed at intervals as the *Zoological Record* never has.

The nomenclators of Agassiz, Marschall, Scudder, Sherborn, Waterhouse, Schulze, and Neave mark the history of the tabulation of the names of animals. They are all still indispensable reference works

[2] The usefulness of these volumes is in large part due to the work of W. F. Kirby, David Sharp, J. A. Thomson, R. Lydekker, R. Bowdler Sharpe, G. A. Boulenger, F. A. Bather, W. T. Calman, F. Silvestri, F. W. Edwards, C. T. Regan, W. L. Sclater, H. M. Muir-Wood, A. K. Totton, and N. D. Riley.

for the serious taxonomist, although only Neave is widely used in America. Furthermore, the extremely useful *Catalogue of the Library of the British Museum (Natural History)* in eight volumes and the more inclusive *Royal Society Catalog of Scientific Papers* in nineteen volumes are only the top level of the many bibliographic works which appeared in the late nineteenth and early twentieth centuries.

The growth in the twentieth century of abstracting services for biology as a whole, such as *Biological Abstracts,* has greatly aided taxonomists in the peripheral aspects of their work, though, in fact, they have never replaced the *Zoological Record* and the *Archiv für Naturgeschichte* in the indexing of new names.

Summary

Taxonomy is an observational science, part of the age-old field of natural history. It is concerned with knowing animals as individuals, kinds, and groups of kinds. It is historical in large part, studying the work of the past, adding to it, and reorganizing it to yield new insights.

Although the goals of individual taxonomists may differ, the science itself has a triple goal: to systematize the knowledge of animals, to add to that knowledge by observation and analysis, and to make this organized knowledge available to all biologists. The data used and stored were once produced almost entirely by the taxonomists, but today, a large number are produced by zoologists who are not taxonomists, such as geneticists. Some of these data may be used in classification, but the rest are merely stored under the names of the kinds. The data used directly by taxonomists are now, as in the past, mostly the product of their own observation.

Specialization has come to be a major feature of taxonomic work, with each taxonomist devoting all his efforts to the study of a group of animals, a class, a family, or even a genus, but the purpose and the basic methods have remained little changed. There is no prospect that taxonomists will run out of work to do or that their results will become any less essential to other zoologists. In fact the increasing demands upon taxonomists for identifications, classifications, and the recovery of stored data may lead to a revival of interest in this ancient but modern science.

CHAPTER 2

The Place and Importance of Taxonomy

By the time a person is old enough to take a course in taxonomy, he will already have been involved in a good deal of such taxonomy. He will know how to tell apart some kinds of animals, even if they are only man and various kinds of domestic animals. He will know that not all members of one kind are alike—that some young animals are noticeably different at first from their parents, that male and female may be clearly different, and that some horses are bay, some chestnut, some white, etc. He will, unconsciously at least, have compared different kinds with each other, distinguishing the features in which they differ from the features in which the members of each kind differ. He will have classified many things. When he says, "Rover is a dog," he is using dog in a group sense and Rover as one of the included individuals. He will doubtless know that dogs are part of the larger group called mammals. He will very likely have applied a new name to some pet. Although taxonomy doesn't usually give names to individuals, there is no difference between the naming of individual groups and the naming of individual animals. He may have made a collection, if not of animals, then perhaps of minerals, fossils, plants, postage stamps, or postcards. Some of the features of collecting, and some of the possibilities for studying the collection, will have been illustrated. He may have identified an unknown bird or other animal by looking it up in a book. Whether this involved use of a key or merely a search through pictures, it was identification.

These and other aspects of elementary taxonomy are part of his everyday life. He cannot exist as a thinking being without using some of the ideas and devices of taxonomy.

All zoology students must learn a classification of animals, even if it is only two subkingdoms and ten phyla. This classification is one of the principal products of taxonomists, who have been kept busy erecting it for two centuries. Of course, the ten-phylum scheme is much simplified. The student then studies comparative anatomy— a study of how certain representatives of major groups differ in their major anatomical features. He studies embryology, in which the development of certain customary species, for example "the pig," "the chick," shows him a little of how species (representing entire groups) differ in the anatomy of early stages and in the details of developmental processes. He studies genetics, in which he learns some of the factors that make the members of each kind alike from generation to generation, and also some of the factors that make individuals within the kind different from their fellows in color, size, ability, sex, etc. He soon comes to know that species receive a dual name and that all groups of species receive a single name, such as the names he learned for the subkingdoms and phyla. He soon finds out that some groups have received more than one name, so that he must choose between Mastigophora and Flagellata, for instance.

What taxonomists do. Inasmuch as taxonomy is defined as the work done by taxonomists, it would seem to be a simple thing to show what that includes. But unfortunately one can only list several major aspects of their work and indicate some of the things involved in each.

(1) *Obtain specimens.* This may be by collecting in the field, exchanging, purchasing, going to a museum, or all of these.

(2) *Obtain literature.* All the literature back to the first description of a species in the group is required in most real taxonomic studies and is usually best obtained in a library, preferably the one attached to the museum containing the specimens. Further, some papers are obtainable upon request, or exchange, from the authors, and some books are obtainable by purchase from book stores or used-book dealers.

(3) *Study specimens.* Under this heading must be included not only the taxonomic study of the particular specimens but also the pertinent background knowledge, listed in Chapter 5, such as comparative anatomy, life history, development, and natural history. The taxonomic study *must* employ all appropriate and necessary methods, which will vary greatly from group to group.

(4) *Study literature.* This goes on simultaneously with the study of the specimens, but there must also be a considerable knowledge of previous work before study of the specimens can usefully proceed. Unless the literature is quite restricted in volume, it will be necessary to organize it in some way, usually by some sort of cataloging; however, if catalogs and bibliographies have already been published, this will help.

(5) *Identify specimens.* After the specimens and the literature have been sufficiently studied, the taxonomist will be able to identify some of

the specimens. He will presumably know whether some of his specimens do not correspond to any of the species in the group and therefore represent new species.

(6) *Publish conclusions.* Having worked out the species and groups, he may prepare descriptions, keys, new descriptions, or monographs for publication. These embody his taxonomic conclusions, whether they are about the newness of a species or group, its position in the classification, or an easier way to distinguish it from other species or groups.

(7) *Propose new names.* If any of the species or groups are new, the taxonomist must propose names for them. To do this he must be thoroughly familiar with the procedures and rules of nomenclature accepted by taxonomists. He must also have a working knowledge of the Latin and Greek employed in scientific names.

(8) *Classify species.* He must classify the new species, at least to the extent of assigning them to genera. Eventually he must go much further and classify the genera into groups and these groups into more inclusive groups—prepare a classification, which is a sort of synthesis of all he knows about the species and the groups.

Although these are the principal scientific tasks of the taxonomist, he must usually perform some other tasks that make the taxonomic work possible. Delegation of these more routine tasks to a non-taxonomist is usually not a satisfactory arrangement.

(9) *Maintain collections.* As most of the work of the taxonomist is based on specimens of the animal species and the data recorded about them, it is necessary to maintain permanent collections so that specimens will be available when needed. Moreover, future taxonomists will use these collections to recheck the work and for later studies. Maintenance of the libraries is equally important.

(10) *Study new methods.* Although conventional methods are usually the basis for any taxonomic study, many cases require development or adoption of new methods before a successful solution can be achieved. It is always desirable to be on the lookout for new techniques that offer better results, and the taxonomist should try to adapt these to his work whenever the advance of the science requires it.

(11) *Study nomenclature.* The highly technical nature of the rules under which zoological names are formed and used makes necessary not only an understanding of their provisions by all taxonomists but also the study of the rules themselves by certain taxonomists, who become in effect nomenclature specialists. These nomenclaturists attempt to make the rules more effective in making names serve the needs of all zoologists. They also serve as consultants to other taxonomists on questions of the use of names.

The employment of taxonomy

Historically, the fields of zoology appeared in about this order: taxonomy (the recognition of kinds), natural history (the life history in general), comparative anatomy, classification, cytology, evolution, embryology, genetics, ecology, speciation, and biochemistry.

The first five of these were comparative. They sought out and recorded the pertinent facts for different kinds of animals. If there had been no ability to recognize and refer to the kinds (species), it would have been impossible to build up any comparative field, so that when a dissection was made, or a life history worked out, there would have been no way to record to which animal it applied.

Embryology, genetics, ecology, and biochemistry, for example, are also partly comparative in nature. Embryologists work out the normal development of each species. Gene maps are comparative, and one of the results of genetic work shows the pattern of characters and determinants of the various species. Part of ecology studies the reactions of individuals to various factors, but another part studies the environmental preferences, ranges, and so on of each species.

Biochemistry is taking a great interest these days in how kinds differ in their biochemical components. It is possible to view most of biochemistry as a prelude to comparison of the species in respect to their components and chemical reaction systems.

Taxonomy not only produces knowledge but organizes it from all fields, thus providing the essential framework needed to make the data widely usable. This framework is the classification of organisms, and the keys to the data in the classifications are the names of the species and groups. All other zoological specialties use this framework and the key names. It is impossible to go very far in the study of organisms, whether one is studying ecological, anatomical, distributional, physiological, biochemical, or genetic factors, without a means of referring to the different kinds and a means of relating the data discovered to previous data. Taxonomy is the means of doing both.

The abilities to recognize kinds, to associate them with names that are distinctive, and to recognize the groups into which they fit with similar species are used constantly by all zoologists. Without these abilities there could be no biological science. It is entirely the result of taxonomic work that these essential tools are available, and it is due primarily to continuing taxonomic work that they are constantly revised and corrected.

Two aspects of taxonomy are constantly depended upon by zoologists or biologists, and most other scientists dealing with animals. The first of these is the ability to recover information from the system by use of the name of the species or group. For example, a person encountering the name *Taenia pisiformis* in a textbook or medical literature or veterinary practice, can without difficulty extract from indexes and publications all that is known about this tapeworm species, including hundreds of reports on its occurrence, development, vectors, and hosts, together with the aetiology of disease it produces. Again, a person studying a little fruit fly would not get very far in genetics unless he found out which species of *Drosophila* he was using. After identification, the name would lead him to all the genetic and other work done on that species. A study made of just "a fruit fly" would be useful only as a training exercise. It could produce useful data only if the identity of the fly were definitely established.

The second aspect used by all zoologists is the association of the species at hand with other species in some taxon of the classification. For example, mention of any species of *Drosophila* immediately associates this species not only with the other species in this genus but also with the order Diptera, the class Insecta, and the phylum Arthropoda, so that it will display the features distinctive to each of these groups. It is not necessary to ask if it has four wings or two, because it is assumed that all normal adult Diptera have only two; nor to inquire if it breathes by means of tracheal tubes, because it is assumed that all active adult insects do; nor to wonder if it has a chitinous exoskeleton and a body divided into regions, because these are characteristics of all adult Arthropoda.

Underlying virtually all important biological work, although sometimes unrecognized, is the knowledge of not only what the species is, but also where the organism fits into the system. Without this, it is doubtful if any of the major advances in biology could have been made. Both evolution and genetics were made possible by the ability of taxonomists to identify species and relate them to a classification.

It is difficult to imagine a part of zoology which is not sooner or later dependent upon this taxonomic capability built up over two centuries by generations of taxonomists. It is *not necessary* to conclude from this that taxonomy is more important than any other subject. It *is appropriate* to recognize that the more "modern" and "glamorous" fields of zoology are possible today because taxonomy did not fail in its task of identifying and classifying animals. It is to be hoped that taxonomy can keep its system adequate to the future needs of all zoology.

It may be thought that this is a personal exaggeration of the importance of one field of specialization, but the importance of taxonomy has been recognized by a variety of modern writers.

In *The New Systematics* at least two writers commented on the dependence of biology on taxonomy: "Systematics, in our understanding, is the basis of knowledge of the plant and animal kingdoms. It was not by mere chance that the greatest evolutionist, Charles Darwin, started his work from systematics" (N. I. Vavilov). Again, ". . . as if biology as a whole could possibly advance when the essential work of systematics is allowed to lag behind" (W. H. Thorpe).

An entomologist primarily concerned with non-taxonomic aspects has written: "Taxonomy is the focal point and basis for all the biological sciences," and "Taxonomy must precede all other forms of biological investigation and furnishes the foundation and frame upon which may be built the results of the researches of all the natural sciences" (E. O. Essig). Another non-taxonomist, forty years ago, saw this relationship and expressed it in direct terms: "It is the systematist who has furnished the bricks with which the whole structure of biological knowledge has been reared. Without his labors the fact of organic evolution could scarcely have been perceived, and it is he who to-day really sets the basic problems for the geneticist and the student of experimental evolution" (Raymond Pearl).

A paleontologist and evolutionist who is also a systematist of note has described the place of taxonomy in these terms: "It is impossible to speak of the objects of any study, or to think lucidly about them, unless they are named. It is impossible to examine their relationships to each other and their places among the vast, incredibly complex phenomena of the universe, in short to treat them scientifically, without putting them into some sort of formal arrangement. . . . Taxonomy is at the same time the most elementary and the most inclusive part of zoology, most elementary because animals cannot be discussed or treated in a scientific way until some taxonomy has been achieved, and most inclusive because taxonomy in its various guises and branches eventually gathers together, utilizes, summarizes, and implements everything that is known about animals . . ." (G. G. Simpson).

A conference on systematics was held in 1953 under the aegis of the National Research Council in Washington, D.C. It was organized by the Society of Systematic Zoology for the purpose of acquainting government agencies with the importance of systematics in biological research. The organizer was Dr. Waldo L. Schmitt, whose principal remarks are quoted in full in a later paragraph. Various other speakers discussed aspects of biology that are substantially dependent upon prior or concurrent systematic work. It was there pointed out by a parasitologist that one who knows the systematics can predict life cycles, vectors, and avenues of infection, and in consequence the necessary preventive measures or treatment. The identification of organisms and the resulting correlation with known data are the keys to the study and control of many diseases.

As to paleontological taxonomy, much emphasis was placed on its dependence upon the systematics of living organisms. Paleontology is largely a science of inference from the known to the unknown. Most of the "facts" about fossils, other than the structure of hard parts, are inferences from their nearest relatives among living organisms. This dependence is so great that some paleontological agencies have had to hire and train taxonomists for work on Recent species, because paleontological investigations could not wait for biologists to study these groups.

In September 1962, a conference [1] was held at Lawrence, Kansas, on Taxonomic Biochemistry, Physiology, and Serology. The necessity was clearly shown, in these experimental fields, for a real understanding of the nature of classification and taxonomic procedures. A large number of the biologists in specialized fields discussed their experimental results on particular animals. They all attempted to show how their discoveries agreed or disagreed with the "accepted" classifications, and it appeared they all wished to show that their data and methods could be of value in taxonomy. Unfortunately, it became obvious that few of these scientists understood the nature of classification and taxonomic procedures sufficiently to show the effect of their own data or judge whether it really was of value. Many speakers refused to accept their own results if they failed to agree with whatever classification they thought was the one accepted in taxonomy. Some highly interesting conclusions were brushed aside because of the lack of this agreement. A taxonomically trained person would have seen the implications and known how to study them further.

Taxonomy has not only been important in biological science but has contributed substantially to the biological arts, which are the fields known as the applied sciences, including economic entomology, parasitology, biological control, conservation, veterinary medicine, and public health. To indicate the scope of the use of taxonomy in applied fields, one can do no better than quote in full the paper read by Dr. Schmitt at the 1953 conference on systematics (distributed in a mimeographed report of the conference and later published in the Annual Report of the Smithsonian Institution for 1953).

[1] Since published as Taxonomic biochemistry and serology, ed. by C. A. Leone. Ronald Press, New York, 1964.

Applied Systematics:
The Usefuless of Scientific Names of Animals and Plants

W. L. Schmitt

"It is an error to suppose as many do that classification is an out-moded phase of natural history. It affords a continuing test of evolutionary doctrine. The increasing refinement of biological study requires greater certainty than ever before of the identity of animals and plants used in experimental work. The fact that all organisms are now considered to be part of one great family tree is a challenge to the intelligence and skill of the classifier who must reconstruct that tree. Actually the business of classification has today greater vitality and significance than ever before . . ."

—*Paul B. Sears*

The field of biological systematics is a broad one, and within it are brought together at least a part of all natural-science disciplines. It represents the orderly understanding and the sum total of our knowledge of the animal and plant kingdoms. I shall confine myself chiefly to the taxonomic side of the subject, so largely devoted to knowing the scientific names of organisms. However, to name animals and plants intelligently you need to know a great deal about them, their makeup, lives, growth, behavior, and geographic distribution; in short, their biology in the broadest sense of the word.

Systematics in everyday life

Everyone at heart is a taxonomist, either by virtue of necessity or because of mere curiosity. From childhood up we want to know the names of things. What is this, that, or the other object—how, where, why, and what? Children at an early age readily learn to distinguish a number of common things—birds, the various wild-flowers, poison-ivy, bees, wasps, yellow jackets—according to their experience.

Every good housewife can identify the tiny moth flitting through the bedroom or the parlor if it be a clothes moth. This knowledge has a dollars-and-cents value, for the name of a beast or a pest indicates the method of control to be applied. With further experience she can distinguish this kind or species from one that may more rarely flutter through the house but in more disturbing numbers—the moth that sometimes appears in your packaged grain or cereals. Or perhaps it is the winged ant coming out from under the house that catches her attention. In mere self-interest she will want to know if it is an ant or a termite, which, by the way, is not an ant but an insect of quite another order and family. There are also wood-destroying ants, the carpenter ants, infesting houses, yet these rarely if ever become serious pests. With the identification comes the scientific name, which is the key, the index entry, indeed the only device which will open up for one the world's literature containing the extant information regarding any object, animal, mineral, or plant. If a name cannot be found for it, the object is probably new and undescribed, in which case the information regarding it is yet to be developed.

Indeed, wherever man comes to grips with the problems of life and living, the importance of the names of things is most vital, whether he be concerned with disease, the production of food, or merely safe drinking water.

The physical fitness of drinking water can readily be determined by chemical analysis, but only by identifying the organisms existing in it, or rather determining the absence of certain of them, can its safety be assured. Among the biological contaminants that need to be distinguished are to be numbered first of all the enteric bacilli and amoebae, the "germs" of typhoid and cholera; copepods, which are the intermediate hosts of the broad tapeworm of Europe now established in parts of this country; and a host of other organisms that vary as to locality. Unknown waters are not safe to drink even in the high Arctic with its extremely low, often killing temperatures, for there the melting ice and snow in the spring expose and redistribute the well-preserved refuse of the long winter months from human habitations. But please do not look askance at the glass of drinking water before you. Our modern municipal waterworks take pains to treat and filter it carefully. Yet accidents happen and plumbing installations have been found faulty, as in Chicago, where carelessness in this respect resulted in 70 deaths from amoebic dysentery a few years ago.

Whether the water be fresh or salt, pollution not only renders it unfit for use, but, if in sufficient degree, will also destroy the inhabitants useful and economically important to man.

Dr. Ruth Patrick, curator of limnology, specializing in diatoms at the Academy of Natural Sciences of Philadelphia, as the result of her investigations in certain Pennsylvania streams, was perhaps the first to stress the importance of the specific naming of the organisms present in the evaluation of stream pollution, its kind or type, and duration. She found that the heretofore frequently tried method of using indicator organisms simply did not supply the data needed to make such evaluations.

All groups of plants and animals living in a stream, particularly the sessile or attached forms, or those which moved about in only a small area, merited serious consideration definitely at the species level. This entailed extensive collecting in relatively shallow water, the area in which the majority of such forms live, and required the cooperation of a number of experienced taxonomists to identify specifically the material collected, especially the algae, rotifers, worms, mollusks, Crustacea, insects, and fishes. Sooner or later we all discover, as did Dr. Patrick, that there is no satisfactory shortcut to the solution of a biologic problem, ecologic, medical, agricultural, or otherwise, that ignores names of the species involved.

Engineering

Ordinarily you would not expect an electric light company to have a biological problem, let alone one in which mollusks were involved. Six or seven years ago a heavily armored power cable lying on the bottom of the bay between Palm Beach and West Palm Beach suffered one of a series of blowouts as the result of the penetration of the outer insulation and the heavy-load casing by marine mollusks. The company's officials naturally wanted to know what manner of shell-fish this was and what could be done to prevent further damage. Though the animal was found to be a new species, which was subsequently described, it was at once recognized by the expert to whom it was submitted as belonging to a genus of boring mollusks which would quickly be suffocated if the cable were buried several inches below the bottom of the bay. This would also prevent further damage of the same sort. In the 8 years preceding, the cable had suffered 15 failures, entailing repairs costing upwards of $12,000. Here an identification solved a costly engineering problem.

The field of ecology

Engineering problems accompanied by specimens are easier to solve than ecological ones unsupported by specimens. Recently an ecologist was discussing the behavior of a green parrotfish in the waters about a tropical island. You can imagine his consternation when he was asked to which of 10 possible species he was referring!

The importance of specific names to ecologists may be illustrated by this excerpt from a letter received by one of our Museum curators from a well-known student of jungle life: "I have all of my voluminous field notes ready and only await the names of the [specimens] which I sent you a long time ago. Have you had a chance to go over them? I have the names of most, but there are still many left and I can publish nothing until I get them."

And Charles Elton, in his book "Animal Ecology," writes: "One of the biggest tasks confronting anyone engaged upon ecological survey work is that of getting all the animals identified. Indeed, it is usually impossible to get all groups identified down to species, owing to lag in the systematic study of some of them. The material collected may either be worked out by the ecologist himself or he may get the specimens identified by experts. The latter plan is the better of the two, since it is much more sensible to get animals identified properly by a man who knows them well, than to attain a fallacious sense of independence by working them out oneself—wrong."

Evolution and genetics studies

The abundance of the pasturage in what are known as "the meadows of the sea" is being evaluated these days by the oceanographers in terms of the chlorophyll collected by their continuous plankton recorders without having to take the pains of identifying the many species of which the plankton mass is composed. At least samplings of the organisms involved should be specifically determined, for there are bad as well as good plank-toners in the sea, just as there are good and bad plants on some of our western ranges. Pasturage in meadows on land, by certain tests, may yield a very high chlorophyll rating, but a lot of it could be locoweed. If the marine chlorophyll ratings are to have real significance, the species on which they are based need to be known.

In evolutionary and genetic studies, it is especially important to know well the species dealt with and the literature about them. Years of effort can go for naught if pertinent taxonomic finds, procedures and discoveries are disregarded.

An unfortunate instance of this sort was a rather impressive report on "An Investigation of Evolution in Chrysomelid Beetles of the Genus *Leptinotarsa*," published some years ago, a 320-page volume illustrated with 31 text figures and 30 plates, some in color. Aside from a number of unnecessary nomenclatorial mistakes, records of distribution and occurrence were far out of line with published work on these beetles. Although the author stated that three species were found in the United States, and that life histories were almost entirely undescribed, actually eight species were known from the States at the time, and seven life histories had been published previously. Several forms which he enumerated as species were invalidated by evidence given in his own work, and to have given it standing he should have supplied or published elsewhere satisfactory descriptions of the new forms he mentioned but concerning which his text was insufficient and unclear.

As the informed entomologist who reviewed this work remarked, "Even a slight acquaintance with the literature of the subject would have saved [the author] from errors which are surprising in a man who claims to have devoted eleven years to his subject." Is it not to be regretted that so much time and money were expended on work so deficient for want of adequate taxonomic background? For "it is the systematist," said Raymond Pearl, "who has furnished the bricks with which the whole structure of biological knowledge has been reared. Without his labors the fact of organic evolution could scarcely have been perceived and it is he who today really sets the basic problem for the geneticist and the student of experimental evolution."

The National Museum's contribution to systematic studies

The U.S. National Museum is one of the world's great centers for systematic research. The studies that the Museum is unable to accomplish with its own staff it tries to encourage others to undertake. That is how it happened that the late Dr. J. A. Cushman became interested in working up the Museum's collections of Foraminifera. In his day he knew more about the classification and distribution in time and space of Foraminifera than perhaps any other man. His great knowledge of these shelled protozoans was derived in great measure from the vast collections that had been dredged up from the seven seas and stored in the National Museum, largely unstudied, before his time.

When these microscopic organisms came into prominence as primary indicators of oil-bearing strata, particularly in the Gulf of Mexico region, Dr. Cushman was the authority to whom the oil companies turned for help in applying this information. His special taxonomic knowledge of the group enabled him to predict from the species brought up in drillings the proximity of a given sample to oil-bearing strata within several hundred feet. His determinations were worth millions of dollars in revenue to the oil companies and in taxes to the United States Government. Though other techniques, electronic and geophysical, are now frequently employed in prospecting for oil, the Foraminifera are still important in identifying and correlating strata and in subsurface mapping in oil-producing areas.

The foregoing is perhaps the most outstanding example of the eventual successful application of purely systematic studies and the naming of species to economic ends. It can safely be said that most, if not all, systematic work has a dollar-and-cents value, perhaps not today or tomorrow, but certainly in time.

Biological controls

In looking over some recent literature dealing with biological controls, I saw reference to the classical example with which I became acquainted in my earlier days in the Government service some 40 years ago. It was the story of the identification of an insect that played the role of a villain threatening the destruction of the sugar industry of Mauritius back in 1910, and how it was circumvented in the best tradition of the popular "who-done-its" by a systematic entomologist. The villain was a destructive white grub that bored in the roots of the sugarcane, killing the plant. It appeared very suddenly in such alarming numbers and spread so rapidly that the threat of the ruination of the plantations of sugarcane, the big money crop of the island, could not be ignored. With such information as was at hand, the best guess was that the borer was the larva of an African genus of beetle represented on the island by two species and the only remedies that suggested themselves were to dig up the root stumps to destroy the larvae or to catch the beetles as they flew about at night in search of food. The invader, lurking unknown in introduced cane cuttings, and finding itself a favorable environment without enemies, in reproductive capacity far outstripped all human efforts to control it despite the fact that in less than 6 months more than 27 million insects were accounted for. Meanwhile, the aid of the specialists in the British Museum was sought. With the extensive reference collections and library there available, it was soon determined that the beetle was not an African one, but a New World form, of which, however, no record or specific description could be found. In an ensuing search through the large collections of that Museum three specimens of this self-same beetle, labeled "Trinidad," turned up. The fact that this native of the West Indies had never been mentioned in literature implied that it was of so little economic importance that it had failed to attract the attention of any entomologists stationed in the islands. What kept its numbers down at home?

With specimens for comparison, a trained entomologist soon located both the beetle and its larval stages in cane roots on Barbados. It further developed that there it had two natural enemies. The only one in evidence at the time was a so-called blackbird which eagerly followed field hands rooting up cane stumps, to eat the grubs turned up, but unable to reach those beneath the ground. The other natural enemy, a tiny, inconspicuous wasp, was discovered by a neat bit of detective work on the part of the entomologists. Attached to one of several Barbados root borers transmitted to the British Museum was observed a tiny white grub. In the manner of its attachment it suggested the larval stage of a small wasp common in Barbados, one of the family of solitary wasps known to parasitize beetle larvae but not heretofore the cane borer. The wasp lays her eggs upon the borers after paralyzing them with her sting so that they will serve as food for her own young on hatching. Introduced into Mauritius, this little wasp soon turned the tables on the cane borer.

One cannot leave the subject of biological controls in the field of agriculture without touching upon one of the most remarkable successes of all time. This particular one was made possible by the taxonomic studies that preceded and were undertaken in connection with it. It was the conquest of the prickly pear in Australia by the cactus moth borer, *Cactoblastis*.

Cactuses are peculiar to the New World. As horticultural curiosities, and also as hosts of the cochineal insect, they were introduced shortly after their discovery into many other lands. The dates of the early introductions of the prickly pears, or Opuntias, into Australia are not known. Some planted as hedges and in gardens escaped to run wild in the surrounding country. As with the cane borer in Mauritius, in a favorable environment and without natural enemies to keep them in check, they spread at a tremendous rate. The rapidity of their increase has been called one of the botanical wonders of the world. In a period of 20 years the land area preempted by these prickly pears increased from 10 million to 50 million acres.

It was imperative that something drastic be done if the Opuntias were not to take over the land. Millions of acres had become veritable wildernesses of prickly pears. The Australians soon discovered that the cost of eradicating cacti by hand, poison, or mechanical means so greatly exceeded the value of the land that it was prohibitive. Some less costly method would have to be employed if the land was to be reclaimed. Biological control seemed to offer the greatest hope. Forthwith, the systematic literature of the world was searched for all pertinent information—the kinds, distribution, and habits of the prickly pears, and especially the literature relating to the animals and plants that have been reported to live in or upon them. Australian entomologists searched the world, so to speak, for the known and yet unknown enemies of prickly pears, studying those preserved in museum collections, as well as the living ones in the field. The most favorable places for these investigations were Argentina and the United States, where the great natural stands of "pears" existed. Some 150 to 160 different kinds of insects injurious to the cacti were found, of which 50 proved to be new to science. Twelve of the most promising ones were introduced, and of these one, *Cactoblastis cactorum,* described in 1885 from South America, proved so successful that further introductions were unnecessary. From an original shipment of about 2,800 *Cactoblastis* eggs in 1927, 10 million were reared in the next 2 years. In the course of 6 years 3 billion eggs were released. The cactuses literally disintegrated before the onslaughts of the *Cactoblastis* grubs feeding within their tissues. Within 15 months after the first trial liberation, huge stands of cactus lay rotting on the ground and in the next 7 years the last large area occupied by the pears collapsed. By 1940 less than 100,000 acres were believed to be infested with patches of dense or moderately heavy cactus growth, whether of regrowth or seeding origin, as compared with the hundreds of square miles of a few years before.

In Queensland, the worst-affected state, it would have cost from four to five hundred million dollars to have cleared the infested areas of prickly pears by poison and mechanical means. Following a careful study of the problem, and, I would emphasize again, especially the taxonomic literature bearing on it, by introducing and distributing eggs of *Cactoblastis,* the Commonwealth Prickly Pear Board accomplished the task at a cost of less than a million dollars. Thus, the formerly useless acreage regained for settlement because of its suitability for grazing, dairying, and agricultural purposes became an asset valued at 40 to 50 million dollars, to say nothing of the worth of new improvements and the future yield of the land, which, in time, will amount to many times its present value. Moreover, the benefits of the Australian experiences extended to South Africa and India, where prickly pears had also been giving considerable trouble.

The Australian entomologists give due credit to the work of the taxonomists who preceded them in the study of cactuses and cactus insects for their share in the accomplishment of this latter-day miracle, but who could have foreseen 65 years ago that the then published description of an insect found to be new by an Argentine zoologist making known to science the animal life of his part of the world would, half a century later, be instrumental in saving a continent from a pest run wild?

Insect quarantine

To prevent such unwitting introductions as this cane borer, the prickly pear, and other pests, our Department of Agriculture has a farflung inspection service at all ports of entry and at border stations. Indicative of the importance that the Department attaches to the necessity of having all "immigrants" of agricultural import promptly identified is the fact that it maintains insect, plant, and phytopathological identification services. The division of insect identification, located in part in the National Museum in Washington, comprises a staff of about 40 entomologists and technical assistants, and elsewhere in the Department an equally alert staff of taxonomic botanists and plant pathologists. They handle many thousands of identifications each year and in the course of making them have detected many harmful insects and other forms of life which might otherwise have become serious agricultural pests.

Epidemiological applications

Malaria ranks as one of the great scourges of mankind. We hear a great deal of the wonder drugs developed to overcome it, but very little of the role that taxonomy played in furthering its control. From the time that Ross first discovered that the causative parasites were transmitted by anopheline mosquitoes, it was thought that the problem could be solved quite simply by a reduction of the mosquito population—by treating their breeding places with larvacides, by introducing the little mosquito-eating fish *Gambusia*, by clearing out aquatic vegetation, and by drainage. The results in the States, and in Panama in the course of the construction of the Canal where expense was no object, were most gratifying, but when the Rockefeller Foundation tried to apply these methods in southern Europe, the same successes were not achieved. The carrier abroad was a different species, to be sure, with different habits and capable of breeding at the edges of running water, where its North American congener was a pool breeder. Nevertheless, it was impossible to establish any correlation between the incidence of intense malaria and the relatively few anophelines found in houses. On the other hand, there were localities with incredible numbers of the anophelines, tens of thousands in a single stable, and no malaria whatever; there were swamps without malaria, and a great deal of malaria without swamps.

It was not until two important and, at the time, unrelated discoveries were brought to bear on the problem that the apparent anomalous behavior of the common malaria mosquito abroad was cleared up. The first was the precipitin test, permitting the exact identification of blood, both human and animal, a serological and purely taxonomic procedure by means of which, no matter where a mosquito was lurking at the time of its capture, its host or hosts could be determined. The other discovery, which to my mind is one of the most important discoveries in the history of malariology, and certainly in its European aspects, was the discovery by Falleroni that females of the apparent European carrier, which he carefully raised, deposited five different types of beautifully ornamented but consistently different eggs, and that a given female always laid the same type of egg. Today it seems incredible that 7 long years elapsed before this significant discovery was properly appraised and applied to the taxonomy of mosquitoes.

By means of these discoveries, what had been formerly considered a single but unpredictable and widely distributed form, was found to be in reality several distinct species and distinguishable races. Thus, the hitherto inexplicable behavior of the European malaria mosquitoes was resolved with the aid of taxonomy, and the way cleared for effective control.

Species sanitation

The exact knowledge of the species of mosquitoes found in any given area is of greatest importance in preventing the waste of effort and funds on unnecessary control measures and permitting full attention to be paid to the dangerous species. Species identification insures a maximum of effective control at minimum cost; we have, therefore, today "species sanitation," as it is called, as the accepted practice in mosquito control.

A notable instance where species sanitation was most successfully carried out was in the Natal, Brazil, area from 1938 to 1940. It was here that the late Raymond Shannon, formerly with the U.S. Department of Agriculture, and, at the time, with the Rockefeller Foundation, made the startling discovery in 1930 that the dread African carrier of malaria, *Anopheles gambiae,* was on the loose in the New World. Probably shipborne, it brought about, in all, what is said to have been the worst malaria epidemic in history— some 300,000 cases, with enormous mortality, in a comparatively limited area.

At once steps were taken to eliminate this exceedingly efficient carrier from the immediate vicinity of Natal. This was accomplished in the next 12 months, but, rather strange to say, no efforts were made to look farther afield for this highly dangerous insect. It apparently made the most of the opportunity so afforded. Nothing is known of its ravages in the interim. In 1938, however, it caused a serious epidemic of malaria some hundreds of miles inland. This time there was no hesitation. All possible means of control were directed against this much to be feared species. Nothing was left undone to completely eradicate it. In 2 years of intensive effort complete success was apparently achieved. No trace of *Anopheles gambiae* seems since to have been found in Brazil. A few airborne individuals, however, have been detected in planes from Africa and promptly destroyed. Here again, species identification proved to be the important thing. It made species sanitation possible, and definitely effective, in a comparatively short space of time, and today enables the Brazilian Government to keep this dangerous enemy out of the country.

The Black Death

The story of the plague—bubonic plague, the highly fatal Black Death of the Middle Ages—and of its spread and control in India, Ceylon, and elsewhere parallels that of malaria, and, like it, turns upon the critical recognition of species of insects—in this case, fleas.

Much of our knowledge of the dissemination of plague by fleas we owe to two men, L. Fabian Hirst, health officer at Colombo, Ceylon, and Nathan Charles Rothschild, an authority on the kinds of fleas, who dis-

covered that the prevalent rat fleas of India and the Orient did not constitute a single species. It was Hirst who first suggested and then demonstrated that these fleas had quite different biting habits and different appe-
tites for human blood, and thus varied in their effectiveness in transmitting
plague from rat to rat, and rat to man. Their discoveries established the
geographic distribution of the different species of rat fleas as one of the
most important factors governing the spread of plague, and for the first
time furnished a logical explanation for the relative immunity of certain
parts of India and Ceylon to both epidemic and epizootic bubonic plague.

This discovery was the natural outcome of the purely zoological researches of Rothschild and others on the systematics of fleas.

In time of war and in national defense

Though not accorded recognition in the headlines of the daily press or
rewarded with oak-leaf clusters, the taxonomists made many note-worthy
but unheralded contributions to the waging and winning of the late great
war with their prompt identification of the many things about which vitally
important information was urgently needed.

In war we have much the same problems in medicine, epidemics, disease, and health as in times of peace, only more intensified and more
urgently calling for solution or alleviation. The immobilization of armies
by attacks of malaria in the European theater and the casualties, if we
may call them that, from the same cause and insect-borne diseases in the
Pacific became so serious that it was of utmost importance that the mosquitoes, fleas, ticks, and other pests or vermin be identified without delay.
Those that could not be named by the sanitary and medical units in the
field were given the very highest priority to Washington for immediate
determination.

During World War II a well-known news commentator, for want of a
more timely subject perhaps, took it upon himself to ridicule a systematic
treatise of the fleas of North America. It is the type of technical work that
is of utmost value to the specialist desiring to make prompt and accurate
determinations. He described this Government publication as a waste of
paper, containing no useful information because it did not tell how to free
your dog of fleas. But it was just the sort of book that would have enabled
Rothschild to distinguish the species of flea that was the chief carrier of
bubonic plague in India from the less harmful kinds. Moreover, this publication has in it the very information which enables one to identify this
particular oriental plague carrier, which, by the way, has become established in this country, but happily, so far as we know, is not here infected
with that most serious of diseases.

A museum friend of mine, though not a scientist, was utterly shocked
by the low regard that the commentator had for work so important. He
wrote the commentator a letter which I believe is still pertinent, and I
quote part of it:

"Having for many years been connected with a scientific establishment, and not being a scientist myself, I have come to realize the real value of such scientific works as you disparaged, and for the first time in my life I am moved 'to write to the editor.'

"This impulse was perhaps strengthened by the fact that the very next morning [after your broadcast] I was pointedly reminded of it by the receipt from the medical officer at one of our outlying bases of a *single* specimen of flea which he particularly desired to have identified with reference to its function as a possible carrier of disease. Only by knowing the exact identity of an insect can information of this character be given promptly, and the scientific entomologist turns instantly to such works as you ridiculed just as you would seize 'Who's Who' or the Encyclopedia Britannica, or some report of the Department of Commerce for data you might need.

"A steady stream of mosquitoes, ticks, and the like is pouring into Washington each day by airplane under highest priorities from our farflung battle fronts, in order that the local specialists may make prompt identifications, thereby furnishing the medical officers in the field the guidance necessary for applying the most effective control measures.

"The mere knowledge of the precise name of 'resident' fleas and other insects will enable the medical and sanitary services of our Armed Forces to quickly ascertain which of several towns in plague-infested areas are the safest for quartering men.

"Such works as the one under discussion are a distinct contribution by the home front to our forces on the battle front. In this connection I am moved to quote a line from one of Kipling's 'Barrack Room Ballads':

> 'Making fun of uniforms
> That guard you while you sleep
> Is cheaper than them uniforms
> And they're starvation cheap.' "

Men, expendable I assume, were landed from submarines on more than one occasion to reconnoiter the places and the islands to be attacked. They were also instructed to bring back what they could of the animal life encountered. As important as was the knowledge of the numbers and disposal of the enemy was the identity of the insect vectors in calculating the risks of attack and casualties from disease, which, more times than we care to admit, laid out more men than the enemy. The identification of dangerous insects in the war areas can be speedily accomplished only because of the stores of knowledge that the taxonomists have accumulated over the years. They supplied much exceedingly valuable information in other directions also.

Something had to be done about floating mines drifting into our coastal waters to menace shipping. Our patrol fleet needed to know the paths they traversed through the sea, so that they might be intercepted before their hostile mission could be accomplished. It was also imperative to determine where the far more dangerous German submarines sinking tons of shipping in the western Atlantic and Caribbean areas had their bases for overhaul, refueling, and the replenishment of stores. From the surfaces of the mines and submarines were scraped marine growths and from the ballast and trimming tanks of the few submarines that were captured intact were recovered traces of bottom mud and sediments pumped into the tanks as the submarines were anchored near, or rested on, the bottom of the shoal bays of their rendezvous. These growths and the sediments, their mineral constituents and contained organisms, were carefully examined and named by appropriate specialists. When the identifications were checked against the known distribution of the various materials it became possible to plot the probable paths of the mines and also to trace the submarines to their bases where they could be destroyed.

You may well remember the paper balloons with which the Japanese so ingeniously took advantage of the currents of the upper atmosphere for dropping bombs on the States during the war with a minimum of effort and cost to themselves. Until recently it was not known that during the 6 months that the Japanese continued this unique barrage over 9,000 such balloons had been launched and evidence had been found that 300, perhaps many more, had reached this country, some traveling as far east as Michigan. These balloons and the bombs they carried might have been frightfully dangerous. They could kill and maim, start devastating forest fires, and, had they been so employed, would have been capable of spreading disease and noxious insect pests. How were we to stop them? The balloons were so constructed that after the last of their bomb load was dropped an explosive charge destroyed the balloon. To keep the Japanese from learning of the success of their efforts, the strictest censorship was imposed, but word was also quietly sent out to appropriate State officials that an intact balloon must be recovered at all costs, so that it could be carefully examined. One was fortunately secured by an Oregon sheriff. The resulting identification of the sand ballast that the balloon carried, along with some of the remains of microscopic plants that were found in the sand, pointed to five possible launching sites. Armed with this information, our Air Force promptly bombed all five sites and, in so doing, must have hit the right one or ones, for soon thereafter this menacing offensive ceased.

Fisheries biology and conservation

I do not have space to tell of many examples in other fields of study in which the name of a species or organism solved a biologic problem. But because of their special pertinence I should like to cite three little instances that bear on the economics of fisheries.

Some months ago an American specialist on sipunculid worms was asked for copies of his technical publications by an Alaskan cod fisherman who had found that where these worms occurred he always made good hauls of fish. He wanted to plot the distribution of the worms in order to do better and to extend his operations.

A matter of weeks ago an ardent sport fisherman brought in a mantis shrimp about which he wanted to know its mode of life, its distribution, and where it could be obtained in quantity. Of course, to make a search for information, the species had to be identified. In turn, we learned something also—that this stomatopod was the favorite food of certain desirable panfish much sought after by fishermen in the Chesapeake Bay area.

In the Carolinas, where shad enjoy a certain amount of legal protection, the State conservation agent must be able to distinguish between four or five species of fish, all superficially more or less alike, if he is to catch violators of the law and avoid congesting the courts with the innocent. So, even in the enforcement of conservation laws, a knowledge of the species involved must be had.

Some botanical applications

Recently I was discussing some of these things with a friend of mine who is a systematic botanist. He spoke of cortisone and yams, and mentioned how much and how often the plant taxonomists are being called upon these days for information regarding not only the names of plants, but also their phylogeny and systematic relationships. If a plant contains a rare alkaloid or drug, what about its relatives? Knowledge of kinship has facilitated many such investigations.

The lore of ancient, primitive, and often unlettered peoples contains much of interest and value to us if only we can find the scientific names for the animals and plants of which they had learned the properties, good or bad, useful or harmful, by long and often sad experience. Curaré is one of these.

Botanists these days work with maintenance crews in keeping clear fire lanes and electric-power and telegraph rights-of-way for the purpose of identifying the plants, so that the appropriate herbicides may be used to kill off unwanted vegetation. The result desired and achieved is a dense growth of low shrubbery that will so occupy and shade the ground that all other growth will be inhibited, yet itself will not hinder or impede the passage of inspection, maintenance, and repair crews. Manual as well as mechanical clearing of ways, uphill and downdale, in these days of high labor and operating costs, is an expensive proposition, which, at best, only temporarily controls the situation. Spraying, too, can be a costly affair, as well as ineffective, if indiscriminate, without regard to the kinds of plants involved.

Again we are moved to remark that wherever one turns, a thorough knowledge of the kinds of organisms, whether of plant or animal origin, sooner or later proves of real value and often of considerable economic importance in the most unexpected ways and places. And finally, may we repeat what George Gaylord Simpson once so well stated:

"It is impossible to speak of the objects of any study, or to think lucidly about them, unless they are named. It is impossible to examine their relationships to each other and their places among the vast, incredibly complex phenomena of the universe, in short to treat them scientifically, without putting them into some sort of formal arrangement. . . . Taxonomy is at the same time the most elementary and the most inclusive part of zoology, most elementary because animals cannot be discussed or treated in a scientific way until some taxonomy has been achieved, and most inclusive because taxonomy in its various guises and branches eventually gathers together, utilizes, summarizes, and implements everything that is known about animals. . . ."

Summary

Biology has many branches and there is much cooperation or overlap among them. Taxonomy is not so much one of the branches as a line of synthesis among them. It is an entire level of approach, the level of the kind, recognizing to what kind individuals belong, grouping the kinds, and storing the limitless data that relate to the kinds. Taxonomy is an activity based on mental processes of distinguishing and grouping used by all biologists and all sentient animals. In addition, the direct or indirect use of taxonomic concepts or data are virtually universal among biologists.

Dr. Schmitt's varied examples of the utility of taxonomic information make it unreasonable ever to say that knowledge of some particular animal or group of animals is not important. We never know when it will be important. Knowledge is good, and there are many ways in which any particular item of knowledge may come to the aid of mankind.

CHAPTER 3

Academic and Professional Taxonomy

Of the twenty-five hundred taxonomists in the United States at the present time, roughly 50% are in academic situations, where taxonomy is their research interest and all but a few are teaching other subjects primarily. About 35% are in positions here described as professional, but only a handful of these are employed for full-time taxonomic research. The remaining 15% are not professionally or academically employed but work on taxonomy as a hobby or avocation.

Taxonomists

From the earliest days of taxonomy, whether or not we take 1758 as the starting point, taxonomists have struggled to make known the kinds of animals that inhabit the earth. The number of people who have been involved in this work runs into the hundreds of thousands. Among them were carefully trained scientists, self-trained professionals, experienced amateurs, and dilettantes. Among these were men of wide biological knowledge and understanding, very one-sided men of strong opinion, and men with no interest in theories or anything except the building of collections.

The number of people in the world interested directly in the taxonomy of some group of animals must be at least ten thousand at the present time. Some spend all their time at taxonomic activities, some use it for relaxation only, and for some it is a research sideline. Both professionals and amateurs have produced excellent taxonomic work, but it would probably be impossible to determine for the entire period of formal taxonomy (a little more than two hundred years) whether the professionals or the amateurs had done the best work. It is certain, however, that in number of workers the amateur group is far ahead, and there seems to be little doubt that the amateurs, who predominated during the first century and a half, have done the larger share.

Even if there has been little correlation between professional or amateur rating and the quality of the output, there does appear to be a direct relationship between the quality of work and the training and experience of the worker producing it. More important than formal training are the ability to receive the ideas of others, a discriminating eye for the features of the animals, an interest in the previous literature, and a knowledge of the animals themselves. Nowadays, it is increasingly hard for a person to become proficient in taxonomy in his spare time; nevertheless, the fact that several respected present-day taxonomists have accomplished the feat shows that it is still possible.

What are the motives that bring zoologists to study the kinds of animals and their classification? The answer varies, but it always seems to involve an interest in the animals and in the systematization of knowledge, together with a satisfaction in assembling, possessing, and studying a diverse collection of specimens representing one sector of the diversity of nature. Some taxonomists in the past may have been trying merely to increase their own importance as authors of new species or owners of large collections, and some may have been only interested in satisfying their employers; but the person who goes into taxonomy these days is almost sure to be interested in the animals themselves. Of course there have been zoologists who felt forced to do some taxonomic work because they were unable to find a taxonomist willing or able to make the identifications for their ecological or evolutionary studies. These men have rarely become taxonomists in the full sense of the term, but many of them have made important contributions.

Training of taxonomists

Taxonomy should, of course, enlist the services of biologists of the highest calibre and with the best possible training. The implication above that a large part of taxonomic work has been done by amateurs may seem to imply that this requirement has not been met. But increasingly in modern times taxonomic research, as distinct from the curating and identification aspects of taxonomy, has been performed by persons with either formal zoological and taxonomic training or extensive practical experience in taxonomy.

The actual training necessary or desirable as a basis for taxonomic work is indicated by the outline of subsequent chapters in this book. *First,* the knowledge of the animals themselves in all their aspects must be substantial. This should involve all the aspects of life suggested in the chapters on Diversity, and it may be gained from the living animals, from preserved specimens, and from the published records of what others have observed. *Second,* a detailed familiarity with the taxonomic literature and the use of such stored data is essential in taxonomy, surpassing the need for this in other fields of biology. *Third,* experience in the methods of collecting, handling, preserving, and studying specimens and kinds of organisms is indispensable. To this may be added the methods of reporting the observations—publication. *Fourth,* a detailed knowledge of the rules of nomenclature and their application to species and other taxa, without which all other taxonomic work can be wasted.

Beyond all this a taxonomist is today expected to be well grounded in all the major fields of biology. He will be expected to deal to some extent with cytology, biochemistry, genetics, evolution, ecology, biogeography, parasitology, behavior, embryology, and perhaps paleontology. To do all this requires the broadest possible background of zoology, in addition to a high degree of specialization in the taxonomic aspects. Almost any zoology department can provide the general training, but the special training in systematics can be obtained at only a handful of universities. But even where there are no formal courses, taxonomic training can sometimes be obtained by a sort of apprenticeship under the influence of an experienced taxonomist whose research is taxonomic even if his teaching is not directly so.

Academic taxonomy

As an academic subject (or course of study), taxonomy is often entirely omitted from the biology curriculum. A random search of college catalogs might well turn up none at all under this name. It has sometimes been said that there are no courses in taxonomy but merely directed learning through experience. This is certainly not so, because many schools offer courses specifically in taxonomic practice and theory; nevertheless, it is unquestionably true that taxonomy is missing from many curricula in which it could make a real contribution to the biological training of both undergraduate and graduate students.

In many schools taxonomic courses are offered in departments other than Zoology. Among those known to be offering such courses at the time this is written are the following:

University of Alberta (Entomology)
Amherst College (Entomology)
University of California, Berkeley (Entomology)
University of California, Los Angeles (Entomology)
Catholic University of America (Biology)
Cornell University (Entomology)
Duke University (Zoology)
University of Florida (Biology)

Fresno State College (Biology)
Harvard University (Biology)
University of Illinois (Zoology)
University of Indiana (Zoology)
University of Kansas (Natural History Museum, Entomology)
University of Maryland (Entomology)
University of Michigan (Museum of Zoology, Geology)
University of Mississippi (Biology)
University of Nebraska (Entomology)
Ohio State University (Zoology and Entomology)
Pacific Marine Station
Philadelphia Academy of Natural Sciences
University of Pittsburgh (Biological Science)
Purdue University (Biological Sciences, Entomology)
University of Rhode Island (Zoology)
Rutgers University (Zoology)
Sam Houston State College (Biology)
San Jose State College (Natural Science)
University of Southern California, Hancock Foundation (Geology)
Southern Illinois University (Zoology)
Stanford University (Biology, Geology)
University of Tennessee (Entomology)
Tulane University (Zoology)
Union University (Biology)
University of Utah (Zoology)
Walla Walla College (Entomology)
State College of Washington (Zoology)
University of Wisconsin (Entomology)

In many other schools taxonomy, although not offered in direct course form, is taught to some extent through the practical means of identifying specimens, seeing how they fit into the existing classifications, and preparing revisions for practice or publication. This may be done under the guise of research, or it may be in a course on a group of animals. There is necessarily a good deal of taxonomy in courses in Mammalogy, Ornithology, Herpetology, Ichthyology, Entomology, Acarology, Protozoology, Parasitology, and Paleontology.

Where formal study of taxonomy is not included at an early stage in a zoological curriculum, it becomes necessary to teach some basic taxonomic ideas in other courses, because much of biology deals with kinds of organisms or with individuals that are referred to particular kinds. It is not enough for comparative anatomy students to dissect "just any animal." It must be an animal of known position in the classification. Its structures are noted and interpreted in the light of its position in the classification, and by means of this position it is compared with other specimens known to occupy different places in the classification. The whole basis of classification thus becomes evident as do the similarities and differences between the successive specimens.

In some courses entitled Comparative Anatomy the comparative aspect is largely abandoned. The animals dissected are looked at primarily as stand-ins for man, and the work is considered to be merely a prelude to the study of medicine and human anatomy. Other courses may also involve recognition or identification of species or more inclusive groups. Names are used, whether for species or for orders. But both classifications and names are taken for granted, and it is frequently assumed that the work of classifying animals is completed. How far this is from the truth should be evident to anyone who probes beneath the surface of the classification of animals as a whole.

Teachers. Taxonomic work is well-suited to research by college professors. It does not require expensive equipment; it can be conducted in short spaces of time between other responsibilities; and it can be published in small segments. Academic taxonomy can still make a large contribution to the taxonomy of animals, directly, in addition to the teaching of taxonomic ideas and methods to potential taxonomists.

The people involved in academic taxonomy, the teachers, are usually not professional taxonomists, in the strict sense of being paid directly to do taxonomic research. Those whose research is taxonomic and who teach only taxonomic courses are few. For the rest the research is something of a sideline, after their real job of teaching other subjects is done. This is no reflection on the individuals as taxonomists or teachers; it merely shows the lack of interest in taxonomy as a formal course subject in many schools.

Professional taxonomy

As a profession, taxonomy has never been attractive to any save the student with a real interest in the ordering or systematizing of knowledge, one willing to work in a glamorless field where rewards are primarily in personal satisfactions.

Some taxonomists work professionally in museums and research institutes. The number of such positions is probably less than two hundred in the United States. In some cases a large amount of time is spent in the more routine tasks of curating collections and identifying specimens. Often this is under the aegis of a public-supported organization, which must perform certain services for the public. Research time may be considerably reduced. A few taxonomists work for large pest control organizations, usually public service authorities such as mosquito-abatement districts. Probably a thousand or more American taxonomists work for oil-producing companies and mineral-prospecting firms. These are paleontologists, who sometimes refuse to be called taxonomists. They study the fossils in order to predict where the oil- or mineral-bearing rocks will be found. A majority of these study the microfossils, Foraminifera, diatoms, and Ostracoda. They are often more highly paid than other taxonomists; but there are usually severe restrictions on their freedom of publication, because of the economic implications of their work.

These are about all the professional jobs in taxonomy proper. The number of such positions, especially outside economic paleontology, is low, and vacancies are not common. There are occasionally attractive temporary jobs on expeditions or in large projects, but these are usually filled by persons on leave of absence from their regular work.

It has become increasingly difficult in recent years for a biologically untrained person to work effectively in this field, although it is still possible for a person to train himself. But taxonomy covers such a wide range of groups, of activities, and of viewpoints that it is seldom hard to find a field of endeavor suited to the means, the time, and the training of any person interested in adding to the knowledge of the kinds of organisms.

The organizations of taxonomy

In this age nearly all human endeavors have been organized in some way. Taxonomy is no exception. In general, the number of organizations is likely to reflect the number of people involved, and the distinctness of the organizations will be related to the isolation of the field from other branches of learning. Accordingly, the organizations of taxonomy are few, and most of them are not exclusively concerned with this one field.

Taxonomic organizations. In the United States there is only a single organization which devotes itself entirely to taxonomy and attempts to cover all aspects of it in all groups of animals:

Society of Systematic Zoology; founded in 1948; publishes *Systematic Zoology, SSZ Newsletter, Directory of Zoological Taxonomists of the World,* and *Books on Zoology.*

In England, a comparable organization exists:

Systematics Association; publishes occasional symposia, including:
Bibliography of Key Works for the Identification of the British Fauna and Flora (1953).
The Species Concept in Palaeontology (1956).
Function and Taxonomic Importance (1959).

Taxonomy and Geography (1962).
Speciation in the Sea (1963).
Phenetic and Phylogenetic Classification (1964).

One supposedly international organization publishes a journal of direct taxonomic interest:

International Trust for Zoological Nomenclature (a private body formed to publish the decisions of the International Commission on Zoological Nomenclature); publishes *Bulletin of Zoological Nomenclature* and various Official Lists of names.

National societies concerned with one or two groups. Many societies have arisen to promote interest in particular groups of animals. At first, most of these were active predominantly in taxonomy, but many later broadened their interests. Among the national societies of this type are:

American Society of Mammalogists; publishes *Journal of Mammalogy.*
American Ornithologists' Union (AOU); founded in 1883; publishes *The Auk* (1883–).
Herpetologists' League.
American Society of Ichthyologists and Herpetologists; founded in 1916; publishes *Copeia* (1913–).
Entomological Society of America; founded in 1953 by union of the former Entomological Society of America (1906) and the American Association of Economic Entomologists (1889); publishes *Annals of the Entomological Society of America* (1908–), *Miscellaneous Publications of the E. S. A., Bulletin of the E. S. A.* (1955–), *Journal of Economic Entomology* (1908–), and monographs under the name of the Thomas Say Foundation (1916–).
Entomological Society of Canada; founded in 1951; co-publishes *The Canadian Entomologist* (1868–).
Lepidopterists' Society; founded in 1947; publishes *Journal of the Lepidopterists' Society* (previously *Lepidopterists' News*) (1947–), and *Memoirs of the Lepidopterists' Society* (1964–).
American Malacological Union, founded in 1931.
Society of Protozoologists; publishes *Journal of Protozoology.*

National societies for mixed groups. Several national organizations deal with a variety of groups brought together by some such aspect as their manner of living:

American Society of Parasitologists; founded in 1924; publishes *The Journal of Parasitology.*
American Microscopical Society; publishes *Transactions of the American Microscopical Society.*
Society of Marine Borer Biologists and Chemists.

Paleontological societies. For those interested in fossils, including their taxonomy as well as other aspects, there are:

The Paleontological Society; founded in 1908; publishes *Journal of Paleontology* (1927–) (jointly with Soc. Econ. Paleo. and Miner., below)

The Society of Economic Paleontologists and Mineralogists; founded in 1927; publishes *Journal of Paleontology* (1927–). (See The Paleo. Soc., above)

Paleontological Research Institution; founded 1932; publishes *Bulletin of American Paleontology,* and *Palaeontographica Americana.*

Society of Vertebrate Paleontologists; founded in 1940; publishes *News Bulletin,* and *Bibliography of Vertebrate Paleontology and Related Subjects.*

Geological Society of America; founded in 1888; publishes *Bulletin of the Geological Society of America, Special Papers of the G. S. A., Memoirs of the G. S. A., Bibliography of Fossil Vertebrates, Treatise on Invertebrate Paleontology,* etc.

Regional societies on one group. There are regional societies, interested in certain groups of animals, often with national appeal and membership:

Wilson Ornithological Club; founded in 1888; publishes *The Wilson Bulletin.*

Cooper Ornithological Society; publishes *The Condor.*

Ohio Herpetological Society.

New York Entomological Society; founded in 1892; publishes *Journal of the New York Entomological Society* (1893–).

Brooklyn Entomological Society; founded in 1872; publishes *Bulletin of the Brooklyn Entomological Society* (1912–), and *Entomologia Americana* (1926–).

The American Entomological Society (Philadelphia); founded in 1859; publishes *Entomological News* (1890–), and *Transactions of the American Entomological Society* (1861–).

Entomological Society of Washington; founded in 1884; publishes *Proceedings of the Entomological Society of Washington* (1884–), and *Memoirs of the E. S. W.* (1939–).

Kansas Entomological Society; founded in 1925; publishes *Journal of the Kansas Entomological Society* (1928–).

Florida Entomological Society; founded in 1916; publishes *The Florida Entomologist* (1920–).

Pacific Coast Entomological Society; founded in 1901; publishes *Pan-Pacific Entomologist* (1924–), *Memoirs of the P.C.E.S.* (1951–).

Hawaiian Entomological Society; founded in 1904; publishes *Proceedings of the Hawaiian Entomological Society* (1906–).

Cambridge Entomological Club; founded in 1874; publishes *Psyche* (1874–).

Entomological Society of Quebec; founded in 1951; publishes *Annals of the Entomological Society of Quebec* (1956–).

Entomological Society of Ontario; founded in 1863; publishes *Canadian Entomologist* (1868–), and *Annual Reports of the Entomological Society of Ontario* (1870–).

Entomological Society of Alberta; founded in 1952; publishes *Proceedings of the Entomological Society of Alberta* (1953–).

Entomological Society of British Columbia; founded in 1902; publishes *Proceedings of the Entomological Society of British Columbia* (1911), and *Occasional Papers of the E.S.B.C.* (1951–).

Connecticut Shell Club.

Hawaiian Malacological Society.

Helminthological Society of Washington; publishes *Proceedings of the Helminthological Society of Washington.*

Regional societies for many groups. There are also a few regional societies interested in a wide variety of groups:

Association of Southeastern Biologists.

Cambridge Society of Natural History.

Buffalo Society of Natural Sciences.

Society of Natural History of Delaware.

Biological Society of Washington [D.C.]; publishes *Proceedings of the Biological Society of Washington.*

Elisha Mitchell Scientific Society; publishes *Journal of the Elisha Mitchell Scientific Society.*

Southwestern Society of Naturalists.

Supporting organizations. Many other organizations are occasionally concerned with taxonomy to the extent of supporting research or publishing its results. Among these are most general science associations, as well as academies of science:

American Association for the Advancement of Science; founded 1848; publishes *Science.*

American Institute of Biological Sciences; founded 1948; publishes *A.I.B.S. Bulletin* (now *Bio-Science*), *Quarterly Review of Biology, In Brief— In Biology,* etc.

Society of the Sigma Xi; publishes *The American Scientist.*

Washington Academy of Sciences [D.C.]; publishes *Journal of the Washington Academy of Sciences.*

California Academy of Sciences; founded 1854; publishes *Academy Newsletter, Pacific Discovery, Occasional Papers of the California Academy of Sciences,* and *Proceedings of the C. A. S.*

New York Academy of Sciences; publishes *Annals of the New York Academy of Sciences* (1877–).

Southern California Academy of Sciences; publishes *Bulletin of the Southern California Academy of Sciences.*

Kansas Academy of Science; founded 1872; publishes *Transactions of the Kansas Academy of Science* (1872–).

Virginia Academy of Science; publishes *Virginia Journal of Science* (1940–).

Museums. Museums are of all sizes and for many purposes. Some have been very active in the production and support of research and in the publication of taxonomic results. Among the more active non-university museums are the following:

United States National Museum (U.S.N.M.); publishes *Proceedings of the United States National Museum, U.S. National Museum Bulletin,* etc. (The U.S.N.M. is part of the Smithsonian Institution which publishes *Smithsonian Miscellaneous Collections,* etc.).

American Museum of Natural History (A.M.N.H.); publishes *American Museum Novitates, Bulletin of the American Museum of Natural History.*

Chicago Museum of Natural History (formerly Field Museum of Natural History); publishes *Fieldiana.*

Los Angeles County Museum.

Royal Ontario Museum.

Several museums important in taxonomy are associated with major universities. At the risk of omitting some of equal standing, we may cite:

Museum of Comparative Zoology (M.C.Z.), at Harvard University; publishes *Johnsonia* (mollusks).

Museum of Natural History, at University of Kansas; publishes *University of Kansas Publication (Museum of Natural History).*

Museum of Vertebrate Zoology (M.V.Z.), at University of California, Berkeley.

Museum of Natural History, at Stanford University.

Museum of Zoology, at University of Michigan.

A list of the museums of North America has been published, with the institutions listed by state, name, and field of interest:

American Association of Museums; published *Museums Directory of the United States and Canada* (1961).

Universities. Several universities sponsor publication of journals in the field of taxonomy:

Catholic University of America; publishes *Coleopterists' Bulletin.*

Pomona College; publishes *Journal of Entomology and Zoology.*

Stanford University; publishes *Microentomology.*

University of Kansas; publishes *University of Kansas Science Bulletin,* and *University of Kansas Paleontological Contributions.*

University of Notre Dame; publishes *American Midland Naturalist.*
University of San Francisco; publishes *Wasmann Journal of Biology.*

Commercial supporters. One biological supply house publishes a small journal with articles of interest to taxonomists:

General Biological Supply House, Inc.; publishes *Turtox News.*

Plant societies. One international plant taxonomy organization deserves to be cited here:

International Association for Plant Taxonomy; publishes *Taxon,* and various directories.

Summary

Taxonomy, more than most sciences, has been a field in which persons without formal training have continued to make real contributions long after a professional level of technicality was attained. Professional training is now required for most employment in taxonomic positions, but it is not essential to producing good taxonomic work.

There are always fashions or popular specialties among the sciences. The top position of taxonomy at the end of the eighteenth century has gradually given way to a series of other fields, until today it seems to be at the bottom of the list, without ready access to support of its research. Yet all other fields of zoology are dependent upon its identifications and classifications.

Few fields of science have been as internally fragmented as taxonomy. Its workers have studied animals so different from each other as to seem to have nothing in common, and some of the fragments, like entomology, have been so large that they were virtually autonomous. There has also been geographical division. Today, the field is so large that there can be no taxonomists experienced in all groups; consequently, specialization is universal. These specialties and regions have for more than a century had their professional organizations: Regional societies and institutions grew up within each field. In recent years there have been a few integrating organizations, but they have had a limited effect. These organizations are mostly not splinter groups or offshoots of more general societies. They originated separately and could be brought together only with difficulty.

Many taxonomists view taxonomy as a research sideline rather than a profession. Taxonomically, their first allegiance is to the society concerned with their group of animals. Only secondarily are they involved with the broader organizations of taxonomy, biology, or science in general.

PART II

Practical Taxonomy

"Taxonomy may be considered as a way of arranging and interpreting information." (Davis and Heywood, 1963)

"I believe systematics to be a way of looking at biology rather than a specialization of it . . . an approach to biology rather than a department of it." (Munroe, 1964)

Chapter 4. The Use of Classification   51
    Storage of data   51
    Recovery of data   52
    Hierarchy of groups   53
    Special terms   54
    Summary   58

Chapter 5. Identification   60
    Background experience   61
    Methods of identification   68
        Direct comparison   68
        Identification from literature   69
            Keys   69
            Pictures   72
            Descriptions   72
        Identification by specialists   73
    Summary   74

Chapter 6. Curating   76
    Collecting   76
        Sampling   78
        Killing   79
        Preservation in the field   80
        Recording field data   80
    Shipment   80
    Storage   82
    Summary   82

Chapter 7. Recording the Data   84
    Field notebooks   84
        Data sheets   85
    Research notes   88
    The cataloging of data   89
        Cataloging of specimens   90
        Cataloging of species   90
        Cataloging of literature   91
    Publication   96
        Types of publications   96
    Summary   100

CHAPTER 4

The Use of Classification

Classification is the grouping of like things. It is a part of the every-day life of every human being. A man classifies his neighbors as children, teen-agers, or adults; he classifies his food as meat, vege-tables, fruit, or dairy products; he classifies the roads as smooth or bumpy; and he classifies his knowledge as science, art, or literature. Every noun and adjective represents a classification of ideas or objects.

Sometimes objects are classified directly, as in the classification yards of railroads, where freight cars are sorted out according to their destinations, or as in a kitchen, where dishes are stacked in a cup-board according to size or shape.

Sometimes things are classified according to features which can be detected with the senses, with the result that people can be classified into tall and short, or fat and thin, without pushing them into groups. Musical compositions can be classified, according to their patterns, as symphonies, songs, sonatas, operas, hymns, overtures, and so on, even though heard in no particular sequence. Trees are classified as evergreen or deciduous, without transplanting them into physical groups. And one may classify ideas that have no physical existence or counterparts, such as democracy and tyranny.

Furthermore, after these groups are made by classifying individual things—the freight cars, trees, and governments—the groups are fre-quently further classified into more-inclusive groups. The very words for all the things of each kind cited above are examples of this. "Neighbors" is a group of people combining the groups "children," "teen-agers," and "adults." "Trees" is a group combining the groups "deciduous trees" and "evergreen trees."

Storage of data

The grouping of ideas or objects under class names or words is a means of storing information, particularly the fact that they are alike in some way. The storage effect is at first in the mind of the grouper, but it may be communicated to others through oral or written language. The storage is done by use of a coding system of scientific names. These names refer to the groupings and serve as a key to the information stored about them.

If the classification is more elaborate than just a few groups at one level, its erection also serves a related function automatically. It shows the existence of groups and of groups within groups. The very existence of the successive groups is new information, frequently of great value.

The non-taxonomist does not normally take part in the synthesis of classification schemes. He does not normally describe species, write revisions, or study the nomenclatural problems. But if he deals with animals in any scientific way, he is almost sure to make use of the classifications of the taxonomists.

Recovery of data

The use of a classification to recover the stored knowledge about a group is its major function. This is done by using the code names. Any name will lead a zoologist to all that is known about that species or group. It will also lead him to information about other species or groups and how they differ from the first.

The name *Drosophila melanogaster* will lead any zoologist to knowledge of this species of animal by some of these steps: The genus name *Drosophila* can be looked up in a nomenclator, where it will be found to be a member of the family Drosophilidae, the order Diptera, the class Insecta, and the phylum Arthropoda. Reference to these major groups in general books will show that there are certain special features characteristic of each. For example, the Arthropoda have jointed legs and a segmented sclerotic exoskeleton; the Insecta have only six legs but also tracheae and three body regions; the Diptera have hind wings replaced by halteres; and the Drosophilidae have aristate antennae, oral vibrissae, and certain wing vein features. All these things apply to *Drosophila melanogaster*.

Knowing the genus, through the first part of the specific name, has now led to knowledge of the family, the order, and so on. In the literature on these, obtainable from the bibliographies in the works already consulted, will be found: (1) the generic revisions or monographs that will provide information on the genus and this particular species, (2) the references to any work on this species in the non-taxonomic fields of zoology, such as distribution, ecology, genetics, and embryology, and (3) the list of other names (synonyms) by which this species has at other times been called and under which other information about it is recorded. When all of these leads have been run down, including all the other leads cited in these, all the known information about the species *Drosophila melanogaster* will have been assembled.

In this particular example, the amount of data is enormous, because of the great interest in this species on the part of geneticists. Every possible aspect of the species is covered, and there are special works referring to it alone. For most species of animals, there is no such large volume of information; nevertheless, most species have been reported on more than once, and it takes a determined search to uncover all that has been discovered and recorded about any one of them.

Classification also provides a means of referring to groups of kinds at one time. Echinoidea refers in one word to all the hundreds of species of sea-urchins, heart-urchins, cake-urchins, and sand-dollars that exist. All that is publicly known about this group can be recovered from the literature. But also the position of Echinoidea in a formal classification indicates that it is, for example, one of the five existing classes in the phylum Echinodermata; that it is grouped with the classes Asteroidea, Ophiuroidea, and Holothurioidea as the more-inclusive group Eleutherozoa; that it is itself made up of seven sub-groups treated as orders. Its place in the Animal Kingdom is thus made clear, so far as it is known.

Hierarchy of groups

Literally hundreds of facts about this species are thus shown merely by its place in the classification (Fig. 4-1). It is obviously impossible

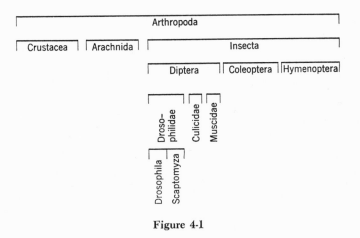

Figure 4-1

to show a million species of animals in such a diagram. It is therefore customary to show only the placement of the one group at each level, in a descending scale of levels or categories, thus:

Phylum   Arthropoda
Class   Insecta
Order   Diptera
Family   Drosophilidae
Genus   *Drosophila*
Species   *Drosophila melanogaster*

It is taken for granted that each level contains more than just the one indicated subgroup. For example, there will usually be other orders within the class and other families within the order (as shown in Figure 4-1).

This descending sequence of levels or categories in the first column is a hierarchy. The sequence of groups in the second column is sometimes called a "classification" of the indicated species. This is an unfortunate use of the word. It would be better to call it the hierarchy of groups of this species.

Special terms

All of these remarks take for granted an understanding of the words species, genus, group, hierarchy, and category. As there is some misunderstanding of these, even among professional taxonomists, it is necessary to define and discuss them. There are two ordinary English words that are at the base of all taxonomy and classification. These are "kinds" and "groups."

It has been known from ancient times that each sort of animal reproduces itself, producing others of the same sort. These are *kinds*, the animals which can interbreed and produce more of the same kind (or produce more by asexual reproduction). A kind includes all the individuals that might have been produced as offspring of one pair. Even before there was much knowledge of heredity and its mechanisms, it was obvious that kinds consisted of the animals that were or could have been related by being descendents of similar individuals. This is the same concept which is now called more technically *species*.

Groups also were recognized in ancient times. The many kinds which had their bodies covered with feathers were different from those covered with hair. Together these feathered kinds formed the group known as birds. The group could be large or small, inclusive or exclusive, natural or artificial. Any two or more kinds that are placed together figuratively because of some characteristic held by all of them are a group. The term is both general and specific, as it can be used effectively to refer to the group of aquatic animals with fins or to the group of kinds of birds we know as sparrows.

Groups may consist of a set of kinds or they may consist of a set of groups. The groups that are called individually sparrows, woodpeckers, warblers, herons, etc., will altogether be the group called birds. And the group of birds, along with the group of mammals, the group of reptiles, etc., will be the group called vertebrates. Thus, group is a noncommittal word, not specifying the number of its members or even whether they are individual kinds or groups of kinds.

If the group is a taxonomic one, as in these examples, then the word *taxon* is a synonym of group. Taxon was first used by botanists and is now frequently used by zoological taxonomists. In taxonomy, any group is automatically a taxonomic one. There seems to be no real need for any more technical word, so taxon is used in this book principally in the technical discussions of nomenclature.

It is possible to arrange kinds (species), groups of kinds (taxa), and groups of groups (also taxa) in an ascending series of ever greater inclusiveness. Several kinds are assembled to form a group, and several such groups into a more inclusive group. Then several of these more-inclusive-groups are assembled into a still-more-inclusive-group. For example, see Figure 4-2. This figure indicates that kind K, for example, belongs to group 3, to more-inclusive-group b, and to still-more-inclusive-group I. There are four levels of these groupings, and the levels are called *categories.* In this arbitrary case the categories are kind, group, more-inclusive-group, and still-more-inclusive-group. This ascending scale is a *hierarchy,* which is a series of levels into which the groups can be arranged.

In zoological classification, the levels in the hierarchy are named species, genus, family, order, class, phylum, and kingdom, from the least inclusive to the most inclusive. It is not correct to speak of some categories as larger than others, because a phylum might consist of a single species and thus be no larger. The categories are properly described as higher or lower. The groups assigned to the categories, however, are not higher or lower. They are simply more or less inclusive, or can even be described as larger or smaller, because they are at least potentially so.

In a practical sense the first level above the kind (species) is the category genus. The groups placed at this level are called *genera*

Kinds	Groups	More-inclusive-groups	Still-more-inclusive-groups
A B C	1	a	I
E F	2		
H I J K	3	b	
M N O	4	c	II
Q R S	5		
U V	6	d	
W X Y Z	7		

Figure 4-2. The grouping of groups.

(singular, genus). This use of identical names for the group and the category is the cause of considerable confusion and erroneous usage. There is no objective way to define genus, but each genus does consist of a group of species sharing certain features. The genus is then the group of the species that show these characters that have been selected to set off this genus. Of course, this group like any other group, may consist of just one object.

The next major level above the category genus is the category family. The groups of genera placed at this level are also called families. Each is simply the group of genera possessing the features deemed to be appropriate to family rank. Above the family are in turn the order, the class, and the phylum levels. The groups of families are called orders; the groups of orders are called classes; and the groups of classes are called phyla.

In addition to these levels, which are almost universally used, there may be any number of additional levels inserted in between. Groups of families may be put at an intermediate level called superfamily; then the superfamily groups make up the orders. Or the orders may be assembled into groups called subclasses, which in turn make up the classes. There are no direct limits to the number of categories (levels) that can be used, but the scale in Figure 4-3 includes those most commonly seen.

```
Kingdom
  Subkingdom
    Phylum
      Subphylum
        Superclass
          Class
            Subclass
              Superorder
                Order
                  Suborder
                    Infraorder
                      Superfamily
                        Family
                          Subfamily
                            Tribe
                              Subtribe
                                Genus
                                  Subgenus
                                    (Species)
```

Figure 4-3. The usual categories of the zoological hierarchy.

There can be no definition of what a family is, or what a superclass is. It is either a level in the hierarchy, defined solely by its position in the series, or it is a group (taxon) placed at the family or superclass level and called a family or a superclass. The only standard to determine the correct level for a group is the agreement of specialists. The group may be quite definite and understood by all, but its level may be subject to much difference of opinion. It may appear in various classifications as a class, a subclass, an order, or even a phylum, according to the views of the classifier.

It must be remembered that each of the categories potentially contains many groups of the next lower level. A class may consist of many orders, or the actual number of orders may be as low as one.

The fact that a certain class contains just one order does not make the class and order levels identical, because the class potentially contains other orders as yet unknown.

For example, the phylum Phoronida consists of just two genera, probably belonging in a single family. It serves no purpose to recount that there is just one class, with one order, one suborder, and one family, because potentially the phylum could include other worms so different from *Phoronis* that they would be placed in a new class. And the same thing can happen at any other level. For this reason the basic categories of phylum, class, order, family, and genus are generally cited, even in groups consisting of only one or a few species.

Summary

Classification is a logical device to deal with many objects under a few headings. The means employed to make the classes (or groups of things being classified) determines what use can be made of the scheme. If the groups are based on the visible features of animals, then the groupings will reflect the fact that the members of any one group are alike in these features. Only kinds showing the features of the group are included in it, so that when the species at hand is *Aurelia aurita* Lamarck, a member of the class Scyphozoa, it is at once apparent that it is a jellyfish, and a coelenterate; possesses nematocysts, oral arms, radial canals, etc.; and is marine. This is the use of classification—to recover information. The recoverable information is that which went into the building of the classification. Other information is also recoverable from the identification of the species.

How such a classification is erected will be studied in a later chapter. The methods greatly affect the usability of the resulting scheme. Some classifications are based on a small amount of information about the kinds, and therefore only a small amount can be obtained from it; but some contain so much correlated information that there is almost no end. For example, the worm class Polychaeta has sometimes been divided into two groups, Errantia and Sedentaria, based on their motile or sedentary habits. A few features of the body combine to make these habits possible, but in general these animals do not separate clearly into these two groups, so that this classification is not very useful. On the other hand, the classification of the phylum Mollusca into classes takes into account many features—type of shell, body arrangement, habits, habitats, major structures, and so on. It is useful for many purposes and represents a variety of information.

A classification is only as good as the information used to produce it. It is useful in proportion to the extent of this information. The factor that summarizes its effectiveness is correlation. The many features used in the scheme are undoubtedly highly correlated, which means that they occur in comparable patterns. It is a mistake, however, to assume that taxonomy needs to spend much time establishing these correlations. The correlations are principally useful when the readily visible features do not serve to show the taxonomic relationships. Accordingly, rather than try to classify a difficult group on the basis of egg characters, electrophoresis patterns, or serological reactions, the taxonomist tries to find overlooked features of ordinary nature that can be shown to be correlated with the unusual characters through which the situation was recognized. Numerical methods or computer manipulation may be helpful in establishing the correlation; however, in practice, such methods have generally not been necessary.

For discussion of more advanced aspects of classification, including data, basis, units, and methods, see Chapter 12, The Practice of Classification. For the theoretical nature of classification, see Chapter 18, The Nature of Classification and of Species.

CHAPTER 5

Identification

Identification has been called the utilitarian side of taxonomy. Classification stores the facts known about kinds of animals, but identification enables us to retrieve the appropriate facts from the system to be associated with some specimen at hand. Thus, identification is better described as the recovery side of taxonomy.

Identification refers to the association of specimens with the correct name for the species, which means association through the name with other specimens of that species. It presupposes that classification has already distinguished the species of the group and that names have been applied to them. In taxonomic practice there is often no sharp distinction between identification and classification, as the latter always involves the former and the former may lead to extension and improvement of the latter.

There are comparatively few groups of animals in which there is no background of classification. It is thus usually possible to undertake to identify specimens by reference to what has already been published. In some groups, the percentage of previously unknown species is high, so that attempts at identification, in the strict sense, fail, because there is nothing with which to identify many of the specimens. In these cases, further description and classification is necessary, so that the unknown or "new" species are added to the recorded classification.

There is no quick and easy road to identification in any group. Even if an identification is as certain today as is humanly possible, it may be quite wrong tomorrow, when someone discovers and publishes a fact hitherto unavailable. There is no substitute for experience. No one should ever expect to make a publishable identification of an isolated individual or population in a group in which he has had no previous experience. The safest identifications are usually those made by a specialist in the group at the time he is identifying many specimens of many species and thus obtains a view of the group as a whole, as well as of its parts.

Background experience

In order to make useful identifications in most groups of animals it is necessary to have experience of several sorts. This experience includes knowledge of taxonomic methods in general, the features and terms for them employed in the particular group, the usual study techniques, the identification literature on the group, and the natural history and comparative zoology of the group.

Familiarity with taxonomic methods and resources. Identification involves a working knowledge of all the other subjects that are part of taxonomy, or even of zoology. It is not possible to identify a particular organism with a particular group without an understanding of how the classification was made and what its limitations are. It is necessary to be experienced in taxonomic characters and terms in the particular group, as well as in the techniques used to preserve the specimens. There must be an understanding of the state of the knowledge of the group and the extent and availability of the literature about it, together with a considerable knowledge of the biology of the group of animals.

If this sounds as if identification of dependable accuracy can be made in most groups only by a specialist with long experience with those particular animals, the point has been successfully made. The point applies, however, primarily to genera, species, and subspecies. Useful identification to group can often be made by the non-specialist, by carrying the identification down the hierarchy only so far as the technical requirements or difficulties permit.

The concepts of taxonomy and classification are discussed elsewhere in this book. It is here assumed that the student is able to work with the ideas of species, genera, homology, similarity, and so on, to extract pertinent features from descriptions, and to handle the aspects of identification represented by the following paragraphs.

Taxonomic characters of the group. Nearly all taxonomy is highly technical and specialized. The characteristics used in distinguishing or grouping birds are so different from those used for nematodes that even a detailed knowledge of one would be of no help in making identifications in the other. Even within a single phylum, the differences between classes or orders may be great. In some groups of insects wing venation is extremely important as a taxonomic feature, but in others venation is seldom mentioned, while in still others wings are absent. In the Coelenterata all the Scyphozoa have the medusa form and are classified on features of the medusa. Anthozoa all have the polyp form and are classified on features of the polyp. The Hydrozoa, on the other hand, exist as either polyp or medusa and also in other forms called polypoid and medusoid; they may have both polyp and medusa forms in one life cycle; consequently, they cannot be classified on the basis of the features of any one body form.

It is, therefore, essential for a person undertaking to identify specimens in a group to have a general knowledge of the nature of the animals and the features that have in the past been used for identification of them.

Special terms in the group. Along with the specialized nature of the study on many groups, there have grown up special terminologies for the features used in classification and identification. In order to use the identification literature, a student must have a familiarity with these terms.

In a recent monograph of the Bryozoa there is a list of three hundred and sixty-one special terms necessary to describe the species, and the special terms used in describing insects have been the subject of several large glossaries. Glossaries of the terms in particular groups are not uncommon, but they usually appear as an appendix to a monograph or general book and are difficult to find. Some of the reference books for definitions of the specialized terms are listed here.

Glossaries of Terms

General

(Anonymous). 1960. Preliminary List of Works Relevant to Descriptive Terminology. *Taxon*, vol. 9, pp. 245–257. (Systematics Association committee for descriptive terminology)

Borror, D. J. 1960. *Dictionary of Word Roots and Combining Forms.* Palo Alto, California; N-P Publications.

Brown, R. W. 1954. *Composition of Scientific Words.* Washington, D.C.; by the author. (Now distributed by the Smithsonian Institution, Washington, D.C.)

Jaeger, E. C. 1950. *A Source-Book of Biological Names and Terms,* 2nd ed. Springfield, Illinois; C. C Thomas.

Melander, A. L. 1940. *Source Book of Biological Terms.* New York; Department of Biology, The City College.

Nybakken, O. E. 1959. *Greek and Latin in Scientific Terminology.* Ames, Iowa; Iowa State College Press.

Woods, R. S. 1944. *The Naturalist's Lexicon. A List of Classical Greek and Latin Words Used or Suitable for Use in Biological Nomenclature.* Pasadena, California; Abbey Garden Press. (*Addenda to The Naturalist's Lexicon,* 1947.)

Protozoa

Campbell, A. S. 1954. In *Treatise on Invertebrate Paleontology*, ed. by R. C. Moore. Part D; Fossil Radiolaria, pp. D11–D163. New York; Geological Society of America.

Campbell, A. S. 1954. In *Treatise on Invertebrate Paleontology*, ed. by R. C. Moore. Part D; Fossil Tintinnina, pp. D166–D181. New York; Geological Society of America.

Corliss, J. O. 1959. An Illustrated Key to the Higher Groups of the Ciliated Protozoa, *Journ. Protozool.*, vol. 6, pp. 265–284.

Invertebrates

Pratt, H. S. 1935. *A Manual of the Common Invertebrate Animals (Exclusive of Insects)*. Philadelphia; Blakiston. (Now Blakiston Division of McGraw-Hill Book Co.)

Porifera

de Laubenfels, M. M. 1955. In *Treatise on Invertebrate Paleontology*, ed. by R. C. Moore. Part E; Porifera, pp. E21–E112. New York; Geological Society of America.

Archaeocyatha

Okulitch, V. J. 1955. In *Treatise on Invertebrate Paleontology*, ed. by R. C. Moore. Part E; Archaeocyatha, pp. E2–E20. New York; Geological Society of America.

Bryozoa

Bassler, R. S. 1955. In *Treatise on Invertebrate Paleontology*, ed. by R. C. Moore. Part G; Bryozoa, pp. G2–G253. New York; Geological Society of America.

Mollusca

Arkell, W. J. et al. 1957. In *Treatise on Invertebrate Paleontology*, ed. by R. C. Moore. Part L; Ammonoidea, pp. L2–L6. New York; Geological Society of America.

Burch, B. L. 1950. *Illustrated Glossary of Gastropoda, Scaphopoda, Amphineura*. Los Angeles; John Q. Burch.

Cox, L. R. 1955. Observations on Gastropod Descriptive Terminology. *Proc. Malac. Soc. London*, vol. 31, pp. 190–202.

Cox, L. R. 1960. In *Treatise on Invertebrate Paleontology*, ed. by R. C. Moore. Part I; Shells of Gastropoda, pp. I84–I169. New York; Geological Society of America.

Arthropoda

Harrington, H. J. et al. 1959. In *Treatise on Invertebrate Paleontology*, ed. by R. C. Moore. Part O; Trilobita, pp. O117–O126. New York; Geological Society of America.

Moore, R. C. 1961. In *Treatise on Invertebrate Paleontology*, ed. by R. C. Moore. Part Q; Fossil Ostracoda, pp. Q47–Q56. New York; Geological Society of America.

Petrunkevitch, A. 1955. In *Treatise on Invertebrate Paleontology*, ed. by R. C. Moore. Part P; Arachnida, pp. P42–P173. New York; Geological Society of America.

Størmer, L. 1955. In *Treatise on Invertebrate Paleontology*, ed. by R. C. Moore. Part P; Merostomata, pp. P4–P41. New York; Geological Society of America.

Insecta

Dillon, E. S. and Dillon, L. S. 1961. *A Manual of Common Beetles of Eastern North America*. Evanston, Illinois; Row, Peterson and Co.

Harrison, R. A. 1959. Acalyptrate Diptera of New Zealand. Christchurch; New Zealand Department of Scientific and Industrial Research. *Bull.* 128.

de la Torre Bueno, J. R. 1937. *A Glossary of Entomology*. Lancaster, Pennsylvania; Science Press. (Supplement A: G. S. Tulloch. 1960. Brooklyn Entomological Society)

Tuxen, S. L., ed. 1956. *Taxonomist's Glossary of Genitalia of Insects*. Copenhagen; Munksgaard.

Usinger, R. L., ed. 1956. *Aquatic Insects of California with Keys to North American Genera and California Species*. Berkeley, California; University of California Press. (Glossary, pp. 483–488, by W. C. Bentinck.)

Hemichordata

Bulman, O. M. B. 1955. In *Treatise on Invertebrate Paleontology*, ed. by R. C. Moore. Part V; Graptolithina and Fossil Enteropneusta and Pterobranchia, pp. V3–V101. New York; Geological Society of America.

Vertebrata

Blair, W. F. et al. 1957. *Vertebrates of the United States*. New York; McGraw-Hill Book Co.

Hass, W. H. 1962. In *Treatise on Invertebrate Paleontology*, ed. by R. C. Moore. Part W; Conodonts, pp. W3–W69. New York; Geological Society of America.

Pisces

Slastenenko, E. P. 1958. *The Freshwater Fishes of Canada*. Toronto; by the author.

Amphibia and Reptilia

Peters, J. A. 1964. *Dictionary of Herpetology*. New York; Hafner Publishing Co.

Smith, H. M. 1956. *Handbook of Amphibians and Reptiles of Kansas*. Univ. of Kansas Museum of Natural History, Misc. Publ. No. 9 (ed. 2).

Stebbins, R. C. 1954. *Amphibians and Reptiles of Western North America*. New York; McGraw-Hill Book Co.

Stebbins, R. C. 1951. *Amphibians of Western North America*. Berkeley, California; University of California Press.

Aves

Van Tyne, J., and Berger, A. J. 1959. *Fundamentals of Ornithology*. New York; John Wiley and Sons.

Mammalia

Burt, W. H. 1957. *Mammals of the Great Lakes Region*. Ann Arbor, Michigan; University of Michigan Press.

Hall, E. R. 1955. *Handbook of Mammals of Kansas*. Univ. of Kansas Museum of Natural History, Misc. Publ. No. 7.

Hall, E. R. 1946. *The Mammals of Nevada*. Berkeley, California; University of California Press.

Study techniques currently employed. The techniques of preparing and studying animal specimens vary greatly from group to group. Birds and mammals are ordinarily skinned and the skin and bones used for study. Fishes are usually preserved whole in a liquid preservative. Insects are either mounted dry on pins, cleared and stained for slide mounting, or kept entire in fluid. All taxonomic work is limited by the condition of the specimens available.

From the early days of taxonomy to the present time there has been a continual change or evolution in the methods of study. Better microscopes provide better visual observation; special media permit the making of microscope slide mounts that are more revealing and permanent, and better storage methods permit direct comparison over a longer period of time. In some groups the workers actively seek out new techniques. In other groups change comes very slowly, perhaps because the old techniques are still giving satisfactory results. It is possible to argue the advantages and disadvantages of both the old and the new methods of taxonomic study. Here we are only concerned with the fact that the methods in use at any period have a considerable effect on the taxonomic work produced in that period. The study of fleas and lice was never very effective until the invention of suitable slide-mount techniques; and the customary ways of studying pinned insects proved to be inadequate to solve the problems of the malaria mosquitoes in southern Europe until the taxonomists began to study the eggs. In order to understand the taxonomic work in some groups it is necessary to know the study methods and their limitations.

The literature of the group. There is a large accumulation of separate publications on most groups of animals. These publications range from short papers in obscure journals or pamphlets to large books and monographs. They are available in widely varying degrees. Few groups are so well cataloged that all pertinent literature can be readily discovered. Most non-specialist identifications are made from the literature, especially from the less technical types of publications. Consequently, it is often necessary to know the general reference literature which will lead one to the specific works appropriate for the problem at hand. (See Chapter 14, The Use of Literature.)

One of the few ways for a beginner to get a start in identifying specimens in a certain group of animals is to use the books which cover (sometimes incompletely or superficially) all groups or at least a wide segment of the kingdom. From these books can be obtained clues to the more technical and restricted works. Some of these general works are listed in a later paragraph.

Familiarity with the animals. The actual mechanics of identifying animals is so different in the various groups that it benefits us little to cite any of the details in a general book. A student must learn by working with the particular group, seeing how older work has been done and what facilities are available. Although it was previously stated that a detailed knowledge of one group will help little in making *identifications* in another group, it is nevertheless true that detailed experience in one group will make it much easier to study a second group.

The number of kinds of animals now known is so great, the diversity among them covers such a tremendous range, and the terminologies in use for describing this diversity are so specialized, that no one can expect to make sound identifications in an unfamiliar group of animals. The problems are very different in widely separated groups of animals. Furthermore, familiarity with the group must be considerable if the group consists of animals with few clear distinctions, has a disorganized literature, requires elaborate preparation of specimens, or has been the scene of contradicting or highly personal monographing. What it means to be familiar with a group in this sense is suggested under several headings below.

Natural history. The most obvious sort of familiarity with a group of animals is to know how and where its members live. The old term natural history is nowadays sometimes replaced by ecology, but the latter is much more formal, implying technical studies of animals under various conditions. Natural history still covers the study of the animals as they are seen in natural surroundings by an interested person. This aspect of knowledge can be of great help in identification, but it is probably of less direct value than the other aspects of knowledge of the group.

What the taxonomist does need to know is where the animal lives, in association with what plants and other animals, in how large aggregations or populations, with what habits and reactions, with what seasonal or cyclic changes in these, and in general all there is to know about its life.

Life cycle and developmental forms. Nearly all formal identification is restricted to adult animals. This does not mean that it is not necessary to make identifications of immature forms of species but merely that this occupies a minor place, especially with respect to the identification literature. It often happens that so far as a given book is concerned only adults can be identified, and this makes it necessary to be able to recognize the developmental stage of any specimen—whether it is adult or larval.

In most sexually reproducing animals, there is little difficulty in recognizing the first normal developmental stage—the egg—and the embryonic stage produced by the early cleavages of the egg. In viviparous animals the foetus or late embryo also is not difficult to recognize, because of our general familiarity with the appearance of unborn or unhatched animals. But in the vast majority of kinds of animals there are stages in the life cycle which are so different from the adults that they cannot be identified, except through use of special methods designed for such larval forms. The variety of these developmental forms is suggested in a later chapter. It is here necessary only to recognize the immature forms as such, and to use the special means to identify them.

In many groups of animals, it is not possible to identify male and female by the same key or the same features. Here again it is necessary to be able to recognize the sex of the specimen. It is quite common to make identification keys that apply only to one sex.

Comparative zoology. It is strange how seldom the literature refers to the fact that the data in all aspects of taxonomy are comparative in nature. Taxonomy has always compared structures, colors, shapes, and sizes. As data became available on physiology, biochemistry, ecology, behavior, distribution, etc., the more alert taxonomists utilized these also, insofar as they were comparative and the procedures of these disciplines could be adapted.

In most groups of animals the standard morphological data have been gradually augmented during the present century by data of other types. It is often not within the capacities of taxonomists to gather these data themselves. They can use it only when it is supplied by specialists in the fields concerned. For example, few taxonomists have the training or the facilities to collect biochemical data about their animals. If the information is made available to them by biochemists or serologists, they generally welcome it and use it to whatever extent is practicable. Whether from biochemistry or serology or from any other field, all data that are comparative are of importance to taxonomy.

The various types of comparative data are discussed in a later chapter. Here we are interested primarily in the data useful in identification. In general, these are the ones used in classification, with one major class of exceptions. Identification frequently depends even more heavily on the fact that many attributes are correlated; for example, that an animal's metabolism, development, and behavior are almost certain to vary in a pattern parallel to some of its external structural features. Inasmuch as classification should already have established these correlations, any of the correlated characters can be used for identification. In this way it has been found that the different species of birds have distinctive songs. These species can be identified by the song alone, even when the bird itself is unseen.

Methods of identification

Assuming that the inquiring person has enough of the foregoing background knowledge to proceed with identification, there are only a few ways he can proceed. If a collection of examples of all the pertinent species is available, he can compare his specimens directly with one after another until he finds the match to it. This method may be one of the most accurate available for zoological work, but it can also lead to serious error.

If specimens are not available for comparison, it may be possible to obtain the help of someone familiar with the group. A taxonomist who specializes on this group is probably the best source of accurate identifications, and this is an excellent way to start building a reference collection by means of which future identifications can be made. Even when a reference collection is available, the major source of identifications is the literature. This may be the basic descriptions of the species, later monographs or revisions, or keys specially made for the purpose.

Direct comparison. Theoretically all identifications are made either by direct comparison of specimens or by comparison of specimens with published statements about the group. In only a few groups of animals has it become possible to identify most species by memory, but even here the reference books are in the background. Identification by direct comparison is practically limited to preserved specimens in museums. In large museums it may be one of the most important means of identification. Familiarity with the classification of the group is still required, but the use of literature is eliminated or reduced.

In practice, direct comparison of specimens may suffice for identification in any situation where identified specimens are available for the comparison. If a herpetologist is studying the frogs of an area and has preserved and identified examples of the five species known to occur there, he can probably make on-the-spot identifications by direct comparisons. This can be done only so long as he doesn't encounter a sixth species, one that has strayed into the area or was previously undetected there. It is this possibility which makes identification by direct comparison alone less than certain. Only a person familiar with the group will be able to tell the meaning of the difference that will usually appear. Only experience will enable one to judge if the specimen at hand does or does not agree sufficiently closely with any of the comparative specimens to be safely identified with that species.

Identification from literature. In associating a specimen with the species to which it belongs, the basic act necessary, if the original specimens of the species are not available for comparison, is to compare it with the published descriptions of the species. In a group of any great size, this would be a very tedious task, involving hundreds or thousands of comparisons. Therefore, it is necessary to narrow the field—to skip over the great bulk of the species which are not closely similar to the specimen at hand. This elimination can be accomplished in some cases by the quick scanning of pictures to eliminate the subgroups that are obviously not appropriate. More effective elimination is possible if there are keys to the group.

Keys. A key is a tabular device that presents alternatives referring to features of the specimens. By comparing the specimen feature by feature with the key couplets, one gradually eliminates all the nonagreeing subgroups and arrives at the only one which agrees. For example:

Key to Writing Instruments

1. Using ink 2
   Using lead 3
2. Using a sharp nib for spreading the ink . . . . *ordinary pen*
   Using a minute ball to spread the ink . . . . *ball-point pen*
3. Wooden body holding a fixed lead . . . . *ordinary pencil*
   Metal body with movable lead . . . . *mechanical pencil*

If one held an unknown type of writing instrument to identify with this key, he would compare it with couplet 1. If it used lead, that line would tell him to pass to couplet 3. Comparing the instrument with couplet 3, if it had fixed lead in a wooden body, the line would tell him that it is an *ordinary pencil*. The instrument has now been identified as an ordinary pencil by use of the key.

It could happen that the unknown instrument uses neither ink nor lead but instead writes with a waxy crayon material. If so, it could not be identified in this key. If the key was complete when written, it included all known kinds of writing instruments, so it would mean that the crayon is a new species. A new key would be made to include it, thus:

1. Consisting of waxy crayon wrapped in paper . . . . *crayon*
   Not as above 2
2. Using ink . . . . (pens) 3
   Using lead . . . . (pencils) 4
3. Using a sharp nib for spreading the ink . . . . *ordinary pen*
   Using a minute ball to spread the ink . . . . *ball-point pen*
4. Wooden body holding a fixed lead . . . . *ordinary pencil*
   Metal body with movable lead . . . . *mechanical pencil*

It could be, again, that during wartime metal shortages, some manufacturers would make a mechanical pencil with movable lead but enclosed in wood. This instrument could not be certainly identified in these keys. There would be ambiguity as to whether one should put importance on the body material or the movability of the lead.

These illustrate two of the common faults of keys: (1) not covering the item being identified, and (2) variability of the item with resulting ambiguity in the key.

Other faults include: (3) a key designed for adult animals may not work for young ones, but the identifier may not know which stage he has before him; (4) a character in a key may apply only to individuals of one sex, but the identifier may have only the other sex or may not know which he has; (5) a key couplet may work well for most examples of an unknown but may fail when an unusual one appears; for example, dwarfs, terata, or hybrids.

Keys are made for several purposes that are somewhat distinct. *First,* there are general keys that are intended to place the specimen in the major group to which it belongs. Such a key should work for all variations, both sexes, and even immature forms. It is intended to put the identifier on the track of the more specific literature that he needs. This it does by giving him the name of the phylum, class, or order, by means of which he can pass on to more specialized keys.

Some of the works suitable for identifying the major groups of American animals are the following:

General Identification Works for Non-specialists

Driver, E. C. 1942. *Name that Animal.* Northampton, Massachusetts; Kraushar Press.

Eddy, S. and Hodson, A. C. 1955. *Taxonomic Keys to the Common Animals of the North Central States Exclusive of the Parasitic Worms, Insects and Birds*, Rev. ed. Minneapolis, Minnesota; Burgess Publishing Co.

Edmondson, C. H., ed. 1959. *Ward & Whipple's Fresh-Water Biology*. New York; John Wiley and Sons.

Jaques, H. E., ed. 1947–1960. *The Pictured-Key Nature Series*. Dubuque, Iowa; Wm. C. Brown Co. (Includes volumes by various authors on Living Things, Insects, Land Birds, Immature Insects, Protozoa, Mammals, Beetles, Spiders, Freshwater Fishes, Water Birds, Butterflies, and Grasshoppers.)

Light, S. F. et al. 1954. *Intertidal Invertebrates of the Central California Coast*. Berkeley, California; University of California Press.

Palmer, E. L. 1949. *Fieldbook of Natural History*. McGraw-Hill Book Co.

Pennak, R. W. 1953. *Fresh-Water Invertebrates of the United States*. New York; Ronald Press.

Peterson, R. T., ed. *The Peterson Field Guide Series*. Boston; Houghton-Mifflin Co. (Includes guides to The Birds of Eastern and Central North America, Western Birds, The Shells of Our Atlantic and Gulf Coasts, The Butterflies, The Mammals, Shells of the Pacific Coast and Hawaii, The Birds of Britain and Europe, Animal Tracks, and Reptiles and Amphibians.)

Pratt, H. S. 1935. *A Manual of the Common Invertebrate Animals (Exclusive of Insects)*. Philadelphia; Blakiston. (Now Blakiston Division of McGraw-Hill Book Co., New York.

Putnam's Nature Field Books. New York; G. P. Putnam's Sons. (Includes field books of Seashore Life, North American Mammals, Wild Birds and Their Music, Birds of the Ocean, Birds of the Pacific Coast, Insects, and Animals in Winter.)

Ward, H. B. and Whipple, G. C. 1918. *Fresh-Water Biology*. New York; John Wiley and Sons. (See also new edition by W. T. Edmondson et al.)

These books are useful for making tentative or field identifications or for non-professional work. They are not intended for taxonomic identifications and usually will not serve for fine distinctions, rare species, unusual individuals, or accurate nomenclatural decisions.

A *second* sort of key might have been prepared by specialists in a group for professional use by non-taxonomists. Field manuals for the identification of parasites or pests of great importance may contain very effective keys of highly artificial nature. They may serve their temporary purpose adequately, but they do not satisfy the taxonomist in his own work because they were not designed to convey much information. For example, two parasites might be adequately and promptly distinguished by the fact that one shriveled up at once when placed in 95% alcohol, whereas the other parasite maintained its size and shape. This would be an effective key character for public health workers, but the taxonomic parasitologist would wish to use a feature of more objective nature, very likely something that could be seen in already preserved specimens.

The *third* sort of key is one that would be drawn up by a taxonomist for his own use and the use of his colleagues in this specialty. Here, the first characters in the key would usually be the ones thought to divide the included group of forms into the most important subgroups. It would doubtless use much more technical terms than the previous two, and it would assume an intimate knowledge of the animals on the part of the identifier.

In some groups of vertebrates, such as the birds, the keys may have been simplified to such an extent that they would fall into our first type. The species are so well known and so few in number that they can be adequately identified without technical keys. In the large phyla, and most invertebrates, it has been found nearly impossible to make good keys to species capable of use by a non-specialist. Sometimes it is even difficult to make such keys to genera or even families.

Pictures. In a few groups of animals, nearly all the species can be identified from pictures of them, especially if the pictures are assembled in atlases that show all the forms. In most parts of the world the mammals and birds can be identified from colored pictures; the butterflies of the world have been illustrated in color to such an extent that most can be identified; and in parts of the world some of the mollusks can be readily identified by colored pictures of their shells. In the rest of the animal kingdom, the use of pictures is limited principally to verification of decisions made from keys and descriptions, but they are very useful for this.

Descriptions. Identification of a specimen with a named species depends on finding that the specimen is identical in pertinent features with the original specimens of that species. In the absence of direct physical comparison, usually not possible in practice, the last resort is to compare the specimen with the published description of the original specimens and the later descriptions of monographers. Very often this is the ultimate source of a specialist's identification. It must be recognized, however, that a great deal of unseen background also goes into this process. There will be detailed familiarity with these animals and the descriptive works of the author of the species. There will be an expert's knowledge of the pitfalls connected with the use of descriptions and the application of names.

It cannot be emphasized too firmly that in most groups of animals it is impossible to make publishable or dependable identifications of species by routine use of non-specialized books or comparisons. A large part of the extant records involve doubtful identifications and are to this extent worthless.

By specialists. Probably the safest single procedure to obtain correct identifications is to solicit the help of a specialist in the group, particularly one who has recently monographed the species of the world. Such a specialist has the maximum of knowledge and experience of the group, and he will be able to make positive identifications better than someone using literature or even identified specimens.

It should not be supposed, however, that a specialist can always make such identifications at a glance. In many cases the specialist is aware of so many unseen problems that he will require careful study of even routine specimens. There are groups in which adequate identifications can be made with little difficulty, but these are exceptional.

It is very common to send specimens for identification to a specialist in the group; however, there are certain matters of manners and professional ethics involved:

(1) Specimens should never be sent without prior permission.

(2) Identifications should never be requested without allowing all the time the identifier needs for the work. After all, he has other responsibilities and interests, and he does not generally owe any such service to you. In case of real emergency, a full statement of the need will probably educe cooperation.

(3) Inasmuch as most identifications are made without remuneration, it is courteous and sometimes expected that the enquirer allow the identifier to keep the specimen or part of the series. This is a slight recompense for his time. If the enquirer is himself trying to build up a reference collection, he may reasonably ask to have examples returned whenever possible. He will often find that the identifier can supply him with examples from other lots that will be at least as useful to him.[1]

(4) Always be sure that specimens sent for identification are properly prepared, preserved, and packed. It is a discourtesy to ask a busy specialist to examine inadequate material.

(5) Never seek a specialist's time for sorting of miscellaneous material. Sort the specimens carefully to species, and keep together all the examples taken at one time.

(6) Give the specialist all possible information about the specimens. Where, when, and by whom they were collected, together with the weather, ecology, parasites, or any other known information.

[1] Many aspects of professional courtesy and responsibility are cited in Mayr, Linsley, and Usinger, 1953.

(7) Send as many examples as possible, as the variation within the series will be of interest and value to the identifier.

It is not intended as a reflection on any specialist to point out that even here the results are not entirely free of the possibility of error. Even a specialist is human; he may underrate the importance of these specimens, he may make a slip-of-the-pen in writing down the name, he may even be ignorant of some new discovery or other factor that would influence his study, or he may even be careless.

Finding the appropriate specialist is not easy. A museum or zoology department may be able to provide a start. A directory of such specialists was published by the Society of Systematic Zoology in 1961. Although its addresses get more out of date each year, it gives lists of the specialists in all groups of animals during the period 1955–1960:

Blackwelder, R. E., and Blackwelder, R. M. *Directory of Zoological Taxonomists of the World.* Carbondale, Illinois; Southern Illinois University Press, 1961.

Summary

Identification is directly involved in several aspects of taxonomic work: (1) When new information is discovered about an animal, an identification is necessary before the new data can be published as referring to that species; (2) when one wishes to learn what is known about an animal, identification provides the key to all the literature about it; and (3) when specimens are to be accumulated or stored for later study, they can be arranged only by identification, which brings together similar specimens and kinds and permits a physical classification.

The practical need for identification of animals extends far beyond zoology, as illustrated by the cases recounted by Dr. W. L. Schmitt in the article "Applied Systematics" quoted earlier. Many activities of man involve this process in both theoretical and practical aspects.

Animals are so diverse and have become so complex in their forms and activities that identification of a given specimen is easily accomplished only in a few well-known groups, such as mammals, birds, and butterflies, and then only in certain parts of the world. In the rest of the animal kingdom reliable identification is much more difficult, especially of isolated specimens. The experience needed for such work includes knowledge of the biology of the animals, of taxonomy in general, of the literature, of the collections and specialists on these kinds, of the terminology of the group, and of the languages in which the literature is published.

Identification can be obtained by three methods or combinations of them: (1) comparison with already-identified specimens; (2) comparison with published descriptions, keys, or pictures; (3) obtaining the help of a specialist (who will use the first two methods).

Further discussion of taxonomic characters and the steps which depend upon identification will be found in Chapter 15.

CHAPTER 6

Curating

Taxonomy deals with the individual organisms, collecting, classifying, identifying, preserving, publishing, and so on. It is therefore reasonable to start a study of practical taxonomy with consideration of the problems involved in having these individuals, usually called specimens, available for study. In a broad use of the term, curating will cover the following activities: collecting, preserving, storing, and cataloging.

Treatment of these subjects will necessarily be brief and general, for no one volume could contain detailed methods for curating all types of animal specimens. The chief purpose will be to emphasize the facts that: (1) taxonomic work of all sorts is dependent upon curating procedures, and (2) the procedures themselves can have a substantial effect upon the taxonomic work.

Collecting

The procedures used for obtaining specimens from living populations cover a wide range of activities. They include picking up the insects that fly to a porch light in the evening, as well as trawling or dredging for deep-sea animals with elaborate and specially designed equipment, used with the aid of an ocean-going vessel, operated by a crew of sailors and scientists. Such direct collection is often supplemented by indirect methods, such as trapping, baiting, employing other animals to do the collecting, and bringing about environmental changes that cause the animals to come out of hiding. Some of the most effective collecting involves unusual methods that are quite ingenious, and there is some justification for feeling that if only standard methods are employed, only the common animals will be obtained.

Nowadays collecting is almost always a highly specialized procedure. By this is meant merely that the collector, to be successful, must concentrate on one group of organisms. He cannot possibly collect effectively all types *at once,* although it is possible for a resident collector to obtain, over a period of years, a wide variety of organisms from his region.

The late William Procter of Bar Harbor, Maine, made a hobby of collecting specimens of any sort from Mt. Desert Island for any naturalist who was interested. Over the years, responding to the interests of various taxonomists, he collected many thousands of species in two major groups of the animal kingdom, first marine invertebrates and then insects. His published lists of these species include nearly 10,000 kinds of animals.

For many years between World Wars I and II, the late D. C. Graham lived in Szechuan Province, in the interior of China. He collected any specimens which were desired by his customers, including insects and other invertebrates as well as vertebrates. He collected most of the giant pandas that are now known, as well as many other novelties. To do this he had to employ a variety of methods. Because he did not always specialize, his collections were far from complete.

Most collectors must specialize whenever they collect. Preliminary surveys can be made by indiscriminate collecting, but most taxonomy requires highly specialized collections that do not come from a quick collecting trip.

It is sometimes overlooked by inexperienced collectors that nothing short of intensive collecting will serve if one wishes to find *all* the species of any group living in a given area. One of the most extensively collected areas in the United States, for insects at least, was New York City and its environs from 1850 to 1920. Many fine collectors took part in a partially organized effort over the years, and a list of the species found was published in 1928 as the New York State List. At about this time a young man living in Brooklyn, named Kenneth W. Cooper, started intensive collecting literally in his own back yard, making an extremely close search for obscure insects the ordinary collector might miss. He promptly turned up a whole series of beetles not previously found in the state, including some completely new to science. It was the care in searching and the concentration on habitats often overlooked which resulted in this discovery of previously unknown animals.

It very frequently happens that the greatest rarities are found long after the more common forms are well known. And they are often found by non-specialists looking for something else. An example is the recent discovery of the living fossil *Neopilina.* All known members of this group of Mollusca were fossil, and all seemed to have lived in shallow water, littoral habitats. A person searching for surviving species would have known enough to search in the shallow parts of the ocean. An expedition that was not looking for these forms happened to dredge up, from very deep parts of the ocean, some shells that proved to be a living representative of the group.

The writer spent nearly two years collecting principally one family of beetles in the West Indies. With stops to be made on eighteen islands, there was not time for concentration at any one place for even a few weeks, and it was obvious that the rather extensive trip might result in the capture of none but the most common forms. To try to counteract this, a day or two was spent on each island simply searching out the spots worth further attention, especially those unusual habitats that could not be found at just any spot. Of the 400 species of the one beetle family that were collected, half were new, and many of these were the uncommon types that would not be taken in a single stop, no matter how carefully it was collected.

On many occasions the desired animals are present in such small numbers, or so widely scattered in space, that the collector simply takes all those that can be found or caught. By adding to these few the ones similarly caught by other collectors and stored in collections, he may be able to assemble a series of the species that gives adequate basis for a variety of studies. If not, he is forced to make the most of the inadequate representation. As a matter of fact, it is still true that a vast number of animals are still known from only a few specimens. This is because no opportunity has ever presented itself for workers to obtain a long series. There are, however, many species present in such large numbers, and so easy to collect, that it is possible for the taxonomist to study as many specimens as he deems necessary. The choice of the ones to be studied, from all those potentially available, is one of the first problems of the taxonomist. The procedure is termed sampling.

Sampling. In recent years there has been increasing biological interest in populations. A population consists, according to one definition, of the individuals present in one interbreeding system in a continuous area at a given time; or, according to another definition, it is any group of organisms systematically related to each other. No matter which definition is accepted, it will only rarely be possible to bring, physically, an entire population into the laboratory or museum for study. To circumvent this difficulty, methods have been developed to obtain a *sample* of the population. The sample must be small enough to be studied in detail and should be large enough to represent the population. This sample then becomes a statistical representative of that population at that time.

As general biological interest in populations has increased, there
have been suggestions that taxonomists must concern themselves primarily with populations. For example, it has been stated that most of
the actual taxonomic work is done with populations and that it is the
basic taxonomic unit. These populations are defined in this first sense,
as a local population; that is, the total sum of conspecific individuals
of a particular locality comprising a single potential interbreeding
unit. Such a population *can* be sampled. However, nearly all taxonomists deal principally with species and genera. The populations may
be of interest, but ordinarily no population is studied extensively
before its species is known to some extent. The interest in the local
populations is likely to be non-taxonomic. In only a few groups of
animals are populations commonly recorded as such. It is the species
that the taxonomist needs to know, and they consist of more than the
sum of all the local populations. He learns about the species from the
individuals, their life histories, and all their natures.

More recently it has been stated that it is not possible to classify
either individuals or groups, that what is classified is always a population, defined in the broadest sense as any group of organisms systematically related to each other. (Local populations are then called
demes.) This is a curious viewpoint. It is not clear at what level the
organisms must be systematically related, but this is cleared up by
the definition that a species is composed of a population. Thus, population and species refer to the same thing in practice; that is, all of
the individuals of one kind. With this definition, one could agree that
taxonomists primarily study populations, but it will be clearer if the
statement is worded that they study principally species.

At the same time, the sampling of populations is described as an
essential preliminary to study. Unfortunately, although sampling can
be carried out on local populations, there are no practical methods for
sampling a wide-spread species. No local population at any one time
is representative of a species or will show the limits of variability;
therefore, this usage of population also is taxonomically useless.

The conclusion drawn here is that populations are rarely of primary
interest to taxonomists. When they branch out to study speciation,
genetics, or ecology, taxonomists may well need to study populations.
The validity of the concept in those studies is not at all challenged
here. It does seem, however, that virtually no published taxonomic
work is based on population studies, although occasionally it is followed or supported by such studies. This follows almost automatically
from the nature of taxonomic study.

Killing. In some collecting methods the specimens are killed in the process of collecting. For example, some steel traps used for small mammals kill them instantly. In most collecting procedures, however, the specimens come to hand still alive. Inasmuch as it is possible to make a specimen useless by subjecting it to the wrong treatment, every collector must be acquainted with the requirements of the animals he is collecting, whether he merely drops them into alcohol or other fluid, uses a quick blow on the head, or carefully anaesthetizes them in a relaxed condition.

Preservation in the field

Like killing, the method of preservation, from the collector's point of view, varies from group to group. Some animals can be dried for shipment; some must be kept in fluid preservative; and some must be frozen entire.

It is essential for the collector to learn in advance the techniques used in the group he is to collect. There have been many short papers published on collecting and preserving methods, but these all apply to individual groups. They can be found listed in the Subject Index of *The Zoological Record* or in *Biological Abstracts,* and a few are cited in our bibliography. The quickest way to obtain adequate information is to ask a museum or a collector that deals with this group of animals.

This does not mean that the previously used techniques are necessarily adequate. Not only must the collector know the standard methods for that type of animal, but he must be alert to see the need for new techniques; his goal should be improvement in the recovery and delivery of adequate material, perhaps in response to more specific requirements. It will sometimes be possible for him to supply the needs of several zoologists at once, one with skins, one with skeletons, and one with blood sera, for example. All of this has to be prepared for in advance.

Recording field data

After the specimens are preserved, or even immediately after capture, a record can be made of data about the capture and the live animal. This is the last time that it will be possible to record the circumstances without recourse to memory, which is notoriously subject to error. It is customary to state that a collector should record all possible information about the place and time of capture, the activities and appearance of the specimen, its relation to other organisms and to the physical world, etc. A more realistic statement would be that for each type of specimen there is (1) a minimum amount of data that must be recorded if that specimen is to have any real value, (2) an additional amount that can be of value in taxonomic studies, and (3) a practically unlimited further amount that would be of value in other ways if it too could be recorded.

For taxonomic collecting, a real effort should be made to fulfil (2), although just what all this taxonomic data will consist of cannot always be known in advance. What constitutes the desirable data or even the minimum varies from group to group. The following represents roughly the priority in some groups:

Geographic locality
Stratigraphic position (if fossil)
Date
Elevation
Food plant (if plant-feeding)
Host (if parasitic)
Colors (if subject to change in preservation)
Collector

Other sorts of data may be very useful to collateral (non-taxonomic) studies.

The most important of these records *must* accompany every specimen or container of specimens. *All* records should also be written in a field notebook, in which are recorded additional details of the precise spot of capture, the nature of the terrain, the abundance and activities of the species, the collecting methods, and many other things. (See also the paragraphs on Notebooks in the next chapter.)

Shipment

Specimens are seldom collected at the spot where they are to be studied. They must be shipped or transported safely to the museum or laboratory where they will be stored and studied. The nature and size of the specimens and the manner of their preservation will govern the choice of shipping containers and transportation facilities. A few general hints may help to avoid certain common difficulties:

(1) Reduce weight by pouring off excess liquid preservative. This may avoid postal regulations against mailing of liquids, and most specimens will remain sufficiently moist for the period of transit.

(2) When forwarding to major cities from abroad, ship in bond so that customs inspection can be supervised by the addressee, who is more familiar with the required handling and repacking precautions than customs inspectors in the ports-of-entry.

(3) Be sure to take care of all postal and customs regulations in advance. Do not try to evade the rules, but do pick the agency best suited to this particular shipment.

(4) Never ship living animals or plants without a prior permit.

(5) If you are collecting for or shipping to someone other than yourself, get in advance his instructions for preservation and shipment. It may well be that care in this will mean the difference between excellent material for study and poor specimens hardly worth the effort of unpacking.

Storage

The storage of a study collection soon becomes a problem for any serious taxonomist. Specimens are subject to damage or destruction by pests; they can be ruined by dust, light, or damp; or they may accumulate until they occupy so much space that this becomes a major problem.

The nature and source of supply of the equipment necessary for safe storage differs greatly from group to group. Much information can be obtained from special papers in the literature of the group, or from a museum or experienced taxonomist.

Equipment can be obtained from biological supply houses, of which the following are a selection:

Ward's Natural Science Establishment, Rochester, N.Y.
Carolina Biological Supply Co., Burlington, N.C.
General Biological Supply House (known as Turtox), 8200 So. Hoyne Ave., Chicago 20, Ill.
Troyer Natural Science Service, Oak Ridges, Ont., Canada.
Ohkura and Co., P.O. Box 2, Shakujii, Nerimaku, Tokyo, Japan.
Standard Scientific Supply Corp., 808 Broadway, New York 3, N.Y.
Bio-Metal Associates, Box 346, Beverly Hills, Calif.

At this point it may be well to insert a postscript about the permanency of collections. Many times a valuable collection built up by much effort on the part of some collector or taxonomist has been kept in private hands after the really interested parties died. In some cases such collections are even in museums, especially local ones. Because of lack of interest, the necessary care is not given to the collection, and it deteriorates. It is not uncommon for such collections to be entirely destroyed. This can be a great loss to science. Every person who collects specimens or assembles a collection for taxonomic purposes should make provision for permanent care of the material. The way to accomplish this is to arrange for it to be given to a major museum which has facilities for this purpose. Most large museums will gladly accept the gift of a collection with the proviso that the donor retain control and possession during his lifetime.

Summary

Curating is an extremely varied business that occupies a large part of the time of most taxonomists. There is no way to relieve the taxonomist of this burden, because much of it is an essential part of his background experience. The curating activities themselves are intimately associated with the taxonomic activities.

With the broad definition of curating used here, there is much diversity in the curatorial activities of individual taxonomists. Some do much collecting; some take care of large collections; and some do a great deal of bibliographic and cataloging work. Nearly all must be involved in all of these to some extent, as well as in the truly taxonomic activities of identification, classification, and naming.

CHAPTER 7

Recording the Data

The results of research are of little use to anyone unless they are recorded. Even the observer or experimenter will find his data useless, unless he keeps a record of his results. For studies to have a lasting effect on science, for them to become part of human knowledge, the results, at least, must be published. Publication is the virtually necessary prelude to use and criticism by other scientists, to repetition of the experiments, to analysis of implications in other fields, and to verification of consistency with other knowledge. The final judgment on a theory may not come for years, but in the end the faults and virtues of every work will be known.

The keeping of notes is not as pertinent to the study of classification as to some other fields, but it can be of very great value. It is beyond the scope of this book to describe the various methods of taking and storing notes, but the value of accurate and detailed records of data is pertinent to several aspects of taxonomy and may appropriately be emphasized here. References to this subject in the literature of systematics are not common.

A distinction will be made here between the notes taken in the field to accompany specimens intended for taxonomic study and notes taken in the course of research (taxonomic or other). The first are intended as permanent records to accompany the specimens from then on. The latter may be intended for permanent record in publication but are more likely to be of temporary nature to record the progress of the study.

Field notebooks

It has already been noted that accurate records of the place, time, and circumstances of capture of the specimens are of great importance. In recent years, the employment of classifications in other branches of biology has become much more frequent, and the amount of data

84

contributed by these fields to classification is becoming substantial. It is therefore desirable for taxonomists to take maximum advantage of information available at the time of collection, even though this information is not the goal of the taxonomic study.

When a taxonomic study is preceded by a period of planned collecting of specimens, thought is usually given to the problem of just what data also need be recorded to give the specimens maximum value. The better this preliminary planning, the better the results in accuracy and usability of the data. However, it is often found that even carefully planned records turn out to be inadequate in some important but overlooked detail, especially when the study leads into unexpected avenues of investigation, as is often the case with good projects. Some examples may serve to highlight both the advance planning and its possible failures.

Data sheets. In many fields of biology, whether experimental or observational, records are customarily kept on prearranged forms that facilitate the recording. Such forms are well-nigh indispensable in keeping the data uniform, complete, and capable of ready storage, but there can be serious disadvantages which must be recognized. They provide spaces for all the types of desired data, and the notes are taken on the spot at the time of collecting. These forms help to ensure uniformity of data for all specimens, and encourage complete recording of data, because the blanks provided serve as reminders. In addition, the use of the forms encourages advance thought and planning and permanent storage of the uniform notes.

These forms have some disadvantages, which should not be overlooked. They tend to restrict the recording of any data not provided for, and they sometimes prove too elaborate for effective use.

Figure 7-1 is a facsimile of one-half of a 4 × 7 inch sheet prepared by a student collector for use on a collecting trip to a new part of the world, where general collecting of insects was to be undertaken. He hoped to make the form so complete that the station number and a few circled words would be about all that was necessary. The form turned out to be totally inadequate, and most records were kept by writing between the lines. (The peculiarities of the capitalization and printing were due merely to the fact that the student set the type by hand and printed the form himself!)

In general, it can be said that the usefulness of such forms in the long run depends on several factors: (1) the care used in their preparation, (2) the extent to which the possible future ramifications of the problem can be foretold, and, especially, (3) the temperament of

Collector   Date   No.

Locality

State, County   Country

El.   ft. meters   Host

Swimming Floating Jumping Crawling Running Flying Burrowing Mining Stribing

Under In Resting on Breeding in Feeding on   BEATING SWEEPING SIFTING

Lake Stream Marsh Pond Puddle Beach Bank Submerged-plant Intertidal rocks

Forest Edge Meadow Cult-field Ground Wall House Fence City Country

Flower Stem Leaf Trunk Bark Cambium Root Petal Needles Bud Cone branch

Logs Stump Fruit Twigs Seeds   Stone Board Leaves Refuse Humus Soil

Dung Carrion Parasitic Predaceous   Dead Hibernating Paralyzed   Larva Pupa

No·   Collector   Host

SWIMMING FLOATING JUMPING CRAWLING RUNNING FLYING BURROWING MINING

Under In Resting on Breeding in Feeding on   BEATING SWEEPING SIFTING

Lake Stream Marsh Pond Puddle Beach Bank Submerged-plant Intertidal rocks

Forest Edge Meadow Cult.-field Ground Wall House Fence City Country Road

Flower Stem Leaf Trunk Bark Cambium Root Petal Needles Bud Cone branch

Logs Stump Fruit Twigs Seeds   Stone board Leaves Refuse Humus Soil

Dung Carrion Parasitic Predaceous   Dead Hibernating Paralyzed   Larva Pupa

Figure 7-1. An entomological collector's record sheet.

the user in relation to keeping records and being conscientious in using the forms.

An example of two detailed forms (Figs. 7-2 and 7-3) prepared after years of preliminary experience and designed to give uniform data for a particular study is illustrated here. They were used by Dr. John C. Downey in a study of the population and variability of a species of butterfly living on lupine plants.

The Population Ecology sheet (Fig. 7-2) was used in the field to record ecological and collecting data. It was designed to encourage the keeping of detailed environment records. It proved to be effective in several ways: In addition to the advantages listed above, it made possible rapid comparison of some particular aspect on a series of the forms. For example, the temperature at the time of capture could be rapidly compared in various populations to show the effect of this feature on flight activity during the day. It often reminded the collector to note a factor that would otherwise have been overlooked.

A subsequent development highlights some factors in the keeping of such forms. A companion on one trip used a bound notebook to keep a journal, a combined diary-and-field-notes. At every stop, notes were made of all the usual features, as well as brief lists of other specimens taken incidentally and external factors affecting the trip, with

POPULATION ECOLOGY

Field No._____
State_____
Locality_____
County_____

LOCALITY
  Description:_____
_____
_____
_____

  Elev. (m):_____   Date:_____

ENVIRONMENT
  Temperature:_____   Time:_____

  Weather:_____
_____

  Soil:_____
_____

  Moisture:_____

  Vegetation:_____
_____

  Condition of lupine:_____

  Abundance: _____ Food plant noted:_____

  Estimate No. Species Lupine_____

  General: _____
_____

BEHAVOUR OF ADULTS
  Nos:_____ Proximity of host plants:_____

  Condition:_____

  Activity:_____
_____

  Altitudinal range in vicinity:_____

  Seasonal status:_____

  Field larvae observed:_____ Eggs observed:_____

  % Parasitism:_____

  Predominant Insects in Area:_____
_____

REMARKS

Figure 7-2. A collector's ecological data sheet.

mileages between collecting sites, names of intervening towns, and
any unusual features or occurrences.

This journal proved to be so helpful that the forms were abandoned
for the rest of the trip. However, the items of information listed on
the data sheet were consciously entered into the journal, so that it
served a dual purpose. In subsequent use on the trip the journal
seemed entirely adequate. After return to the laboratory, when a
specific feature was to be compared for each locality, it was found that
the journal was more difficult to use than the forms, because the
arrangement varied. It was also found that the journal sometimes
missed a pertinent fact that would have been recorded on the form
because of the place reserved for it.

This fault of the journal could be eliminated by any of a variety of
means, such as use of a large rubber stamp at each new locality. The
stamp would have the desired features of the form and would thus
help to combine the good features of both types of notebooks. If the
journal were loose-leaf, one of the original forms could be inserted for
each collecting station. One other advantage of a journal is that it
can be illustrated by means of pasted-in maps, photos, or other material. It would thus be possible to have the rubber-stamped form on
gummed labels that could be pasted in where appropriate. The advantage of this would be use of a bound rather than a loose-leaf notebook, the latter being subject to loss or disarrangement of pages. There
is no end to the possibilities. Imagination and foresight are the in-
gredients necessary to devise an effective note-taking system.

Research notes

The keeping of notes is necessary not only in the field to accompany
the specimen collected but also in the laboratory, where the specimens
are later studied. Here again, uniformity is an important benefit be-
stowed by pre-arranged forms. This results in direct comparability of
the observations. These research notes are often the result of assembling the data for report or publication. They are not only written but
may be drawings, photographs, recordings, graphs, diagrams, etc.,
and merge into the permanently stored dissections, preparations, and
specimens themselves.

A good museum has not only the specimens but also the collecting
notes, the research notes, and the resulting publications.

An example of a prepared form for research notes in the same project as described above (the butterfly taxonomy and ecology) is
shown in Fig. 7-3. Its principal advantages are permanence, completeness, and comparability.

SPECIMEN DATA RECORD

Sex: _____ Genitalia Dis. No. _____

Population
  State: _____ County: _____ Locality: _____

Wings
  Length forewing _____ Length hindwing _____

  Macules
    all present.

    missing macule no.: _____
    additional macules: _____
    fusions: _____
    spot ratio, fore to hindwing: _____

  Constellations
    Forewing value: _____ hindwing value: _____
    Placement: _____
    Ground color: _____

    Termen in male: _____
    Blue of male: _____
    Orange in female: _____
    Blue of female: _____
    Brown of female: _____

Body
  length: _____
  setae: _____

Genitalia: Male
  F _____ H _____ U _____ E _____
  Valve length _____ Valve width _____
  D _____
  Ratios: F/U _____ Forewing/Valve _____
  Other: _____

Figure 7-3. A research data sheet.

The cataloging of data

The verb to catalog means rather different things to different taxonomists: (1) the cataloging of specimens, (2) the cataloging of species and groups, and (3) the cataloging of literature. They are discussed separately below.

Cataloging of specimens. In some groups of animals, the specimens in the collections are recorded individually or in lots in what is called cataloging. This is sometimes in the nature of an accession record, showing how the animal was received, the data of its collection, its identification, and the place of storage in the collection.

In the higher vertebrates, this generally takes the form of assignment of an individual number to every specimen and the recording of the relevant information on a card to be filed in a card catalog either numerically, systematically, geographically, or all of these.

In collections stored in fluid preservatives, such as those of many aquatic invertebrates, the individuals may not be cataloged but merely the lots. Each collection will receive a number, or each species in each collection. In insect collections such cataloging is extremely rare, because the large numbers of both species and specimens make it entirely impractical. It is sometimes possible to include on the pin labels a reference to a lot number or catalog number, but this has generally proven to be of temporary value at best.

As a means of keeping extra data associated with specimens, such procedures as cataloging of specimens are highly desirable. In practice, however, the time and cost have proven so high that cataloging has often been deemed unjustified. As detailed data becomes more necessary to fill the demands on taxonomy, methods must be found to keep track of data in an economical manner.

One class of specimens is, however, almost universally cataloged. These are the type specimens on which the species are based. In major museums these type catalogs are usually bound books in which the types are serially recorded by number. The number on the specimen refers back to the notes on how, when, and by whom the specimen was made a type.

Cataloging of species. Checklists and catalogs are often physically somewhat different, as explained in a later section, but they are similar in being basically lists of species. Such lists of the species, either of one group or of one region, are sometimes made on cards primarily for personal use, but they are often prepared with publication as the goal. They may list the species of a given region, the species of a given group for the world, or the species or specimens in a particular museum or collection.

When such a catalog is made on cards for personal use, it is not clearly distinct from the next type. A card catalog of the species in any group from any region may simply consist of a card for each species, with the original author, date, and reference, the formal synonyms (other names for that species), and notes on any other information desired by the collector. The cards serve as a list of the known species. For publication, the cards are carefully arranged and transcribed into a list for the printer. The published catalog may contain only the names themselves, or it may contain a great deal of additional information about each one.

Cataloging of literature. In general, taxonomic cataloging of literature is the assembly of an annotated bibliography of the previous work on a group of organisms. It is usually more than a mere list of the published books and papers, as its purpose is to arrange taxonomic and nomenclatural information, not library information. Because flexibility in arranging the data is essential, cataloging is nearly always done on cards or sheets that can be arranged as desired or handled with the aid of mechanical devices.

The catalogs of *species* are frequently not prepared for publication. Complete catalogs of the *literature* are also first of all for personal use of the taxonomist, but they are sometimes published. They are extremely valuable, if well prepared, but they are so expensive to publish that at best they are usually combined with the catalog of species.

The catalog is made from the original books and papers. If these works are then sent back to a library, the value of the catalog will be greatly reduced. The catalog may contain an abstract of the information in each publication, but it serves primarily as an index to those publications. The works will have to be consulted again on many occasions, if the catalog is in active use. It is impossible to foresee all the problems that will necessitate reference back to the original. Therefore, a taxonomic catalog should be assembled in conjunction with a set of the pertinent papers and books, so far as possible.

Cataloging of the literature is essential for serious taxonomic work. On many groups of animals, the literature is so diverse, so voluminous, so scattered, that the taxonomist can assemble it only through years of effort. His catalog must be highly accurate and complete, unless it is to be misleading, because many taxonomic decisions are based on the data in the catalog. The extent to which the quality of detailed taxonomic work depends on the knowledge of previous work is sometimes not fully realized. The failure to uncover one pertinent paper published at any time since 1758 might quite possibly render an entire analysis useless and produce a whole series of conclusions that would be logical but erroneous. This has happened many times. It can be prevented only by the most thorough bibliographic work and properly designed cataloging of the historical data.

It is said that there are almost as many methods of filing reprints as there are scientists filing them. Something of the same sort could be said about the methods of cataloging. This is not entirely bad, because cataloging should be carefully planned to fill the needs of the situation at hand and therefore should vary with the circumstances.

The first step in any sort of literature cataloging is to determine the exact nature of the data desired and the manner in which it is to be filed. This will determine the course of the rest of the work. The exact limits of coverage must also be decided at this time. Inadequate planning at the beginning can result in the necessity to repeat the early work when a change in coverage or data becomes necessary. It is always best to plan a more detailed and complete catalog than the minimum which at first seems to be all that would be required. For example, when the author of this book started to catalog the family Staphylinidae, he spent several years covering the eighteenth-century books and then worked on publications of the early nineteenth century. From the beginning, he included several groups that had sometimes been placed in the family and sometimes not, such as the Micropeplidae. Had these been omitted and later found to belong in the family, much of the search would have had to be done over. After a large part of the cataloging was complete, it was discovered that the genus *Inopeplus,* always placed in a distant part of the Coleoptera, probably belonged in the Staphylinidae. Only by reexamining much of the early literature on beetles could this group be brought into synonymic harmony with the rest of the Staphylinidae. This job is too large to repeat, and the genus will probably remain for a long time imperfectly cataloged.

The second part of the cataloging work is bibliographic—to determine all the publications that have any bearing on the problem at hand. In a synonymic study this means every publication in which any of the members of the group have appeared. It is not enough in synonymic work to study only the major works. In many cases every use of each name must be examined and evaluated. This job continues as long as taxonomic work is being done. Unknown older papers are forever turning up, and new publications are appearing daily in many groups. There is no cut-off point between 1758 and today.

The difficulties encountered in the bibliography depend on the previous cataloging, the existence of published bibliographies, the extent and completeness of recent monographic works, the size of the group, the volume of the accumulated literature, the geographic spread of the literature, the number of languages involved, and other factors. A good catalog cannot be produced without excellent bibliographic foundation, and good taxonomic work cannot be done on a monographic scale without adequate cataloging.

The third step is to obtain each of the appropriate publications and extract from it all the data pertinent to the catalog. In a small catalog, the entire pertinent section may be copied onto the card. Or a reprint or photostat may be cut up and pasted on the cards. In an extensive catalog, the best that can usually be done is to make the card serve as an index to where the data is to be found, with annotations as extensive as possible of the data itself.

When the group being cataloged is a large one, the catalog will have to be little more than an index to the literature—an index to all the names and to every reference to each. Two cards from such a file are reproduced as Fig. 7-4. They are both from one work, showing an important reference (to a new species description) and a minor reference (a species merely compared to a new species). The first card shows that this species was described as new in the cited publication, was placed in a particular subgenus, and came from a certain locality. This is about all one could expect to find about this species in this publication.

The second card represents a reference to the species *P. planus* that some catalogers would consider too trivial to record. However, in comparing this species with the new one, some previously undescribed features may have been mentioned. Furthermore, this may be the first time that the species has been placed in this particular subgenus. This emphasizes that no citation can be considered too trivial in cataloging. Generic names, particularly, are governed by such strict rules, that every time they are printed, and every spelling variation, may prove to be of the highest importance in nomenclature.

Some catalogers try to catalog the relevant works in chronological order. If one knows in advance all the relevant works, it will hardly be necessary to catalog them at all. Furthermore, it is in practice simply impossible to do this, and the cataloger will save much time and effort by taking books as they come, searching in all possible places for additional references to his species, and keeping a meticulous record of what he has already cataloged.

A bibliography of the publications cataloged is an important part of the catalog. It should be in standard form, complete, accurate, and annotated to show the extent of the data therein, the library where the book is obtainable, and the date on which it was cataloged. In large catalogs, it has been found extremely useful to keep a duplicate bibliography in chronological order. Not only taxonomic data and nomenclature can be organized but geographical and bibliographical

Phloeonomus pinicola n.sp. 1920

Champion ---- Ent.Mo.Mag.,56,p.241,242.

Subg. Phloeostiba
Desc., notes
Kumaon

Phloeonomus planus Payk. 1920

Champion ---- Ent.Mo.Mag.,56,p.242.

Subg. Phloeostiba

D.f. P.pinicola n.sp.

Figure 7-4. Examples of literature catalog cards.

data also. Taxonomy involves all of these, and simultaneous use of all data is essential to sound results.

Since animals can only be referred to by names, nomenclature is a key factor in cataloging data about animals. Although we say that the "correct" name of a certain species is *Ocypus olens*, this name is correct only at the present time. We must recognize that the species was originally called *Staphylinus olens,* that it was later called *Goerius olens,* then *Dinothenarus olens,* then *Ocypus (Goerius) olens,* as well as *Ocypus major, Ocypus maxillosus,* and *Ocypus unicolor.* It must be cataloged under each of these and several others, although it would be possible to file them all at one place. (This latter is not practicable or desirable in a large catalog.) The following synonymy gives the principal names under which the data on this species would be filed.

Staphylinus olens Muller (1764)	(1)
Goerius olens (Muller) Westwood (1827)	(2)
Ocypus olens (Muller) Curtis (1829)	(3)
Emus olens (Muller) Dejean (1833)	(4)
Physetops olens (Muller) Motschulsky (1858)	(5)
Anodus olens (Muller) Motschulsky (1858)	(6)
Ocypus (Goerius) olens (Muller) Mulsant and Rey (1876)	(7)
Dinothenarus olens (Muller) Heyden (1887)	(8)
Staphylinus (Goerius) olens Muller (Ganglbauer, 1895)	(9)
Staphylinus (Ocypus) olens Muller (Fauvel, 1897)	(10)
Staphylinus major Degeer (1774)	(11)
Staphylinus maior Degeer (1781) (emendation of *major*)	(12)
Goerius major (Degeer) Stephens (1829)	(13)
Ocypus major (Degeer) Bertolini (1872)	(14)
Staphylinus maxillosus (Schrank, 1781) (not Linnaeus, 1758)	(15)
Ocypus maxillosus (Schrank) Gemminger and Harold (1868)	(16)
Goerius maxillosus (Schrank) Reitter (1909)	(17)
Staphylinus unicolor Herbst (1784)	(18)
Ocypus unicolor (Herbst) Gemminger and Harold (1868)	(19)
Goerius unicolor (Herbst) Reitter (1909)	(20)
Emus morosus Dejean (1833)	(21)

Cross-referencing is essential. Attention to spelling is necessary not only to avoid new errors but also to prevent omission of data published under a misspelling. For example *Goerius* has also been misspelled *Georius* and *Coarus.* In the latter case, the data would be lost in a distant section of the file if it was not recognized and cross-indexed.

The mechanics of filing the data require careful attention also. Ordinary cards and guides are usually not adequate. Guide cards can be obtained with three tabs across the top, and ones with five tabs may then be used as subguides. Other subguides can be made by having cards cut from stock in a size ⅜-inch higher than the catalog cards, so that a typed line at the top serves as a guide. Cards of various colors can be used, and there are clip signals that can be attached to show special features. These must be worked out to serve the needs of each catalog, but the mechanical devices should aid the user, not merely add to the work.

Catalogs are the basis of most taxonomic work, and without published catalogs the bibliographic work-load of many taxonomists would be greatly increased.

Publication

No research has much impact on science until it is made available to other scientists. This can be done orally—in conversation or by presentation of reports at meetings of scientists—or it can be done by correspondence or the circulation of printed reports. Only the latter means is customarily implied by the word publication, and it is also the only effective means of disseminating taxonomic information.

In using taxonomic knowledge, it is often the most recent monograph which is consulted first, because it will be the most complete. Nevertheless, the opinion of the monographer must be in conformity with the earliest work, the original publications, or it will be set aside. In the final analysis, all taxonomy and all nomenclature must be consistent with original sources, and intervening work, even of monographic nature, will be of secondary importance. If the monographic work is well done, it may supplant the earlier work to a large extent, but this happens only in the better-known groups, where a high degree of taxonomic stability has been achieved.

Taxonomic papers retain their value or interest for the specialist for decades or even centuries. In many cases a publication will remain the only source of data on the subject for fifty years or more. This makes publication of data, analysis, and conclusions of prime importance in this field.

In taxonomy, publications may take the physical form of books, pamphlets, journal articles, symposium chapters, etc.; they may be primarily factual, theoretical, methodological, essayistic, or critical; they may be analytical or synthetic; and they may range in size from a one-paragraph note on a new observation to a multivolume monograph. The major types of taxonomic publications are listed below, with notation of some of the features and purposes of each.

Publications presenting new zoological data. The largest number of taxonomic publications present the results of studies (observations) on the comparative attributes of animals. These publications include all the ordinary descriptive papers, whether the subjects are new species, life-history stages, distribution, or biochemistry.

Comparative studies. The basis for all taxonomy and classification is comparative data. Therefore the first type of study required for systematic work is the study of the diversity of the group and the variation of the included species. The more that can be learned about these two things, the better the taxonomic work can be.

Comparative anatomical studies have been made in several parts of the animal kingdom, but the lack of them in many groups is a serious deterrent to sound classification. In the vertebrates, where comparative anatomy seems to be far advanced, there are many conspicuous gaps in the knowledge, so that sound classification has been delayed. In some phyla and classes of the invertebrates, nearly all the available anatomical data stands by itself, not effectively compared with data on the other groups. Even in such a largely comparative study as Hyman's compilation on *The Invertebrates,* there is little attempt to tabulate or compare item-for-item the numerous data presented.

To be effective these studies must be analytical, determining the nature and extent of variation. Frequently, they must explore new aspects, as new techniques become available. Their purpose is to discover and record all the ways in which species or individuals are similar or unlike, and the extent to which the data can be reasonably used in segregating and combining groups.

These comparative studies may be taxonomic contributions. Among them are life history studies, embryological studies, zoogeographies, stratigraphies, ecological studies, and behavioral studies, as well as comparative anatomy, comparative physiology, and comparative biochemistry.

Descriptions of new taxa. The simplest type of direct taxonomic paper is the description of one or more new species or other taxa. Although some writers hint that such single contributions are to be frowned upon as inadequate, they sometimes include all the requirements for making a contribution to science.

Descriptive revisions. Also called synopses and reviews, these papers summarize some aspects of the taxonomy of a group of species and incorporate the writer's views on the classification. They also enable others to identify the species. In extent of coverage, they range all the way from the previous type up to monographs. They may deal with species, or they may be limited to a study of the genera.

Monographs. These are the most complete systematic works, dealing with all the known facts about the species covered, usually not restricted so much as revisions in coverage. They have been described as "complete systematic publications (involving) full systematic treatment of all species, subspecies, and other taxonomic units and a thorough knowledge . . . of the comparative anatomy of the group, the biology of the species and subspecies included, the immature stages in groups exhibiting metamorphosis, and detailed distributional data." This is an ideal. It is sufficient for a good monograph to bring together all that is known and add the writer's analysis and systematic conclusions. However, a monograph is necessarily exhaustive in its bibliographic background, because this is essential to assembly of all the existing information.

Faunal studies. Merging with revisions and monographs are detailed studies of the fauna of a single region. Monographs are restricted to a single group of animals and can be prepared only by an experienced specialist on that group. Faunal studies cover all animals of an area or at least all those in a major group. The author generally is not a specialist but has studied the local fauna, made keys to the species, or tabulated their occurrence.

Atlases. These are comparative studies in picture form. They present anatomical data by means of illustrations.

Classifications. The studies which deal only with the grouping of animals and the arrangement of the groups into categories of the hierarchy are not very numerous, although much of this type of work at lower levels is included in monographs and even revisions. These papers may cover all the genera in a class (Simpson's *Classification of Mammals*), all the families in a class (Wetmore's *A Classification for the Birds of the World*), all the phyla in the Animal Kingdom (Hyman's *The Invertebrates*, vol. 1, chap. 2), or the subgroups within a group at any level.

Publications presenting new studies of names. Nearly all taxonomic revisions and monographs deal with names extensively, but a few papers deal only with the solution of the problems involved in using the correct name for each taxon.

Nomenclature studies. These are studies of the names themselves, their derivation, orthography, validity, typification, synonymy, homonymy, and so on. They are not to be confused either with nomenclators (which list names) or with studies of rules of nomenclature. Examples are the following extensive generic name studies.

Knight, J. Brooks, 1941. Paleozoic Gastropod Genotypes. *Geol. Soc. America, Spec. Pap.* 32, 510 pp. 96 pl., 32 figs.

Blackwelder, R. E. 1952. The Generic Names of the Beetle Family Staphylinidae with an Essay on Genotypy. *United States Nat. Mus. Bull.* 200, 483 pp.

Publications presenting methods of study. The working out of new methods to learn more about animals is a very important activity, one of those tending to keep the field up-to-date in techniques and outlooks. The study of rules of nomenclature amounts to study of methods of perfecting the system of naming.

Methods. Reports on the methods found useful in any aspect of taxonomy are valuable aids to other taxonomists. They may be methods of collection, preservation, storage, observation, examination, preparation, cataloging, filing, classification, measuring, comparing, recording, or any other aspect faced by taxonomists.

Rules of nomenclature. These are studies of the nomenclature rules, especially those analytical papers that offer solutions to problems of nomenclature. These papers are generally short and cannot be listed here. They will be found listed in the subject indexes of the *Zoological Record* and in *Biological Abstracts*.

Reference publications. Works of reference are usually major works, requiring years of work and resulting in lightening the work load of all other workers in the field. Some works listed above could also be included here, such as classifications, because they are much used for reference by non-specialists.

Catalogs and checklists. Both of these words have been used for rather different things. A checklist is literally a list of names prepared for the purpose of checking off certain ones. In the groups that have been popular with collectors—birds, butterflies, and beetles—these have been used to record the species in the collection. If the species were numbered consecutively, the numbers could be used as a shorthand notation when referring to the listed species. Through extension of meaning, the word checklist has come to refer, in work with vertebrates especially, to any tabulation of species, even if it includes most of the features usually found in catalogs.

A catalog is a tabulation of species which also cites some of the following information about each: (1) the original description reference, (2) later references, (3) synonyms with references, (4) range, (5) type locality, (6) genotypes of generic names, (7) annotations of various sorts, and (8) other pertinent data. Few catalogs are able to present all of these data. The circumstances dictate what will be shown in each case. The extreme use of the term catalog is the descriptive catalog, in which the species are actually described.

Keys. Most revisionary or monographic works include keys for the identification of the genera and species. In some cases such keys may be published separately, either singly or collected into a volume that covers a major group. An example is:

Bradley, J. C. 1930. *A manual of the Genera of Beetles of America North of Mexico. Keys for the Determination of the Families, Subfamilies, Tribes, and Genera . . .* , Ithaca, N.Y.; Dow, Illston and Company.

Handbooks. Books designed to enable the layman or non-specialist to identify the more common local species may be a product of taxonomic endeavor, but they usually do not form a part of the real taxonomic literature. They may be in the form of field guides or in the form of student manuals. Occasionally they are scholarly volumes of relatively complete taxonomic treatment.

Bibliographies. Inasmuch as nearly all taxonomic work is largely dependent on previous literature, guides to this literature are of universal importance to taxonomists. Most taxonomists must do a great deal of bibliographical work in their own research, and some of the results are published for the use of all. There is no end to the variety, extending from large summaries through annual lists to studies of some one work or serial. (See Chapter 14.)

Nomenclators. Lists of names, usually generic, intended for reference to show what names have been used and where, are called nomenclators. They differ from checklists in being alphabetically rather than systematically arranged.

Critiques. Less strictly taxonomic are certain papers that discuss the attitudes or work of taxonomists. These are very important in directing attention to goals, pitfalls, and inadequacies. They are not very numerous but may be cited under two headings.

Book reviews. Reviews of new books in taxonomy are not different from reviews of any other sort of book. It is not common to review monographic works, but brief notices are sometimes published to announce such works. The few general books that are published on taxonomy are reviewed in appropriate journals.

Essays. Many of the types of works cited above are primarily concerned with the presentation of taxonomic data. It is sometimes believed that such works should not contain discussions of methods, concepts, theories, the relation of taxonomy to other fields, or critical analysis of other work. Consequently this discussion material is frequently published in the form of essays.

Summary

Recording the data about kinds is an activity closely related to curating. It is in great part a mechanical activity but has close connections with taxonomic procedures of utilization of specimens and classification of taxa. Scarcely any taxonomy can be accomplished without the help of these recording procedures, and the results of any work can become part of science only when publicly recorded.

The judgment of the taxonomist in distinguishing and grouping taxa will be largely the result of his experience with the data of taxonomy, as recorded in notes, labels, catalogs, and publications.

For further discussion of recording by publication, see Chapter 16, The Publication of Data.

PART III

The Diversity to be Classified

"The zoologist whose work lies in one of the greater museums inevitably comes to see the problems of taxonomy somewhat differently from those of his colleagues who study animals in the field or in the laboratory. In the British Museum, for example, where the staff are, year after year, constantly unpacking and studying great collections from the uttermost parts of the earth and from the depths of the seven seas, one gets an impression, more vivid perhaps than can be gained anywhere else, of the unending diversity of animal form." (Calman, *New Systematics*. 1940)

Chapter 8. The Diversity of Individuals   105
   The kinds of diversity   105
   The variety in the diversity   106
   The diversity within a species   109
         The causes of diversity   110
         The forms of diversity   111
      Polymorphism   112
         Functional   113
         Sexual   114
         Developmental   115
      Variation   116
         Variation in development   116
         Sexual variation   116
         Genetic variation   116
         Environmental variation   117
         Willful or accidental variation   118
   Diversity of individuals in taxonomy   119

Chapter 9. The Diversity of Kinds and Groups   121
      Limitations   122
      Universal features   122
   The extent of the diversity   123
   General features of group diversity   123
      Diversity in group size   124
      Diversity in breadth of groups   125
   Differences between groups   126
      Chemical diversity   127
      Cell diversity   127
      Organ diversity   128
      Diversity in organ systems   128
         Variety of hard parts   128
      Diversity in reproduction   129
      Diversity in development   130
      Other group diversities   130
   The diversity of kinds   130
   The diversity of taxa   131

CHAPTER 8

The Diversity of Individuals

The kinds of diversity

It is an evident truism, scarcely ever put into words, that animals are not all alike; they exist in a variety of forms, sizes, and colors and perform a variety of activities. It is also evident that individual animals do not occur with random combinations of features, but that there are large groups of individuals with substantially the same features, clearly set off from other groups of individuals possessing different sets of features.

These facts have long been tacitly assumed in the recognition that there are different *kinds* of animals. The word kind sometimes signifies groups of considerable size, as birds are recognized as being different *in kind* from fishes, but it is more often applied to the narrowest or most restricted concept; for example, a cottontail is not the same *kind* as a jack rabbit. These kinds are in general what the scientist calls species.

The members of each kind are recognized because they share certain features. This does not mean that they are exactly alike, however—one recognizes different house cats by minor features in which they differ from each other, just as one recognizes different human individuals. It is common knowledge, too, that within a given species there may be a difference in appearance between the young and the adult individuals, and between male and female. There may also be differences due to the season of the year, the amount of food consumed, or the activities of the individual.

The preceding remarks have suggested several of the ways in which animals are diverse—different kinds, developmental forms, seasonal forms, growth differences, functional forms, and all the differences between individuals and between parts of individuals. These cover a tremendous range of variety, at many levels. This variety is oftentimes forgotten in the current tendency to enlarge upon the unity of all living things. There *is* basic similarity, but the similarities do not extend as far up the scale of organization as is sometimes implied, and the dissimilarity is frequently substantial at several levels. Some of the diversity is tabulated in this chapter and the next.

All animals and plants consist of protoplasm and its products, and this protoplasm consists basically of immensely complex molecules of a few types. These molecules are principally chains of carbon atoms with side chains and a variety of prosthetic groups. But different organisms possess many complex molecules that differ from those of other organisms, and they combine to form cells, structures, and organ systems of an almost endless variety. Unity or similarity is interwoven with multiplicity and diversity. Taxonomy is based on the fact of diversity, and it records both similarities and diversities.

Up to this point the term diversity has been used in a broad sense to cover all of the ways in which two individual animals may be different, whether they belong to the same species or not, or even the same group. It is useful to make a distinction between the diversity shown by the million kinds and the variability shown within one kind. The word diversity implies both of these, so it is appropriate to refer to the first as *diversity of groups* (or group diversity). The word variation does not generally cover features of group diversity, so it can be restricted to the variability within the species. However, variation will not readily include the gross forms in which animals appear (here called polymorphs), so the general term for all the diversity within the species must remain the *diversity of individuals* (or individual diversity). Although many of the same features are involved in these two levels of diversity, there is generally a marked difference in the expression at the two levels. They can thus be discussed separately in large part.

The variety in the diversity

The following list shows some of the ways in which the diversity appears. The numbered forms of diversity are stated in general terms, some more technical conditions are cited, and sometimes examples for clarification. The items are not mutually exclusive but are intended to suggest different aspects of diversity.[1]

---

[1] Part of this diversity is tabulated in quite different manner in Mayr, Linsley, and Usinger (1953, p. 82) and in the excellent article on *Taxonomy* by Linsley and Usinger in The Encyclopedia of the Biological Sciences, edited by P. Gray (1961, p. 995).

A. Two individual animals (specimens) may be different although belonging to the same species:

1. Different *age* (even among adults, age alone may produce differences).

2. Different *sex* (male or female; see also item 16 below).

3. Different physical *castes* (queen, drone, worker, soldier, replete, etc.).

4. Different phases of a *life cycle* (developmental stages; e.g., egg, larva, cyst, embryo, juvenile, nymph, pupa, adult).

5. Differing *body forms* (polyp, medusa, medusoid, dactylozooid, gonangium).

6. Were in differing positions in a *colony* (terminal individuals or basal ones, performing different functions, differing in structure).

7. Were living at different *seasons* or in different climatic cycles (spring and summer forms, and cyclomorphosis).

8. Were living in different physical *habitats* (ecophenotypes; arctic and temperate individuals).

9. Had responded in color to differing *backgrounds* (color changes produced by integumentary chromatophores in response to environment).

10. Were living in different *hosts* (in absence of complete host specificity, or in presence of host variability, a parasite may differ in or on different hosts).

11. Were feeding on different prey or plants (*food*) (difference similar to that in item 10 above).

12. Were living under different *crowding* conditions (density-dependent variation, sometimes related to availability of food).

13. Were differently *parasitized* (mechanical distortion response to presence of a parasite, as in stylopization).

14. Differ in *karyotype* principally (diploidy and haploidy, homozygosity and heterozygosity with dominance; sex chromosome differences would normally also result in sex differences as in item 2 above).

15. Had developed disproportionately in *size* (size alone would be a difference; sometimes large size is correlated with excessive growth of one part; size may be due to differences in food supply, age, etc.).

16. Differing in *sex combination* (gynandromorph, intersex; in some species there may be male, female, hermaphroditic, and sexless individuals).

17. From different extremes of continuous *character expression*.

18. From differing sectors of a *discontinuous character expression*.

19. Were *deformed* at birth or any time during development (normal vs. teratological; at least sometimes due to factors in item 20 below).

20. Had suffered an *accident* (leaving permanent physical damage).
21. Was *diseased,* in the sense of infection by germs or poisons (normal vs. pathological but usually temporary).
22. Underwent *post-mortem changes* (due to preservation of specimens or lack of it—color, size, oiliness, etc.).

B. Two individuals, while belonging to different species, may differ in any of the ways 1–22 (specimens of different species may differ in any of these ways but real differences between species may be found in the ability to differ in these ways, and in the limits or frequency of these differences) and the following:

23. Differing in one or more relatively *minor feature,* structural, physiological, behavioral, or other (these are the customary specific characters).

C. Two individuals, while belonging to the same group but not the same species, may differ in any of the ways 1–23 (specimens of different groups may differ in any of these ways but real differences between groups may be found in the ability to differ in these ways, and in the limits or frequency of these differences) and the following:

24. In *most gross features* (as any two phyla differ).
25. In *some gross features* but not most (as any two classes in a phylum, or orders in a class).
26. In a *major structural feature* or combination (as any two families or genera).
27. In a basic *body arrangement* (symmetry, segmentation, strobilation, cephalization).
28. In a *primary ecological adaptation* (fins for swimming, wings for flying; air-breathing vs. aquatic).
29. In an important *metabolic process* (use of different hydrogen acceptors in muscle metabolism).
30. In a major *activity capability* (winged, sessile, sense organs).
31. In being *solitary or colonial* (earthworms and bryozoans, jellyfish and corals).
32. In having a drastically different *life history* (such as juvenile-to-adult in Nematoda, or miracidium-to-sporocyst-to-redia-to-cercaria-to-metacercaria-to-adult in Trematoda, or larva-to-pupa-to-adult in Insecta).
33. In basic *cleavage pattern* (the direction of the first planes of cleavage and distribution of egg contents thereby).
34. In any of the multitudinous body systems, functions, activities, distributions, etc., not cited above.

This list is by no means exhaustive, especially B and C. There are innumerable variations of these diversities, intermediates between them, and combinations of them. (Items 1–22 include polymorphism and individual variation (within the species). Item 23 consists of the usual features distinguishing species.

The diversity within a species

Before a taxonomist can classify species, he must know these species. Whether he believes that he can classify individual specimens into species or that he must infer the species from the sample (specimens) before him, there are many differences between individuals which will not be used to distinguish the kinds. These are the differences found between individuals of the same species. If these characteristics are not to be used in classification, they must be recognized and consciously rejected.

Man happens to be a species that occurs in relatively few forms. Thus, there are no normal physical coloniality, no functional hermaphroditism, no distinct seasonal forms, no parthenogenesis, no asexual generations, and no physical castes. Nevertheless, the variation from individual to individual seems to us to be great, involving male and female, babies, youths, and adults, tall and short, hirsute and bald. We tend to forget the more extreme examples such as dwarfs, megalocephalics, and Siamese twins. That these all belong to one species might not be obvious if it happened not to be the one most familiar to us.

Although such differences between individuals will not be useful in classification, they must be recognized and understood, else all the males will be classified together and all the females. Many errors in classification have been made because of failure to recognize the variation within the species.

The diversity within a species, often called variation or individual variation, can be in the form of continuous change, such as growth, or of discrete types, such as sexes or castes. Either of these may be part of a sequence of forms (growth or alternating generations) or not sequential (climatically induced or castes existing together).

The word variation has been used in various ways. It is restricted here to the difference (whether structural, functional, behavioral, chemical, or other) that exists between the offspring of the same parents or between individuals of the same species. These are the differences that might appear among the progeny of one asexual parent, or one parthenogenetic parent, or two parents that could themselves have been offspring of another such pair.

The amount of variation within different species is extremely diverse, but no species are known in which there is no detectable variation (except where there are too few specimens for comparison).

Theoretically, the nearest approach to this is found among the asexually produced populations (clones), in which the basic gene complements may at first be identical and environmental pressures relatively uniform.

The causes of diversity. There are many brief statements of the causes or sources of variation. None of these seems entirely adequate to explain all of items 1–21 in the previous tabulation. These causes or sources have been classified as intrinsic or extrinsic, inherited or non-inherited, continuous or discontinuous, and constant or sporadic. However, the two most useful ways of distinguishing causes or sources of variation seem to be the following: *First,* the distinction between those that occur normally in the life cycle and those that occur only sporadically or accidentally; *second,* the distinction between those that are produced by genetic factors and those that are determined by the environment (within limits set by the genes).

There are five main subdivisions of variation: polymorphism, other normal cyclic changes, mutations, normal genetic differences, and externally induced differences.

Examples or a brief discussion of each of the causes of variation are given here:

(1) *Normal developmental processes* produce the stages in the life cycle (developmental polymorphism), growth and aging, alternating generations, and obligate seasonal forms (including cyclomorphosis).

(2) *Sexuality* refers to sexual polymorphism (male, female, hermaphrodite, neuter), sequence of sexes (protandry and protogyny), sex-linked characters (those of one sex only, such as male plumage and ovipositors of females), and some castes.

(3) *Colonial division of labor* refers to the polyp and medusa, and the polymorphic colonies (such as *Obelia, Physalia,* and the Bryozoa).

(4) *Genetic recombination* accounts for a large part of general variation but can seldom be recognized except by genetic analysis. *Ploidy* involves the haploid and diploid conditions as well as the rarer polyploids. Gynandromorphs and intersexes are also due to genetic aberrations.

(5) *Mutations* may produce either apparently continuous variation or discontinuous variation. The number of mutations normally present in populations of most species is unknown, but they are doubtless a real part of the individual variation.

(6) *Response to environment* results in a variety of sorts of variation:
    (a) habitat-induced;
    (b) host-determined;
    (c) density-dependent (crowding);
    (d) climatically controlled (e.g., arctic forms);
    (e) food (prey or plant food, some castes);

(f)   disease (response to pathogenic organisms);
(g)   parasite-induced;
(h)   color changes;
(i)   seasonal forms (if not obligate); and
(j)   terata and accidental mutilations.

There are also some that can scarcely be classified. One of these is autotomy, self-mutilation, in response to accident but determined by intrinsic decision. Another case is the effect of domestication. Still another is the effect of fossilization and preservation in sediments and rocks.

The word variation is most likely to bring to mind the small differences, particularly those that are continuous or not obviously discontinuous. This variation is accounted for in large part by the following items: growth and aging, sex-linked characters, recombination, mutations, habitat and food effects, color changes, and very diverse mixtures of these, affecting diverse features of the animals.

The forms of diversity. Some of the kinds of diversity may be tabulated thus:

A. Inherent diversity
    1. Part of the life cycle
        Age variation and change
        Growth (normal stages or heterogonic growth)
        Seasonal (including cyclomorphosis)
        Alternating generations
        Sex arrangement (protandry, protogyny, extra form of neuter or
            hermaphrodite)
        Developmental stages
        Body forms (polyp and medusa, functional forms)
    2. Genetically produced (but not in usual life cycle)
        Castes
        Sexes
        Gynandromorphs
        Intersexes
        Genetic polymorphism (recombination)
        Ploidy
        Mutations (discontinuous)
        Variation of any feature
B. Externally induced
    3. Environmentally induced
        Ecological
            Habitat
            Host-determined
            Density-dependent

Climatic (arctic forms)
Food (prey or plant food)
Disease
Parasite-induced
Neurogenic or neurohumeral color changes
C. Occasional and abnormal
    4. Autotomy
    5. Teratological changes
    6. Accidents
    7. Post-mortem changes
    8. Fossilization
    9. Artifacts
D. Indistinguishable combinations of these
    10. Continuous variation of any features or of the combination of
        many features

In discussing below some aspects of this variation, it will be helpful to use still another arrangement of the factors. Among the causes listed above are several which produce strikingly different body forms (correctly called polymorphism in the classical sense). There are others which produce continuous change during the life of the individual (*development*). There are some that involve *sexuality* and the distribution of the sex organs. There are some that are clearly an aspect of the normal *environment*. And there are some that consist of *willful or accidental* events. These are not clearly distinct, and this overlap is one of the factors that makes this variation hard to recognize and make allowance for.

Polymorphism [2]

Five forms of diversity produce sharply different forms within some species: developmental stages, alternating generations, body forms, castes, and sexes. These occur regularly in the normal life of the species, which is to say in the lives of the individuals which make up the species.

[2] In some recent books the term polymorphism is used for the relatively minutely discontinuous diversity produced by mutation and recombination. These are distinct at their level, but they never involve the entire form of the individual. They appear as occasional sports, albinos, and phases, in unpredictable but very low proportions. To use the name polymorphism for these is to misappropriate a term used for a century for the obvious body forms that occur regularly and are often so different as to be placed initially in widely separate groups. It has been suggested that the term *polyphasy* can be applied to the production of the obvious forms. This term is not inappropriate, but there is little reason to abandon the equally appropriate term polymorphism just because a different implication is nowadays applied to it by one group of zoologists.

Polymorphism is defined here as the existence of individuals of more than one form within a species. This situation is very nearly universal among animals, for different forms may be assumed by males and females, by young and adults, by successive "generations," by individuals receiving different food, and by others. In the zoological books which refer to polymorphism in detail, sexual and developmental forms are often omitted, but if we exclude them by definition we would have to set up a new term to include all the regular *intraspecific* but discontinuous diversity.

All polymorphism is genetically controlled in the sense that nothing can take place which is not provided for in the genetic constitution. Most polymorphism is directly produced by genetic means, such as sexual, developmental, alternating, and allometric. Some is produced or controlled by the environment, such as nutritional, seasonal, and occasionally sexual.

It has not been found useful to classify the various types of polymorphism. There is too much overlap in causes, too little known about how the various types arise, and little real difference in the manner of expression. It is possible to group some types as involving two or more forms simultaneously as against those appearing successively. Or those that perform different functions as against those that do not. Or those that exist in a united colony as against those that are completely separate individuals. But no useful classification results, so the different types will here be discussed separately under the most appropriate headings, beginning with those that occur simultaneously.

Functional polymorphism. In several groups of animals there is a real division of labor between individuals. This is seen among bees, ants, and termites, where individuals may be specialized for breeding, tending the young, protecting the colony, gathering food, storing food, and so on. These forms are called *castes*, and the individuals are, of course, completely separate.

In the colonies of connected individuals, such as occur in Coelenterata, Bryozoa, and Tunicata, the individuals may be very highly specialized, with as many as six different kinds of individuals in one colony. Again these specialized individuals may each perform one function, such as grasping food, reproduction, anchoring or floating the colony, nursing other individuals, feeding, swimming, or protection.

In each of these types of colonies its very existence depends upon the presence of these specialized individuals. We can thus look upon this as obligatory polymorphism. Sexual dimorphism, which is of course also functional, is sometimes obligatory, and it is sometimes partly facultative in the sense that it is not essential to species continuance, although the individuals do not have any choice in the matter.

Sexual dimorphism. Substantial difference in appearance between the two sexes in any species is a form of polymorphism. It is almost universal among sexual animals. In a few cases, there are more than two sexual forms, described below, and some forms of general structural polymorphism are correlated with the sex of the individuals.

The simplest animals to show clear dimorphism of the sexes are Acanthocephala, Nematoda, and Rotifera. In some of the echiuroid worms, the difference between male and female reaches such an extreme that the male becomes an internal parasite in the body of the female, bearing almost no resemblance to her in form, size, or activity. In many groups of Arthropoda there is a marked size difference between the sexes, or evident modification of some external feature in one sex. In some insects the ovipositor of the female is very distinct. In some the antennae or mouthparts may be very different in male and female. Sometimes color pattern is strikingly different.

There are all degrees of difference, ranging from sexes which differ only in the cryptic fact that the gonad produces sperm in one and ova in the other to the almost complete dissimilarity between the sexes in scale insects and Strepsiptera (with parasitic females). Perhaps mainly because of greater size, the differences between the sexes in many vertebrates are well known. Probably none are so extreme as *Bonellia* or *Lecanium,* but they are none-the-less obvious. In many birds the dimorphism is striking.

Sexual polymorphism, in the sense of more than two sexual forms, occurs in Protozoa, Coelenterata, Gastrotricha, Calyssozoa (Endoprocta), and such colonial forms as may possess neuter individuals (Coelenterata, Bryozoa, Pterobranchia, and Tunicata).

It is easy to let sexuality get beyond the proper range of polymorphism, as defined above. Mere separation of sexes does not produce polymorphs, unless the individuals are noticeably different. In the Gastrotricha at least one species can have males, females, or hermaphrodites, but striking differences in appearance seem to be lacking. In the Calyssozoa at least one species occurs as male, female, or hermaphrodite, but again no great differences are found in appearance. None of these, therefore, is truly polymorphic.

In social insects there may be males, females, and neuters of several sorts. These may be clearly distinguishable and may be considered polymorphic. (They are also properly treated as castes.)

In ciliate Protozoa, the mating types described by Sonneborn seem to be very similar in nature to sexes. There may be as many as six mating types in one species. However, the absence of differences in external appearance would make it inappropriate to call this polymorphism.

Developmental polymorphism. Most familiar animals originate from zygotes or fertilized eggs formed in each case by the union of sperm and ovum from its parents. We cannot say all, because there are two classes of exceptions. In a wide variety of groups, some individuals originate by asexual means, and in other groups unfertilized ova may develop into new individuals. (There are even groups in which males are not known to exist at all.)

In animals that originate from ova, either fertilized or parthenogenetic, there is always a considerable change of form during development. Even if the egg gives forth a miniature of the adult, which simply grows larger to become mature, there are two forms of the individual—the egg and the adult.

In most such animals, the individual which hatches from the egg is somewhat different, perhaps almost entirely different, in appearance from the adult (as well as from the egg). In these animals the newly hatched individual is called a larva. The change from the larva to the adult is sometimes called metamorphosis. It is part of development.

The occurrence of these structurally, and sometimes functionally, different stages in the development of an individual is polymorphism in a successional sense. It is obligate in the sense that the stages cannot be avoided by the individual, which must pass through the sequence to attain adulthood (usually sexual maturity).

If the animal is viviparous, that is if it retains the egg within its body until after the egg hatches, or if there is no egg covering after cleavage begins, the form is judged upon birth. There appears to be no difference between this and oviparity so far as developmental polymorphism is concerned.

There are three principal types of developmental polymorphism, arising from the general habits of the animals. Marine organisms frequently hatch from the egg as a motile larva different in appearance from the adult, which may change into another larval form or directly into the adult. Internal parasites frequently produce eggs which develop through a succession of larval forms, often with asexual multiplication at one or more points, and with the final larval form developing into the adult. Terrestrial Arthropoda usually lay eggs from each of which hatches a grub-like larva, which may later enter a different resting or pupal stage before metamorphosing into the adult form. In all of these, the egg, larva, and adult are distinct in form, and additional larval forms or a pupal form may be interpolated.

Polymorphism in development thus can involve either sexual or asexual processes. In some animals it may involve both in turn, in an alternation of generations. (Some of the ways in which these alternations can occur are diagrammed in the next chapter.)

Variation

Variation in development. In addition to the polymorphic forms cited above in the life cycle of many animals, other kinds develop with little or no evidence of such forms. In Nematoda a juvenile hatches from the egg and grows into an adult with no stages or quick changes in form. The only distinct stages left are egg and adult.

In asexually reproducing forms, such as *Hydra*, by the time the new individual is separated from the old it is already an adult. It never passes through any larval or egg stage.

In these, there is gradual change—in size, proportions, external features, internal organs, and so on. This is developmental variation. It occurs in all forms even when there are polymorphic stages. Other difficulties appear to perplex the zoologist: when adults of some species are larviform, when larvae become sexually mature before assuming the adult form, when regeneration produces structures different from those lost by mutilation (heteromorphosis), and when there are alternations of asexual and sexual forms.

Sexual variation. As indicated above, sexuality does not always result in dimorphism. The sexes may look alike; hermaphrodites may look like unisexual individuals; and even neuters may be indistinguishable. On the other hand, there may be numerous sex-linked features which show any amount of variation due to genes or environment. This is a major source of the continuous variation.

Genetic variation. Although all variation, like all diversity, is either controlled directly by the genes or limited by them, much small variation is the sole result of various genetic processes. Although this can be confidently stated to apply to all animals, it applies much more cogently to sexually reproducing ones, and it is possible to identify this variation as genetic only when some genetic analysis has been made. It thus forms a major source of the minor variation of animals, but a cryptic source to the taxonomist in most cases.

The processes are: *mutation* (of genes, which produces new features), *hybridization* (which introduces new genes into the species), *recombination* (which shuffles the gene expressions into new combinations), *ploidy* (in which diploid and haploid individuals may differ, or polyploids occur), and *chromosome aberrations* (acting much like mutations). These are not entirely distinct and may act in combination.

Environmental variation. Although genetic variation is universal, and although the gene complement serves to channel and limit all other types of variation, by far the largest source of minor variation in most animals is the effect of environmental factors. They can act on almost any part of the body, at any time during its life, in any part of its range. They can cause temporary, cyclic, or permanent changes in the individual. These variations may be due to any of the following:

Climate, as distinct from seasonal change, may produce many variations within a species. The populations of *Rana pipiens,* extending from Canada to Panama, show many differences, apparently due primarily to climate working through the physiology of development.

Elevation and latitude may cause variation, such as size, within a species. Although the difference may be actually caused by some other factor such as temperature, it shows up clearly in relation to altitude or latitude.

Seasons, as cyclic phenomena, frequently induce changes that lead to spring forms and summer forms, or to cyclomorphosis—a succession of shapes due to a succession of seasons.

Background color is an environmental stimulus. Many animals respond to the color of their surroundings by changing their own color. The effect is generally to make them less conspicuous by blending into the background. There are differences in ability to change color from species to species and group to group, but the interest here is in the differences in color between individuals of one species, or changes in one individual through time.

The only groups in which color changes commonly occur are the Cephalopoda, the Crustacea, and the cold-blooded vertebrates. They occur infrequently in the Hirudinea, Gastropoda, Insecta, and Echinoidea.

There are, of course, other types of color diversity in animal kinds; for example, genetic color phases, colors influenced by the food, or color influenced by the color of the material on which development takes place.

Density-dependent phenomena. The density of a population may directly or indirectly produce variation among the individuals. Some insects develop differently if grown under crowded conditions, even if food is plentiful. Many animals show differences when food is scarce or when population pressures force some into marginal habitats.

Food. As just mentioned, the amount of food may affect certain individuals. More obvious variation is caused by differences in the food utilized; for example, where two host-plant species are utilized or where prey is largely of different types for different individuals.

Host-determined. Very similar to the last case is variation among parasites that utilize different hosts. The variation in parasitic insects may appear in size, proportions, cocoon color, and structural features such as presence of wings. The food of the host may affect the parasite, as shown in some parasitic wasps and some nematodes.

Parasite-induced. Such parasites as gall-wasps are well known to evoke changes in the tissues of their plant hosts, and gall-like effects of parasitization are not unknown in animals. Not only do some nematodes produce galls in plants, but some cause gall-like alterations in parasitized ants.

Microorganisms may also be considered parasites, and the diseases produced not only affect the individual during the course of the disease but may leave permanent effects. If the disease occurs during larval stages, there may be drastic effects on development and the appearance of the adult.

Willful or accidental variation. *Autotomy.* A crayfish with one cheliped much smaller than the other may have suffered an accident to a cheliped and rid himself of it by autotomy, i.e. self-mutilation. While a new cheliped is being regenerated, there will be distinct difference from the usual appearance.

It is not certain that all autotomy should be classed as environmentally stimulated. The evisceration of sea cucumbers is in response to some outside stimulus, but the autotomy of the starfish may be due to internal stimuli. In any case, it may result in a four-armed starfish and a comet-star, through regeneration of both fragments.

Accidents. It would be difficult to define accidents, but any physical damage is likely to leave scars, if the damage is not fatal. If it occurs in larval or embryonic stages, it may have drastic effects on the structure of the adult. Some of the "monsters" or terata are produced by physical damage, although genetic factors may sometimes be involved. Even damage to the egg may produce abnormal embryos and adults. The general subject of teratology, regardless of causes, is discussed further below.

In some other Crustacea, if an eye is lost through accident, regeneration may produce not another eye but an antenna in its place. This is heteromorphosis, which leads to a variant abnormal condition.

Teratology. One of the kinds of diversity among the individuals of any species, frequently overlooked, is teratology. In the development of individual animals, it is not at all uncommon for disturbance in either the spatial or the temporal synchronization of the many processes to lead to abnormal organs or individuals. In many animals, these deviations are difficult to detect, because no standard or norm is available. The gross "monsters" such as Siamese twins and two-headed snakes, as well as all the malformed foetuses, fall under this heading. The exceptional features which they show must be recognized as such by taxonomists, if they are not to be used as the basis for spurious groups erected for these non-recurrent forms.

Some living terata have presented features, such as two heads, nowhere else encountered in the animal kingdom. Insects with biramous antennae and double legs are among these. If the deviation affects only one member of a pair of structures, as it apparently usually does, the condition is likely to be readily recognized. Many minor terata have been reported in animals such as insects, where the exoskeleton sometimes makes such features obvious. Without reference to the cause of the deformation, the following types of monstrosities do occur:

1. Giants.
2. Dwarfs.
3. Defects in any structure.
4. Embryonic irregularities:
    a. Incomplete development
    b. Non-synchronous development
    c. Abnormal cleavage.
5. Abnormal hermaphroditism.
6. Cyclops or siren.
7. Reversed organ position.
8. Siamese twins.
9. Two-headedness and other duplications.
10. Heteromorphosis.

Summary: Diversity of individuals in taxonomy

Taxonomists have always been concerned about "individual variation." What this term means and how to recognize the variation between individuals of one species has led to difficulties. It is no easier to identify this sort of diversity today, but at least its nature and causes are better understood.

If taxonomists are to segregate, name, and classify the existing species and if they are to do this on the basis of the attributes of the individuals, they must find the attributes which all corresponding members of a species have in common and they must avoid attributes which are lacking in a segment of the species. They must recognize that most species consist of two sexes, of two or more developmental stages, and of large and small individuals; that some species also consist of several color phases, of diverse structural forms such as polyp and medusa, of several physically different castes, and of seasonal forms; and that in some species, the individuals appear very similar in most detectable ways, but in other species there is a wide range of difference between individuals in obvious features.

How to tell the variation within the species from the differences between species has always been the chief difficulty of taxonomy. The recognition of the hereditary nature of the ideal species characters does not solve this problem, inasmuch as some intraspecific variation is also hereditary and seems to be controlled by the same genetic mechanisms. Theoretically, it is easy to make a distinction between those attributes that are shared by all members of a species and those that are found in only part of the members, but there is no simple way to distinguish these in practice. The comparative methods by which this is done are discussed in a later chapter.

CHAPTER 9

The Diversity of Kinds and Groups

There is diversity of kinds, of groups of kinds, and of groups of groups, as one goes up in the hierarchy of taxa. Kinds are often diverse in the details of their features; groups of kinds are diverse in combinations of the details or in features thought to be less superficial; and groups of groups are diverse in major adaptations—in organ systems, in major diversities in body arrangement, developmental patterns, etc.

The existence of a vast amount of group diversity has been recognized for centuries. In recent years, emphasis has frequently been put on the much smaller amount of unity or basic similarity among all animals. To emphasize that diversity is still the most obvious feature of animals, some of it is suggested in this chapter.

Most of the diversity cited is at group levels fairly high in the hierarchy. In contrast, species within a group would differ in less obvious ways: in proportions of parts, in the number of multiple structures, in coloration (pigment difference or arrangement), in specific protein molecule components, in slight behavioral differences, in definite but slight genetic differences, in slight differences in development, in different responses to environmental influences, in possession of different parasites, in use of different food, in dependence on different hosts, in occupation of different geographical areas, and so on.

All of these can be differences between species, but of course they can also be differences between populations or individuals within the species. If the differences are within the species, they are individual variation, discussed in the previous chapter. How to distinguish which level a difference represents is the first problem to be solved in either classification or descriptive taxonomy. This problem will be discussed further in later chapters, but there is no easy solution or standard formula. Only the broadest possible knowledge of the organisms will serve to make the distinction, and this is usually applied through a series of refinements in successive monographic studies.

121

Limitations. There are limits to diversity, although the nature of the limiting factor is seldom understood. It is certain that the Second Law of Thermodynamics limits what evolution can produce. This is, of course, involved with the very nature of the molecules of protoplasm, as well as with the atomic structure and energy relations of matter itself. On a much more visible scale, one can imagine many things that animals do not produce, such as three pairs of wings in insects, vertebrates with separate heads for feeding and a sensing-control center, or parasites requiring five successive hosts. These things would not be any stranger than some of the things that do occur, such as the variable number of legs in Pycnogonida, the multiple-purpose life histories of some parasites, or the unique systems such as the Aristotle's lantern in Echinoidea.

In the past, there may have been some things tried that are outside the limits appearing today. In fact some extinct groups do show unexplainable unique features. If geologic time is long enough for all possible things to have been tried, given physical conditions then existing, we can assume the diversity that has existed is roughly what could exist. Paleontologists surely do not know all that has existed, very likely not a major part of it.

Universal features. It has been stated that all metazoan animals are sufficiently alike for concepts that work in the Mammalia to work in any other group as well. Unfortunately this statement is clearly due to lack of familiarity with the diversity of invertebrates. Many groups lack the genetic effects of outbreeding, affecting the application of many evolutionary concepts. In many groups, sex is determined in radically different ways. Development at all stages from fertilization on varies in many basic ways. Provisions for so basic a function as absorption of food by digestion are so different in Cestoda, Pogonophora, and Mammalia as to be similar only in the fact of exchange of metabolites with the environment. There are metazoans without germ layers; there are some without separate cells (at least in parts of the body); there are ones that lack major marks of animals, such as locomotion, or possess major features of plants, such as production of carbohydrate skeletons; and there are ones in which there is virtually nothing that can be called behavior, not even coordinated reactions.

It is surely extremely risky to assume that *any* generalization based on higher vertebrates will apply even in principle to all animals, even all metazoans. If protozoans are considered to be animals, then nearly all rules are certain to break down. Even to say that all are composed of protoplasm controlled by DNA systems is scarcely meaningful because of the known diversity in that group of molecules, and little generalization above that level is consistent with the known but sometimes forgotten diversity that exists.

The extent of the diversity

The presence of diversity in the animal kingdom is proverbial, but the extent of it is nowadays greatly understated. A catalog of all the diversity would be an encyclopedia of all zoology. It could not yet be written, because new diversity is still being discovered at a rate higher than generally realized. The present generation has seen the discovery of a new phylum, discovery of living representatives of at least two classes thought to be long extinct, discovery of unknown organs in common kinds of birds, discovery of an entire level of diversity (biochemical) previously not detectable, and recognition of a variety of genetic mechanisms previously undreamed of.

Because the extent of this diversity is too often overlooked, and because it is the basic realm of study of taxonomy, it is thought to be worthwhile to illustrate it here. Examples of diversity have been chosen to illustrate the fact that the diversity is itself diverse—there are a wide range of features which themselves show a wide range of development or expression through the animal kingdom.

General features of group diversity

Textbooks of college zoology usually mention as few as 10 or as many as 20 phyla of animals, often with no indication that there are about this many more groups of animals that cannot be placed for sure in any of the 10 or 20. This is excused by reference to space limitations and the assumption that some groups are more important. Even the recognition of 35 or 40 phyla does not acknowledge the diversity at this level. Evolutionists generally deny that any of the existing phyla arose directly from any other existing ones, which means that all the common-ancestor groups are not only extinct but unrecognized as fossils. Yet these groups must have existed, and there may have been quite a few of them. It is just possible that they are not entirely unknown.

In the more detailed paleontological works, there is frequent mention of unusual fossils which cannot be assigned to known phyla. Some of these are imperfectly known—not enough remains to tell whether they are different from living animals. Others seem to be quite possibly the remnants of distinct groups. For example, conodonts present several features not duplicated in any living group. See Figs. 11-1 and 11-2 in Chap. 11.) There is no real basis for assigning them to any phylum. The criconarids and hyolithids are conical or pyramidal shells of unique types, made by animals of unknown nature. The peculiar fossil *Amiskwia* bears some resemblance to the Chaetognatha, but over a gap of 450,000,000 years, the differences leave some doubt as to the relationship. The little fossil cone shell *Matthevia*, which has the inner chamber divided into two by a thick transverse septum, gives little basis for assignment to any phylum. And there are many others.

Instead of assuming that there are only a limited number of phyla, it seems to be necessary to admit that minor groups and still unrecognized ones very likely occur or have occurred in considerable variety. It is unlikely that evolution produced the present diversity without many developments which eventually proved to be incapable of survival. It may thus be expected that the number of recognizable phyla and classes will increase considerably, as the unique features of these isolated groups come to be recognized. A reasonable prediction would seem to be 50 phyla as a minimum.

In assessing the variety of the animal kingdom, or even in summarizing its aspects for beginning students, one of the most important facts is the diversity in size of the groups which occur. A group (perhaps a phylum or class) which consists of a few species in a genus or two is no less important biologically than another group at the same level in which the number of species now living is vastly greater. In fact, for studying the diversity of animals the odd or peculiar groups, representing blind alleys or relics or recent developments, are in many ways the most interesting ones.

Diversity in group size. Although most of the lesser-known groups are of small size, they do not of themselves show the diversity in size mentioned above. At every level in the classification, there is a wide range of actual size, as the following examples will illustrate.

The phyla range in size, regardless of whose classification is chosen, from such a small homogeneous group as the Chaetognatha, with about 50 known species, to the Arthropoda, with over 800,000 known kinds.

Although the phylum Mollusca contains some 130,000 species, when these are divided into seven classes, they vary in size from the Monoplacophora with two living species (and a handful of fossil ones) to the relatively enormous class Gastropoda with its 80,000 or so living forms and thousands of fossil ones. In the Arthropoda the classes Pauropoda and Symphyla with about 50 species together stand next to the Insecta with more than 700,000 species.

At the level of order the mammalian Monotremata consists of three species, whereas there are some 3,000 species in the order Rodentia. In the Insecta a recent monograph of the order Zoraptera includes

16 species, whereas the order Coleoptera contains about 300,000 kinds.

At the family level the ratio can be as high as 50,000 to 1 within a single order. There is an isolated family of beetles (Platypsyllidae) consisting of a single species, and the family of the weevils (Curculionidae) contains 50,000 species and is still growing as many new ones are discovered.

The final level in this series is that of the genus. Great diversity in size exists, but some writers have suggested that all large genera should be broken up into smaller ones; for example, "There is every reason to believe that these giant genera will eventually be broken up into a number of smaller ones." This remark is based solely on ignorance of these large genera, because the specialists on some of them know that they are truly homogeneous and cannot even be divided into subgenera except on an arbitrary basis. Some of these genera contain several thousand species, while many insect genera are known from only one species of isolated character.

There is not a shred of uniformity in the size of the taxa at any level in the hierarchy of classification. Taxonomy is not concerned with the reasons for this diversity in size; it merely records the facts and tries to uncover patterns, uniformities, or discrepancies among them. The processes that are believed to have produced the many kinds of animals (evolution) are not such as to lead us to expect any uniformity in the size of taxa. Instead, extinction due to various causes would almost necessarily lead to differences in size. It is only the extent of the extremes which might be unexpected.

Diversity in breadth of groups. Some groups of animals, even fairly large ones, are homogeneous, whereas some others are heterogeneous to a substantial degree. By homogeneous is meant consisting only of animals all very much alike in a large number of obvious features. In contrast, some groups contain quite diverse subgroups having in common substantially less than all the obvious features.

For example, the Pelecypoda or Bivalvia among the mollusks are almost all immediately recognizable to anyone familiar with a few of them. The class Gastropoda, on the other hand, includes not only the readily recognized snails but also the land slugs, the sea slugs, the pteropods, the abalone, the limpets, and other forms of quite diverse appearance and structure.

In the Coelenterata, the Scyphozoa form a homogeneous unit, varying only in the details of a single basic pattern. The Hydrozoa, on the other hand, are extremely diverse. They include solitary polyps, solitary medusae, colonial polyps, and complex colonies of polypoid and medusoid members combined. Considering the relatively simple structure of the organisms, the diversity in form within the Hydrozoa is extremely high. If the millepores, the stromatoporoids, and the grapto- lites are included, as is still a common practice, the intra-class diversity becomes much greater, probably greatly exceeding that of any other group of animals.

Two more or less comparable groups of annelid worms are the terrestrial earthworms (Oligochaeta) and the mostly marine Poly- chaeta. The first is a uniform group, doubtfully divisible into sub- groups above the family level, whereas the Polychaeta consist of active swimming forms, as well as less active burrowing forms and sessile tube-dwelling forms. Most writers also include in the Polychaeta a very different sort of animal, the ectoparasitic Myzostomida.

A general measure of the diversity within the various phyla of ani- mals is provided by the number of classes recognized in each. In one classification which includes 40 phyla, the 107 classes are distributed as follows:

Phyla of 1 class each 21		6 classes each 1
2 classes each 7		7 classes each 1
3 classes each 4		8 classes each 1
4 classes each 2		11 classes each 1
5 classes each 1		16 classes each 1

Thus, half the phyla consist of a single type of animal not divisible into classes, but a third of the phyla consist of three or more subgroups diverse enough to be listed as classes. The range of subgroups from one to sixteen represents the variation in internal diversity of the phyla. There is little correlation between this and the diversity in size of the phyla. Although the largest phyla all consist of several classes, some of the single-class phyla are substantial in size and diverse at the ordinal and family levels (Nematoda and Graptozoa), as well as sev- eral of the two-class phyla (Bryozoa and Brachiopoda).

Differences between groups

Differences between groups may be as great as the difference be- tween unicellular and multicellular, between solitary and colonial, between presence or absence of a digestive system, or between having an exoskeleton over the body or being devoid even of an epidermis; they may be as slight as the number of serial organs, the presence of such a structure as a feather, or the position of the anus with respect to the tentacles. Such differences do not automatically qualify to dis- tinguish groups. Any pair of them can occur within one group at a level above the one in which they are distinctive.

The range of features in which such differences can occur is too great to be tabulated here. It can occur at any level of organization or activity, as illustrated in the following paragraphs.

Chemical diversity. Some general biology books imply that there is great unity in the chemical basis of the different forms of life and groups of organisms. At the same time it is held that there is biochemical uniqueness in all kinds of organisms if not actually in all individuals. It is a trend of our times to emphasize the unity—the similarity.

This aspect is almost unique in being summarized in an elaborate ten-volume work entitled *Handbook of Biological Data,* consisting of thousands of tables on many aspects of biochemistry. (It is seriously misnamed, inasmuch as it deals almost exclusively with biochemistry and physiology to the total exclusion of large aspects of biology.) It would seem to be an easy task to summarize the diversity at the biochemical level from these numerous and elaborate tables. The biologist soon finds, however, that this is really a medical reference book in major part, with only occasional references to invertebrates, and even these are usually included as if in concession to those few who might notice the omission. A few pieces of pertinent information can be gleaned from this source. There are some other publications of interest, particularly on insects, but no survey of the biochemistry of the Animal Kingdom is available. Such a work could not be other than sketchy, at best, because almost nothing is known of the biochemistry of many groups, and too often data are not comparable from group to group.

The present-day concepts of genetics require that some of the molecules of the genes be different in each kind of animal. The extent of the difference may be surmised in some cases, but the actual structure of any protein molecule of this sort is still not positively known. Inasmuch as the instructions coded in the gene structure are translated into specific molecules, cells, and organs by enzyme systems, there must also be some differences in enzyme complement between different kinds. It is fairly certain that differences in molecular structure exist in all parts of all living things.

Cell diversity. Cells illustrate well the dual viewpoint of unity and diversity. Basic cell types are found throughout the animal kingdom, but many unique cell types occur in almost every animal group, performing unique functions or variations of widespread functions.

Nearly all animals have some amoeboid cells that are motile. They seem to be similar in structure in a wide variety of groups. There is diversity, however, with specialization in function: Some produce skeletons, some are phagocytic, some transport materials in the body, and some are capable of differentiation into other types for replacement. Amoeboid cells are among the less specialized cells of the animal body.

Nearly all groups of animals possess unique cells found nowhere else. For example, only Ctenophora possess colloblasts (lasso cells) and only fishes possess mormyomasts—the electroreceptors in the lateral lines.

Organ diversity. In addition to such ubiquitous organs as mouths and glands, many groups have unique organs. Among these are the rhynchocoel or proboscis cavity of the Nemertinea, the corona of Rotifera, and the Aristotle's lantern of Echinoidea.

Diversity in organ systems. Almost any general zoology textbook will give a list of the organ systems of animals. The list will consist of the systems common to vertebrates, and there will be little mention of any diversity in different groups. The list will probably closely parallel the list of basic functions or reactions of protoplasm—irritability, contractility, digestion, excretion, respiration, etc.

It is of course true that most animals, vertebrate or invertebrate, do have sets of organs of similar sort for the performance of some of these functions, but the diversity among these is actually much greater than is usually stated. There are even a few unique systems found only in one group and not usually mentioned. The following are examples:

demanian system, Nematoda; probably an accessory reproductive system
water vascular system, Echinodermata; for hydrostatic manipulation
tracheal system, Arthropoda; for breathing air
Malpighian tubule system, Arthropoda; for excretion
lateral line system, fishes; composed of sense organs
rectal respiratory tree system, Holothurioidea; for respiration

In addition to these unique systems, there is substantial diversity among animal groups in the "standard" systems. In fact, uniformity throughout a major part of the kingdom is the exception rather than the rule.

Variety of hard parts. Most textbooks refer to various hard parts of animals, particularly the bones of vertebrates, the exoskeleton of arthropods, and the shells of mollusks. These do represent the three most common types of skeletal structures, but they are far from being all the hard parts produced by animals. There are also animals which produce tests, spicules, thecae, loricae, calyces, zooecia, jaws, cuticles, chitinous exoskeletons, external tubes, shells, coenecia, dermal ossicles, plates, and scales, fin-rays, beaks, feathers, hair, claws, teeth, hooves, antlers, and horns.

Hard parts may be secreted in four ways: First, on the outside of the body by the outer layer of tissue (epidermis) or by the surface layer of a protozoan cell. These include the tests of Foraminifera and other Protozoa, some of the coral of Coelenterata, cuticles in all animals, the zooecia of Bryozoa, the shells of Mollusca and Brachiopoda, the exoskeletons of Arthropoda and Vertebrata, and the dwelling tubes of Phoronida, Pogonophora, insect larvae, etc. Second, hard parts may be secreted on the inside of the body by mesenchyme amoeboid cells (Porifera). Third, they may be secreted by the dermis or middle layer of the body wall—mesodermal in origin (Echinodermata and Vertebrata). And fourth, by connective tissue—also mesodermal in origin (Vertebrata).

The skeleton of many invertebrates and protozoans is external, a lifeless secretion, forming a hard covering over the body. The exceptions include the echinoderms and the sponges. The vertebrate skeleton is almost invariably cellular, composed either entirely of hardened cells or of cells and cell products. These may be either epidermal, dermal, or mesenchymal in origin. Exoskeletons may be epidermal or dermal. Endoskeletal structures may be dermal or mesenchymal.

Diversity in reproduction. It is sometimes implied that animals typically reproduce by sexual means and their development from the fertilized ovum always follows much the same course. This is not merely an oversimplification, because sexual reproduction is by no means universal, and the details of development are extremely diverse. There is little in common among all animals, except that part of the parent (or of both parents) goes to produce a new individual, and that mitosis is always involved in some manner.

If sexual reproduction be defined as the production of new individuals by the union of sperm and egg from the two parents, then there are many animals in which sexual reproduction is not known to occur. In fact, it may be that Sonneborn [1] was correct in stating that not over half of the kinds of animals are produced by a bisexual process involving two parents. It is probable that non-bisexual processes (division, fragmentation, multiple fission, budding, sporulation, polyembryony, and parthenogenesis) produce a large majority of the individuals of all animals together. At least nine forms of asexual reproduction occur in at least fourteen phyla of animals. At least fifteen forms of sexual reproduction occur, of which eleven are in the Protozoa.

[1] T. M. Sonneborn. 1957. *Breeding systems, reproductive methods, and species problems in Protozoa*, pp. 155–324. In E. Mayr [ed.], *The Species Problem*, American Assoc. Adv. Sci. Publ. No. 50.

Diversity in development. The development of animals from the beginning of their separate existence until their death is generally assumed to include always an embryology—a period of egg cleavage, gastrulation, and embryo formation. This is, of course, a gross misconception. The individuals produced by six of the asexual methods develop without any of these stages.

Among those animals that do arise from an ovum, there is much diversity at several points: There may or may not be fertilization; the first cleavage plane may be meridional or equatorial or superficial; cleavage may result in any one of four types of blastulae; gastrulation may occur in any of nine ways; the three possible germ layers may be present in any of six combinations; the blastopore may become the mouth, the anus, or neither, or may never exist; mesoderm may form in any of six ways from either ectoderm or endoderm; any of four types of body cavities may occur; and larval stages in almost infinite variety may be present singly or in series in one life cycle.

Other group diversities. Few of the aspects of behavior have been cited above. There is diversity also in everything animals do. The study of comparative behavior is a relatively new field, but there is a wealth of diversity to be recorded. Comparative psychology would of course be included here, along with such features as consortism and constructions.

Every kind of animal occupies a certain normal range, which it seldom or never escapes from. A variety of factors account for these ranges, and kinds differ extremely in the size of the range, the limiting factors, the density of the occupation, the contact with similar kinds in other ranges, and so on. Although the ranges of animal kinds do usually fit into a general pattern over the entire earth, they vary without end in all aspects of distribution, as well as in the extent and means of their migrations, their capacity for transport by man or other means, and the changes they undergo through geologic time. A whole series of diversities are involved in this neglected aspect of comparative zoology.

The diversity of kinds

All of the preceding examples illustrate primarily the diversity among groups, especially at high levels in the hierarchy. It is more difficult to illustrate the diversity at the level of species, partly because it is much more extensive.

The kinds of *flies* may be distinguished by minute differences in the chaetotaxy (placement of bristles), the shape of cells between the wing veins, the slight differences in shape of antennal or leg segments, the proportions of any body sclerite or appendage, details of the genital armature, and arrangement of colors.

The kinds of *mammals* may be distinguished by pellage color, slight differences in bones including muscle scars and shapes, relative size of body parts, shape and color of teeth, and details of special hair arrangements.

The kinds of *jellyfishes* may be distinguished by the position of buds, the length of the manubrium, the shape of the umbrella, the position of the gonads, the number and arrangement of nematocysts of each type, and the number and shape of the tentacles.

The kinds of *mites* may be distinguished by bristle arrangement, size and proportions of body parts, the details of genital armature, the shape and vestiture of mouthparts and other appendages, the details of the openings of the tracheae or air tubes, and the body markings.

The kinds of *sporozoans* may be distinguished by the size of the sporozoites, the color of various spots, the presence of microscopic hairs, the number and size of glycogen granules, the size of the nucleus, the number of spores and their size groups, shape and extent of syzygy, the number of schizonts and their nuclei, the presence of a polar filament, and the location within host tissues.

In short, species in every group are distinguished by details within the general pattern of the group. That species do differ in such details is unquestioned after two hundred years of descriptive taxonomy. How to tell in advance what the details will be in any given group remains a secret, but in retrospect it is possible to generalize that they will be features that do distinguish species, usually the various expressions of some feature found in a group of related species.

Species are seldom distinguished by completely unique features, such as an organ known nowhere else. They are more likely to be distinguished by some detail of an organ, such as some difference in position or color.

A difference which is found to distinguish two species in one group may not serve to distinguish a third species. Features which yield useful distinctions in one group may be useless in another group. There is no way to define species characters or class characters in general, but it is possible to tell whether they are effective in any given case.

The diversity of taxa

These examples may suggest the almost limitless variety of features among the groups of animals. Although most of the examples are cited at the class and species levels, they exist at all levels.

Although the diversity seems to be without end, when one starts to tabulate the features of the different kinds of animals, there are very definite limits to most of the variety beyond which animals just do not go. No insect ever has more than four wings, although there are three thoracic segments which could conceivably all have produced a pair of them. Apparently no animal is based on a grouping of three like parts around a central axis, although four, five, six, etc., are common, and bilaterality can be considered to be two such parts. No animal is adapted to live its life continuously floating in the atmosphere, although this is conceivable in the same manner as floating in water.

The diversity is also limited by universal occurrence of some features. Apparently all protoplasm consists basically of carbon-chain molecules. Respiration always utilizes oxygen and produces carbon dioxide as a waste product. So far as known, all kinds pass hereditary determiners (genes) on to the next generation. And so on. But each of these very soon shows diversity if one carries the description a little farther, just as the sex of animals is *usually* determined by the genes but is *sometimes* determined by one of a variety of other mechanisms. The carbon-chain molecules are extremely diverse, as are the resulting structures and functions.

It is the presence of this diversity *and* the presence of uniformity within each kind that makes classification necessary and possible. The rest of taxonomy is largely the prelude to, or mechanical operation of, the classification system.

PART IV

Classification, Naming, Description

"Classification, as a process, is a fundamental necessity in human life. Indeed, we can see that this ability to classify is necessary to the ability to communicate. Nouns and adjectives, our chief classifiers of the world about us, are an absolute necessity for the exchange of information." (Burma, 1954)

"A plant's name is the key to its literature." (Van Steenis, 1957)

"It is a major thesis of this volume that the proper describing of species in order that they may really be known is the most important practical task which the systematist has to perform. Second only to the matter of determining what are and what are not species, it constitutes the basic portion of all systematic work and upon the way in which it is done depends the validity of everything which follows it." (Ferris, 1928)

Chapter 10. Comparative Data 137
Resemblance 139
Hereditary resemblance 139
Convergence 139
Parallelism 139
Environmental resemblance 140
Chance resemblance 140
Difference 140
Correspondence 141
Homology 141
Diversity vs. taxonomic features 143
The data of taxonomy 144
Kinds of data 145
Taxonomic characters 148
Diagnostic characters 149
Weighting 149
Qualitative vs. quantitative
characters 151
Comparative zoology 151
Comparative anatomy 152
Comparative physiology 153
Comparative biochemistry 154
The methods of comparing 155
Graphic methods 158
Statistical comparisons 159
Summary 161

Chapter 11. Species and Subspecies 162
Species 162
Species in taxonomy 164
Terms descriptive of species
165
Typification of species 165
Non-taxonomic species 169
Subspecies 171
Definition 172
Taxonomic nature 172
Pseudotaxa 174
Other subdivisions of species 175
Parataxa 176
Summary 180

Chapter 12. The Practice of
Classification 182
Purpose 182
Types of classification 184
Natural vs. artificial 184
Evolutionary or phylogenetic
classifications 186
Phylogenetic vs. phenetic
classifications 187
Typological or archetypal
classifications 188
Horizontal vs. vertical
classifications 188
Larval vs. adult classifications
188
General vs. special classifications
189
Modern omnispective
classification 190
Basis of classification 191
What is classified 193
Existence of groups 194
Differences and similarities 195
Stability of classification 195
Types in classification 196
Genotypes 196
Family types 199
The data of classification 200
For fossil animals 201
Appropriate data 202
Selection of features 202
The categorical level 203
Good and bad characters 203
The units of classification 204
Methods of classification 206
The method of association 207
The method of subdivision 208
Summary 209

Chapter 13. The Use of Names 210
Vernacular names for sexes, age forms,
and groups 211

Standardized common names 213
Scientific names 214
    Historical background 215
  Names of phyla, classes, and
  orders 216
  Names of the family-group taxa
  220
  The names of genera 224
  The names of subgenera 226
  Specific names and subspecific
    names 228
    Citation 228
    Meaning of the name 229
    Spelling 229
    Change of name 230
    Citation of rejected names
    231
Pronunciation of names 231
Summary 238

Chapter 14. The Use of Literature 239
Recovery of data from the literature
  239
  Through the name of the taxon
  239
    Groups above the genus
    level 240
    Groups at the family level
    241
    Groups at the generic level
    242
    Groups at the species level
    244
  Through the name of the author
  245
  Directories 246
  Personal bibliographies 248
  Through a general subject 251
  Methods and equipment
  251
  Collections 252

Expeditions 252
Localities 253
Institutions with taxonomic
  interests 253
Periodicals 253
Books 254
Nomenclature 255
Interpreting the literature 256
  The physical book 256
    Title page 256
    Issuance in parts 257
    Reprintings and new editions
    258
    Authorship 259
    General 260
  The author's intent 261
  Language forms 261
    Languages 262
    Diacritic marks 262
    Ligatures 263
    Abbreviations 264
    Translation 266
    Transliteration 266
    Latin phrases 266
  Interpretation of the contents
  267
    Identification 268
    Synonymy 268
    Bibliographic references
    276
Summary 277

Chapter 15. Descriptive Taxonomy 279
  Observation of features 282
    Preparation of material 283
    The series or sample 284
  Taxonomic characters 285
    Selection of characters 286
    Good characters 286
    Measurements 287
    Statistical characters 287

Description 288
    Graphic description 291
Types in descriptive taxonomy 292
    Type-locality 295
Descriptive publications 296
    Diagnoses 296
    Revisions and monographs 297
    Keys 297
    Classifications 299
    Phylogenetic trees 301
    Synonymies 302
    Bibliographies 306
Naming the taxa 306
Summary 307

Chapter 16. The Publication of Data 308
    General problems of publication 308
        What to publish 309
        Where to publish 310
        Authorship of taxonomic papers
            311
        Ethics in publication 312

Technical problems of publication
    314
    Descriptions 314
    Keys 315
    Classifications 315
    Synonymies 316
    Bibliographies 316
    Nomenclatural aspects 316
    Languages 317
    Historical documentation 317
    Spelling of names 317
Preparation of papers 318
    Manuals of style 318
    Terms 318
    Technical writing manuals 319
    The use of words 319
    Preparation of illustrations 319
    Common errors 320
The distribution of publications 326
    Dealers in secondhand books
        327
Summary 327

CHAPTER 10

Comparative Data

The diversity of animals, described in Part III of this book, is the basis for all classification of animals. Indeed, the existence of this diversity is the reason why classification is necessary. The word classification implies that there are "classes" of things.[1] If we start with individual things and compare one with another, we can unite similar ones in classes based at first thought on similarities. If we start with many things at once and divide them into classes and subclasses, we will have used differences as our first criterion. As methods of classifying, these two approaches will be discussed in the next chapter. Here we are interested in the fact that both approaches to grouping are based on comparison of the animals or their attributes. In the first we compare to find what there is that is similar. In the second we compare to find what there is that is different. The key word and concept is *comparison*.

Two things can be compared even if they have nothing in common, but there may seem to be no purpose to the comparison. To group or classify things, there must also be some similarities among them. This is *resemblance*. We know that resemblance may be produced by several processes or circumstances, but it is the existence of resemblance that allows comparison to recognize groups, and it is the existence at the same time of *differences* between animals that allows us to classify these groups by comparing them among themselves.

Even where there is difference between two kinds, there is always some *correspondence*. This correspondence may be functional; for example, nearly all animals have digestive tracts which correspond in gross function. The correspondence may be functional but not structural, as the wings of birds and insects serve the same locomotory purpose with no similarity in structure or development. The corre-

[1] The word *class* is here used in its general sense, the one used also in logic, rather than in its technical meaning of a hierarchical level, Class.

137

spondence may be structural, as the forelimbs of bats and of men are composed of similar bones in the same relative positions, even if their shape is different and they are used differently. The nature of this correspondence is as important in taxonomy as the existence of the diversity, because some correspondence, like some diversity, is of no direct interest to taxonomy yet must be recognized to be avoided.

Zoological classification is based entirely on data about the attributes of the animals. To be useful in classification, the data must be comparative in nature—either animals agree in possessing an attribute, or they differ in its possession or extent. The differences may be such as to attract our attention through any one of the senses, but most commonly it is through direct vision. On the other hand, the difference may be such as to appear only after the application of special techniques—microscopy, experimentation, analysis of measurements, instrumental recording and interpretation of invisible attributes, chemical analysis, and so on.

In general, to be used in classification, the data must reveal discontinuities in the attributes. Satisfactory taxonomic groups cannot be made in material which shows continuous variation, because taxonomic groups must be clear-cut. Currently accepted ideas of evolution seem to deny the existence of clear-cut discontinuities between groups as a natural feature, but this must be a false appearance for we cannot deny "the simple fact that readily recognizable and definable groups of associated organisms do really occur in nature" (Simpson, 1961). This was also the conclusion of Robson and Richards in 1936, who wrote: ". . . throughout the animal kingdom there is a tendency for individuals to be capable of arrangement in a hierarchy of groups. . . ."

Whatever process gave rise to these groups was a natural one, whether it was genetic or simply due to extinction of intermediates. It is, then, not quite correct to say that taxonomists "construct" the classes in a classification. In general they recognize and select the groups which already exist or which appear in the samples studied. The origin of these groups is of great interest to biologists, but the taxonomist can proceed with his work without regard to the theories of origins, if he so wishes. It will very likely make no difference in his results.

Five relevant questions may be asked: What sorts of resemblance and difference are there? How is correspondence recognized and used? What sorts of data are available? How are the taxonomic data distinguished from the individual variation? How are the kinds compared? These require separate discussion.

Resemblance

When two things are alike in any feature or group of features, we say they resemble each other to that extent. Thus, resemblance is simply likeness. Similarity is another synonym.

In zoology there is always resemblance. The most completely different animals, such as an eagle and a paramecium, are similar in being composed of protoplasm made largely of carbon-chain compounds operating through enzyme systems, in requiring water for their very life, in being subject to the vicissitudes of the environment, in carrying on cell division to reproduce their kind, etc. The similarity may be sharply limited and overshadowed by the differences, as in this case, or the similarities may involve most features of the animals. The latter would be the case if we were comparing a bald eagle with a golden eagle.

The resemblance that interests taxonomists is, of course, the resemblance in what are called the taxonomic features. These are simply the ones the taxonomist decides to use in his taxonomic work. (The basis for this choice is discussed later in the chapter.) But this definition is a little too easy. It hides the fact that there are several kinds of resemblance of very different nature which must be distinguished by the taxonomist to some extent.

Hereditary resemblance is that which is produced by similar genes passed on from parent to offspring. It will involve most parts of the body and all stages in the life cycle. Intermixed with it will be a relatively small amount of difference produced by recombination, mutation, and other genetic factors. The great majority of all resemblances are of this hereditary nature, but in the outward appearance of the animal they may be overshadowed by other factors.

Convergence, or convergent resemblance, is similarity in two genetic lines or lineages not due to a common gene heritage but to adaptations in one or both lines. The adaptation presumably is in response to ecological pressures and is an evolutionary process. Convergence generally involves limited features of the organisms, but it must be recognized by the taxonomist to prevent grouping the distinct forms because of this real but non-taxonomic resemblance. Mimicry is a form of convergence in which the general appearance of the two forms are similar because of the convergent similarities. Likewise, warning coloration, protective resemblance, and obliterative coloration may be forms of convergence.

Parallelism or parallel resemblance is usually treated with convergence by taxonomists, but it differs from convergence in that the development of the similar features is the result of and is channelled by a common ancestry, not principally by the environment. The common ancestry is so remote that lineages of evolving species are involved, not merely a few generations of individuals as in hereditary resemblance.

Environmental resemblance is of two sorts, of which the first has already been treated as convergence. In that type the resemblances are produced by the genes which were screened by the selective processes of the environment. But there are also resemblances of temporary nature that are not impressed on the genes and appear only when the environment acts directly on an individual. A furry animal may have longer and denser hair in cold areas or seasons than the same kind has in warm areas or seasons. In some animals, size is largely due to the amount of food available to each individual. Environmental resemblance may become fixed in the genes by selection and thus become convergent resemblance.

Chance resemblance would be difficult to define and to identify. All of the above types involve chance to some extent. It is not always possible to see the environmental influence, if it exists, and, in the absence of direct hereditary control, we can only say that chance is the major factor. This can best be illustrated where large numbers of species are involved. Among the thousands of kinds of flies and bees, it is not strange that some one of each could be found that looks very much like the other. If they occur in widely separate regions, none of the preceding forms of resemblance is convincingly applicable. We might even postulate that with this number of kinds involved, some similarity would be expected by chance without the influence of any other factors.

Difference

Although there are some similarities between any two animals and many similarities between some of them, the predominant fact of life is that there are great differences between most kinds of animals and some differences between even the most similar ones. Difference is dominant over similarity, because it is present at every level we can conceive. Resemblance is always present at some levels but there may be none at certain levels of consideration.

It would be possible to classify differences in the same manner as done above for resemblances; in fact, the classes would be more distinct. However, there seem to be no terms of similar nature for the kinds of difference. Surely there are differences caused by gene influence and environmental influence. The taxonomist must try to distinguish between these two, but what interests him most is whether they are individual differences, specific differences, or group differences. This is the central problem of the comparative part of taxonomy.

In making comparisons, another problem occurs. In things as diverse as animals, there is always a wide variety of differences. In comparing two animals it would be meaningless to say that the fur of one is longer or whiter than that of another if the tail fur of one specimen were being considered and the belly fur of another. The taxonomist must compare things that correspond—that are comparable. This problem is considered next.

Correspondence

The chief rule to be followed in comparing any two things—structures or individuals or taxa—is to compare only things which correspond. We must not try to compare a genus with a species, a fingernail with a feather, or a distribution pattern with a stratigraphic sequence. But it is by no means always obvious what things are comparable, what things correspond in the necessary way.

There are a variety of causes which make structures appear to correspond. These are largely phylogenetic or evolutionary causes. As such they are not objectively recognizable but must be used on some subjective basis. They include homology, parallelism, convergence, and mimicry. These may or may not involve correspondence in any given sense, but they are more reasonably spoken of as causes of resemblance. Resemblance may also be caused by two other factors, analogy and chance similarity.

Homology. In comparing individuals—specimens—zoologists have for a hundred years been ostensibly guided by the concept of homology. Certain structures are "homologous," such as the wing of a bird, the foreleg of a horse, and the arm of a man, and these can be compared usefully with each other. Structures on two individuals are not comparable unless they are homologous. For this it is obviously necessary to be able to recognize homology between two structures on separate animals.

Definitions of homology commonly relate the correspondence to common ancestry. Simpson in 1961 gave this simple and clear definition: "Homology is resemblance due to inheritance from a common ancestry." The criteria for recognizing homology are cited by Simpson and paraphrased here: (1) Minuteness of resemblance; (2) multiplicity of similarities (it being a "sound principle" that conclusions on affinities are stronger the more characters are involved); (3) the probability of homology is greater the more intricate are the complexes of characters; (4) correlation of characteristics of radically different kinds is unlikely, so homology is indicated by concordance of such characteristics; (5) developmental patterns are inherited and so give evidence of homology; (6) evidence of chromosome and genetical identity; (7) continuing adaptation to environment; and (8) "the most direct criterion of homology is of course the discovery of the common ancestry or of fossil lineages clearly converging backward in time toward that ancestry."

Simpson believes that these criteria "almost always can suffice to recognize homology with a satisfactory degree of probability."

In practice, any student will find that only the second of these criteria is of any direct use to him, unless he is lucky enough to have also evidence of development (5). All the other criteria are substantially subjective, subject to varying interpretation, and requiring more or less rigid application according to circumstances. The student will find that he can judge the presence of homology *only* on the basis of experience with previous work and the results of embryological, anatomical, genetical, or physiological studies on supposedly related species.

The fact is that the ancestor of one or two individuals (the prehistoric ancestor, of course) can be guessed at exclusively on taxonomic evidence, on the basis of similarities of attributes. If this is turned around to say that the features of two animals are comparable because they are "known" to have had a common ancestry, this is proving a proposition by its own conclusions.

In some cases there is "evidence" of correspondence from several viewpoints. How far these separate and partly subjective evidences bolster each other is the crucial question. There is no way to answer that question objectively. This leaves us with no direct way to test for homology and no way of judging the accuracy of the subjective conclusion we may reach.

This view of the basis of homology is not the view accepted by most writers today, but it seems to be a necessary conclusion even from the arguments of its opponents. Recognized homologies, which are ones generally accepted at one period, have many times been shown to be erroneous at a later period. This can happen to virtually any homology determined under the usual criteria.[2]

It might seem to be a reasonable conclusion that this point of view destroys entirely the usefulness of homology as a criterion of com-

[2] For additional arguments regarding the basis of homology, see the papers cited under this heading in the Bibliography.

parability, but this is not at all the case. Taxonomists have never had an objective basis for homology, and yet they have used this concept for a hundred years as the basis for all classification. It is, in fact, probably true that no classification would be possible without this concept. Taxonomists have compared structures which seemed to them to be comparable, and when they erred it was often because the organisms were even more complex and diverse than they had realized. The correction of these errors, upon later evidence, is one of the continuing jobs of this science, just as such corrections are a major part of all science, where verification is an essential procedure.

Taxonomists cannot do without the concept of homology. Unfortunately they cannot at present give it any objective basis, even though it is a logical necessity in the evolution of animals.

Two animals that do not seem otherwise to be closely related in a phylogenetic sense may have certain structures that are very similar. These similar structures may be the result of environmental pressures, phylogenetic lack of change, or chance. Those which result from descent without change are called *homogeny,* and the only example is parallelism. Those which are not due to inheritance from a common ancestor are called *homoplasy,* and include convergence, analogy, mimicry, and chance similarity.

Diversity vs. taxonomic features

In Part III of this book the diversity of animals is shown to be of numerous sorts. The two chapters therein foreshadow a distinction of the utmost importance in taxonomy. Both descriptive taxonomy (based on differences) and classification (based on resemblances), are founded on features which members of a species or group have in common, in which there is little or no individual variation.

The first job of taxonomy, therefore, is to distinguish between individual variation and the diversity which individuals share or which species share. This is usually stated in the form that taxonomy is based on *taxonomic* characters and is not concerned with individual variation. Inasmuch as there is no definite way to distinguish between the two offhand, taxonomy must concern itself with both until it has definitely determined which is taxonomic.

A taxonomist must distinguish between taxonomic features and individual variation, but there are few general rules to make this distinction easy. A feature that distinguishes species in one genus may vary within the species of another. The only way to distinguish them is to test them individually in every case. This is done almost automatically by an experienced taxonomist, who knows so much about the total biology of his animals that he develops a feel for the taxonomic features.

There are some formal methods of studying variation to see if it is individual or holds good for all the specimens of a species or even of a family. Some of these are mentioned later in this chapter. In general, this is accomplished by comparing specimens and groups of specimens, and these aspects are dealt with below. The *data* of taxonomy are the usable diversities. The *characters* are the ones chosen for making distinctions or groupings.

The data of taxonomy

The things that can be known about kinds and groups of animals are the potential data of taxonomy. Taxonomy and classification have always been overwhelmingly structural in their data, primarily because these were the only data consistently available. Classification serves to store all kinds of data if these are associated with known species or other taxa.

Some writers have recently stated that taxonomy cannot be adequate if it uses only anatomical or structural data (usually called morphological), but this involves at least two important misconceptions. *First,* it is implied that other types of data are largely available. But it is probable that there is no genus of animals about which there is enough non-structural data for classification of the species to be accomplished by its use alone. The genus *Drosophila* may be an exception, but there seem to be, even in this much-worked group, some species which are known only from structural features. Even if such exceptions were fairly numerous, the great bulk of animal species cannot at the present time be compared on the basis of any data other than those obtainable from preserved specimens. *Second,* it is implied that use of non-structural data is a recent development that has been neglected by taxonomists. The fact is that non-structural data have been used since the time of Linnaeus in increasing amounts as such data have become available. But, because such non-structural data are not *even now* ordinarily available in sufficient completeness, these data are and can be used only in rare cases to solve problems that have proved puzzling. Furthermore, in most cases where such data are available and are studied, they merely corroborate the conclusions from structural data.

A later section discusses at some length the theoretical basis of classification and the nature of the attributes by which groups are recognized. There has been no great change in the practice of taxonomy in recent years, in spite of some statements to the contrary.

There is no prospect of change in the future and no reason why one should expect it. Animals are still classified almost entirely on the basis of comparison of their attributes. Whereas a few score years ago the data generally available were structural, today taxonomists frequently know something also of the physiology, ecology, distribution, and even biochemistry of some of the animals studied. (The very substantial limitations on the use of this new data will be discussed later.)

The newer types of data or methods of study have sometimes been presented as if they were panaceas for all difficulties in taxonomy. They have not proved to be any such thing. There is no reason to believe that these newer features are any less subject to the difficulties that beset the use of the older data—irregular variability, convergence, parallelism, difficulty of observation, and so on. There is no reason to believe that they are as a group any more surely related to the basic nature of the species or the hereditary processes. *All* characters are correlated to some extent in the gene complement. Some are controlled directly and absolutely by the genes; some are merely initiated or limited by the genes. It does not appear that any one type is more reliable than the others.

The attributes of animals, when recorded in terms that are at least potentially comparative, form the data about kinds. Although the data available in Linnaeus' time were almost exclusively the physical features of the specimens, after two hundred years of zoological study there is available now a large range of attributes, including much beside the physical structure. As it is sometimes misunderstood how wide a variety is represented in this range of attributes, a few remarks will be included for the major types.

Kinds of data. In talking about the features in which certain organisms differ from others, taxonomists commonly refer to them as *characters*. These are simply the specific features chosen to make a comparison. These characters may come from any kind of data, but usually a character is a visible feature of some sort. In the present section, it is intended to discuss the various sorts of data that can be used comparatively, whether or not they are customarily a source of characters.

At the outset we should understand that comparative data can be physical, chemical, spatial, or temporal. They can relate to any part of the life cycle of the animal or to any of the phases in which the kind appears. They can relate to any level of organization, from the atomic and molecular to the community. They can represent shape, material, arrangement, activity, or posture. Some of the aspects have so far yielded few examples of taxonomically useful data. Some are coming into considerable prominence at the present time. It is necessary to keep in mind that the subjects discussed below are those which have yielded usable data, but they are not necessarily fields whose data *must* be used by taxonomists. The distinction should become clear in later paragraphs.

Morphological data implies to many people the data of organs, anatomy, the structures of the body. This is a misuse of the word morphology, which actually refers only to form or shape. In this more restricted sense there are morphological data, but they are seldom used in taxonomy except at the highest levels of classification.

The diversity of form among animals is suggested in several places in Part III. It relates to symmetry, repetition of parts, gross features of body arrangement, and polymorphism—the form of the body, and rarely the form of parts of the body. Data of this type are useful in classifying the major groups of animals, principally the phyla and classes.

Structural data are those usually referred to as morphological, especially external anatomy. It surely also includes internal and developmental anatomy. The data of osteology, haematology, conchology, and the structural aspects of paleontology would be included.

Here, of course, are all the normal facts of classical taxonomy—the descriptive data about specimens. These may be from preserved specimens which are adult individuals or immature specimens; they may be anatomical (about the internal organs), histological, or cytological. A special case of this is what is called cytotaxonomic data, the number and shape of the chromosomes or karyotype. Going still lower in the scale of size, they may be at the molecular level, at which structure is one aspect of biochemical study.

Physiological data are those relating to the functioning of the parts of the organism. These are derived from adults and from all immature stages. Under this heading would also be classed data from endocrinology, serology, and much of neurology. At the protoplasmic level, these data begin to merge into biochemistry, in which the reactions of the molecules and substances yield physiological data.

Ethological data relate to the differences in behavior. It is often believed that differences in behavior must be correlated with differences in structure at some level, but the behavioral differences may be easier to use in some instances. Aspects of behavior for which data are most often available are reproduction, vocalization, and responses to various stimuli.

Genetical data would be listed by many zoologists, perhaps together with the cytotaxonomic data. In reality the genetical data are structural, physiological, or what we may call hereditary—the sequence of other data in the succession of individuals. Only the hereditary aspect is really unique. Nevertheless, there is a large number of data from genetics research, and it certainly is part of the data available to taxonomy.

Parasitological data consists of information on the hosts of each parasite species and the parasites of each host species. Either of these can be of taxonomic value, and both have been cited as characteristic of species. With the so-called micropredators, or ectoparasites, this subject comes close to some aspects of ethology and ecology. Predation supplies similar data that may be distinctive.

Ecological data are akin to ethological data as well as parasitological. Animals differ in their response to the environment, their preferences, tolerances, peaks of abundance, seasonal development, etc.

Geographical data have been used as long as structural data, inasmuch as Linnaeus recognized the source of the specimens as pertinent to taxonomy. Sometimes the geographical distribution seems to be dependent upon ecological factors, but the former is easier to detect and record.

Stratigraphic data are similar to geographical data in representing the chance location of the individuals at a particular geological time. They are part of the paleontological data, of which the balance is mostly structural.

The taxonomist may use these data in classifying the animals—in fact he certainly will use some of them. The rest of the data will not be useful to him for a variety of reasons, such as: (1) the data are incomplete, being available for only a part of the specimens, species or groups being classified; (2) the data are difficult to present, as may be the case with complex three-dimensional shapes; (3) the data may be so numerous that they are hard to interpret, even when tabulated, as is the case with much of the numerical data used in computer analysis; (4) the data in question may be fully correlated with some other type of data that is easier to handle, as is the case with much genetic data which directly produce visible features that are more readily compared; or (5) the data may be produced by procedures normally beyond the facilities of the taxonomist, who thus has difficulty in repeating, verifying, and adding to the results, as is frequently the case with serological and electrophoretic data. Whatever type of data is used, they must at the outset be interpreted or adapted for this particular purpose. The taxonomist must select the specific items he will use, the particular forms of these that will be compared—in the modern phrase, the parameters of the classification he is producing. These individual features to be employed are the taxonomic characters.

Taxonomic characters

Some recent books have used the term *character* in much the same general meaning as used above for *data*. For example, Davis and Heywood (1963) in an excellent plant taxonomy book write: "Taxonomic evidence (in other words, characters) can be drawn from any part and phase of development . . ." (p. 142). They define a character for practical purposes as: "any feature whose expression can be measured, counted or otherwise assessed . . ." (p. 113).

This is a reasonable usage and one consistently followed in that book. It requires that each character "be divisible into two or more expressions or states (including here presence or absence in qualitative featives)" (p. 113). For example, a character might be the number of segments in the tarsus of a beetle. It can be expressed as 1, 2, 3, 4, or 5 segments. Or it might be a certain spot of red on the wing of a bird, expressed as presence or absence.

Approximately the same usage is represented by the definition by I. M. Newell (1953) of characters as "any . . . feature" (p. 120). The particular expression of the character was then called a variant or character variant.

Neither of these usages conforms to the common usage of "character" in animal taxonomy. For example, Mayr, Linsley, and Usinger (1953) remark that "it must be determined how constant a given character is" (p. 121). In the system of Davis and Heywood, one could not use a constant character but would want to know how constant is each expression of the character. Again, "taxonomic characters should not be drawn from single representatives of populations. . . ." Only one expression of the character could be so drawn in the multiple-expression system cited above.

It seems more useful to give a slightly different definition to the word character, which would provide a clearer distinction from the general term taxonomic data, as used in the earlier chapter. This is to apply the word character to the *expression* of the feature in the individual. We would not then say that a certain seta is a character but that the presence of that seta is a character and the absence of that seta is another character. This seems to be more in keeping with the common practice of saying that the character which separates species A from all other species is the presence of a certain feature. It is true that this implies that it is absent from all the others, and thus that two expressions are involved, but it is equally reasonable to say that two characters are involved or one character and a lack of one character.

The new *International Code of Zoological Nomenclature* (1961, Art. 13), while not pretending to deal with the subject of taxonomic characters, does contain this passage: ". . . a statement that purports to give characters differentiating the taxon. . . ." It would take only one phase of the dual-type character to differentiate one taxon, but it would take two phases to distinguish between two taxa. The use of the singular "taxon" seems to show that the writers were thinking of a character as one expression of a feature.

In several recent works G. G. Simpson uses the phrase "characters in common" when speaking of the similarities between two groups. If he were using the dual-type character, it would have served to show the dissimilarities between groups, but to show similarities or characters in common one must be using character as a single expression.

This distinction really reflects no fundamental difference in classification or philosophy. It must be understood, however, which of these two meanings is intended, whenever the word character is used. The above quotations are intended merely to indicate that in work on animals a usage different from that in plants is common.

Characters are the features of the individual animals. Taxonomic characters are the ones chosen by the taxonomist for use in discrimination or grouping. Some of the problems in the use of characters in taxonomy are suggested below, whereas others, including the determination of which characters are taxonomic and which are not, are discussed in Chapter 12, The Practice of Classification and Chapter 15, Descriptive Taxonomy.

Diagnostic characters. These are taxonomic characters of limited occurrence which are selected because they can be used alone as distinguishing features. They are sometimes called **key characters.** They are useful in distinguishing between taxa and in identification. They are always presumed to be "good" characters; at least, they are the best ones in the opinion of that taxonomist at that time.

Diagnostic characters are discussed at length, with reference to their origin and use, by Davis and Heywood (1963).

Weighting. A problem brought to the fore by the recent emphasis on numerical or phenetic taxonomy arises because of the tendency of some workers to place more confidence in certain kinds of features. "Weighting means, in taxonomic usage, giving greater importance to one character than to another, for any reason whatever" (Davis and Heywood, p. 48). These authors list four main reasons why characters are weighted by taxonomists: (1) Because they are most easily observed, the enormous number of characters making it virtually impossible to describe them all; (2) because they show the highest correlation with others in a natural group (*correlation weighting* or *a posteriori weighting*); (3) because the character is believed to be of particular importance for reasons other than those deduced by correlation studies, such as supposed "conservative" or phylogenetic nature (*a priori weighting*); and (4) because they are the characters left when we have rejected others (*rejection weighting* or *residual weighting*). Only item (2) represents a process that *should* be used for classification, but the other three have most often been the actual reasons.

Weighting appeared in the early days in the form of belief that, for example, mouthparts of insects were more useful than other parts of the body at certain levels of classification. Much more recently there has been heavy weighting of genitalic characters in insects, in the belief that these are less subject to variation. At one time the structure of the feet of birds was thought to be an especially useful feature for classification.

It seems to be a safe general rule that reliance on any one character or type of character is likely to give unsatisfactory results in natural classification, if only because the factors molding this feature are many and largely unknown and there is thus real danger that the groupings will be unnatural. No one has demonstrated that *any* one type of feature is always of most significance, and it is generally believed that the synthesis of many features is the safest procedure.

It has been argued, especially by the numerical taxonomists, that the logical goal should be equal weighting of all the characters selected. Inasmuch as selection is itself a form of weighting and the operations of (1) and (4) above are well-nigh universal, it appears better at this time to caution against excessive weighting rather than attempt to eliminate it. Weighting has not yet been proven to be inherently disadvantageous, and there is some possibility that it may be practically unavoidable. It has been noted repeatedly that any effect of weighting is likely to be reduced or eliminated by subsequent revisions by other workers. This may be the best or even the only way to prevent weighting from interfering with the effectiveness of classification.

For further discussion of this problem see Davis and Heywood (1963) and papers cited therein.

Qualitative vs. quantitative characters. There has also been considerable discussion recently of the relative value or importance of qualitative and quantitative characters. There is no theoretical difference in value of characters which can be expressed in numbers and those which can be readily expressed only in descriptive words. This statement is true in the logical type of classification assumed in this book, although it has not always been considered true by scientists in other fields.

Mathematics is simply one of the forms of symbolic logic. Whether or not there is some theoretical advantage to the use of this particular form of logic or any other is not of so much interest to us at this moment as whether or not it is possible or practicable to express all of biology in these symbols. It is obviously not possible at the present time and will not be in the immediate future. As to its advantages, there can be little question that there are some things that are better described in numbers than in words—the size of a circle, the length of any object, its weight, and the number of discrete objects involved. But try to put into numbers the shape of a fly's wing, the arrangement of the veins and cross-veins, and the manner in which these wings open out when the insect first emerges.

As a result, it is not appropriate to urge all taxonomists to use numerical characters exclusively. In some situations quantitatively expressed characters can be more definite and more easily compared. The taxonomist should be alert to recognize these situations. He should be equally alert to use verbal characterizations when appropriate and graphic characterization when appropriate. He should not feel that there is any inherent difference in the various ways of expressing his observations. He will, however, find it advisable to make his records comparable to those of other workers in his field. If they are not, he will very likely find his work unused and unaccepted. This actually happened to Adanson more than a century ago, when he urged a radically new method of classification which could not readily be incorporated into the ways of his time.

Comparative zoology

Taxonomy is entirely comparative, at least in its two major aspects—descriptive segregation and classification. We compare animals to see which are alike, in order to group them, and we compare animals to see which are different, in order to distinguish them. The data of taxonomy are all comparative. It must be emphasized also that any comparative data are of interest to taxonomy, whether or not they are of a conventional nature. Taxonomy may not make use of all such data in classifying, but whatever part is not directly used is stored in the system.

Unfortunately, the producers of these data are often too little familiar with taxonomic practice to produce their data in taxonomically usable form. There is now coming to be a large enough body of these new data to be usefully discussed in relation to classification. It should be clear that no data can be *the basis* of classification until that type is available for all the forms to be classified. The new types are not yet able to attain this position.

Of the numerous fields which can supply comparative data to taxonomy, only three are well known in combination with the adjective comparative. These are comparative anatomy, comparative physiology, and comparative biochemistry. They are discussed briefly under separate headings.

For the rest, any field which can present its data comparatively, or in such form that it can be compared, can contribute to taxonomy. For example, there are several fields that deal with the mechanisms and sequences of reproduction and development; these give rise to *comparative genetics* and *comparative embryology.* There are two fields that deal with distribution of animals in time and space; these give rise to comparative aspects of zoogeography and stratigraphy. There are two fields that deal with the behavior and mental processes of animals; they give rise to *comparative ethology* and *comparative psychology.* And there are several fields that deal with the reactions of animals to their environment and the interactions of different kinds of animals; these give rise to autecology, which is potentially comparative, and to parasitology.

Parasitology has a truly comparative aspect which deals with the parasites of each kind of animal, the host of each kind of parasite, the specificity of the arrangement, and the other forms of individual-to-individual contact that are usually not called parasitism but consortism, epizoism, symbiosis, commensalism, predation, micropredation, or phoresy.

Comparative anatomy. The study of the differences in the structure of animals is comparative anatomy. The extent, variability, and even origin of the differences are all involved. The structure can be at any level from body plan of the individual and arrangement of a colony to the architecture of the cell and the molecular structure of protoplasm. Because any one scientist can be a specialist in only a limited number of these aspects, it has been customary for those who call themselves comparative anatomists to restrict their study largely to the body plan, organs, and tissues of some particular group of animals.

Comparative anatomy has come in recent years to mean the study of vertebrate anatomy as a prelude to the understanding of the anatomy of man. The comparative aspect is largely lost, with almost no emphasis on the nature of diversity, the methods of comparing features, or explanations of the meaning of the diversity in animal structure. The very retention of the term comparative anatomy indicates historical influences, because anatomy or structure has long since ceased to be the only comparative field. Comparative zoology is a term rarely seen, but it is much more appropriate for indicating the breadth of the data which can be compared. Included in this are the animal aspects of comparative physiology, comparative parasitology, comparative ethology, and all the comparative aspects of biochemistry, embryology, geographical distribution, and ecology.

Lankester's 1878 definition of comparative anatomy expresses the true basis for its study but a basis mostly forgotten in modern usage. He wrote: "Comparative anatomy . . . a work in which the comparative method is put prominently forward as the guiding principle in the treatment of the results of anatomical investigation." Just how this principle is to be put into the work is not quite clear. It involves philosophical questions as well as problems of taxonomic procedure. It involves homology and all other forms of correspondence between kinds. It involves the nature of the diversity of structure, as well as its extent and distribution. It involves all the basic aspects of classification and is absolutely indispensable to that science.

The data available from comparative zoology have been outlined above and the diversity exemplified in the preceding Part III. Some of the branches of anatomy have received names and are comparative insofar as they deal with animals other than man or the laboratory animals which act as stand-ins for man in experimentation. These include osteology; its derivative vertebrate paleontology, which is substantially osteological and largely comparative; cytology; histology; developmental anatomy; and organology. Even physical anthropology is largely comparative, dealing with the similarities and differences among the races of man.

Comparative physiology. A great deal is known about the physiology of a great many animals. Recently there has been published a series of books dealing with the physiology of such groups as fishes, insects, crustaceans, and mollusks. These books are a mixture of information about particular species and summaries of the usual situation in the group as a whole. They are potentially comparative, but they are not arranged to be directly useful in most comparative studies.

A large compilation entitled *Handbook of Biological Data,* edited by W. S. Spector, was issued in 1956 by the W. B. Saunders Co. It consists of hundreds of tables of physiological data. In general, the information is comparative, but on a level so high in the hierarchy as to be of little use in taxonomy. Moreover, the tables include only data thought to be of widest interest whereas comparative zoology is equally concerned with the rare and unusual forms. Companion volumes are entitled: *Standard Values in Blood* (1951), *Standard Values in Nutrition and Metabolism* (1953), *Handbook of Toxicology* (5 volumes proposed, 1955–1959), and *Handbook of Respiration* (1958).

Recent textbooks of comparative physiology provide a great deal of data on a variety of vertebrates, but for invertebrates the coverage is so sparse as to be merely suggestive. Physiological data such as could be used in species taxonomy are presented only in the numerous papers published in the experimental journals, in which the comparative aspects are usually subordinated to description of the results of observations on the one species.

Comparative physiology is an immense field for comparative data. Potentially, it far surpasses comparative anatomy, at least in its grosser forms, but it is largely undeveloped at the level that could be useful in taxonomy. Comparative biochemistry is an offshoot which has outstripped the parent in current taxonomic interest.

A branch of comparative physiology is represented by the term comparative endocrinology. There are also physiological aspects to several others, such as genetics, serology, and parasitology. Functional anatomy is a hybrid field, but it usually leans heavily on physiology and may be comparative.

Comparative biochemistry. Only in the last decade or two has enough been known about the biochemistry of animals for there to be anything substantial in the line of comparative biochemical data about kinds. Even now this data is restricted almost entirely to the presence or absence of materials; very little is known comparatively about metabolic pathways, differences in enzyme chain actions, and in general the functioning at the molecular level, as related to particular species.

The biochemical studies were at first almost exclusively serological. These included precipitin reactions that showed the presence or absence, and relative abundance, of antigenic materials. More recently one of the common types of study has been the separation and identification of amino acids or other molecules by electrophoresis or by chromatography. The requirements and usefulness of biochemical data in taxonomy are substantially the same as for other types of data. The limitations and the difficulties of use are also of the same general nature but may be somewhat different in detail.

To be useful in taxonomy, any data must be related to known kinds. In most cases this means that the species must be accurately identified. Data obtained from "the rat" or "the frog" will be of use only in the beginning steps of perfecting techniques or in training technicians. Taxonomy can use the data only if it knows exactly what species are involved. A minor exception to this can sometimes be justified where comparisons are being made between groups rather than species. It is possible to get some useful data about the comparative biochemistry of two orders by comparing unidentified examples of each, but it must be emphasized that the usefulness of such data is very limited. It cannot be assumed that *any* species is representative of a large group, and it cannot be assumed safely that there is not substantial diversity within the group.

For example, it has for some years been known that in the metabolic processes of muscle action there are two compounds that may serve as hydrogen acceptors. These are phosphocreatin and phosphoarginine. One was found in all the vertebrates examined, as well as in the Echinodermata. All other invertebrates employed the other, so far as known. Conclusions as to phylogenetic relationships of vertebrates and echinoderms were drawn on the basis of this data, and these conclusions have been quoted in many books. As a modern tabulation would show, this simple dichotomy is entirely spurious. More thorough coverage of groups, more examples of different kinds within the groups, show that this is not a simple situation but a complex one. It is not at present possible to see any phylogenetic implications in this feature.

For further discussion of comparative biochemistry consult the recent multivolume series, the following general works, and the additional references in the Bibliography.

Florkin, M., and Mason, H. S., eds. 1960–1965. *Comparative Biochemistry. A Comprehensive Treatise.* New York; Academic Press. (7 volumes)

Bier, M. 1959. *Electrophoresis: Theory, Methods, and Applications.* New York; Academic Press.

Boyden, A. A. 1943. Serology and Animal Systematics. *American Nat.*, vol. 77, pp. 234–255.

Micks, D. W. 1956. Paper Chromatography in Insect Taxonomy. *Ann. Ent. Soc. America*, vol. 49, pp. 576–581.

Sibley, C. G. 1962. The Comparative Morphology of Protein Molecules as Data for Classification. *Syst. Zool.*, vol. 11, pp. 108–118.

The methods of comparing

The absence in recent books of direct discussion of methods of classifying may be due in part to the fact that it is a procedure of great subjectivity. It may also be partly due to the fact that in the better-known groups of animals classification is much less frequently practiced than identification. In the view of classification adopted in this book, comparison can involve either specimens or the conceptual groups which are the taxa—the species, families, and orders. The methods of comparison are not always the same for specimens and taxa, but in both cases a major factor is the previous experience of the taxonomist in the particular group of animals, and to a lesser extent his experience in the taxonomy of other groups.

Neither specimens nor taxa are compared in all their attributes. One knows from experience that there are classes of attributes which yield no useful comparisons. A taxonomist working with species soon learns to be on the lookout for any of the forms of infraspecific variation listed in Chapter 9. When working with genera, he avoids the attributes which were found to differ at the species level. And when working with families, he avoids the ones which differ at the genus level. But there are many pitfalls.

Four things are basic to taxonomic work as it is approached here. As discussed earlier in this chapter, these are: similarities, by which like things are grouped; differences, by which kinds are distinguished; correspondence, by which is known what to use in grouping and distinguishing; and comparisons, by which the actual similarities and differences are discovered. These are inextricably inter-related phenomena. All of these aspects have now been discussed except the actual comparison and the methods by which the taxonomist decides which features to compare.

The features of use to a taxonomist are those that vary between the groups being studied but not within them. For example, in classifying a series of species within a particular genus, what are needed are features which every member of species A has in a certain form and which every member of species B has in a different form. The form of appearance of this feature then serves to distinguish the two species and its presence in some form or other serves to unite them in one group. Any feature which is variable within the species will fail to distinguish them. Any feature which fails to appear in some individuals will be useless in uniting the individuals of the species. From this it is seen that the chief problem of the comparative aspect of taxonomy (classification) is to determine the variability of all possible characters *within* the groups being classified and *between* those same groups.

Some of the means for determining this variability and stability are as follows: If a taxonomist has two lots of specimens and wishes to determine if they belong to one species, he must first select specimens from each lot which are comparable—adult males, or first instar larvae, or the medusa stage (correspondence of individuals). He then examines these specimens for features in which they are not all the same (internal diversity). In examining these features on every specimen, he will eliminate any features which are not represented by states that are clearly recognizable (distinguishability of characters). He will also eliminate any features which are not detectable on all the specimens (uniformity of occurrence). Among the things so eliminated, there will be many which, by his knowledge of the biology of these animals, he knows without examination are not appropriate. These are the things listed in Chapter 8. They are individual variation and polymorphism. The former are eliminated because they are inconstant; the latter because they do not correspond.

If no features can be found which differ consistently between the two groups of specimens, it will be concluded that they represent two samples of the same species. If any features appear which differ consistently in the corresponding individuals of the two groups, the taxonomist is assured that he has two groups that differ, but he still has not answered the question of whether they represent two species. They may be merely two artificially separated lots from the same species.

It is now necessary for the taxonomist to decide what value to assign to these differences. Do they distinguish species or not? The only ways to solve this problem are: (1) comparison with the features used to separate other species in the same or nearby genera, and (2) experimental testing of the genetic distinctness of the two groups. In the Animal Kingdom as a whole, it is rare for such a problem to be approachable by experimentation, because most animals cannot readily be bred and reared in captivity. If they cannot, then the only possible direct solution is through comparison with what has been found to work with neighboring species. (And in the rare case where there are no precedents or experience to help, the taxonomist must guess. The resulting groups will eventually be tested in use by taxonomists and corrected if necessary.)

In general this comparison is made automatically and perhaps even unconsciously. To do sound taxonomic work of this sort requires a detailed knowledge not only of the animals themselves but also of previous work on this group. The qualified taxonomist will therefore immediately know which features are suitable for use in this case. This subjective and ill-defined method of decision is subject to error, but every successive taxonomist on the group will weed out the poor characters, find more stable ones, and correct the misplacements.

Although the method appears to be unscientific and unreliable, in practice it is highly effective.

As for the possibly unscientific nature of it, there is some reason to believe that it conforms closely to the usual idea of the scientific method. On the basis of his information, the taxonomist postulates that certain features will separate the animals into workable groups. If he can experiment with breeding the animals in question, his judgment may be at once vindicated or challenged. If not, the judgment waits until he or someone else finds other data which corroborate or disagree with his conclusions.

It has recently been demonstrated that computers can mathematically choose the feature with the most distinct differences from a group presented to it. The computer, however, cannot search the specimens for these features but relies on a taxonomist. The computer cannot know which features are due merely to sex, caste, season, environment, or any other variable except as the taxonomist gives it this information. Thus, the taxonomist must choose the characters and interpret the results. The extensive work done by the computer in a short space of time is thus merely a mechanical aid to the taxonomist. It does not by itself perform any real taxonomic activities. The computer's analysis can, however, be valuable to the taxonomist. There are also ways of studying the variation of animals by graphs, scatter diagrams, etc. However, in the end the taxonomist must make the decisions on the basis of his experience and knowledge of these particular organisms. And his decisions will be the result of comparative studies, even if this comparison is largely unconscious.

Graphic methods. In reporting numerical or statistical data it is sometimes possible to use visual means of presentation. These include such things as graphs, histograms, tables, population range diagrams, curves, scatter diagrams, and various kinds of maps. It is sometimes implied that such diagrams are readily applicable to taxonomy, yet it is rare that one is seen presenting taxonomic data. Most often they are presenting data on populations, a subject which can be taxonomic but which is generally more involved in speciation studies, population genetics, and ecology (see next section).

For discussion of this subject and references to its literature, consult the following:

Mayr, E., Linsley, E. G., and Usinger, R. L. 1953. *Methods and Principles of Systematic Zoology.* New York; McGraw-Hill Book Co. (pp. 148–151, 295).

Simpson, G. G., Roe, A., and Lewontin, R. C. 1960. *Quantitative Zoology,* revised edition. New York; Harcourt, Brace & World.

Statistical comparisons. About thirty-five years ago, there arose a branch of biology called biometry—the statistical treatment of measurements of animals. This field has grown tremendously, but it has mostly abandoned its original name in favor of biostatistics. In the beginning this field was intended to aid taxonomists, among zoologists in general, in the comparison of things which could be expressed numerically, by measurements, variation patterns, ratios, etc.

As statistics in general became better known and more used in other branches of knowledge, it came to be held by some people that only statistical or numerical statements have any validity in science, and that only problems that can be attacked statistically are really worthy of investigation. This came to be coupled with the view of the physical scientist that science recognizes no other source of data than experimentation, because statistical treatment is generally possible with experimental data. It is very obvious that statistics and accurate measurement are helpful in most branches of zoology, including taxonomy. It is equally obvious that statistical methods are only one of the tools available, and that even when statistical treatment is possible, it is not always necessary, practicable, or desirable.

The main cause of the current trend toward statistical treatment in taxonomy is the belief of some taxonomists that all taxonomy is based on populations. Few taxonomists will deny the importance of populations, but many will deny that populations are of prime importance in all parts of the field. In the study of the higher vertebrates, the taxonomy has been so nearly completed at the levels of species, genera, and more inclusive groups that most taxonomists working on these animals are now studying the structure within the species. This is population dynamics, a subject of great importance to evolution, ecology, and genetics. It has been mistakenly assumed that it is the principal "modern" business of taxonomy.

Among the more than one hundred classes of animals, there are probably less than a dozen in which population studies are possible on any substantial scale. One of these is the Insecta, the largest of all classes. The number of places in the Insecta in which population work has been done to any large extent are entirely insignificant compared to those in which it has not. This does not prove that population work is not important, needed, useful, or scientific. It merely shows that *most* taxonomy is still not based on populations as such.

It has recently been stated that what is classified is always a population, and that specimens are used merely to give a statistical inference to the population. There are many thousands of species known only from a single specimen. It is likely that any taxonomist would admit that there are surely other specimens belonging to the same species, and if such other individuals do exist, it is possible to say that the individual at hand is a member of the population, but it does not follow that taxonomists at large consciously use anything except the attributes of the specimens and the knowledge and experience which they already have of this group of animals.

Normally the taxonomist compares the specimen directly or indirectly with other specimens. He also compares it with the background data of the variation of the known species. This variation may be sexual, developmental, seasonal, environmental, geographical, or other. It would perhaps be possible, with modern computers, to make these comparisons statistically, but few taxonomists are in a position to use statistics or machines for more than a fraction of this comparison.

In practice in the study of many groups, there is little conscious use of populations or statistics. That this statement is less applicable among vertebrates merely emphasizes that vertebrates are a small minority of animals and that their special problems are not the major ones faced by taxonomists.

All of this assumes that taxonomy deals only with species, which *are* made up of populations. This tacit assumption is false. In many aspects of zoology the classification of animals into genera and groups of genera is of more immediate value than the discrimination of species. Much is to be learned about animals by systematizing them by their attributes—their structure, organs, functions, behavior, development, origin, location, and so on. Only individuals have these characteristics, although populations may also have some characteristics.

The characteristics of the populations are generally statistical. Thus statistics is said to be essential in taxonomy primarily by those who deal with populations. In the many groups in which populations are not a major object of study, statistical approaches assume only a minor role. This role may be of considerable importance, however, in aiding the taxonomist in the subjective judgments on which all classification rests at the present time. Each worker should make such use of statistics as he can, but the value of his work will not necessarily depend on the amount of statistics employed.

The expression Numerical Taxonomy has recently come into use among the statistically minded taxonomists. The extent to which this is used in the study of specimens or species rather than of populations is not entirely clear. Only the former would be part of taxonomy. For further discussion see the works of Sokal, Sneath, Cain, and others in the Bibliography.

Summary

Our knowledge of the existence of diversity is derived from comparison of individuals and groups that differ among themselves. All of taxonomy and classification are based on such comparison, which puts this activity at the very heart of all systematizing procedures.

The comparing may be deliberate and formal, or it may be almost unconscious. It may be direct, between objects put side by side; or indirect, through pictures or descriptions; or assisted, by instruments ranging from microscopes to computers.

The comparisons show the distinctive features of each individual or group and also the features shared by several of them. The distinctive features are the characters of taxonomy. The shared features are the characters of classification. There is no real difference between these two, as one feature can unite species into a genus and simultaneously distinguish the genus from other genera.

The features may be of any sort—structures, shapes, sizes, activities, functions, occurrence, etc. Virtually any feature of any animal is a character of some sort, uniting like animals into a group or distinguishing unlike animals from each other. Every step involves comparison.

One basic rule of comparison requires that one compare only corresponding things. They may correspond in any of several ways, such as origin, history, location, or activity. Correspondence itself cannot be determined objectively but is one of the conclusions from comparison. The two are thus interdependent.

The comparing of corresponding features of different individuals or groups by a variety of methods gives the taxonomic characters of resemblance or difference on which discrimination and classification are based.

CHAPTER 11

Species and Subspecies

Although species and subspecies are linked together in nomenclatural problems, and in the informal speech of taxonomists, they seem to be taxa of substantially different nature. They are, in fact, more different from each other than species are from genera. There is some possibility that subspecies are not taxa at all but merely concepts.

Some of the ways in which taxonomists deal with species and subspecies involve identical actions and procedures. In this chapter these species and subspecies will be discussed separately, with occasional references to their similarities and differences.

Species

It is possible to make an elaborate problem of the question "What is a species?" In biology in general this may be necessary, as there are many ways of looking at the manner in which animals occur in nature. In taxonomy, however, it is neither necessary nor appropriate, because taxonomic species are simple and easy to work with. This is not the same thing as saying that they are objective or easily defined or simple to discover, for they are none of these. It may be partly because of the subjectivity and lack of definition that taxonomists have been able to build up a functioning system of species and groups of species, flexible enough to represent growing knowledge yet accurate enough to be effective.

There is an apparent paradox in the statement that species, which cannot be defined, can be classified. This is due entirely to the mixture of two meanings of the word species. It is not possible to define what a species is, or what species are, in general. They are the taxa placed at the species level, but there is no real definition in this statement. Species in this conceptual sense cannot be classified either. Just as a chemist probably could not rigidly define just what is "a chemical" and what is not, so the taxonomist cannot define what is a

162

species and what is not. The chemist, however, *can* define each one of the chemicals known to him and can distinguish them, and the taxonomist *can* define each species known to him.

Thus, definition, which is frequently thought of as a necessity in science, is really a necessity only at certain levels. It is not necessary to be able to define rigidly such words as zoology, psychology, and so on, but it is necessary to define some of the units which are employed. Taxonomy has no difficulty in doing this, even though the actual understanding of these individual units is constantly increasing.

Different species are different "kinds" of animals. The members of each kind are not always alike, but we gradually learn which differences distinguish kinds and which differences merely distinguish some of the individuals within a kind. When somewhat different individuals live together normally, it is assumed that they are the same kind. When somewhat different individuals develop from one batch of eggs or in one litter, it is recognized that their differences are those between children of one family. The young are generally somewhat different from their parents, but they are still not different species. If there are two parents, they may be different in one or more ways, and these differences too are recognized as not showing difference in kind (species). There are many of these differences between individuals within a species; for example, differences due to nutrition; malformation in development; mutilation, accidental or otherwise; environmental conditions; or dominance and recessiveness of genes. In some circumstances little variety is to be noted among the individuals of a species, but the possibilities listed above generally produce a very substantial variety if all individuals are considered.

There is no objective way to tell whether a difference is merely a variation within a species or actually a distinction between two species. It is the basis of good taxonomy that the taxonomist develops the ability to recognize the value of most differences. When he errs, a later reviser with more relevant knowledge corrects his error. This continuous correction not only serves to perfect the system but constantly improves the accuracy of the alert taxonomist in making his judgments. This is the very essence of taxonomic work.

In modern terms, a species consists of all the individuals with a common inheritance back to the point where the ancestors differed in enough features to be considered a distinct species. There are immense philosophical difficulties in the way of translation of this into taxonomic discrimination, but these philosophical difficulties do not generally prevent the taxonomist from making judgments which yield a useful set of species that can be described and distinguished by differential features. These are the species of the taxonomist; they are not necessarily the species of the geneticist or the evolutionist.

Some taxonomists, including all those whose opinions are known to the writer, believe that when the species are well enough known and biological concepts in general are sufficiently well understood, the species of the taxonomist will coincide with those of the geneticist. At the present time, however, the species definitions of the geneticist and those of the evolutionist are almost completely unusable by the taxonomist.

Species in taxonomy. Formal taxonomy deals only with the species of the taxonomist. These may be called taxonomic species, a term not overworked but nevertheless highly appropriate. The possible relationship between this kind of species and the kinds used by evolutionists, ecologists, and others is suggested in the next section (Nontaxonomic species) and discussed in Chapter 18, The Nature of Classification and of Species.

In this taxonomic sense, a species consists of all the specimens which are, or would be, considered by a particular taxonomist to be members of a single kind as shown by the evidence or the assumption that they are as alike as their offspring or their hereditary relatives within a few generations. When there is no evidence of the hereditary relationship, the taxonomist will rely on distinctions that have been found to be effective in segregating species among other animals. These distinctions were discussed in Chapter 10.

Most of this book is concerned with species. Virtually all aspects of taxonomic theory and practice involve them, either directly, as groups, or as specimens. The following aspects of species are dealt with directly as shown:

Species in applied taxonomy (Chapter 2)
Species in the hierarchy of classification (Chapter 4)
Identification of species (Chapter 5)
The diversity of kinds (Chapter 9)
The characters of species (Chapter 10)
Classifying species into groups (Chapter 12)
The use of the names of species (Chapter 13)
Finding the literature on a species (Chapter 14)
Describing species (Chapter 15)
Publishing descriptions of species (Chapter 16)
Controversies over species concepts (Chapter 18)
The rules governing names of species (Chapters 20, 26, and 29)

These aspects are not re-discussed in the present chapter, but certain other points relating to species are cited briefly. The non-taxonomic usages of this term are then cited for the purpose of distinguishing them from the taxonomic uses.

Terms descriptive of species. Some of the terms employing the word species are descriptive of the species of the taxonomist, whereas many others are used to refer to evolutionary or ecological ideas. The latter are cited in a later section, and the former are discussed briefly here.

Cryptic species are ones which are hidden, whose distinctive features are not evident under the usual procedures. Not usually a technical term.

Sibling species—a term applied to pairs or groups of very similar and closely related species. In practice this term can be applied only to similar species, when it becomes a synonym for cryptic species. When applied to closely related species (in a phylogenetic sense) it becomes a hypothetical situation, which cannot be dealt with in taxonomy but can be useful in speculations on evolution. (See non-taxonomic sibling species, below.)

Physiological species are those distinguished by activities but not readily by structural features. There is no real distinction, as structure is always involved at some level. These are also cryptic species.

Sympatric species are ones normally occupying the same geographical area.

Allopatric species are ones normally inhabiting completely different areas.

Continental species are those that live on the large land masses, as distinct from the insular species.

Insular species are those living on isolated islands which owe their fauna to dispersal methods other than overland migration. This isolation affords opportunity for study of immigration, speciation, etc.

Cosmopolitan species. This expression is used for species that occur widely over the earth, in all major regions. The Greek base of this word implies universality. No species occurs everywhere in the world because none can live on land and in deep water alike. This term is meaningless if used carelessly or taken too literally.

Tropicopolitan species. Literally, a citizen of the tropics. This ill-formed word is used to mean found throughout the tropics. (Also called *pantropical species.*)

Montane species are those found only at higher elevations on mountain ranges, isolated by the surrounding lowlands.

Morpho-geographical species are those of the ordinary taxonomist, from Linnaeus to modern times. Although data other than "morphology" and geography have been increasingly used in recent years, these still remain the basic species of taxonomy.

Typification of species. In modern taxonomic practice, all species (and subspecies) must have type specimens. In some cases the type may never have been selected or labeled, or the type specimen may have been lost. A proposal of a species without a type, or at least a specimen that *could be* the type, would not be accepted in taxonomy, and the taxon could not receive a name acceptable in zoological nomenclature. This has been a basic rule in taxonomy for more than a century.

Types are used in a variety of ways. These may be either taxonomic or nomenclatural in purpose. They include the following:

(1) In early knowledge of a species the type is usually the chief source of unchallengeable characters of the species. It will not show all the characters of the species, but it will show some which can be expected to appear on all comparable specimens of the species.

(2) It continues to show, for every sort of diversity, one point that does unequivocally occur within that species. Every feature of the type falls within the variation range of the species.

(3) It serves as a check on the accuracy of the published descriptions and their completeness.

(4) It provides an anchor for the name, indicating the point in the diversity to which that name is forever attached.

Each type thus serves as basis for description, as a standard for identification, and as name-bearer. It is evident that these uses are not entirely distinct. It is impossible to separate a species from the characters of its members or the names applied to it.

It will be noted that these functions of types do not include "showing what the species is," or "showing the limits of the species," or being the "basis for the definition of species." No taxonomist believes that the type can do these things. It is surprising, therefore, that some recent writers have felt it necessary to deny at length that a type can serve all these purposes. Taxonomists use the type as the primary basis for the original description. To the description of the type is properly added the range of features of the other known specimens thought to be part of the species. In the course of time, data accumulate until a monographer can more fully describe the species—all the specimens believed to belong to the same specific taxon as the type. Taxonomists do not use the type as basis for *definition* of the species. No specimen can possibly show limits of variation or the variety of forms assumed within the species. No taxonomists attempt such an obvious misuse of types.

Simpson (1961) lists the three supposed functions of a type: Basis for the description and definition of species, standard of comparison, and vehicle for a name. He states categorically that "No one specimen can possibly fulfil them all properly." This is true, but only because of the inclusion of the word species. No specimen of any sort can ever serve for description of a species, which always includes variation of several sorts among individuals. Taxonomists have not tried to describe the species from the type. They describe the type as one specimen that unequivocally belongs to that species. As taxonomists have usually proceeded, the type thus can be used for description, for comparison, and for anchoring the name. Types are universally used in all these ways, except where the specimens are consistently fragmentary. The fragments are inadequate for description of even one individual animal; they fail in comparison unless corresponding fragments happen to be available; and they even fail as name-bearers if later specimens cannot be identified with the type because of lack of correspondence of the fragments. It is thus primarily among vertebrate fossils that any difficulty of this sort occurs.

Although Linnaeus did not use types in the modern sense, the need for them arose gradually and has resulted in continual refinement of the concept. Whether they are called types or not, the specimens on which a species was based will inevitably be used to settle questions of ambiguity in description, of mixed species in the original lot, and as a source for data not recorded by the original describer.

Simpson, in denying the effectiveness of types for all non-nomenclatural purposes, has suggested the term *hypodigm* for "all the specimens personally known to [the taxonomist] at that time, considered by him to be unequivocal members of the taxon. . . ." This hypodigm, rather than a type, would be the basis of the species. This would be satisfactory at that time, but when later authors learn more about the supposed species, this hypodigm will be of little use to them in separating the mixed data. The word is acceptable as denoting the entire understanding of a species (or group) by a taxonomist at a particular time, but it cannot replace the type in subsequent study.

The hypodigm might possibly include more than one species, both for the original describer of the species and for subsequent classifiers. The type cannot be composite by definition, although there can be errors in association of fragments from several individuals. As a concept the hypodigm always exists, the sum of all that is known by a taxonomist about what he deems to be one species. As something that a later worker can refer to, or use, there is no such thing as a conceptual hypodigm of an earlier author. Simpson did not use the term as a concept but to refer to the specimens themselves. This specimen-hypodigm is therefore of little taxonomic value to future students and is no different from the expression "original series."

Simpson also argues that only populations can be classified and that therefore a single specimen is not so useful as the sample of the population. The difficulty here is that "population" is an abstract term; a congregation of animals is not necessarily a population but frequently a mixture of several populations. The taxonomist is interested in species. The only animal that can be positively known to belong to a given species is the one specimen on which the species was based. All others are merely *identified* with it, believed to be the same species, even believed to be part of the same population. (This statement is actually inapplicable to asexually reproducing animals, where the members of a clone still physically connected are certainly the same. At the extreme in sexual animals, there is no real assurance of identity even in litter-mates or parent-offspring sets, because hybridization is always a possibility and usually cannot be excluded in practice.)

In connection with typification and the hypodigm there should be noted another viewpoint forcibly expressed by Simpson. He considers the hypodigm, the original specimens, to be a sample from which the taxonomist draws inferences as to the variation in the entire population. One judges whether a specimen belongs to a particular species not by comparison with any one specimen (type) but on whether it falls within the limits of variation that have been inferred from the hypodigm. The inferences are probably derived by statistical procedures. He even claims that if only one specimen is known, inferences are drawn from it, and these inferences are the basis for comparison and identification.

It seems unnecessary to challenge the logic of these statements because they simply do not apply to taxonomy. Populations are indispensable to the evolutionist. They are unquestionably features of species occurrence, because the term merely describes the manner in which animals are known to occur. But populations are not generally important to the taxonomist. He is not usually interested in statistical probabilities, but in actual situations. He can describe specimens, and he can group them in species. He may eventually be able to verify this grouping by various biological evidence. He can group the species in a classification. Populations do exist, but they are part of a different approach to animals. Populations do not represent levels in the taxonomic hierarchy, nor units of classification, nor anything that the taxonomist is likely to know sufficiently well to use in taxonomic work.

Typology. For at least a century and a half taxonomists have made use of the device of typifying a concept with a selected specimen or an included subconcept. These "types" have been in use for species, for groups of species, and for subdivisions of species. The types serve a variety of purposes and have been indispensable in taxonomy.

In recent years evolutionists have introduced the idea that the use of types is what is called typology. Typology and neo-typology have come to be terms of derogation, used by evolutionists to emphasize the supposed unscientific nature of taxonomy. For example, the basic concept of typology is described by Simpson (1961, p. 46): "Every natural group of organisms, hence every taxon in classification, has an invariant, generalized or idealized pattern shared by all members of the group." The taxonomists would agree with Simpson that the concept he describes is not a useful one in taxonomy or evolution, but they would challenge him to show that this is really how types have been used.

In Hansen (1961, p. 11) we read that "The Linnaean type concept led taxonomists up a blind alley and is no longer accepted; the idea that drove it out is that of biological evolution." It is not certain that it is correct to label the type concept of the last hundred years as "the Linnaean type concept," but the fact is that in the decades after Darwin there was no change in the use of types by taxonomists in general.

Hansen further states that "evolutionary theory . . . destroyed the type concept . . . , for the central idea of evolution is change, which is the antithesis of static types." In the first place the type concept has not been destroyed but is in daily use by virtually all taxonomists. It is, furthermore, completely false to imply that there is anything static in the taxonomic use of types. As pointed out by Muesebeck (1942, p. 753), "Around this (type) there needs to be built the species concept . . . ," and this species is not static; it encompasses much diversity not shown by the type and may change in time.

The typology of the evolutionists is not the type concept of taxonomy, whether we think of pre-Darwinian or post-Darwinian times, or of current practice.

Non-taxonomic species. As new branches of biology have become interested in aspects of animal existence, they have frequently used the species of the taxonomist in conceptual framework quite different from the subjective and continually corrected system he employs. Some of these uses are based on facts of the nature of the animals, and some are based on hypotheses of the history of the species, or the capabilities of the animals. By grouping these uses under the heading Non-taxonomic Species, it is not intended to imply that none of them are of interest to the taxonomist or relevant to his work. It is not intended to give rigorous definitions of each of these kinds of species. It is enough for our purpose to point out that these are, in general, within the purview of biologists other than taxonomists—that when a taxonomist says species, he probably is not referring to any of these but rather to what Davis and Heywood refer to as the morphological-geographical species, herein called the taxonomic species.

Among these formal and informal terms or expressions there are some that relate to some feature of the species, some that represent hypothetical conditions, others that are not formal terms but mere adjectival expressions, and still others that serve only to reflect unfavorably on certain workers.

Genetical species are groups of interbreeding populations, which are reproductively isolated from each other. They are thus the same as biological species.

Biological species are usually defined as groups of actually or potentially interbreeding natural populations, which are reproductively isolated from other such groups. This gives theoretical groups, which can seldom be distinguished in practice. As Simpson has pointed out, *all* definitions of animal species give us *biological* species; he therefore prefers the name genetical species for this and cites also biospecies. (It should be noted that populations do not interbreed, only individual animals.)

Biospecies (see Biological and Genetical Species).

Agamospecies are ones consisting of uniparental organisms. They may produce gametes but there is no fertilization. Such a distinction cannot really be made, because fertilization may be facultative, and many animals can reproduce either uniparentally or biparentally either as species or individuals.

Sibling species—a term applied to "pairs or groups of very similar and closely related species." When applied to "closely related species" (in a phylogenetic sense), this expression refers to hypothetical species; these cannot be dealt with in taxonomy, but can be useful in speculations on evolution. (Compare under taxonomic species.)

Polytypic species are those which consist of two or more subspecies. This is the original definition of Huxley (1940); according to later writers this usage "has now been almost universally adopted." Nevertheless, there have been many uses of this term to cover variable species in which no subspecies have been explicitly recognized, where the populations differ among themselves but insufficiently to be recognized as subspecies, in the opinion of the reviser. The term *polymorphic species* has also been used in these same ways, but the term polymorphism is used in such diverse ways in biology that it would be best avoided as a general modifier of the word species.

Monotypic species consist of a single subspecies. This and the preceding would be considered taxonomic concepts if subspecies were considered to be taxa.

Evolutionary species are lineages (ancestral-descendant sequences of populations) evolving separately from each other and with their own unitary evolutionary roles and tendencies. (Simpson, 1961, p. 153.)

Transient species are the ones existing contemporaneously, as a cross section of the lineages of evolutionary species.

Successional species are temporally successive species in a single lineage.

Paleospecies (see Successional Species).

Paleontological species have no representatives now living, all known examples being fossils.

Panmictic species are those in which a single interbreeding population occurs. Therefore these are the theoretical counterparts of monotypic species, theoretical because panmixia can only be postulated.

Philopatric species are ones which show no tendency to extend their range. (Again, this is a hybrid term, formed from two languages.)

Incipient species are geographical subspecies or other segregates, which, it is presumed or postulated, will become isolated and then be distinct species.

Morphospecies are ones "established by morphological similarity regardless of other considerations" (Simpson). They are not genetic groups but groups of like objects. (These include the "form species" and "paraspecies," following.)

Form species are groups of fossil objects not identifiable as any particular biological species, such as fragments or isolated parts.

Paraspecies are the parataxa at the species level. (See Parataxa, below.)

Non-dimensional species. The taxonomic species have been described as lacking dimensions of space and time. With this as a premise, they are called non-dimensional, applicable only to non-evolving animals, and therefore not biologically acceptable. In reality both space and time are present in the use of taxonomic species, but only a short segment of time (the span of a few generations) is available for consideration. This expression serves only to derogate the real basis of taxonomy; it is used only by the speciationists who attempt to force taxonomy into evolutionary studies.

Subspecies

Much has been written in recent decades about subspecies and their use in taxonomy. There are strong feelings that they are usable, useful, and desirable. There are also strong feelings that they are not really relevant to taxonomy and are an unnecessary encumbrance to classification and nomenclature.

Subspecies are widely used by zoologists who work with vertebrates, perhaps consistently used. They have been used in many invertebrate groups but in most of them only occasionally. In work with invertebrates, in fact, the concept of subspecies often merges into that of varieties, with no uniform basis for discrimination.

The problem of terms discussed above for species applies in considerable part also to subspecies. There is a level in the taxonomic hierarchy in which are placed groups of animals (taxa) called subspecies. The level is the category subspecies; the taxa are the subspecies taxa. The term thus refers both to a level in the hierarchy and to the taxa placed at that level.

The principal theoretical question to be asked about subspecies is seldom discussed and has not been answered effectively. It is this: Are the segregates called subspecies actually of the same nature as the taxa of species level and above? If this question cannot be answered in the affirmative, there will be serious doubt as to the relevance of subspecies in taxonomy.

Definition. Recent definitions of subspecies are in agreement on the two basic features: (1) a distinct geographical area is occupied, and (2) there are structural features partially setting off the subspecies. Other ideas included in some definitions are that it consists of local populations, that it is set off also by ecological features, and that its distinctive features will hold for about 75% of the specimens but not necessarily for all of them.

A recent definition by Mayr (1963, p. 348) incorporates most of these ideas: "A subspecies is an aggregate of local populations of a species, inhabiting a geographical subdivision of the range of the species, and differing taxonomically from other populations of the species." This emphasizes the population nature of the subspecies and also the fact that the distinctive features need not be "morphological" but must be "taxonomic." It should be noted here that the subspecies cannot be known until the species is known. Only after the limits of the species are set, for that moment, can the variants be seen to represent what might be called subspecies.

Some definitions of subspecies imply that they are "incipient species" or at least populations that are nearing specific status. It is certainly theoretically possible for a subspecies to become isolated and evolve into a distinct species, but the ordinary subspecies is distinguished by its present features and distribution only, not by its future possibilities. In pointing this out, Simpson states that subspecies are taxa of a markedly different kind than species.

Taxonomic nature. It is sometimes assumed that there is no basic difference between subspecific and specific characters, either morphologically or in mode of inheritance. It is taken to follow from these presumed facts that there is therefore no conceptual difference between species and subspecies. It seems to have been forgotten that these are not the only ways in which species and subspecies can differ.

In nearly all definitions, species can be distinguished because of gaps in the variation of their features. Subspecies as usually defined cannot be so distinguished, except in some percentage of cases, a figure often placed at 75%. Both species and subspecies consist of the same basic materials, the individuals, but there is a complete break between species and subspecies conceptually in the distinctness of the former and the overlapping of the latter.

The entire taxonomic hierarchic system of zoological classification is based on the fact that species and other taxa are presented to us as distinguishable entities. The individuals can be identified with one species or another in practice. Whether or not this is theoretically to be expected among evolving organisms is not relevant—the system is based on the assumption that it can be done, and the elaborate classification now in use is evidence that it has been done successfully in the eyes of most practicing taxonomists. But this system may not be capable of classifying taxa that are not distinct—that are therefore not really taxa.

It was noted by Borgmeier (1957) that while two species are essentially different (in the features taken to be specific), two races or subspecies are essentially alike "because they agree in all basic structures and are linked together genetically." Because not all species form subspecies, this level is not one of universal significance as is the species level. Borgmeier concludes that it is only a partial subcategory of the species level.

It was stated as long ago as 1940 that "there are relatively few 'good' species that are not actually composed of groups of 'subspecies'." This was, of course, an extreme overstatement, as the structure of the vast majority of species had never been (and still have not been) examined for this purpose. The statement may apply in the bird and mammal groups, and there is little doubt that there will be an increasing number of cases discovered in other groups. Nevertheless, this does not justify the enunciation of a new law of nature.

Hubbell (1954) concluded that "population analysis below the species level is too complex to be bound by formal taxonomic and nomenclatorial rules. . . ." He proposed six "precepts" which serve to summarize his views of subspecies: "(1) Clinal variation is variation of characters in populations, not of populations. (2) Clines in themselves therefore cannot be taxonomic units. (3) In a population showing only a stepped cline in a single character or in a group of correlated characters it is feasible and may be convenient to treat the segments between steps as subspecies. (4) In populations in which clinal change is gradual or in which two or more non-coincident clines exist it is inadvisable to try to separate subspecies. (5) In general, infraspecific variation is best treated by description, graphic presentation, and non-technical names. . . . (6) Nothing should ever be named for the sake of naming it, but only in order that something may be said about it."

In the large recent literature on subspecies in taxonomy, it is evident that nomenclature is a major factor in the controversy that has arisen. Many papers that seem to be addressed to the problems of subspecies turn out to be concerned almost exclusively with the question of whether such segregates should be named. This is unfortunate, because the decision on whether or how to name any biological objects should not affect discussions of whether there is something biologically worthwhile to be recorded about them.

The problems involved here are (1) whether there is in nature enough diversity within some species to be usefully studied; (2) if so, whether this diversity can be treated in the taxonomic system; and (3) if so, whether the segregates can or should be named in the formal system of nomenclature. The first question is generally answered in the affirmative. The second question has scarcely ever been faced; it is the crux of the present problem and is here believed to be likely to be eventually answered in the negative. The third question has clouded the second and is answered either negatively or affirmatively according to the experience of the speakers. The Rules of Nomenclature have for a half-century permitted such naming.

It is possible to look on subspecies as pseudotaxa. A taxon is a classificatory unit of any rank, but it is also a group of individuals. If the group cannot be circumscribed, at least in practice, it can scarcely be classified. It is doubtful if subspecies are ever classified in the same manner as species. They are populations recognized within the species, not groups of individuals assembled to produce a taxon. They are thus not classified and so are not taxa.

Pseudotaxa would thus be segregates of a taxon which appear to be subordinate taxa but in reality cannot be segregated and so cannot be classified. They are thus false taxa from a taxonomic point of view. They may, however, be acceptable isolates in the eyes of other zoologists. In any case they may be of much interest to taxonomists even if they cannot be dealt with as taxa are.

It seems quite likely that the study of subspecies is the aspect through which taxonomy became involved in recent decades with population studies. The subspecies are populations, presumably ones with a certain amount of unity. This cannot be said of species, which are (potentially, at least) groups of populations, and which are by definition fully set off from other such groups of populations by the features which are therefore called taxonomic (or specific).

These general conclusions have resulted in there being few references to subspecies in this book. Their treatment and use is often part of other aspects of systematics. References to recent discussions of subspecies, their problems, and their misuse are given in the Bibliography. Their names have been regulated by all modern codes of nomenclature, and this aspect is dealt with in Part VI of this book.

Other subdivisions of species

From the very beginning of formal taxonomy, it was recognized that some species include groups of individuals that can be readily distinguished. The basis of the distinction was varied, but the cause was usually unknown. Some individuals were of a different color, had larger body parts, or differed in other features to a lesser degree than was elsewhere accepted as of specific value.

As knowledge of animals increased, it was recognized that some of these variations were geographically isolated. These were often called *races*, but in later years such geographical segregates became universally known as *subspecies*. Where there were color differences, the term *color form* was often used. Sometimes *phase* or *color phase* was used for this also. In a few cases, such as butterflies, specimens so rare as often to be unique were termed *transition forms*, supposedly intermediate between two species or subspecies (but not hybrids). Similarly, single specimens with unusual features, sometimes probably teratological, have been termed *aberrations.*

Specimens or groups of specimens have also been described as *forms* (*formae*), *Rassenkreise, Formenkreise, natios,* and *colonies.* Other terms, such as *blastovariations, morphae, mutations,* and *variants,* have been used, both in zoology and botany, but they are not always intended as formal taxonomic categories. They have had little use in animal taxonomy and have no standing in the system of nomenclature.

There is no reason why any taxonomist should not use any idea or term which he finds useful, whether or not it has been widely adopted. He will have a responsibility to define and use it consistently. But none of the possible levels below species are to be named within the system of nomenclature, except subspecies.

An example of an elaborate system of terms for minute segregates within a species was published in 1938 by Bright and Leeds. In recording the aberrations in color pattern of a certain species of blue butterfly (Lycaenidae) these writers use seven hundred and thirty-seven names for the aberrations of this one species, and they also propose a system of terms for identifying similar aberrations in other species. These names do not enter into zoological nomenclature. They were evidently found useful to these writers in publishing the results of their study of the variability of these insects.

Parataxa. When the material (specimens) with which a taxonomist works is fragmentary, it may consist solely of isolated parts of animals —parts not now and not likely to be identifiable to species because in the various species they were too much alike. If these fragments are of use in stratigraphic correlation even without specific identification, there arises a need for names to supplement the system of specific names, so that the fragments can be referred to even when their species cannot be determined.

Paleontologists have sometimes named these fragments as if they were genera and species, in the usual manner. If, later on, the fragments are associated with other parts of the animal which can be identified with species, there is a conflict of names. Furthermore, the fragment-species may include objects that actually belong to several species, so that synonymy will not suffice to show the situation. To circumvent this problem, it has been proposed to name the fragments as *parataxa* under separate rules of parataxal nomenclature. The following paragraphs are taken from a preliminary draft of such a proposal, by R. C. Moore and P. C. Sylvester-Bradley, with their kind permission.

"Discrete parts of various kinds of animals, chiefly skeletal parts, occur commonly in nature; more especially they are represented by abundant fossils in sedimentary strata of all geological ages from Cambrian to Recent. Examples are isolated coccoliths; spicules of sponges, octocorals, and holothurians; ossicles of crinoids, cystoids, blastoids, echinoids, and asterozoans; annelid jaws (scolecodonts); radular elements and opercula of gastropods and cephalopods (aptychi); and the abundant fossils of unknown zoological affinities called conodonts. [See Figs. 11-1 and 11-2.] A large majority, if not all, of these bodies are usefully classifiable within the groups to which they belong, even though the genera and species of animals from which they were derived is almost universally unknown. Such discrete fragments of animals constitute a special category of zoological entities which, though classifiable in varying degrees of detail and precision, offers critical problems in nomenclature.

"There is little need for the classification and nomenclature of fragments when whole specimens of animals are available for study. This applies to virtually all work by neozoologists on living animals and may be accepted also for most work by paleozoolo-

gists on extinct animals because the fossils on which many thousand taxa have been recognized and named are judged adequate
for discrimination of various genera and species of whole animals.
In addition, there are multitudinous dissociated fragments of animals which are far from sufficient for identification of the whole
animals that produced them and yet these are so distinctive in
themselves as to have great usefulness for identifying the sedimentary strata containing them. These fragmentary paleontological materials are indispensable for correlations of many rock formations in the earth's crust and for aid in establishing a trust-
worthy geochronology of the post-Precambrian part of geological
time. However, in order to make use of such fragments, they must
be classified, named, described, figured, and recorded as to occurrence. When this is done, many prove to be invaluable. For example, the dissociated fossils called conodonts have been demonstrated to constitute the only reliable means for determining correlations and relative geological age of various strata containing
these fragments. Other highly fragmental remains of animals,
especially echinoderms are similarly useful, but so far have been
little studied because no satisfactory means of naming them in
accordance with zoological rules has been available. When suitable procedure is provided for applying names to discrete animal
fragments without reference to the whole-animal species which
they represent, this will encourage greatly the study of such fragments, making them useful in stratigraphical palaeontology.

"The taxonomic arrangement adopted in by far the greater
majority of fossils studied is exactly comparable to that which
would have been proposed if whole animals had been available
for study. In many cases, if a fragmental specimen is at first inadequate for the identification of the whole animal from which it was
derived, evidence may accumulate later which will establish its
identity. In these cases the normal operation of Art. 27 of the
Règles (which states that the Law of Priority applies when any
part of an animal is named before the animal itself) takes care
of the nomenclatural situation. In a certain number of cases, however, the stratigraphical importance of the fragments far tran-
scends their importance as biological entities. In these cases a dual
nomenclature has grown up, one providing names for the fragments, the other for the whole animals. Such dual systems are
contrary to the present provisions of the Règles, but they have
great utilitarian value and are currently employed in the taxonomy
of conodonts . . . , ammonoid aptychi . . . , holothurian spic-

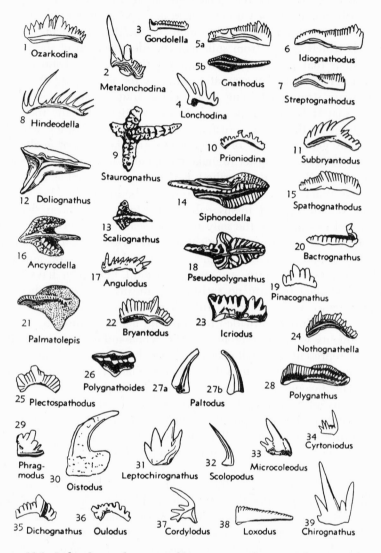

Figure 11-1. Isolated conodonts, jaw-like structures known only as fossils and produced by animals of unknown nature. These can be identified and correlated from stratum to stratum. They have been named as genera and species. From: Moore, R. C. et al., *Invertebrate Fossils.*

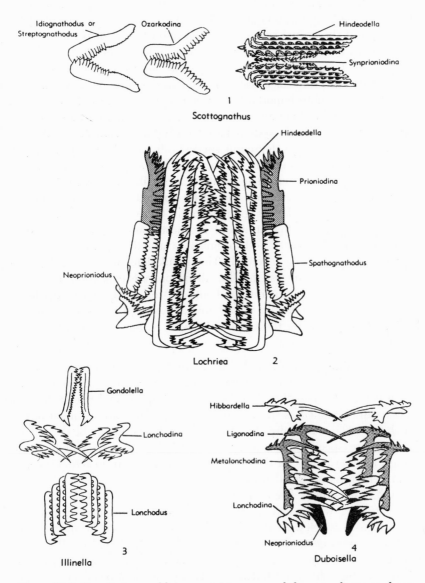

Figure 11-2. Conodont assemblages containing some of the same forms as shown in Figure 11-1. Because each assemblage evidently came from an individual, each is now believed to represent a species, and the isolated objects in Figure 11-1 are now thought to represent something less than species, a level called parataxa. From: *Treatise on Invertebrate Paleontology*, Part W, Fig. 42.

ules . . . , and, to a somewhat lesser extent, in a number of other groups. This application seeks to regularize the establishment of certain of these dual systems by the establishment of parataxa as a special category for the classification and nomenclature of the specified fragments. In a sense a parataxon is a taxonomic category, but, as Professor Chester Bradley has pointed out to us, in a zoologically more important sense it is outside of taxonomy. The study of parataxa might even be termed 'parataxonomy.' Zoological taxonomy is a single system based on natural relationships into which, with varying degree of success, all animals can be fitted. It is just because fragments of the type here described cannot be fitted into that system that parataxa are called for. It might be argued that if these names cannot be applied in ordinary taxonomy, then they are better ignored; to this there is the very forceful counter argument that it would be most confusing to have the same name applied to both ordinary taxa and to parataxa. Such homonymy must be avoided. The regulations we here recommend therefore suggest that for all purposes except those of the Law of Homonymy, parataxa should be regarded as not coordinate with corresponding whole animal taxa. To this extent they may be ignored by the taxonomist who is only concerned with zoological taxonomy."

This proposal by paleontologists at the London Congress was rejected as far as the rules of nomenclature are concerned. Parataxa as such cannot be named within the system of binominal nomenclature. They can be named as species, genera, etc., if they are assumed to be such taxa. The names already introduced will presumably be treated as if they are specific and generic names. Paleontologists will have to resort to other means for distinguishing these "taxa" from actual species and genera based on identifiable specimens.

It has sometimes been suggested that this problem exists also in parasitology, where certain larval forms are found that cannot be identified to species. At this level, the problem would be nearly universal, as unassociated stages or fragments can be found in all groups of animals. The unusual feature among the particular fossils concerned is the stratigraphic need for names. This same need has been claimed in parasitology, but parasitologists are by no means agreed that the use of parataxa is a reasonable solution in their field.

Summary

The idea of species is all-pervading in biology. The word is used with various meanings in different contexts, but it generally refers to a group—the individuals which are related by comparatively recent common ancestry and who could produce more of the same kind by some process of reproduction. In taxonomy these individuals are recognized and brought together because of their common possession of some features that are not possessed by other kinds.

A considerable amount of confusion can arise through any attempt to impose on taxonomy the theoretical concepts of species used in evolution or ecology. The confusion would disappear if different terms were used for the different concepts, but this has not yet been acceptably done.

If the individuals of the species occur in separate populations, these populations *may* develop differences among themselves. These are often called subspecies. It is not supposed that all subspecies *are* evolving into separate species, but some of them *may* be doing so. The species themselves are taxa (taxonomic groups) placed at the species level, and subspecies may also be taxa at the lower level called the subspecies level. There is, however, some doubt whether subspecies are really taxa, because the group usually cannot be completely distinguished from other groups.

Both species and subspecies receive formal names. The validation of both requires selection of a single specimen, or type, to serve both as an anchor for the name and as a source of unchallengeable taxonomic information.

For discussion of the describing of species, see Chapter 15, Descriptive Taxonomy. For discussion of the theoretical nature of species, see Chapter 18, The Nature of Classification and of Species. For problems of the names of species, see Part VI, Zoological Nomenclature.

CHAPTER 12

The Practice of Classification

Classification in the technical sense consists of five activities: The objects are grouped and the groups must be distinct (I, grouping, and II, distinguishing). A hierarchy of levels must be chosen (III, choice of hierarchy) and the groups assigned to the levels (IV, assignment to categories). Finally names must be applied to the groups (V, naming). Each of these activities is discussed below, except V, which is left until a later chapter.

It must not be assumed, however, that classification usually starts out with ungrouped objects, no hierarchy, and no names. This happens so rarely after two hundred years of classifying animals that we can dismiss it as not occurring at all. Classification always proceeds to build upon previous classification; there are always groups, levels, hierarchy, and names. Today, classification is largely the re-evaluation of earlier classifications with addition of new species, new groups, new data, and new understandings.

Although there is almost always previous work to build upon, it is a mistake to think that classification of organisms is complete or nearly so. The remark of Davis and Heywood (1963) about plants, that "the classification of a large part of the world's flora is still in its early or pioneer stages," applies equally to animals (fauna). The well-classified parts of the animal kingdom form no more than a hundredth part of the known fauna and probably only a thousandth part of what actually exists.

Purpose

The principal purpose of classification is to provide a simple practical means by which students of any group may know what they are talking about and others may find out. This statement [1] suffices for

---

[1] Copied from a book on anthropology but apparently originating in Simpson, 1945, p. 13.

very simple classifications, but for a system that must record a million things there is an even more cogent purpose. No person can keep track of a hundredth part of a million things without help of some kind. With an electronic computer he might do so, but it can readily be done by the simple device of classifying the things into groups and the groups into groups of groups.

By this means we can keep track of one aspect of 180,000,000 human beings merely by classifying them as "Americans." Instead of stating for each of the 180,000,000 that he or she is an American, we can by grouping them, show that they all belong in the group "Americans" rather than any other corresponding group such as "Italians."

By grouping Americans, Canadians, Englishmen, Australians, etc., as "English-speaking peoples," that is, by making a group of groups, we can keep track of the native tongue of at least 300,000,000 persons, without obscuring the fact that some of them are Canadians, some of these are Torontoans, and some of those are the Jones family on Queen Street. Classification thus enables us to keep track of many objects or groups of objects or groups of groups without losing track of the individual groups or objects.

The basic purpose of zoological classification is the same—to enable us to keep track of more than a million kinds of animals represented by many millions of individuals. Classification forms the framework in which these may be grouped, and the groups also grouped, but it also provides a system by which we can keep track of the hundreds of facts discovered about each of the kinds. Inasmuch as these facts are often the result of work in other fields, embryology, ecology, physiology, zoogeography, and genetics, classification aids these fields by helping to organize the data which they employ.

A second purpose of all biological classification, one of more scientific nature, is the discovery of new knowledge. Classification is always the result of observation of attributes, and it always results in new knowledge of the variation and distribution of these attributes. It frequently uncovers unsuspected patterns, and these patterns are far more important to biology than the isolated facts.

These purposes do not complete the story. In addition to (1) recording data about kinds, and (2) producing new data through correlation, observation, and analysis, it makes possible (3) identification, (4) prediction of the state of unrecorded features, (5) conclusions on phylogeny, and (6) explanations of the diversity of animals. Successful use of a classification for any of these from (2) to (6) depends upon the extent of the first; maximum information content will permit maximum deductions to be drawn and will also insure maximum stability of the classification.

The principal task of a systematist is to make the species of nature known, as stated by Borgmeier (1957), and the second task is to arrange them into a classification, which will then yield the six benefits listed above.

Types of classifications

The arrangements which we call biological classifications have been described under various terms, sometimes with contrasting meanings. Among these are: natural vs. artificial; evolutionary or phylogenetic; phylogenetic vs. phenetic; typological or archetypal; horizontal vs. vertical; larval vs. adult; general vs. special; and omnispective.

Some of these expressions involve ordinary words in ordinary meanings, such as larval and adult classifications which are based on the characters of larvae and adults respectively. General classifications are simply those intended for general use in various ways, whereas special classifications are designed only for particular purposes.

A definition of each of these, with comments on the implications, should serve to orient a consideration of the basis actually used in classification at the present time.

Natural vs. artificial classification. These are the types most often referred to, and they are also the ones most subject to obfuscation through lack of definition. The implications of this contrasting couplet are discussed in Chapter 18, The Nature of Classification and of Species, in which the theoretical aspects are considered in the light of the recent controversies over the basis of biological classification. It is here necessary only to show how the terms are used, when the distinction is appropriate, and the extent of its use in this book.

Various authors have defined a natural classification to be: (1) one that relies on phylogeny; (2) one that groups so as to place together those with the maximum possible number of attributes in common; (3) one that has a basis in phylogeny; (4) one that conforms to the available knowledge of genetic relations; (5) one that groups those which are related to each other through common ancestry; (6) one that seeks to show the blood-relationships; (7) one that reflects the objectively ascertainable discontinuity of variation; (8) one that reflects the objective state of things; (9) one that enables us to make the maximum number of prophecies and deductions; (10) one that more truly reflects nature; (11) one based on the probable evolutionary relationships, and (12) one in which the objects resemble each other in a multitude of particulars and appear to be grouped together by Nature. These are approximate quotations, and the list could be extended almost indefinitely. These authors had only one thing in common—they all believed that biological classification should be natural, as they understood that word. If the list were extended at random, it would become obvious that recent writers overwhelmingly include phylogeny as the basis in their statements. Number (5) probably represents the commonest statement.

Only a few of these writers are dealing with theoretical aspects of taxonomy. Most of them are quoting in their textbooks the statements which they have read in other works. But an increasing number of thoughtful systematists are questioning these requoted statements and asking for convincing argument that provides a logical basis for the definition. Such argument is simply not forthcoming, and many biologists are trying to turn to a more logically defensible definition. Number (11) shows the nature of the challenge in admitting that the ancestry is never known but only inferred. Number (10) begs the question by depending on the indefinite word nature, and number (8) employs the same device. Assuming that these brief comments show fallacies in certain of these definitions, there are left as justifiable only (2) and (3), grouping on the possession of attributes in common —similarities, preferably many of them, and (7) grouping based on discontinuities in the diversity.

All modern philosophers in the field of biology, as distinguished from taxonomists themselves, seem to be agreed that there is no possible basis in common ancestry (which common ancestry is never known as a fact), that correlation of attributes is the only natural method of classifying, and that this method will prove to be the most useful in taxonomy. They do not say that a classification should not be natural. They are saying that the most natural classification possible is one based on many direct attributes of the specimens or taxa.

It is further denied by some recent writers that it is even possible to make an independent classification of any animals based on phylogeny. It is impossible, first, because phylogeny is not known but merely hypothesized and is therefore not an attribute but only a guess at an attribute, and, second, because the phylogeny (whether real or imaginary) is based on previous classifications. Inasmuch as all animals being classified now are known to some extent, the supposed phylogeny is based on the known facts. It would be a circular argument to then say that a classification based on the supposed phylogeny is independent of the same characters that were used directly in the conventional classification.

No one denies that evolution produced the kinds we are now classifying; that the sequence of forms through which a particular species arose is in fact a phylogeny of that species; that some of the features of the animals are the result of the phylogeny—the result of descent from an ancestor which possessed those same features; and that some of these features are good characters for classification. What is denied, here and in several recent books, is that the phylogenies are ever really known. They are only postulated. The best of them have been changed recently and are apt to be changed again. And there are no examples of classifications actually based on what were taken to be phylogenetic facts as distinct from the features of the animals.

With this situation, there is only one way to define a natural classification consistent with the known facts and useful in systematics. This is a combination of numbers (12) and (2), with recognition also of (7) and (9). A *natural classification* is, therefore, one in which the groups are recognized by having a maximum number of attributes in common, with their limits set by discontinuities in the diversity, and capable of yielding the maximum number of correct deductions about correlations of other features. It is possible to express this less explicitly by saying that a natural classification is one which reflects most of the various natures of the objects.

The artificial classification in the couplet above was simply one that was not based on evolutionary relationships. The usual example cited is the classification of books in a library. This example is entirely irrelevant to biological classification. Furthermore, a classification of books by subject could be natural, because the subject matter of books is a basic feature of their individual natures. With the definition accepted here for natural classification, *artificial classification* becomes simply one that is based on so few characters or such illusory discontinuities that it is not capable of producing acceptable deductions. (See also Special Classification.)

Evolutionary or phylogenetic classification. These two terms are practically synonymous and represent classification based on features derived from a common ancestor. The features are thus held in common by the members of the group. The identifying aspect of this concept is that the features are chosen *because* they were also present in the supposed common ancestor. As a necessary result of our ideas of evolution and knowledge of hereditary mechanisms, there is nothing wrong with this belief in the origin of the features used in classification. The proponents of this way of stating the basis of classification feel that this is a necessary conclusion to the belief in evolution of animals. The opponents believe that the actual phylogenies are an inevitable product of evolution but that we cannot definitely know the details of these phylogenies and so cannot use them as data in any classification.

It has been stated that all classifications, since the acceptance of Darwin's theories, have been based on attempts to recognize the characters that were derived from the common ancestor—that all classifiers have consciously sought out such characters and used them in classifying. There may have been cases in which this was done, but these cases are so rare that none have been cited as examples. It may be that in the best-known groups, such as the higher vertebrates, some evolutionists can speculate on the connection between the working classification and the postulated phylogeny of the group, but in animals in general this is out of the range of feasibility. There are no real phylogenetic data with which to work.

There is no occasion to use either the term evolutionary classification or the term phylogenetic classification in this book, because the natural classifications described here, like all those published on animals, are based on features in common, not on supposed ancestors.

Phylogenetic vs. phenetic classification. The most recent denials of the phylogenetic basis of "natural" classifications have come from the "numerical taxonomists." They see the fallacy of postulating a phylogenetic basis where there is in fact no separate knowledge of phylogenies. These biologists believe that statistical data manipulated by computers can actually serve to classify organisms, and they believe that the use of a large number of data will ensure a classification that will have the maximum usefulness.

Definitions of *phenetic classification* have been evolving rapidly during the decade since the word was introduced. It was originally defined as arrangement by over-all similarity, based on all available characters without any weighting—employing all observable characters. In a more recent book, it is defined as based on over-all affinity (resemblance) as judged by intuitive means and using as many characters or as much evidence as possible. It is currently used loosely in referring to any classification based neither on phylogeny nor on some special artificial feature.

Unfortunately this word still has implications that quantity of features is more important than quality. In practice the features selected are those believed to be suitable for taxonomic purposes. It is sometimes forgotten that even one character is sufficient, if that one is known to be correlated with many others of varied nature. If the correlation has been established, the use of the one feature will result in just as natural a grouping as the use of all these same features together.

This type of classification is sometimes termed *neo-Adansonian,* after the eighteenth-century taxonomist, Michel Adanson, who first advocated the use of a maximum number of characters.

Typological or archetypal classification. The basic concept of typology is given by Simpson (1961, p. 46): "every natural group of organisms, hence every natural taxon in classification, has an invariant generalized or idealized pattern shared by all members of the group. The pattern of a lower taxon is superimposed upon that of a higher taxon to which it belongs, without essentially modifying the higher pattern. Lower patterns include variations on the theme of the higher pattern and they fill in details, different for different taxa at the same level, within the more generalized, less detailed higher pattern." This concept is a very old one and undoubtedly has some effect on modern classification. In most discussions, it is confused with the use of a type concept represented by type specimens of species and type species of genera. It does not really concern taxonomists.

The term *archetypal classification* is simply a synonym of typological, representing the idealized pattern cited. If there are zoological classifications that can properly be described by either of these terms, they have not been identified in recent statements on this subject. If there are not, then it is hardly worth our while to consider them in a study of the practical basis of taxonomy.

Horizontal vs. vertical classification. As applied to animal groups, this is an evolutionary concept, not a taxonomic one. When a paleontologist thinks he recognizes a sequence of species—a lineage—he may classify the sequence as a genus. A neontologist, not seeing the lineage but only the end species of this and other sequences, will classify these terminal species as a genus. The paleontologist's classification is vertical, along the time axis of evolution. The neontologist's classification is horizontal, across the lineages that exist at the present time.

Although there is a reasonable theoretical problem here, in practice the lineages are not known but only postulated. The taxonomist still has to select his groups largely on the basis of the attributes in common of the species involved. In a phylogenetic study, this will be a real problem, but in a taxonomic study, there is no difficulty not encountered in any horizontal classification. The stratigraphic succession of the fossils does show that a time factor is involved, but it doesn't show what the evolutionary sequence or lineage was.

Larval vs. adult classification. This couplet does not represent the same sort of distinction as the others. It is included here to emphasize one of the things frequently forgotten by taxonomists, other zoologists, and students. Although a substantial number of animals are produced by asexual means and do not pass through a period of embryonic development, all sexually produced animals start as an ovum either fertilized or activated to development. From this point until death, the animal is an individual and a member of one species and various more inclusive groups. The planula larva of a jellyfish, barely at the blastula stage of development is no less a member of the species *Aurelia aurita* Lamarck than were its parents. It is no less a member of the Scyphozoa and the Coelenterata.

In spite of this obvious fact, that all forms and all stages belong to the species, it is common for writers and speakers to forget all but the adults. Species are almost universally described from adults, often of just one sex. Keys and classifications are generally based on only that one stage.

There are practical reasons why this is sometimes necessary. However, it should not blind the student to the facts that all forms are potentially of interest to zoology and therefore to the data-filing aspect of taxonomy. Furthermore, the unusual forms may become of direct use in taxonomy, as was the case with the malaria mosquitoes in Southern France described in the section headed Applied Systematics.

Because the forms and stages are often very different in physical features, behavior, and occurrence, it is often necessary to study the larvae, for example, separately. It then sometimes happens that a classification of the larvae is prepared. This classification may or may not agree with that based on other stages such as the adult. It should be clear that the term classification is not well used in this case. To classify the animals, the features of all stages should be considered, and the groupings selected should be those that are shown by all or most of the stages. In this way a true classification of the species will result, not merely a key to the adult males of the species.

In one sense, then, larval and adult classifications are not really classifications. In another sense, however, if the two stages show groupings which cannot be harmonized, two classifications will in fact result, but it is reasonable to believe that this happens because of inadequate knowledge at some point. Not only can there be larval or adult classifications, but egg classifications, male and female classifications, or even ones based on head features as contrasted to genitalic features, for example. But none of these are classifications of the species.

General vs. special classification. A special classification is one designed for a particular purpose, often based on one attribute or one type of attribute only. It would in some sense be an artificial classification. Some of the types described above are special classifications, including larval and some evolutionary ones. The deme and ecotype classifications mentioned later are special classifications.

A general classification is merely one that is not special—one whose purpose is general in taxonomy. It will be natural, in the sense of the definition given earlier, and it will be horizontal, because no other basis is directly available.

On the whole, the best classifications of the last hundred years were all general. It is probable that successive classifications become ever more general as taxonomists broaden the base of knowledge about the animals. These are the classifications called natural by Gilmour and by Davis and Heywood. They tend to become the all-inclusive classifications recently referred to as omnispective by Blackwelder (1964). The phenetic classifications of Davis and Heywood would qualify as general classifications, but it is not so clear whether the phenetic classifications of the numerical taxonomists could qualify, since their features are selected for the requirements of the machine and are judged by quantity of correlation rather than quality.

Modern omnispective classification. It has been correctly said by Davis and Heywood that "different classifications are needed for different purposes." This accounts in part for the variety of classifications described above. Any of these may have a legitimate use, and the remarks under these heads are not intended to deny these legitimate functions.

Only taxonomic classifications are of concern here, to be used for purposes decided upon by taxonomists (or at least designed to serve only these purposes). If they do not serve the purposes of evolutionists, or ecologists, this merely shows the need for other types of classifications in those fields. (Such other classifications have already been proposed, based on the concept of deme in evolution and that of ecotype in ecology.)

The taxonomy of the past has sometimes been denounced as inadequate or obsolescent. The sources of such charges are generally nontaxonomists. Experienced taxonomists recognize that for the purposes and times for which it was designed, classical taxonomy has been highly successful. It is used continually but often unconsciously by most zoologists, who may fret over its imperfections but still are able to use it effectively. This classical system is based on comparative data drawn from individual organisms—the only basis so far available. The primary feature of this customary system is the use of all available data as far as necessary. This availability is limited to cases where comparative data of the particular sort are available for all objects or taxa being classified. The necessity is fulfilled when a satisfactory classification is produced.

In this system the extensive background of the experienced taxonomist enables him to pass over, almost without conscious thought, all the non-varying features of the organisms, as well as the ones due to sex, age, pathology, and so on, and to use a workable number of features that are evidently correlated with many unmentioned ones. In this system, then, all features of the organisms are considered, so far as they are available, and only those are employed which are necessary to show the groupings and distinctions that occur.

This, then, is an all-seeing or all-considering system. It does not pretend to *use* all features, because it decides against the use of some. It knows of the probable existence of other features which it does not yet have access to. This has been called the omnispective system [2] and is the one in current use by most taxonomists. It is at present the only workable system, and the reason why it has changed so little is that it was very largely adequate to the purposes for which it was designed.

Basis of classification

Because several recent books have stated categorically that classification *must* be based on phylogeny in some sense, it is necessary to open the discussion of the basis of classification with a statement which anticipates the treatment of this subject in Chapter 18, The Nature of Classification and of Species.

If there are any zoologists today who deny that animals have evolved and that the kinds observed now are the result of evolution, taxonomists as a group are not among that minority. They are too much aware of the wide variety of diversities of animals. It is certain that there is difference of opinion on what processes of evolution resulted in these diversities, but the question of whether change has occurred is not at issue.

Taxonomists classify the kinds of animals produced by evolution, whatever the processes were. It is correct to say, then, that classification is *possible* because evolution has produced diverse kinds. Furthermore, classification is *necessary* because evolution has produced such a diversity of kinds. From this it is possible to generalize that classification is *based on* evolution, but it must be recognized that this means based on the results of evolution.

When one speaks of a particular scheme of classification, the arrangement is based on the fact that the kinds can be arranged in groups in a hierarchy on the basis of the features with which evolution endowed them. These features are the real basis of the classifica-

[2] Blackwelder, R. E., 1964. Phyletic and phenetic versus omnispective classification. Syst. Assoc., Publ. No. 6, pp. 17–28.

tion; evolution is the process that determined that these kinds do possess these particular features. Evolution, successive change, can alter the features of a kind until it would be called a different kind. After the successive changes have acted for long enough, it is evident that there has been a sequence of kinds. This is called a lineage. If one starts at the recent end of the lineage and looks back, along the lineage, he "sees" the history of the present kind. This he calls its phylogeny.

There is reason to believe that some phylogenies, at least, represent very complex histories. It is easy to talk about a phylogeny as if it were simple, and as if one could actually see it. In reality, it cannot be seen even when it is simple. No one can know anything directly about it, although it can be theorized about from several sorts of evidence.

A statement that taxonomists can base classification on phylogeny can only mean on the supposed-phylogenies, because that is all any of them are. Even in groups with the best fossil record and the best evidence of a phylogeny, such as the horses, no proposed phylogeny has lasted fifty years. The one that is accepted today probably won't last either, because it's not *the* phylogeny of the horse but only the best current theory as to the course of its phylogeny. (It must be confessed, however, that the present equine phylogeny is a convincing one.) There are no such supposed-phylogeny-based classifications in taxonomic use. There may be some that are useful in evolutionary studies as working hypotheses, but that is not taxonomy.

Some writers say that classification cannot be **based on** phylogeny but must be consistent with it. Here again, it would merely be consistent with a theorized phylogeny. This is no basis for a working classification that must distinguish kinds and groups and make possible identifications and recovery of information. The proper statement is that classification should be consistent with everything that is known about the kinds. If it is also consistent with the supposed-phylogenies, that would be fine except that it would merely help to bolster the supposed-phylogeny instead of the classification!

Classification deals with the results of evolution. It is necessary because of evolution. It is not affected by the nature of the evolutionary processes. The processes produce phylogenies of the kinds, but these phylogenies cannot be definitely known. The same processes produce features which the animals possess because they are phylogenetically related to others which also possessed them or had the potential to possess them. These features can be seen and recorded. They are definite, objective, and discoverable and do serve to unite animals in groups that are useful. There is a real possibility that these groups are in some way the result of the phylogeny, but this has not yet been proved.

Animals cannot be classified on the basis of phylogeny or evolution. They can only be classified on the basis of features that were produced by evolution and appeared sometime in the phylogeny, or on the basis of features that were made possible by evolution but are actually produced by influences of the environment. These are the only possible bases for grouping.

What is classified. Taxonomy really encompasses two things that involve classification. First it classifies the diversity found in nature, giving us data on the diversity—its forms, extent, sequence of appearance, and relation to other aspects of the organisms. Some of these data will be useful in understanding the diversity used in the second step. Second, taxonomy classifies the animals into kinds and the kinds into groups of kinds. This means into species, genera, etc.

Some writers assume that classification can deal only with species and more inclusive groups—that individuals cannot be classified. Whatever the theoretical basis for this idea, it cannot be justified, for some taxonomists *do* classify individuals, building an entire scheme before making any attempt to discriminate species. It may be that this is not a good way to classify, but there has been no evidence adduced that the classifications so produced are less adequate or useful than any others. In a logical sense there can be no question that taxonomists can and do classify both objects and groups. They usually classify the concepts that represent the objects or groups, by listing the names or symbols that have been assigned to them, but they also deal directly with both objects and groups of objects, physically sorting them into the prearranged system.

In this connection the apparently contrary view of Simpson (1961, p. 18) requires comment. Simpson here states categorically that an individual never is and cannot be classified. What is involved when an individual is assigned to a particular species is not classification but identification. This appears to be the result of a mammal-centered viewpoint. All the species of mammals are known and a system of classification is in existence, so an individual can be placed in the system by showing its identity with one of the previously described species. This situation is by no means universal. In many groups of animals monographic work can and does proceed without first identifying species. The entire classification of a family of twenty-five thousand species can be accomplished with specimens alone, without early reference to whether they belong to named species or not.

It is true that identification and classification are often intermixed in taxonomic work, and it is certainly true that specimens can sometimes be identified with known species without specific reference to classification. This would need to be identification by comparison. A more scientific and accurate method of identification is to classify the specimen—to place it in the classification scheme step by step in descending order of levels, before trying to guess what species it belongs to. Both systems are used, depending on the circumstances, but the latter is the more accurate and is in fact to some extent almost universally employed.

Existence of groups. Most of the evolutionary processes which are known or suspected to be effective in producing new kinds, are believed to produce small changes that only gradually accumulate to produce a recognizable difference. From this it might be assumed that kinds grade into other kinds, being indistinctly separated at the edges. Theoretically this is to be expected, and there are cases in which it has been thought to occur. In general, however, to the surprise of many evolutionists, there can be no question that kinds and groups of animals are clearly different from one another. This might be due to extinction of intermediates through time, so that the present-day fauna represents mostly remnants of the continuum.

The existence of these gaps is so widely accepted, by evolutionists as well as by taxonomists, that it seems to be redundant to emphasize the fact. Robson and Richards (1936), Schindewolf (1950), and Davis and Heywood (1963) all have statements that are neatly covered by Simpson's reference to "the simple fact that readily recognizable and definable groups of associated organisms do really occur in nature." Thus, the groups are accidental, in the sense that they are produced by accidents of extinction. They are not predictable, although they may not be entirely random, because the causes of extinction may vary. The groups will be of a wide variety of sizes, and the gaps between them will range from only slightly more than species differences to the very substantial differences between the phyla.

The accidental origin of the groups does not make them any less real. If a continuous series of numerals is taken, such as

$$1 \quad 2 \quad 3 \quad 4 \quad 5 \quad 6 \quad 7 \quad 8 \quad 9$$

and any one of the numerals is removed:

$$1 \quad 2 \quad 3 \quad . \quad 5 \quad 6 \quad 7 \quad 8 \quad 9$$

two clear-cut groups are produced. It may seem as if the two were once part of a single series, but at this time there is no direct connection. The two groups are definite, objective, recognizable, and usable. If the "extinction" had been more extensive, any connection might not be so apparent:

$$1 \quad 2 \quad 3 \quad . \quad . \quad . \quad 7 \quad 8 \quad 9$$

Thus the gap can vary in size without requiring new factors to produce it.

In speaking of these groups and the gaps between them, it is possible to be thinking of a single feature instead of the animal as a whole. A gap between the forms of one structure *may* be significant, but if it exists simultaneously in a variety of features it is almost certain to be significant. This is the basis of the belief—erroneous in this context—that classification should be founded upon a wide variety or all of the features. In this connection it is sometimes forgotten that keys are not usually classifications. The use of single obvious features may be entirely satisfactory in a key, but it would not give confidence in a classification.

It was remarked by Simpson (1961, p. 71) that "it is an axiom of modern taxonomy that variety of data should be pushed as far as possible toward the limits of practicability." It has also been demonstrated, however, that there are no benefits to be derived from further data or variety of data after the classification reaches the point of maximum effectiveness. One feature, correctly chosen and stated, may serve to classify the members of a group with as much effectiveness as a thousand characters. The point is that when the taxonomist does not know in advance which character is the all-effective one, and in practice this is virtually all the time, he can increase the probability of finding effective groups if he utilizes a variety and quantity of features.

Differences and similarities. Up to now, this discussion has emphasized differences between the groups—the gaps in the range of features. These differences by themselves will not serve to classify, because classification is grouping and to have groups there must be similarities among the members. It is thus gaps in the similarities which are used to distinguish groups. And particularly it is correlations between the gaps in several different similarities which give the taxonomist confidence in the groups.

It has been generalized that "taxonomy emphasizes the differences between organisms." This is only a half-truth, because it must emphasize equally the similarities among them. Neither difference nor similarity alone will produce a zoological classification.

Stability of classification. Much has been written about stability in taxonomy, but most of it refers to stability of names. Obviously, it is easier to teach or to learn the classification of a group if the scheme remains unchanged for a long period of time. Thus there are advantages to stability.

As the taxonomist learns more about animals, he finds ways in which the classification can be made more effective, and this results in a change in the scheme. In the long run, any change which is an improvement is advantageous, and the temporary disadvantage will have to be borne. It is possible, however, to make changes before it is certain that they will be advantageous. If these have to be changed back later on, a double disadvantage will result.

It must be remembered that taxonomy is an active, living subject, growing and developing as other fields of zoology unearth new facts about kinds of animals. To suppress change in taxonomy would result only in forcing its replacement with some system that would permit growth. The responsibility of taxonomy is to grow with zoology as a whole but to resist abortive or ill-considered change.

Types in classification. In classification, types are used at the species level and for grouping species into taxa. The use of species types has already been described in Chapter 11; the specimens which are believed to be conspecific with the type are said to belong to that species.

For building taxa, types are used at the generic levels and at the family levels. The use is a mixture of identification, nomenclature, and grouping. Only the last is dealt with in this chapter.

Genotypes. The type of a genus (or subgenus), variously called genotype, generitype, generotype, and type-species, is the means of attaching a generic name to a particular species, as an anchor similar to the type specimen of a species. This species will tell us a great deal about the genus, about one part of it, but nothing about the limits of the genus. The genus must always include at least that one species.

The present rules of nomenclature state that the type-species is type of the *genus,* but in practically all respects it is found to be type of the generic *name.* (See Chapter 25, Genotypy and the Types of Genera.)

Problems of synonymy and of dividing a genus into two genera both involve genotypes in relation to the names, and it is impossible to speak understandably about a certain genus without being certain which species is its type and therefore which species is inescapably a member of it. For example, to say that the species *Planaria tigrina* belongs in the genus *Dugesia* is meaningful only if we know with certainty what other animals also belong in that genus and are correctly grouped with *P. tigrina*.

In our catalogs and revisionary studies, we frequently find generic synonymy comparable to the following:

Hypotheticus Linnaeus, 1758
Subsequens Smith, 1908

We understand this to mean that these two names have been applied to the same genus and are therefore synonyms. According to the dictionary, synonyms are two or more names for the same thing, with all of them being called synonyms. By this definition *Hypotheticus* is as much a synonym as *Subsequens*. The important point is that it is the older one that is the proper one to use. It is best to distinguish this correct one as the senior synonym, with the unaccepted one as a junior synonym.

In biology there are two quite distinct kinds of junior synonyms. There are some which were clearly proposed for the same genus (new names, stillborn synonyms, and names based on exactly the same species) and are therefore absolutely synonymous; they can never be separated by any means. These are called absolute synonyms, objective synonyms, or nomenclatural synonyms. There are other synonyms that are synonyms only in the opinion of one or more students. One may lump the two genera into one, making the names synonyms; the other may split them into two genera, making both *names* correct. Synonyms that are thus based on opinion are called conditional synonyms, subjective synonyms, or zoological synonyms. The synonymy can be denied and removed at any time.

The first type includes what are here called absolute synonyms. Their identity can never be questioned. The two names can never be correctly applied to different genera under any circumstances, and there are elaborate rules to determine which one will be the name that is to be used for the genus. Once a case of absolute or objective synonymy is recognized, there are few problems, because opinion and taxonomic conclusions have little effect. The two names have identical standing, except for priority and homonymy. These latter determine which one will be used. The second type includes what are here called conditional synonyms, because anyone may challenge the identity. Their synonymy depends on the opinion of each taxonomist as to the zoological status of the species included in the genera. The synonymy is therefore subjective and based on zoological considerations.

In stating that these two names in the example above are conditional synonyms, the reviser is stating in effect that the species previously included under both are congeneric and therefore belong in a single genus. However, it may be that some of the species previously included under *Subsequens* belong in *Hypotheticus* and some do not. The reviser cannot apply the name *Subsequens* to both groups —as a synonym of *Hypotheticus* for one and as a separate genus for the other. There must be some means to determine which species the name will follow. This is accomplished under the present rules of nomenclature by the use of the genotype principle. Every generic name is tied to one species, so that the assignment (on zoological grounds) of that species will determine the fate of the generic name. In the example, if the type of *Hypotheticus* is species 1 and that of *Subsequens* is species 7, it is merely necessary to determine whether these two species belong in one genus or not. If they do, then *Subsequens* is a junior synonym of *Hypotheticus,* regardless of the status of the other species assigned to either genus (or generic name). Thus, generic synonymy cannot be determined without use of the genotypes.

In dividing genera, some means of determining which group shall retain the name is again required. For example, when it is found that the genus *Compositus,* with ten species, actually contains two groups of species which deserve to be recognized as distinct genera (species 1 to 5 and 6 to 10), the original name must be retained for one of the groups.

Previously: Spp. 1–10 = *Compositus* (in broad sense)
Now: Spp. 1–5   = ? (*Compositus* s. str. or new genus)
         Spp. 6–10 = ? (New genus or *Compositus* s. str.)

Under the Rules *Compositus* must be used for the group containing the particular species which is established as its type species or genotype. If species 4 is the genotype, then the name *Compositus* must be retained for the genus including species 4, or group 1–5.

It is thus necessary, and the Rules require, that each generic name have a genotype, to fix the application of that name. Then, whenever any writer places a species in a particular genus, he is in effect stating his belief that that species is congeneric with the type species of that genus—that the two species belong in the same genus. Otherwise he would not put them together. In order to state that his species belongs with the type species of this genus, he must know what the type species is and what its characteristics are. No generic transfer or assignment of a new species to a genus is much more than a guess unless the genotype is known and considered. For example, a writer describes a new species *albus,* and he places it in the genus *X-us.* If he does not know the genotype of *X-us,* his assignment of *albus* is little more than a guess.

If a writer believes that two genera are the same and cites the name of one as a synonym of the other, he is in effect stating that the genotype of one belongs in the same genus as the genotype of the other, because otherwise they cannot be synonyms. It is impossible to prove the synonymy of two names without using the genotypes of both, since the application of each name depends entirely on its genotype. For example, if a writer states that *X-us* and *Z-us* are synonyms because he has a species that is called *X-us* which he discovers belongs with a species that is put in *Z-us,* his conclusion will be completely wrong *unless* it happens that his species of *X-us* really does belong with the type species of *X-us* and his species of *Z-us* really *does* belong with the type species of *Z-us.* If either of these should happen not to be so, then his conclusion on generic synonymy will be worthless.

No writer should ever cite a generic name as a synonym of another generic name until he, or some previous writer whose conclusions he accepts, has determined the genotypes and found that they belong in the same genus. Any synonymy proposed on any other basis is worthless so far as the generic names are concerned. Furthermore, no writer should ever describe a new species in a genus whose genotype has not been determined and is not believed to be congeneric with the new species. Otherwise the new species will later have to be restudied to see if it really does belong in the genus indicated.

Since there is little in taxonomy that can be safely done without use of genotypes, this is one of the most important subjects in nomenclature, as well as one of the most neglected. It is one of the nomenclatural subjects whose effect upon classification is very great.

Family types. The idea of typifying the groups in each level of the hierarchy is an old one in taxonomy. Because types are directly involved in nomenclature, the idea of types has generally been discussed in connection with the rules of nomenclature. As a result, types have been formally used only at the levels directly covered by the rules. Until 1948 this meant at the specific and generic levels only.

In the 1961 Code, rules were extended to names of all the groups in any of the family-group categories, family, subfamily, tribe, etc. Such rules were not applied to groups in the categories of order, class, etc.

Because each family-group name must be formed from the name of an included genus, there are few problems of determining the type genus of a family name. (Here again, the Code treats the type as the type of the *family,* but in nearly all respects it is the type of the *name.*) The type of any family name is the genus (or generic name) on which that family name is constructed. The family must always include that genus. The same problems of synonymy and homonymy occur as with the generic and specific names, but problems of determining the type seldom occur.

The data of classification

It was mentioned above that all the data useful in classification are comparative in nature and that any comparative data may be of such use. The subject of comparative data was discussed at length in the previous chapter, and the selection of data to be treated as taxonomic characters was cited as the first step in both classification and discrimination. It remains to discuss the type of characters most useful in classification and the means by which they are selected.

It is essential to make a clear distinction between the basis of classification and its purpose. Certain data are *used* as basis of a classification, and by means of the classification a great many other data are *recorded.* The data which are recorded in the classification are not always those used in making the system. For example, it has recently been found (Downey, 1963) that all known pupae of the butterfly family Lycaenidae possess unique abdominal structures that enable some to make a grating sound. If it develops that all Lycaenidae possess this feature in their pupal stage, the feature could readily be used in the primary classification of butterfly families to distinguish this family. It is unlikely, however, that it would be so used, because features of the more commonly obtainable adults will segregate the same family for a wider variety of purposes.

If it is found that the pupal structure and the usual adult features are completely correlated, each one always present when the other ones are, then either can be used alone, and nothing is gained by using them both. The unused aspect is stored in the classification but may not be part of its actual basis. It is not correct to say that taxonomists must use biochemical or any other kind of data just because such data are available. They should make an effort to determine whether the new data are correlated with the old. If they are, they are merely stored. If they are not, then their possible contribution to the classification should be studied. The results may or may not produce a change in the classification. The change may or may not be ostensibly based on the new data, because after the change is decided upon it may become obvious that unnoticed data of the older type are easier to use than the newer.

There is nothing different taxonomically about the newer types of data. They are comparative; they are subject to the same faults and difficulties; and they give the same sort of results in classification. There is one thing unique about the older types, however; they have been in use for so long that they have been recorded uniformly for all or most of the species. They are thus in fact usable throughout the group being classified. The newer types are almost universally not so widely available. When they become so, they will be equally likely to be effective in the construction of a classification.

For fossil animals. One special case has been outlined recently, that fossils have data characteristically different from those in neontology and presenting many special problems. The specific points listed are nearly all relative, not fundamentally different, and these do not justify the conclusion. It will be instructive to discuss them individually.

(1) *The available anatomical data are always incomplete.* This is also true for all animals. For Recent animals it is possible to claim that there is a possibility of possessing all the anatomical data, but in practice it can only be said that fossils frequently present less complete anatomical data than Recent animals.

(2) *Few or usually no direct physiological or behavioral data are available.* Here again, for Recent animals there is always the possibility of obtaining such data, but in fact, for the vast majority of animal species, no substantial amount of such data is actually known. One can only conclude that fossils usually present a minimum of such data.

(3) *Ecological data, although available in significant amounts, are generally also much less complete than for Recent animals.* No claim is made here of difference in nature of the data. They are merely available in less profusion, even potentially.

(4) *Large samples are less often obtainable.* Invertebrate fossils are often available in unlimited numbers. More and more kinds are being recovered by new techniques in numbers as large as the taxonomist can handle. Although many Recent species exist in widespread profusion, many others are extremely rare and cannot be obtained in quantity by any known means.

(5) *Large samples may be prohibitively expensive.* Large samples of whales or alligators would surely also be expensive. The expensive samples would frequently be those of large or rare animals, regardless of whether they are fossil or Recent. It is true that fossils sometimes require more elaborate preparation than most other animals, but the difference is only relative. The isolation and preparation of microsporidian protozoans, for example, are more elaborate than those of most other Recent animals.

(6) *The samples may be biased in numerous and sometimes peculiar ways.* Of course, all sampling is biased by a variety of factors, differing in different sorts of animals. The possibility of "peculiar ways" of biasing with fossils is the only thing in these six items which can possibly justify the claim that there are characteristically different data from fossils. There certainly are some unique factors in the sampling of colonial animals, parasitic and symbiotic animals, migratory animals, and in many less obvious samples.

There is a suggestion in (4) and (5), above, of why this unjustified claim should have been made by an eminent paleontologist. Very large animals are mostly vertebrates. Large samples of fossil vertebrates are almost unknown. The cost of obtaining, preparing, and storing such large specimens is exceptionally high. Nevertheless, the general claim is surely unjustified for fossil animals as a whole, and there is no real support for the claim even from the vertebrate viewpoint among the factors that were cited. (These statements were all directly quoted from a recent work by a *vertebrate* paleontologist.)

It is, of course, true, even if not included in the above list, that time is a special factor in paleontology. It has little effect on data, however, although there is a special set of data relating to occurrence in the stratigraphic sequence. The time factor may influence the judgment of correspondence of individuals, and it does give slight differences in the location of specimens a great possible effect on judgments of identity and contemporaneity.

Appropriate data. Data in general have been discussed in Chapter 10, Comparative Data. In descriptive taxonomy and identification, the data used are called characters. In classification, this latter term is appropriate, but it is not often used. In its place will be found simply data, criteria, features, or resemblances. In classification, the characters come from the individuals, just as in discrimination of species, but they are now the characters of groups of individuals and may be expressed differently.

Selection of features. There is no question that all characters to be used are chosen by the taxonomist for that particular use. This is a personal decision, regardless of what basis he uses for his choice. The choice is therefore highly subjective. If the characters are effective ones, the resulting groups and distinctions will be clear-cut. When this happens, the characters can be described as objective, or at least the results can. The choosing of the characters is one of the major functions of the taxonomist.

The taxonomist will examine many features, many potential characters. Probably a majority of them will be found not to be differential—they are the same in all the groups or they are inconclusively variable. These will not be listed as characters. This assessment is made intuitively or consciously, sometimes after tabulation or analysis of the variation. The test is the permanence or uniformity of the feature.

The categorical level. There has been discussion for two centuries over the possibility of defining what features are appropriate for use at each level—species, phylum, etc. No satisfactory solution to the problem has been found, although in retrospect it is possible to cite what *has been* used at each level. In different parts of the Animal Kingdom the features used at the class level, for instance, may all be related to the condition of organs and major structural features, but they differ so among themselves that there is no real uniformity, and we cannot say what types of characters will distinguish classes, or groups at any other level.

A generic character is obviously one that characterizes a genus, but there is no way to tell in advance what sort of character will do this. It is possible to know in a general sense that it will be a character that occurs widely or universally in the members of the genus and one that does not normally occur in nearby genera. However, in a series of similar characters, even though one is found to characterize a certain genus, it may happen that the others will be useful in characterizing only a species, or perhaps an entire family. It is not possible to decide *a priori* at what level a given sort of character will be found useful. Furthermore, a particular feature may be highly uniform in its expression in one taxon and appear to be a source of good characters there, but the same feature may be found in other taxa to be highly variable and therefore not suitable to be a taxonomic character in that group.

Good and bad characters. A good character at any level is one which "works," which produces a useful distinction or grouping. This means that it is (1) readily detectable, (2) clear-cut in its segregation of the group possessing it, (3) complete in its uniting of the members of the group, and (4) evidently correlated with other aspects of the life of this group, so that the grouping itself is useful in a variety of situations.

When the taxonomist is looking for characters, for distinctions between groups, the choice is based solely on his judgment, experience, and analysis. In later usage, the judgment of good or bad is based on whether or not it has proven to be effective, whether or not it produced a useful and workable classification.

Davis and Heywood (1963) emphasize that a good character will be one that is not influenced by environmental factors. They list the theoretical requirements in this way: (1) not subject to wide variation within the samples being considered; (2) not having a high intrinsic genetic variability; (3) not easily susceptible to environmental modification; and (4) showing consistency. They add: "Good characters will have a narrow range of expression and will be easily recognizable."

A bad character would be one which did not serve well to make the distinction or grouping desired. Such a judgment could be made only in a particular case, as it is not possible to prejudge one type of character as to its effectiveness in cases not yet studied. Wide experience with characters and animal diversity will help a taxonomist to make good preliminary choices. Only later examination of that character in the particular group will show whether it is effective in that one case.

In theoretical terms it is possible to define bad characters. First and foremost, they do not meet the criteria set for good characters. They are, moreover, those that make distinctions and groupings that do not correspond to fundamental features of the organisms. They are thus those that are readily affected by the environment, that vary irregularly in the group, that are not shown by all the individuals being classified, or that otherwise fail to distinguish or group. These can be identified only in practice, by trial and recognition of failure.

The units of classification

There has been substantial and continuing difference of opinion as to what is the basic unit of biological classifications. On the one hand there are those who believe that the species, defined on the basis of breeding populations, is the unit that taxonomists classify. On the other hand there are those who believe that it is individuals that are arranged into species and orders.

It is not easy to see how there can be such a sharp difference of opinion. It has been pointed out repeatedly by logicians that it is possible to classify any subjects, whether they are concrete objects, simple concepts representing sense data, or compound concepts of which the members are other concepts. Thompson (1937, p. 160) reiterates this for biologists. Anyone can classify species on any basis he wishes, but to do so he must have the species before him, at least figuratively, so that he can arrange them into genera according to the selected criteria. Likewise anyone can classify individuals in the same way. He can arrange them into species or he can dispense with that level entirely and arrange them directly into genera or groups at any higher level. Groups at any level in the hierarchy can be classified into groups in the still higher levels, unless they are already at the top.

There is no way to distinguish between individuals, species, or genera on this basis.

To say that the species or population is the basic unit in classification would seem to mean that it is the smallest unit involved in the operation. If this is true, then these species must exist in nature as separate entities in such a way that the classifier does not have to construct them or assemble their parts. Such a situation actually exists in a large degree in the case of already known species. It is possible to classify the concepts that have been patiently assembled by previous taxonomists. Where previous work has not delineated the species, there is no possibility of classifying the species directly.

Species can be classified (grouped) only after the individuals have been classified into species. Whenever a taxonomist is studying a group in which the species are all well known, he can almost forget the individuals and proceed to classify the species that were previously made known. Here it is possible to say that one is classifying species, but to say that the species is therefore the basic unit of the classification is a *non sequitur*. Some fictitious species (A, B, C, etc.) could be classified without reference to their components if one wished to do so, but natural species, no matter how that word is defined, do not present any features on which to base a classification until at least one individual has been identified as belonging to each. The individuals individually do not furnish all the data that can be known about the species, but with no individuals there is nothing that can be known about the species.

There is philosophical basis for denying even that the individuals are the basic things in a classification. Gilmour (1940, p. 466) points out that our sense-data derived from these individuals are the only objective materials available for study. So it is merely necessary to understand that the word individual means "as it appears to our senses." It is clear, then, that taxonomists can classify species, individuals, or subfamilies. It is also clear that the individual is a more basic unit of classification than the species, being essential to the very existence of the species.

It must not for a moment be construed that this means that an individual represents a species. A male and a female, a caterpillar and a butterfly, several strains, seasonal forms, polymorphs, and examples of all the range of variation of every sort—these together represent the species. The species is quite different from any possible individual, because its attributes are not absolute; they are ranges within which individual attainments lie. The individual no more represents the species than the species represents the phylum; it is simply one of the members which contribute to the more inclusive concept.

These remarks are not intended to make a distinction between such terms as deme, species, and population. The remarks apply regardless of what basis is adopted or what term is used. It makes no difference at this point whether the species is defined as a group of specimens having some character in common or as a potentially interbreeding series of populations or as a gene pool. If there are no individuals there are none of these.

The idea that species are the basic unit of classification seems to have originated with the workers on higher vertebrates. In the birds, for example, it is now possible to classify species, because they are all well known. It should have been obvious that this is a form of parasitism, in which the descriptive work of previous taxonomists enables the current ones to forget about the individuals. There are classifications of animals, particularly fossil animals, in which species are simply not employed. The individuals are grouped into genera, which are the units most useful for the purposes at hand.

Methods of classification

There are few subjects in systematics so difficult to write about as the actual procedures employed in classifying or reclassifying a group of animals. As a result, almost nothing has been written on the subject. Even the most recent book on taxonomy, which deals with many philosophical aspects of classification, has little to say on the actual methods of classifying. It is evident from published classifications that there are no well-known procedures in use in all groups of animals. It seems extremely likely that clear-cut concepts and procedures do not have any significant influence on classifiers in general.

It seems to be necessary to point out again that the level of knowledge of the species and the extent of classification differs widely from group to group. In the best-known classes, such as the higher vertebrates, discrimination of species and genera has virtually been completed. Classification of these into families and orders is still a matter of study, although there is really very little classification being done. In lesser known, and sometimes much larger groups, only a fraction of the existing species have been described. The arrangement into genera is tentative at best, there being no adequate monographs. The result is that when vertebrate taxonomists speak of classification, they are thinking of quite a different thing from the invertebrate taxonomist.

There are two basic methods of classifying animals, or anything else. Both of these have been used in zoology, but one necessarily preceded the other in history.

1. *The method of association.* Historically, when the first taxonomists sat down to classify, they had before them what was actually, as well as in our perspective, a small number of specimens of a small number of kinds. In studying these specimens they found that some of them had many features in common and some only a few. Knowledge of these animals was adequate to show some resembled each other about as much as the members of one brood, or the offspring of one pair of parents. These were assembled into a group called a species. It was believed that these were the individuals which could reproduce their own kind, with reappearance of the same minor features indefinitely in various combinations. It took no great effort to assemble all the available specimens into a very few thousand kinds of species. This activity was classification, at that time. It was, in fact, the only classification possible in the beginning.

It was then seen that certain of the species were so similar to each other in so many ways that it was possible, and useful, to bring these similar ones together in a group of species that was called a genus. Similarly, some of the genera were so much alike that they were grouped into orders. The orders were then grouped into classes. In all of this the essential feature was recognition that because of differential similarity, certain distinct entities could be brought together as a definable group made up of several of the lesser groups. The arrangement of the groups into more inclusive super-groups in a hierarchy of ascending levels was a device well known in logic long before zoological taxonomy became a formal study. *Grouping* of a comparatively few objects or groups of objects thus started the system.

Once such a system was started, by Linnaeus in the mid-eighteenth century, new discoveries were simply inserted into the existing groups. For a time, most new specimens could be assigned to existing groups at most levels. If they differed from all known species in ways believed to be outside the range of filial diversity, they were segregated as a new species. This process did not really involve much classification. It is still a very common procedure, often denoted by the expression *descriptive taxonomy.*

As the number of species increased and were assigned to the known genera, these genera came to contain species more and more diverse. Finally, students divided the genera by recognizing that the included species could be grouped into two or more clear-cut groups that could themselves be called genera. The same thing happened to orders and classes, so that the number of groups increased at all levels. As more and more specimens were collected in all parts of the world, many were at once seen to be so different from all known species that they couldn't be put in any known genus. At the time these new species were described, they were grouped apart in new genera, and even in new orders and classes. All of this involved primarily recognition (1) of groups of specimens as probably being within the range of variation of a single kind, (2) of groups of these kinds or species under the term genus, and (3) of groups of these groups as orders and classes. Grouping predominated all this activity, and characters held in common were the basis of the grouping.

2. *The method of subdivision.* When a large number of species had become known, and they had been grouped into genera, orders, and classes, it was possible for taxonomists to use a different approach. When a sufficient number of the species in one order were available for study, the taxonomist could disregard all previous classification and work by subdividing groups instead of grouping them. For this purpose he would, in effect, start by treating all the specimens as comprising one group, an order perhaps. Examination of these specimens would immediately reveal that they differed from each other in various ways. The taxonomist, being familiar with these animals, knew much about the variety that occurred among these specimens. He could, usually without any difficulty, divide the specimens into two lots on the basis of features that seemed to him to be the least likely to be the result of common heredity, environment, or the diversity usually associated with species or genera. Each of these segregated groups could again be subdivided, until finally all the known specimens were distributed into genera and then into species on the basis of the extent and manner in which they were different.

It was of course found necessary to insert additional levels into the system, families, suborders, and many others. At each level the goal was to have a comparatively homogeneous group—a genus containing only species very much alike, and a family containing only species more like each other than like those in other families. Groups were divided and subdivided whenever appropriate features provided a distinction. In this way *subdivision* was the keynote, although of course grouping was also involved.

This method of subdivision is quite often the principal approach in a monograph, where a reclassification is undertaken. Previous groupings may be temporarily or even permanently discarded, while the actual specimens are being sorted by subdivision of the entire lot into sublots.

When a group has been well classified previously, such reclassification by subdivision will probably end up with much the same grouping. If the previous classification has been based on poor work and faulty decisions, the new one may be substantially different.

Summary

Classification is a highly practical activity that organizes things or ideas. Science may be defined as organized knowledge, and classification is the principal form of organization in biology. This activity may also be appropriately termed systematizing, the production of a system of knowledge arranged to facilitate the use of that knowledge.

The process of classifying thus stores the knowledge. This is its chief practical purpose. The act of classifying produces new knowledge about the groups. This new knowledge is also stored and is often more useful than the original knowledge on which the system was built.

It is possible to classify because the immense diversity of organisms is not a continuous sequence of minute differences but a series of discrete groups. Gaps in the diversity do exist and make possible the definition of the groups. A variety of aspects could be used for the grouping, but the necessity in zoology to work at times with preserved specimens has channelled the grouping on the basis of physical features of the body which can be preserved.

Because the groups discovered are sometimes later found to be composite, each group needs to be anchored to a single specimen or subgroup which gives an objective base to the group. This helps to solve many later problems of splitting groups, applying names, and identifying the correct place in the classification for a given specimen. These anchors are known in general as types.

The actual classification can be done by grouping like things starting at the bottom, or it can be done by dividing the entire complex into groups and each of these into subgroups. These methods are, respectively, association of objects and subdivision of groups. Both are accomplished by comparison of corresponding features of the specimens or groups. They differ only in the direction of approach.

The difficulties in classification are those of understanding the nature of the animals. The data are very largely objective. The selection of the data to be employed is entirely subjective. Only detailed comparative knowledge of the animals, the methods of classification, and the past studies on these animals will produce effective classifications.

Further discussion of the theoretical basis and nature of classification will be found in Part V, Chapter 18, The Nature of Classification and of Species.

CHAPTER 13

The Use of Names

"For the experienced and productive scientist concern with problems of nomenclature will have its practical reward in preventing the early and unwarranted assignment of his work to limbo upon his demise." (Lamanna and Mallette, 1953.)

This point of view quite closely represents the practical importance of nomenclature. No matter what is thought of it as a scholarly study or as an activity on the part of biologists, it cannot be denied that names are necessary in biology and that a system of naming is necessary for effective employment of the names. Names are inextricably tied up with the system used for grouping organisms. The groups may be taxonomic, hereditary, or ecological, but without names for the animals zoologists would be limited in communicating their groupings and the data about the groups. The people who could assemble and look directly at the groups of specimens would be the only ones that could be reached.

There are a number of unfortunate ideas that cloud the use of names and may be looked upon as myths. They add to the confusion caused by the often unavoidable changes of names. Perhaps if some of the faults can be corrected, nomenclature will be seen in its true light as one of the most important tools in all of biology.

One of these ideas is that there can be stability of names at a time when there is no corresponding stability of taxonomic grouping. Stability of names is said to be a primary goal, but at the same time there is assurance in the 1961 Code that taxonomy must be left free to grow. The two are inextricably tied together; the goal cannot be reached without sacrificing the freedom. Another idea is that the "common names" used in insects and vertebrates are something quite apart from zoological nomenclature and are the same as the names used by farmers, sportsmen, and the man-in-the-street. In reality the common names that are widely used have been officially adopted like

210

the Latin names. They do not correspond to such down-to-earth names as pill-bug, night crawler, or gopher.

It is in part a myth that zoological nomenclature is binominal—a two-name naming system. Zoological names are used for groups at twenty or more levels in the hierarchy. Only one of these levels uses binominal names. Actual trinominals of two types are very common; quadrinominals are correctly used at one level; and quinquenominals have been used extensively in some groups. At the remaining levels, all names are uninominal. The rules of nomenclature do not count the name parts in this way, but a majority of the rules apply to names that are, in at least one sense, not binominal. It is also a myth that citation of the author of the name as a bibliographic reference serves to eliminate the potential difficulties. It does serve a useful purpose, but the difficulties of applying names effectively are legion, being both nomenclatural and taxonomic. It is a widespread myth that in such an expression as *Felis leo*, *Felis* is the name of the genus and *leo* is the name of the species. The very nature of binominal nomenclature is misunderstood here.

It is a myth promulgated by international officials that zoological nomenclature rules are the result of the agreement of the majority of taxonomists. In reality, the majority of taxonomists never have a real opportunity to take part in such studies. The rules are always made and adopted by a very small minority, probably less than 5%, and the balance of taxonomists have no real opportunity to vote on acceptability of the results. (This is not necessarily a bad situation. It cannot honestly be described as democratic in action, but it may well be more effective than a truly democratic system would be.)

Vernacular names for sexes, age forms, and groups

Proper names are the labels for people and places. Nouns of all sorts are the labels for objects, attributes, and ideas. The names for groups of animals are the subject matter of zoological nomenclature, but most individual animals are also represented by some of the labels that denote sex, age, and the number in the assemblage. For example, gender is denoted by such names as *male, female, stag, rooster, doe, mare*. The young of animals are denoted by such names as *baby, kitten, chick, colt, whelp, fingerling*. The more formal stages in the life cycle of some invertebrates are denoted by such names as *larva, pupa, adult, planula, redia, cercaria, pilidium, hexacanth, trochophore*. The customary aggregations of animals are denoted by such collective names as *herd, flock, pack, flight, army, school, colony, swarm, covey, string, pride*.

These names are not the province of zoological nomenclature. They are simply part of the common language or part of the technical language with which the ordinary person and the scientist refer to types of animals.

Lists, of the names showing the gender of the animal, the names for the young of animals, and the names for assemblages of animals, are given under the heading "Names" in *The Encyclopedia of the Biological Sciences*, by Peter Gray, 1961.

Zoological nomenclature

Classification or any other organization of the kinds of organisms would be well-nigh useless unless some way were found to designate the kinds and the groups into which the kinds are assembled. It has been found that the use of names for the groups of organisms is more effective than numerals or formulae but is also accompanied by some difficulties. The field of nomenclature is largely the study of ways to circumvent these difficulties.

Nomenclature, pronounced nō'-men-clā'-chur, is literally the calling of things by name. The word is used to refer to the general field of naming animals as well as to the system of names employed. Its adjective form is nomenclatural, referring to nomenclature. The word nomenclatorial is sometimes used in this sense but entirely erroneously. Nomenclatorial is the adjective form of nomenclator, a person or book which deals with names. There are zoological nomenclators, but they are not to be confused with the system used for naming animals.

It is not usually noted that there are several systems of nomenclature in current use in zoology. These systems supplement each other, although they may also overlap. The first is the binominal system for naming species, the Linnaean system (see page 227). In this system, the first part of the name denotes the genus and the second part the species. The combined name is the name of the species. These names are subject to careful regulation by international agreement. The second system (page 216) is that employed for all groups above the species. It is uninominal and is in part based on international agreement and in part upon custom. The third system (page 213) is the one of standardized common names. In some groups of great importance to man, the best-known species may receive specially coined common names for the benefit of the non-taxonomist. There is no international system, but sometimes the names are chosen by governmental agencies.

All of these systems have their own uses, their own faults, and their own basis for use and control. The student must study them all to understand their proper application and to know how to propose, use, and record the names in each. The binominal system is very much the most elaborate, and it embodies many international agreements as well as many technicalities. This dual-name system will be discussed in detail, after consideration of the common-name system and the uninominal names of groups.

Standardized common names. Names in the ordinary language of the people of a region or country are called common or vernacular names. They simply appear in everyday speech. It is very common for a widespread species to receive many such names, especially in different parts of its range. The European white water lily has 15 English common names, 44 French, 105 German, and 81 Dutch—a total of 245 different common names. There may be no "system" for such vernacular names. They appear in the local language and may attain some stability and definiteness in application. There is no way to tell for certain to what species they apply, to prevent the giving of many such names to one species, or to prevent the use of one name for quite different organisms in separate regions. For example, the name mahogany has been applied to at least 300 different kinds of trees.

There are, however, systems of common names which have at least some of the desirable qualities of uniqueness, distinctness, and definiteness. These are mainly the lists promulgated by organizations concerned with the application of taxonomy. The following ones have been published for North American animals:

Mammals: *Vernacular Names for North American Mammals North of Mexico,* by E. R. Hall et al. University of Kansas, Museum of Natural History, Misc. Publ. No. 14, 1957.

Birds: American Ornithological Union, *Check-List of North American Birds.* 5th ed. 1957.

Amphibians and Reptiles: Common Names for North American Amphibians and Reptiles. *Copeia,* 1956, pp. 177–185.

Fishes: *A List of Common and Scientific Names of Fishes from the United States and Canada,* 2nd ed. American Fisheries Soc., Spec. Publ. No. 2, 1960.

Insects: Common Names of Insects Approved by the Entomological Society of America, by J. L. Laffoon. *Bull. Ent. Soc. America,* vol. 6, pp. 175–211. 1960.

In addition to these, there are in some groups similar names proposed and used by individuals, employing the same style and having much the same features as the names in the "official" systems. They form a sort of appendix to the official lists, applying to lesser-known species or to ones without great economic importance. For example, among butterflies: The Common Sulphur, The Pink-Edged Sulphur, The Giant Sulphur, Booth's Sulphur; The Long-Tailed Skipper, The Silver-Spotted Skipper, The White-Tailed Skipper, The Coyote Skipper. These would not be used in vernacular language, but they have not been fully standardized by agreement.

Scientific names. Neither the common names of these systems nor the true vernacular names in daily use by laymen have been found satisfactory for use in science. To be most effective, scientific names must be of world-wide application, recognized by taxonomists everywhere, and permanent. The solution to these problems is the system of Latin names called the Linnaean system, now regulated by international agreement. These agreements amount to a code of laws, called rules of nomenclature. They have produced more stability than is usually realized, but they have inevitably "overlooked" or even produced some instability. Changes in the rules from time to time are attempts to decrease the instability.

The scientific or Latin names of animal groups include names for species and subspecies, genera and subgenera, and all taxa in the family, order, class, and phylum groups. A few of the general problems are discussed here, with special problems treated in later sections.

(1) The names of animal groups are independent of the names of plant groups (which are governed by a separate code of nomenclature). A botanical name at any level does not prevent there being an identical zoological name.

(2) Scientific names of animal groups are either Latin or latinized, or they are considered and treated as such in case they are not of classical origin.

(3) All names are written in the characters of the Latin alphabet, but this particular Latin alphabet includes some letters not used by the Romans. These are *j*, *k*, *w*, and *y* and the letters now found in European languages that are distinguished by diacritic marks. All of these were acceptable in scientific names until 1961, but the XV International Congress of Zoology ruled that diacritic marks may never be used in scientific names.

(4) The original spelling is to be preserved except in rare cases which are provided for in the rules.

(5) Since for reference purposes it is often desirable to know who proposed a name, the author of the name may be written immediately following the name without interposition of any mark of punctuation. The author is that person who was responsible for the first publication of the name in a manner conforming to the pertinent rules.

(6) All names in zoological nomenclature are written with a capital initial letter. (The second word of the name of a species is not a name by itself and is not capitalized.)

Most rules, precedents, and proposals for the regulation of scientific names apply only to one of the levels of taxa; for example, the family group, or the species group. These rules must be discussed separately under those headings, but it need not be concluded that there is no similarity between them, as the goals of stable reference often lead to similar solutions at the different levels. In order to understand these rules and conventions, it is necessary to know where they came from and how they are administered. A brief account is inserted here.

Historical background. There have been systems or rules for naming animals for just over two hundred years. Various taxonomists from Linnaeus on have suggested ways in which names might be promulgated and used with the maximum benefit to science. Some of these people proposed entire codes, dealing with many aspects of nomenclature. The history of the regulation of names of animals is roughly paralleled by the similar history of names of plants.

With more than five thousand taxonomists in all parts of the world dealing with names, rearranging, studying, and reclassifying groups, using a score of different languages and a variety of biological training, it is not only understandable but inevitable that there should be instability of names, produced by differences of opinion, differences in the attention to various factors, and actual errors on the part of the investigators. These are what the rules of nomenclature are designed to reduce.

By far the greatest instability during the last twenty-five years has been in the rules themselves. Several substantial revisions have been made during this period. The latest complete revision was published in 1961, but it may take several years before its provisions are widely understood. It has already been amended in several respects.

In spite of the recent publication, the rules are at the present time still in a state of flux. The *Règles Internationales de la Nomenclature Zoologique* were in effect since about 1901, when they were adopted by the Fifth International Congress of Zoology (Berlin). The only authority of the Congress was the weight of opinion of those who attended the meeting. The committee which drafted these rules became the International Commission on Zoological Nomenclature, empowered to interpret the Règles and, in a few cases, to make decisions about names under suspension of the Règles.

These rules remained in effect until recently. They were amended a few times by the International Congress. In 1948, a complete revision of the Rules was authorized by the 13th International Congress (Paris). By 1953, this work was so far from complete that a Colloquium of about forty taxonomists restudied most of the code and made numerous decisions for revision. These revisions were approved by the 14th Congress (Copenhagen) that year and were published as *Copenhagen Decisions on Zoological Nomenclature* (1953). The new set of rules which was supposed to result from these decisions was still far from complete in 1958 at the time of the 15th Congress (London). A second Colloquium restudied many of the decisions and set up machinery for the completion and publication of the new code. This entirely revised code was published in 1961 under the title *International Code of Zoological Nomenclature.*

Thus, from 1948 until the present time there have been several substantial revisions of the code. A complete official code is now available for the first time in many years, but the changes made at Paris, Copenhagen, and London have left many zoologists in doubt as to its permanence. It is certain that additional changes will be made at forthcoming congresses.

Some of the provisions of all these codes, and probably of the ones to come, are arbitrary, based on popularity votes rather than logic. A really sound nomenclature must rest upon logical analyses and clear agreements on all points. The 1961 Code is based in part on such careful studies, but there is still much of very dubious quality. The basis and procedures of nomenclature are discussed here as problems in reasonable and logical cooperative behavior and point out the provisions of the present code which bear on each problem. There are still many problems not covered by the rules.

Although species are named in a unique fashion employing a two-part name (binominal nomenclature), all groups composed of species are denoted by a single name (uninominal). These single names are regulated by the Code within the genus group (genus and subgenus) and the family group (superfamily, family, subfamily, and any others between subfamily and genus). The Code now contains no provisions for regulating names in the order group, class group, or phylum group. There is nevertheless a substantial amount of agreement in the employment of the names for groups at these higher levels. These unregulated names are considered first.

Uninominal Latin names

Names of phyla, classes, and orders. In speaking of the names of taxa placed at levels above the family group it is necessary to recall two things. First, there are no fixed or indispensable levels, but each one is used or not according to the needs of the classifier. Second, the placing of a taxon (group) in any level is entirely a matter of opinion.

There is no correct or standard position, even though there may be a historically customary position. (Both of these matters are discussed in Chapter 4, The Use of Classification.)

For convenience, these categories or levels have recently been separated into two groups, known as the Phylum-group and the Order/Class-group. There is little to be gained by combining orders and classes in this manner, so three groups will be discussed here in the aspects which they do not all share.

General considerations. There is little in the rules of nomenclature that applies to the names of taxa at these levels. The general employment of names of taxa above the family group will be described briefly here. The technical details will be discussed in Part VI.

All of these names consist of a single word, such as Animalia, Arthropoda, Coleoptera. They are generally taken directly from Latin or latinized Greek and are treated as plural Latin nouns. They are one of the kinds of scientific names or Latin names applied to animal groups. In general these names are proposed in the publications in the same manner as other names, but there are no formal requirements for availability.

The names of these groups are not usually based upon the names of included genera. For example, there is no genus of arthropod, such as *Arthropus*, upon which the name was based. Instead the names are descriptive, being based on some attribute or feature of the included organism. The name is a noun, referring in a general way to the state of possessing that attribute. The Arthropoda are the animals with jointed feet (legs). The names are always plural, because even if they actually contain only one species, they potentially contain all species which meet the definition.

Saying that the names are not based on the names of included genera is the same thing as saying that they do not have types. The Copenhagen Congress recommended the use of type genera for all these names, but the name of the genus was not necessarily to be part of the group name. For example, the type of the phylum Arthropoda might be *Musca*, a genus of flies which are arthropods. These recommendations from Copenhagen were discarded entirely in the 1961 Code, where there are no specific provisions for names at these levels.

Change of level or category. When a taxonomist finds it to be necessary to change the level of a group, from one category to a higher one (for example, subclass to class), or to a lower one (for example, order to suborder), there is to be no change in the name, unless he is using one of the uniform-ending systems. The level adopted is referred to where necessary by using the category name; for example, "The *class* Myzostomida of earlier workers is herein raised to the phylum level and is called the phylum Myzostomida."

Synonyms. There are hundreds of cases in which groups in these higher categories have received more than one name. There is no rule for rejection of any of these synonyms, so it is up to each taxonomist to choose for himself. This is nearly always done by accepting the names used by a previous worker. Sometimes names are rejected because of inappropriateness, because part of the group is no longer included, or for some other reason. In general, it is best to follow established usage unless there is clear reason for deviating. Almost any change will arouse criticism. Unless it is carefully handled, it may also result in instability of name for that taxon. The change should be made in an appropriate publication.

Many rejected names are not really synonyms in a strict sense. They are the names for groups substantially the same but differing in inclusiveness. For example, the names Coelenterata and Cnidaria are currently used by different workers for exactly the same phylum of animals. They are not really synonyms, however, because Coelenterata did, does, or may include also the Ctenophora (Acnidaria). If one uses Coelenterata in the more-inclusive sense, then Cnidaria is not a synonym but the name for one of the included subgroups. If one uses Coelenterata in the restricted sense, then it is a synonym of Cnidaria. As priority does not apply to these names, either one can be used. The taxonomist will decide between them on any basis he chooses—priority, appropriateness, extent of prior usage, or personal preference.

Homonyms. There are many cases in which two or more names of identical spelling have been used for different groups on these higher levels. There is no rule to prevent use of these homonyms. Although it has seemed unreasonable to some taxonomists to have an order Decapoda in the Crustacea and another order Decapoda in the Cephalopoda, no real harm is done.

At present, there is no basis for abandoning a well-known name because homonymy exists. However, if it becomes necessary to choose between two synonyms in selecting a name for a taxon, it would be wise to avoid selecting one which is also a homonym.

Subdivision of a taxon. If a taxon of ordinal rank or above is divided into two taxa, the original name cannot be used for both. Some zoologists have believed that the name cannot be used for either of the segregates but must be rejected altogether. This view is not widely held, simply for the practical reason that the contents of most taxa are changing constantly, by additions or deletions, and this basis for names would force the change of names every time a change was made in contents. As this would be intolerable, involving at the present time a change in nearly all major phylum names, it is not widely employed and should not ever be used as the basis for changing names. Should a case arise involving the division of a fairly large phylum into approximately equal groups, so that there would be a real question as to which group should continue to be denoted by the original name, there might be an excuse for abandoning the name altogether. However, apparently all the cases that have appeared so far involve the removal of only a relatively small part from the main group. Here, there should be no question of abandoning the well-known name for the group.

For example, the well-known phylum name Coelenterata (for the three classes Hydrozoa, Scyphozoa, and Anthozoa) was recently changed because the Coelenterata had originally included also the Ctenophora and other groups since removed from the phylum. In place of that name was put the junior (that is, younger) synonym Cnidaria, which had usually been applied only to the three classes cited. Actually Cnidaria has not always had the same contents, so it too would not be acceptable under this system.

There is some logical basis for the arguments in favor of this basis for name choice or change, but in practice it is simply unworkable. It should not be used as reason for changing any group name or for adhering to any recent change of this kind (Cnidaria for Coelenterata, Ectoprocta for Bryozoa, etc.).

The categories. It must be remembered that, although the names of groups in these higher categories are divided into three groups, they are treated alike in nearly all respects. An order Cyclostomata in the Bryozoa is a homonym of an order Cyclostomata in the Vertebrata, but likewise a subclass Articulata in the Brachiopoda is a homonym of a class Articulata in the Echinodermata. The names thus belong in a single name-pool, where their current category assignment has little bearing on any of the problems of names.

It is sometimes said that there are certain obligatory categories. These are listed as Kingdom, Phylum, Class, Order, Family, Genus, and Species. There is some justification for this statement, because it is a widespread practice to use at least these seven in all classifications. There is not the slightest justification, however, for any implication that these represent groupings of primary importance in all groups, or that it is incumbent upon any classifier to use them all in every case.

There are phyla in which the subclass is the most indispensable category. And there are small phyla in which the family level is completely useless.

These categories, even with the super and sub forms of each, will not always be adequate to show the groupings which exist. What will work in a phylum of less than 50,000 species, like the Vertebrata, cannot be expected to be adequate to show the major groupings of a phylum of more than 800,000 species, like the Arthropoda. The one order Coleoptera consists of close to 300,000 species; suborders, superfamilies, families, and subfamilies are totally inadequate to classify this large group. Additional categories are inserted as needed.

Names of the phylum-group taxa. This group includes names for taxa at the phylum, subphylum, and all higher levels. Usually these include:

> Kingdom
> Subkingdom
> Phylum
> Subphylum

It is not uncommon to interpolate such levels as Branch, Grade, and Series between Subkingdom and Phylum. It is also possible to use Superphylum or other prefixed names.

Names of the class-group taxa. This group normally includes:

> Superclass
> Class
> Subclass

It may also include Infraclass or any level between Subphylum and Superclass.

Names of the order-group taxa. This group normally includes:

> Superorder
> Order
> Suborder

It is common in the mammals to use also Infraorder as a level below Suborder. And there would be included here also any other levels between Suborder and Superfamily.

Names of the family-group taxa. It has sometimes been said that the families are the most important and widely used of the taxa above the genus level. This is true in some groups, such as the Insecta. There are other groups, however, in which families are seldom used, where orders or even classes are the subgroups of most utility. In any case, there is nothing about the family level which sets it apart in the hierarchy, except that there are rules governing the names of the family-group taxa. Family names are of quite sufficient importance, however, to justify using them in such a way that they are stable, invariable in spelling, universally recognized, and recognizable as family names by uniform endings.

Formation, spelling, rejection, and basis on a type species are all technical aspects of family-group names that must be studied with the legal requirements of nomenclature. Here it is sufficient to point out that (1) all names of family-group taxa (superfamily, family, subfamily, tribe, and any other between family and genus) are based on the name of an included genus, (2) uniform endings are required at each level, (3) changes of name are carefully regulated and not left to the discretion of the taxonomist, and (4) if two or more families are united, the combined family takes the oldest of the names, not the one based on the oldest genus.

There is no theoretical limit to the number of levels or categories in this group. The most commonly used throughout the Animal Kingdom are family and subfamily. In such large orders as those in the Insecta, five or six levels are sometimes employed as follows:

Superfamily
Family
Subfamily
Supertribe
Tribe
Subtribe

These are treated alike so far as the Code of Nomenclature is concerned. Proposal of a name at any of these levels makes it available in all of them. However, homonymy has some misleading aspects.

If a family consists of several subfamilies, there will always be one subfamily which contains the genus which is the type of the family. This subfamily will also have this same genus as its type, and its name also will be based on the name of this genus. If the subfamily is divided into tribes, one of these will have a name based on the same genus as the name of the subfamily. Thus, every level except the top one used (family or superfamily, usually) will have one taxon with name similar to that of the next taxon above. See Fig. 13-1.

The author of a name in the family group is usually not cited. He is the person who first proposed the name for a group above the genus level, regardless of the level used or the spelling employed. The date of publication of the name is the date on which it was published as a name for a group above the genus level, regardless of the level or the

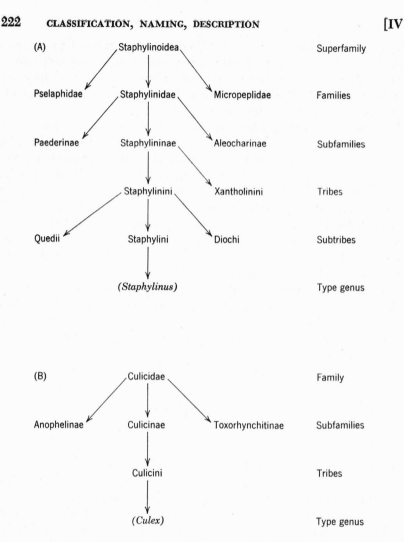

Figure 13-1. The hierarchic sequence of similar names.

ending. No one but an experienced taxonomist should undertake to propose or change a family name. There are many technical requirements which must be understood and met. The ordinary user of family names would be well-advised to follow the decisions of a recent monograph or textbook. When there is more than one name available for a taxon at this level, the non-specialist cannot go wrong in using the best-known one, regardless of what may be the outcome of a detailed nomenclatural study of these synonyms.

Homonymy. In the family group, if two writers propose names based on the same type genus, and therefore presumably of identical spelling, the more recent author is simply presumed to have been using the older name. The "homonyms" cannot apply to different things and are therefore complete synonyms as well. The younger one is simply forgotten.

The possibility of other types of homonymy occurring is remote, but the Code provides for the rejection of the younger name.

Synonymy. Although a family-group taxon formed by the union of two or more taxa takes the oldest valid family-group name among these, there are exceptions which must be worked out by a specialist, preferably a nomenclaturist.

Superfamily names. Names of superfamilies are not directly regulated as to the form of the ending. For many years entomologists have standardized this ending as *-oidea* and have urged adoption of this form in the Code. In the 1961 Code there is no ruling, but it is recommended that *-oidea* be adopted for superfamily names. Nothing is said in the Code about the *exclusive* use of this ending for this level, although to be really useful a uniform ending such as this must not be used elsewhere in the system. Unfortunately, it happens that words ending in *-oidea* have been very commonly used at the class and phylum levels in many groups of animals. Inasmuch as the endings of names at these higher levels are not regulated by the Code, it is unlikely that all these forms will soon be changed. The following are some of the better-known *-oidea* names in the phylum and class groups:

Planuloidea	Nautiloidea	Stelleroidea
Moruloidea	Ammonitoidea	Asteroidea
Stromatoporoidea	Belemnoidea	Ophiuroidea
Cestoidea	Sipunculoidea	Echinoidea
Bdelloidea	Echiuroidea	Holothurioidea
Macrodasyoidea	Trilobitoidea	Planctosphaeroidea
Chaetonotoidea	Sagittoidea	Branchiostomoidea
Priapuloidea	Cystoidea	Petromyzonoidea
Nematoidea	Blastoidea	Aphetohyoidea
Gordioidea	Crinoidea	

In spite of this usage at the class and phylum levels, the use of *-oidea* at the superfamily level is sufficiently widespread, at least in entomology, to be worth continuing. There is not usually much confusion between superfamilies and classes. Most zoologists are sufficiently familiar with the hundred or so classes to have no difficulty in this regard. There have also been some uses of this ending with the names

in the ordinal group. Some of these have recently been changed to
-*ida* or -*oida,* but some remain in use.

Family names. Family names must end in -*idae,* which is an ending
added to the stem of the name of the type genus. Determination of the
stem was formerly based on the Latin genitive, but as an increasing
number of generic names are of non-Latin origin, there are now rules
for determining the stem in all cases. (These rules are listed in an
appendix to the 1961 Code.)

Subfamily names. The names of subfamilies must end in -*inae,* which
is an ending added to the stem of the name of the type genus. These
are treated exactly like family names, except for this ending.

Supertribe names. This level in the taxonomic hierarchy is not directly recognized in the 1961 Code. Uniform endings would be desirable in any work which employed taxa at this level, but none have
been widely adopted. Indeed, the level itself is seldom used. It is
definitely available, however, if needed in any extensive group.

Tribe names. Tribes are specifically included in the family group in
the 1961 Code. Uniform endings for their names are not specified, and
various systems have been proposed. It appears that -*ini* has been more
used in entomology than any other ending, and it is principally in some
of the large families of insects that taxa at this level are employed.

Subtribe names. Subtribes are not directly mentioned in the 1961
Code. There is no generally accepted ending for these names. So long
as this is the first level of taxa above the genus, it is reasonable to
employ a simple -*i* or -*ae* ending, analogous to a plural of the generic
name. For example: *Osorius,* Osorii; *Lispinus,* Lispini; *Aleochara,*
Aleocharae. Uniform endings other than these have also been proposed.

Names of the genus-group taxa. The genus group includes only the
genus and the subgenus. These two categories are the only ones between the family group and the species. Their names are regulated in
detail by the 1961 Code, as they were by previous codes. All of the
technical aspects of formation, gender, publication, synonymy, homonymy, and genotypy are discussed in the section on Nomenclature.
Therefore only the general employment of these names will be discussed here.

The names in this group are uninominal—they consist of a single
word. This word must be a noun in the nominative singular or be
treated as if it were. Until 1961, a generic name could consist of a
hyphenated compound word, for example *Embrik-strandia.* The 1961
Code prohibits such use of the hyphen and requires that all such names
already published be changed to one word without the hyphen.

The spelling of generic names is, of course, subject to error, in both the original publication and subsequent ones. In general, the original spelling must be maintained, except in very unusual circumstances. The new Code, however, forces changes in spelling of several sorts: (1) deletion of any hyphens, (2) deletion of all diacritic marks, (3) insertion of an "e" after a German vowel from which an umlaut has been removed, (4) deletion of an apostrophe from such names as O'Toole, and (5) change of all such names as M'Coy and McCoy to maccoy. For the ordinary user of names, the best procedure is to follow the usage of a standard monograph or reference work. Even if the name used is not exactly right in every aspect, it will be understood, and that is the only function of a name.

The problem of what name is to be applied to a given genus of animals is affected by such technical matters as synonymy (the presence of more than one name for that genus), homonymy (the existence of an older name of the same spelling for some other genus), and genotypy (the identity of the species which is type of that generic name). These are technical matters, but the choice of name to be used is also affected by the presence of subgeneric divisions and differing opinions of possible union of related genera. There are thus both nomenclatural and zoological considerations involved.

The zoological aspects, for example whether a given species "belongs" in a given genus or not, are problems of classification. Identification will determine its species (X-us albus), but whether this species is correctly placed in the genus X-us is determined by comparison of its features with those of the other species in the genus X-us and those in the other genera in the same family (or tribe). Furthermore, a species correctly named as X-us albus today may become Y-us albus tomorrow simply through the published opinion of a reviser that X-us is a subgenus of Y-us. The subgeneric name may be cited, Y-us (X-us) albus, or it may be omitted (Y-us albus).

The most-used names in the zoological system are the generic names. The genera themselves are important objects of discussion, and there are places where species names are little used, as in some fossils. Furthermore, it is common to refer to a species by the generic name alone, when the specific identity is of no immediate concern. One can, for example, refer to a living specimen being studied as *Hydra*, without raising the question of what species is involved. In the varied business of taxonomy, generic names are thus used (1) to denote a particular group of species, (2) as a stand-in name for an undetermined specimen of any one of its species, (3) as part of the name of each of the included species, and (4) as the basis for the formation of all names in the family group.

The species in a genus are often so much alike that it is possible to discuss them all together under the name of the genus. *Amoeba* is spoken of as moving by means of pseudopodia. Obviously this means that the individuals of each of the species in this genus move in this manner.

The genus. There is no way to define the concept genus except to say that it is the level at which is placed the group of species whose names have their first part alike. The actual genus is the group of species. This is of course a definition arrived at from behind, because the genus cannot wait until some species are grouped together. The genus is involved from the naming of the first species, of whose name it forms the first part. Again, it is possible to say that a genus is any group of species included under one generic name by any taxonomist. This is completely subjective, but it is approximately the working definition which is the basis of most taxonomic work. The genus cannot properly be described as the next higher level above the species, because it is common and always possible to use subgenera between genus and species, and to use also sections or other informal categories.

It is sometimes said that a genus is a group of species holding certain features in common. It should be obvious that this definition applies to genera, families, orders, and all groupings. Presumably the species in one genus have *more* features in common than do all the species in the family in which that same genus is but one of several, but how much more cannot be stated in advance. There is no definition that will distinguish genera from tribes (for example). The level above the species which is to be called the genus level has become fairly well established in most groups through the influence of effective monographs. Taxonomists can only continue to use this subjective understanding which has developed over the two hundred years of the binominal system.

The names of subgenera. In non-taxonomic or non-monographic studies, the subgenus is the least used of all categories. Even when a genus has been thoroughly and satisfactorily divided into subgenera, there is no reason why anyone should refer to them unless he finds it desirable to do so. Subgenera are useful to the specialist; their use conveys some additional information about the species cited. For example:

genus	*Anopheles*
subgenera	*Stethomyia, Anopheles* s. str., *Myzomyia*, etc.
species	*Anopheles (Myzomyia) gambiae*
or	*Anopheles gambiae*

The citation of this species as *Anopheles gambiae* is entirely sufficient in nearly all circumstances. The addition of the subgeneric name in non-taxonomic literature has the appearance of pedantry. It is unnecessary and actually undesirable. It should be used only when comparison is implied between two or more species representing different subgenera within the same genus.

So far as their status as zoological names is concerned, subgeneric names are coordinate with generic names. Thus, all names in the generic group are treated together in the rules of formation, spelling, publication, synonymy, homonymy, and genotypy. But subgeneric names do differ in several ways from generic names:

1. They are used only with the generic name, when referring to a particular species (they may of course be used alone when referring to that subgenus).
2. They are always cited in parenthesis between the generic name and the second part of the binomen.
3. They must be younger than the generic name; i.e., their publication date must be subsequent to that of the name of the genus to which they are assigned.
4. They cannot be the basis of family-group names so long as they are treated as names of subgenera.
5. They may be treated or considered as generic names whenever the subgenus denoted is treated as a genus.

Binominal Latin names

Names of the species-group taxa. The taxa of the species group are the species, the subspecies, and any recognized groupings within the subspecies. Although none of the latter are now recognized under the rules of nomenclature, many of them are in the literature and must be recognized and dealt with. They include the names of varieties, races, forms, parataxa, transition forms, and aberrations.

It is not possible to go very far in studying the names of these taxa without running into technicalities of several sorts. For the technical solutions, the student must be referred to Part VI, Zoological Nomenclature. Only the general features of the employment of names will be discussed here, particularly as they are to be used in non-taxonomic literature, because employment of them in revisionary taxonomic studies requires a knowledge and understanding of the numerous legal requirements of the Code now in effect and of the codes that have been in effect in the past.

As pointed out in detail elsewhere, it is intended in this book to explain the requirements and procedures as agreed to by zoologists in the 1961 Code. In addition, it is necessary to explain the requirements in previous codes, because there is much literature that was governed by those earlier agreements. It is further deemed appropriate to point out alternative solutions to some of the problems, and even to point out when the 1961 Code is inadequate, misleading, or illogical.

Specific names and subspecific names. Specific names require many rules to assure uniform employment by workers all over the world. Only the aspects of the day-to-day use of names which are already established will be discussed here. For this purpose the pertinent aspects are: (1) how to cite the name, (2) what the name means, (3) how to spell it and why there are sometimes alternative spellings, (4) why names are sometimes changed, and (5) how to cite the rejected names. All of these refer also to subspecific names.

Citation. A species is always referred to (cited) by its specific name, the binomen or two-part name. For example, the domestic horse is referred to as *Equus caballus*. If the genus has already been clearly indicated, it may be abbreviated to its initial letter, giving *E. caballus*. It cannot be referred to simply as *caballus*, because there are other animals with *caballus* as part of their binomen. There is only one *Equus caballus*.

Inasmuch as the status of species and subspecies, and even synonyms, is usually a matter of opinion, it sometimes happens that two taxonomists use different names for the same species. Each may be using the correct name from his point of view. In this way a species may have two or more current names neither of which can be said without question to be the only correct one. When citing the name of a species in any but the most informal circumstances, it is wise to give some indication of the source of the name. This is done by appending to the name the surname of its original author and the date of the original publication. The author's name follows the specific name without intervening punctuation, and the date follows the author's name in parenthesis or set off by a comma; for example, *Staphylinus maxillosus* Linnaeus, 1758.

If the species name is not in the original form (the generic name has been changed), then the name of the author is enclosed in parentheses: *Creophilus maxillosus* (Linnaeus, 1758). The sole purpose of this convention is to call attention to the fact that if one searches for this species in the indicated original publication (Linnaeus, 1758), he will find it listed under *some other* generic name. The 1961 Code leaves this citation of the author as optional. It is elsewhere generally considered desirable to cite the author's name in all formal publications. Another convention, recognized as optional in the 1961 Code, and widely practiced, results in two authors' names, with the first in parentheses. The first is the name of the original author of the original combination; the second is the name of the writer who placed the species under the new generic name; for example, *Creophilus maxillosus* (Linnaeus) Samouelle. Still another convention is to put the original generic name after the author's name to show in which genus the species was originally described; for example, *Creophilus maxillosus* (Linnaeus) (*Staphylinus*).

Meaning of the name. A name is simply a means of referring to something. The name of a genus refers to a particular genus, a group of animals consisting of one or more species. Its name consists of a single word; it is uninominal. The name of a species refers to a particular species, a group of animals judged to be a taxonomic species. Its name consists of two words; it is binominal. It is this feature of the name of a species that gives rise to the term binominal nomenclature. In this two-part name, called a binomen, the first part denotes the genus, and the second part denotes the species within that genus. The generic part can stand alone, referring to the genus. The specific part cannot stand alone; it requires the generic name to be present before it has any meaning.

This second part of the binominal name of a species is, by itself, just a word, treated as if it were Latin. There can be many such words, of identical spelling, in nomenclature. They acquire the status of names only when combined with generic names, only when each is made into a binomen. These second words have been called trivial names and, in the 1961 Code, specific names, but they are not names in any sense. They are herein referred to as epithets or words. (For further discussion, see Chapter 26, The Names of Species.)

Spelling. There are very definite rules about the spelling of all names. There can be only one correct spelling at any one time, but the spelling may change from time to time under certain common circumstances. (The spelling of generic names was cited under Names of the Genus-Group Taxa.)

Specific or subspecific epithets may be adjectives, nouns in apposition, possessive nouns, participles, or gerunds. All of these are adjectival in nature, modifying the generic name. As in Latin grammar, the true adjectives must agree in gender with the modified generic name, as must some participles and all gerunds. (The other kinds of epithets always keep the same endings.) This agreement is generally evident in the terminal letters of the ending. Thus: *Asilus punctatus* (masculine) becomes *Dasypogon punctatum* (neuter) if it is transferred to *Dasypogon*.

The rules provide for the possibility that an epithet may have been spelled erroneously when first published. Certain types of original errors are to be corrected when discovered. This also produces acceptable spelling variations.

In addition, alternative spellings may be due to typographical errors, lapsus calamorum (slips-of-the-pens), intentional changes that would not be permitted by the rules, errors in determining gender, and so on. It is always necessary to be on the lookout for such alternative spellings, which are not to be accepted but may need to be recorded.

It should also be pointed out that rejection of a name or epithet or of a spelling form does not mean that zoologists must also reject the data published under that name. Much valuable contribution to zoology has been recorded under incorrect names; if the error is recognizable and corrected, then the recorded information is still available to science.

In general, however, subsequent writers must retain the original spelling, except for the gender endings. The detailed rules for spelling and emendation are discussed in Part VI, Zoological Nomenclature.

Change of name. Names are changed (1) because it is discovered that an error had been made earlier, so the wrong name is in use; (2) because increased knowledge of the animals shows that two supposed species are actually the same, or a supposed single species is actually two or more; (3) because a forgotten earlier name is discovered; (4) because it becomes known that the species is in the wrong genus; (5) because it is discovered that there exists an older name of identical spelling, or (6) for certain more technical reasons dealt with in the chapters on nomenclature.

Necessary changes must be accepted by all users of the names. Without this, knowledge of the animals would stagnate and names would lose their ability to denote one species effectively. But it is not always necessary to abandon completely a name which has been suppressed. In a few cases one name has become so widely known among non-taxonomist biologists that we are justified in continuing to use the suppressed name in non-taxonomic situations. For example, *Schistosoma mansoni* is an important parasite of man which is the cause of a disease called schistosomiasis. When taxonomists decided that the name had to be changed to *Bilharzia mansoni*, a tremendous medical literature was affected. Many persons felt it necessary to change also the name of the disease, to bilharziasis. It gradually became clear that no one would misunderstand *Schistosoma mansoni* in medical use, whereas *Bilharzia mansoni* did not add to the usefulness of the work done on the species and the disease. Many persons continued to use *S. mansoni* in non-taxonomic literature.

Citation of rejected names. Whenever any name has been changed since its original publication, there will be two names or name forms for that species; for example, *Staphylinus maxillosus* Linnaeus and *Creophilus maxillosus* (Linnaeus). These two forms are synonyms, regardless of which is the correct name. The citation of synonyms is usually necessary only in revisionary works, classifications, and other formal taxonomic studies. The ordinary users of names are generally not even aware of the existence of synonyms.

There are several methods for indicating names that are not accepted:

1. The generic name of a previous combination may be cited in parenthesis after the accepted name and its author.

Creophilus maxillosus (Linnaeus, 1758) (*Staphylinus*)

This tells the reader that the combination *Staphylinus maxillosus* represents the same species as *Creophilus maxillosus.*

2. The rejected synonym may be cited in parenthesis after the accepted form with an equal sign to show that it is equivalent although rejected.

Creophilus maxillosus (Linn.) (= *C. villosus* Grav.)

3. In a formal presentation, the synonyms may be listed beneath the accepted names, on separate lines.

Creophilus maxillosus (Linnaeus, 1758) Samouelle, 1819
   Staphylinus maxillosus Linnaeus, 1758
   Staphylinus villosus Gravenhorst, 1802
   Creophilus villosus (Gravenhorst, 1802) Laporte, 1835
   Creophilus fasciatus Laporte, 1835

Pronunciation of names

The pronunciation of scientific *terms* derived from Latin or Greek words (or both) is almost hopelessly confused by differing interpretations of the ancient pronunciations and the fact that different systems of pronunciation have been taught in different parts of the world, and at different times.

When it comes to *zoological names,* the situation is not hopeless, although the same factors lead to some substantial difficulties. It would be possible to set up rigid and detailed rules for pronouncing zoological names, although these would inevitably conflict with the usage of some zoologists. No such set of rules is available, and, in its absence, students can only imitate the pronunciations they hear. This is an indecisive and incomplete method. Some of the features that might be part of a system of rules can be cited here, in order to give the student some basis for working out consistent pronunciations.

Pronunciation depends upon at least five factors: (1) combination of letters into diphthongs with a single sound, (2) syllabification, (3) length of the vowels, (4) accent, and (5) the actual sound given to the letters and combinations. Some discussion or tabulation of each of these follows.

Diphthongs. In Latin the diphthongs are *ae, au, ei, eu,* and *oe.* They are pronounced as a single sound (see table, below). Whenever these letters stand together in this order they are diphthongs. (Note: names from non-Latin sources may contain these letter pairs as distinct letters but very seldom do so.)

Syllables. A Latin name has as many syllables as it has vowels and diphthongs, including final ones.[1] The final syllable is called the ultima; the one before the ultima is the penult; the one before the penult is the antepenult.

Dro so phil a	*-so-*	antepenult
	-phil-	penult
	-a	ultima

Vowel length. All rules for pronunciation of terms or names presuppose that the student knows whether the vowels are long or short. Of course, in Greek this was sometimes easy, as there were two letters for each, one short and one long (*eta* and *epsilon, omicron* and *omega*). To determine the length of the Greek vowels from a dictionary requires a knowledge of the language. The following rules will sometimes help determine the length of the vowel in Latin names:

1. All diphthongs are long.
2. Final vowels are short, except for the final *i* of tribe names (Staphylininī), which is long, as *ee.*
3. Penult vowels are short when
   (a) derived from short Greek vowels;
   (b) followed by two consonants (except any of these: *b*, hard *c, d, g, k, p, q, t, ch, ph, th,* followed by *l* or *r*);
   (c) in family names;
   (d) in a Greek plural such as Echinodermata.
4. Penult vowels are long in
   (a) subfamily, tribe, subtribe names;

---

[1] The vowels are *a, e, i, o, u,* and *y*; the diphthongs are *ae, au, ei, eu,* and *oe.*

(b) names derived from Latin past participles and ending in *-ata, -atus,* or *-atum;* such as plicāta;

(c) names derived from Latin adjectives and ending in *-alis;*

(d) names ending in *-ina, -ica, -ana, -anus, -anum, -ura, -odes, -soma;*

(e) words in which it is *u* (except when followed by *l*).

5. An antepenult vowel is long when

(a) it is followed by another vowel (in the penult);

(b) it is *a, e, o,* or *u,* followed by a single consonant and two vowels, the first of which is *e, i,* or *y;*

(c) it is *u,* followed by a single consonant.

6. Antepenult vowels are short except as in (5) above.

The length of the Latin vowels is: short *a* as in "fat," long *a* as in "father"; short *e* as in "met," long *e* as in "prey"; short *i* as in "pit," long *i* as in "machine"; short *o* as in "con," long *o* as in "hole"; short *u* as in "full," long *u* as in "rule."

Accent. The last syllable is never accented; if the penult is long, it is accented; if the penult is short, the accent falls on the antepenult; but there are many exceptions.

Pin′ us	pon der o′ sa		For am in i′ fer a
Mam mal′ i a		Lum′ bri cus	ter res′ tris

1. There is no accent when the name consists of a single syllable.

2. Accent the penult when

(a) the name contains only two syllables;

(b) it contains a diphthong;

(c) the vowel in the penult is followed by *x* or *z;*

(d) the vowel in the penult is long (see Vowel Length, above);

(e) the name is a subfamily or tribe name;

(f) the vowel in the penult is short but is followed by two consonants (except for the list in 3(b), below).

3. Accent the antepenult when

(a) the vowel of the penult is *u* (except when followed by *l*);

(b) the vowel of the penult is followed by one of these, *b,* hard *c, d, g, k, p, q, t, ch, ph, th,* followed by *l* or *r;*

(c) the vowel of the antepenult is long and that of the penult is not;

(d) the vowel is *a, e, o,* or *u,* followed by a single consonant and two vowels, the first of which is *e, i,* or *y;*

(e) the vowel is *u* and followed by a single consonant;

(f) the vowel is a diphthong;

(g) the vowel is short but that of the penult is also short.

Letter pronunciation. Although there are several systems of pronunciation of Latin letters, it is desirable to give some help to those who do not have a mastery of Latin. A choice had to be made. The following table gives one of the systems, largely consistent with English pronunciation of terms of Latin origin.

This table is for biological *names* only. It should not be used to determine pronunciation of English terms.

a   initial *a* long, as in "Aves"
     final *a* has sound of *ah* or *uh*
     when short, as in "cat"
     before *r* as *ah*
     in Greek plurals, as *ah*. Ex.: Echinodermata.
     otherwise long, as in "gate"

ae   a Latin diphthong, has sound of *ee*. Ex.: caenogenesis.
     if followed by two or more consonants, often shortened. Ex.: septum (saeptum).
     often coalesced to *e*, as in "hemal" (haemal)

ai   as *i* in "mine"
au   as in "fraught." Ex.: caudal, glaucoma.
aw   as *au* in "caudal"
ay   as long *a*

b   as in English

c   soft before *e, i, y, ae, oe*, as in "see." Ex.: coelom, cyanide.
     after an accented syllable, before an *i* followed by another vowel, it is pronounced *sh*. Ex.: social, species.
     hard before all other sounds. Ex.: carpal, costa, cunicular.
     double *c* follows above rules, first *c* is hard, second one hard except before *e, i, y, ae, oe*. Ex.: coccyx, flaccid, vaccine (hard and soft); coccus (both hard).

ch   (from the Greek chi) is a double consonant that is generally hard, as *k*. Ex.: chromosome.
     (from English) would be either *k* or *ch* as in "chair," depending on the source. Ex.: chamberlini (*ch*); reichenbacheri (*k,k*).

ck   as *k*

d   as in English "did"

e   if from Greek eta, long as in "prey"
     if from Greek epsilon, short as in "met"
     final *es* as in "ease." Ex.: crises, phalanges.
     if derived from Greek *ae* or *oe* then long, as *ee* in "knee." Ex.: ameba, hemoglobin.

ea   as in English "beat"
eau   as long *o*
ee   as in "deed"
ei   as *i* in "pine"

eo   as in "leopard"
eu   as in "neuter." Ex.: neurology, pharmaceutical.
ew   as in "new"
ey   as in "obey"

f   as in "clef" or "few"

g   soft before *e, i, y, ae, oe* as *j* in "joy." Ex.: gingivitis, genus, gyroscope.
    hard before other sounds as in "gun." Ex.: guttural, gonad, gland,
      grandis.
gh   as in "ghetto"
gn   as *ny* in "canyon"
gu   as in "guest"

h   always pronounced when initial; often in combination with *c, g, n, p,*
      r, s, t, w

i   final *i* is long, as "eye." Ex.: *smithi*, stimuli, radii.
    otherwise it is short as in "pit" or long as in "machine"
    with *es* in the plural (following *t* after an unaccented syllable), *ies* is
      like a long *e*, as "tease." Ex.: cavities, verities.
ie   as *ee* in "knee"
ieu   as in "lieu"
iew   as in "view"
io   as in "question"

j (consonant) as in "jug"

k   as *k* or *ck* in "kick"

l   single or double, always as in "lull"
ll   as in "yellow"

m   as in "mum"

n   as in "nun"
ng   as in "English"

o   long in an accented final syllable. Ex.: coelom.
    usually short in an unaccented syllable
    usually long in the final syllable of first part of a compound word.
      Ex.: Corrodentia.
    short in an accented antepenult. Ex.: Loligo.
    if from the Greek omega, always long as in "go"
    if from the Greek omicron, always short as in "optical"
oa   as in "roar"
oe   a Latin diphthong, has sound of *ee* in "seen." Ex.: amoeba, foetid,
      coelom.
    often coalesced to *e*, as esophagus (oesophagus)

oi   as *oy* in "boy"
oo   as in "book," "door," "zoom," "blood"
ou   as in "out"
ow   as in "crow"
oy   as in "boy"

p   as in "papa"
ph   from the Greek phi, as *f*. Ex.: pharynx, phloem, phylogeny, Phoronida.
ps   from the Greek psi, as initial letter of a Greek root, has sound of *s* in "ess." Ex.: psychic, pseudopod, Psocoptera.
     in middle of word or root, is not from the Greek and has sounds of separate letters *p* and *s* in "ipso," generally in separate syllables

qu   as *kw* sound in "quick"

r   as in "run"
rh   as in "rhetoric," "rheumatism"

s   as in "song"
sc   as in "science" or as separate letters
sch   as *sh* in "ship"
sh   as in "ship"
ss   as in "miss"

t   if preceded by an accented syllable, *t* before an *i* followed by another vowel has sound of *sh* in "ship." Ex.: ratio, but verities (*-tease*).
th   as in "thin"

u   if accented, generally long as in "unison," but sometimes short as in "uncus"
     if not accented, generally short as in "serum" but sometimes long as in "uletic"
ua   as in "suave"
ue   as *ee* in "feel"
ui   as in "built"
uoy   as in "buoy"
uy   as in "buy"

v   as in "valve"

w (vowel) as in *ow* and *ew*
w (consonant) as in "water"
wh   as in "when"

x   if from the Greek chi as an initial of a Greek root, has sound of *z*. Ex.: xiphoid, xeric, pseudoxeric.
     if from the Greek chi, not as an initial, has sound as in "wax." Ex.: oxygen, taxonomy, hexaphene.

y (vowel) in combination with other vowels, as in "Zoysia"; or alone, as in
"Zephyranthes"

     short before or after *s* as in "syringe" or "physics"

     otherwise long as in "zygote"

y (consonant) as an initial, as in English "yolk," "yucca"

z   as in "zany" or "Zululand"

General pronouncing rules. The following rules cover some of the most frequent problems of pronunciation.

Rule I. Determine the source of the name—the language from which it came and the type of word on which it was based.

A. If it is a Latin word or a latinized Greek word, apply the rules of Latin pronunciation.
   1. It has as many syllables as it has vowels and diphthongs.
   2. Final vowels are always pronounced.
   3. A syllable is long if it contains (a) a long vowel, (b) a diphthong, or, (c) a short vowel followed by two consonants or by either *x* or *z*.
   4. The last syllable (the ultima) is never accented.
   5. If the next-to-last syllable (the penult) is long, it is accented.
   6. Otherwise the accent falls on the preceding syllable (the antepenult).
B. If it is a latinized word from some modern language, it is pronounced as if Latin and may be spelled only with letters of the neo-Latin alphabet.[2]
C. If it is an arbitrary combination of letters, it is to be pronounced as if Latin.
D. If it is based on a geographical name, the root should be given the local pronunciation and the Latin termination given the Latin pronunciation.
E. If it is based on a proper name (name of a person), the proper name portion is to be pronounced as the person's name is pronounced, with the ending (*-i, -ae, -orum, -arum* in trivial names, *-ius, -ia, -ium, -us, -a, -um* in generic names) as in Latin.

Rule II. Pronounce the root elements clearly, distinguishing all syllables (one for each vowel or diphthong—a, e, i, o, u, y and ae, au, ei, eu, oe). The purpose of pronunciation is not to be sonorous but to be understood; clarity may be promoted by enunciating the simple components of the names so that they will be heard as distinct etymological entities. (After R. W. Brown, 1954.)

[2] The same as the English alphabet—the Latin alphabet plus the letters *j, k, w,* and *y.*

Rule III. Pronounce correctly if you can, but do not correct the pro-nunciation of your colleagues. If they say AH' care instead of A' sir for Acer, go along with them in conversation.

There are a few books from which the pronunciation of some zoological names can be determined. These include:

Century Dictionary, 6 vols. New York; Century Co. 1889–1891. (The most complete of all pronouncing guides.)

Jaeger, E. C. *The Biologists' Handbook of Pronunciations.* Springfield, Illinois; C. C Thomas. 1960. (A convenient handbook but one apparently inconsistent. Only a comparatively few zoological names are included. The basis of the system is not clearly stated.)

Nybakken, O. E. *Greek and Latin in Scientific Terminology.* Ames, Iowa; Iowa State College Press. 1959. (A recent book of unusual nature, including a short chapter on pronunciation of terms.)

Woods, R. S. *The Naturalist's Lexicon* . . . Pasadena, California; Abbey Garden Press. 1944. (In general restricted to Latin and Greek words before they are made into names.)

Textbooks in several zoological fields, in which the pronunciation of included names is indicated. Such as: Borror, D. J. and DeLong, D. M. *An Introduction to the Study of Insects.* New York; Rinehart and Co. 1954.

Summary

The system of classification would be ineffective unless there were some means of referring to the groups produced. An elaborate system for naming the groups has developed alongside the classifications, being almost indistinguishable from them. The naming system could have taken a variety of forms, but the technical names employed are unique single, double, or triple names tied irrevocably to the taxa of the classification by means of their types.

The names are applied under several conventions agreed to by zoologists at international meetings, but these conventions are not the same at all levels in the classification. Somewhat different sets of rules apply to species, genera, families, and the levels from order to kingdom. The application of these conventions, codified in sets of rules, is the most detailed and difficult aspect of taxonomy. It occupies nearly a third of the text of this book.

For all the technical aspects of the use of names see Part VI, Zoological Nomenclature.

CHAPTER 14

The Use of Literature

Taxonomy is a highly historical subject. The taxonomic work of the past is recorded in the literature. This record of what has been discovered is of primary importance in most taxonomic work. When new facts are discovered about any kind of animal or any group of animals, these will become part of the science only if they are recorded in the literature. Thus, literature enters into taxonomy very substantially, as the record of past studies and as the means of adding to the available knowledge of kinds of animals.

The main problems connected with the recovery of what has already been published are (1) finding the appropriate publications, and (2) interpreting their contents.

Recovery of data from the literature

Through the name of the taxon. In setting out to find the publications that contain information about any group of animals or plants, a taxonomist is faced with an incredibly large literature to be searched. There are probably ten thousand periodicals in which information on animals *may* occur, as well as tens of thousands of books and other separate publications of all varieties. No taxonomic search of this literature could possibly be effective without the help of indexes, bibliographies, and other reference works. The number of the available reference works is not very large, but the variety is considerable.

In general, the data to be covered are the comparative data about that group or kind. This means the descriptive information in the original publication plus everything that is recorded by subsequent writers, whether structural, behavioral, biochemical, or of any other aspect. In some cases, a recent summary or monograph will serve our purpose, but in a serious taxonomic study it will be necessary to see every publication, so far as possible. Of course, there are a few species that have such a large literature that it is impracticable to examine every item. Here it is necessary to examine only certain items, but the choice of these is never easy.

It is possible to discuss the search for the literature on species separately from that on groups of species. The problems are sometimes different, and the literature on a group, such as a family, or an order, actually includes all the literature on every one of the included subgroups and species. It would also be possible to take each major group of animals separately, but this would result in much duplication. The nature of the literature seems to make it more practical to approach it from the viewpoint of the breadth of coverage in the Animal Kingdom.

Groups above the genus level. There is no single reference work that covers all animals at all levels in the hierarchy. None of the major reference works dealing with genera or groups at higher levels for animals as a whole gives any help with the species in those groups.

The nearest thing to a universal primary reference is a work almost completely unknown to taxonomists and never listed as a source book.

The Century Dictionary, 6 volumes. New York; The Century Co. 1889–1891.

This dictionary lists thousands of zoological names, including not only names of common genera but all the known names for more-inclusive-groups as well. It is the only real source for identification of names of families, orders, classes, etc., for the kingdom as a whole and during the early periods of taxonomy.

All other major zoological indexes refer only, or principally, to generic or specific names. Recent lists of zoological source books (for example, Smith, 1962) do not even refer to the problem of groups above the generic level. One recent book lists most of the names of the groups above the family level, so far as they have been used in the major classifications of animals within the last half century; one encyclopedia includes a review of various classifications that cites a great many names of taxa at these levels; and one recent dictionary lists the commoner names at all levels:

Classification of the Animal Kingdom, by R. E. Blackwelder. Carbondale, Illinois; Southern Illinois University Press. 1963.
Encyclopaedia Britannica, 11th ed. (1911–1912) or 12th ed. (1921–1922).
Collegiate Dictionary of Zoology, by R. W. Pennak. New York; Ronald Press. 1964.

Within certain groups, especially the better-known ones, there are sometimes lists of the taxa in the levels above genus. These are incidental to some other purpose and do not usually show up in the title of the work. The following are examples:

Protozoa: Copeland, *Classification of Lower Organisms.*
        Foraminifera: *Treatise of Invertebrate Paleontology* (Vol. C).
Porifera: *Treatise* (E).
Coelenterata: *Treatise* (F).
Platyhelminthes: Trematoda: Dawes, *The Trematoda.*
Bryozoa: *Treatise* (G).
Mollusca: *Encyclopaedia Britannica,* 11th ed.
Arthropoda: Trilobitomorpha: *Treatise* (O).
        Chelicerata: *Treatise* (P).
        Acarina: Baker and Wharton, *Acarology.*
        Araneida: *Encyclopaedia Britannica,* 11th ed.
        Ostracoda: *Treatise* (Q).
        Diplopoda: Chamberlin and Hoffman, *Checklist of the Millipeds of North America.*
        Insecta: Brues, Melander, and Carpenter, *Classification of Insects.*
Pogonophora: Ivanov, *The Pogonophora.*
Echinodermata: *Encyclopaedia Britannica,* 11th ed. Hyman, *The Invertebrates,* volume IV.
Vertebrata: Romer, *Vertebrate Paleontology.*
        Pisces: Berg, *Classification of Fishes Both Recent and Fossil.*
        Mammalia: Simpson, *Classification of Mammals.*

Of course, the major catalogs of various groups usually deal with the names of the included subgroups, as well as with the genera and species. Some of these are listed in a later paragraph.

Groups at the family level. There is a peculiarity of zoological nomenclature which puts family names in a unique position. All names of families (and other taxa in the family group—superfamilies, subfamilies, tribes, subtribes, etc.) must be formed by adding a specified ending to a generic name. This means that all such names can be identified to some extent in the nomenclators of generic names. For example, Discinidae must be based on a genus name whose stem is Discin-. Neave's *Nomenclator Zoologicus* shows that this is *Discina,* a genus of Brachiopoda.

There are two considerable difficulties, however. First, there may be several generic names that have the same stem. For example, Tachinidae, as an unknown family name, would be based on the stem Tachin-, but this could be derived from *Tachina* (Diptera) or *Tachinus* (Coleoptera). There is no way to tell which, except by further investigation of the special literature of these two groups. Either one could easily be missed in consulting the nomenclator, because they are separated by many other names in the list.

The second difficulty is that there may be more than one genus of the same spelling. For example: Diprionidae could be based on *Diprion* a coelenterate or *Diprion* an insect. In this case one of the generic names must be older (the insect), and it is the only one on which a family-group name can legally be based. Therefore, a taxonomist might feel safe in assuming that Diprionidae referred to the insect. There is nothing, however, to prevent the introduction of a family name based on the coelenterate genus, even though the generic name is an invalid homonym and the family name would be unacceptable under the Règles. In the nineteenth-century literature especially, before the Règles were adopted, this name might have been used.

Groups at the generic level. At the generic level, there are a wide variety of reference works. From them can be determined the place of any genus in the classification, the original publication of any generic name, and often the list of included species.

At various times during the past one hundred and thirty years, major works have been published, listing all the generic and subgeneric names published up to a specified date. These works are known as nomenclators. Several were intended to cover only a short span of years, but others were supposed to be complete for all time (from 1758) and to include all the names in previous nomenclators. It is thus commonly supposed that the latest nomenclators are complete and are thus the only ones that need to be consulted. For routine identification of generic names, this belief will cause little difficulty. However, in one major study of generic names it was recently found that no one of the "complete" nomenclators listed all the names from all its predecessors. If this is true for one group, it may be true for others, so it must be concluded that, in any exhaustive study of names, *all* the nomenclators must be consulted.

It is not hard to understand how this has come about. Each nomenclator—this word refers to the compiler as well as to the compiled book—adopted rules for the inclusion and exclusion of names, these rules differing in different works. Changes in the rules of nomenclature affect the acceptability of some old names. Together these two factors may make a nomenclator incomplete in the eyes of later workers.

The ten major nomenclators are listed here in the order of their approximate dates of coverage:

[1758–1800] Sherborn, C. D. 1902. *Index Animalium,* sect. 1. Cantabrigiae.

[1758–1842] Agassiz, L. 1846. *Nomenclatoris Zoologici. Index Universalis.* 12 fasc. Soloduri. (393 pp.)

[1758–1842] Agassiz, L. 1848. *Nomenclatoris Zoologici. Index Universalis.* Soloduri. (1135 pp.)

[1758–1873] Marschall, A. 1873. *Nomenclator Zoologicus.* Vindobonae.

[1758–1882] Scudder, S. H. 1882. *Nomenclator Zoologicus. United States Nat. Mus. Bull.* 19, 376 + 340 pp. Washington, D.C.

[1758–1926] Schulze, F. E. et al. 1926–1954. *Nomenclator Animalium Generum et Subgenerum,* 25 Lief., 5 vols. Berlin.

[1758–1935] Neave, S. A. 1939–1940. *Nomenclator Zoologicus . . . ,* 4 vol. London. (With Addendum in vol. 4)

[1801–1850] Sherborn, C. D. 1922–1933. *Index Animalium,* sect. 2, 33 pts. London. (Poche, F. 1938. Supplement zu C. D. Sherborn Index Animalium. *Festschr. 60. Geburtstage E. Strand,* vol. 5, pp. 477–615.)

[1891–1900] Waterhouse, C. O. 1902. *Index Zoologicus,* 421 pp. London.

[1901–1910] Waterhouse, C. O. 1912. *Index Zoologicus,* No. II, 324 pp. London.

[1936–1945] Neave, S. A. 1950. *Nomenclator Zoologicus . . . ,* Suppl. vol. London. (With Addendum)

In using any reference book, the nature of the listing and the mechanics used by the compiler *must* be understood or ascertained. For example, in the Neave *Nomenclator* listed above there are four volumes in alphabetical order and a fifth or supplementary volume. Obviously both lists must always be consulted. Not so obvious is the fact that both volume 4 and the supplementary volume 5 have small appendices at the back. Both of these must also be examined. Thus, a reference to Neave may involve checking in four lists.

In addition to nomenclators, or for the period since the publication of the last one, there are annual listings of new generic names in various forms, but only the first three listed are currently useful as a supplement to a recent nomenclator.

Zoological Record, 1864–
Biological Abstracts, 1928–
Archiv für Naturgeschichte, 1832– (Now as *Zeitschrift für Wissenschaftliche Zoologie,* Abteilung B.)
Concilium Bibliographicum (cards), 1896–1940.
International Catalogue of Scientific Literature. N. Zoology. 1904–1916.

If a generic name cannot be located in any nomenclator, the *Zoological Record,* or the most recent years of *Biological Abstracts,* the searcher is at a stalemate. If the name is not very recent, it is probably a *nomen nudum*—an invalid name, not properly published.

As such it would have been omitted from most listings. This is no guarantee that it will not prove to be acceptable, when studied under later conditions. Should a taxonomist desire to find out about such an invalid name, it can be run down only in the literature of its field, by meticulous search. If the name is probably recent, then it may be too recent to be in the most recent list. Search of the recent volumes of the most likely periodicals might turn it up, but there is no direct way to find it except by waiting for its listing. It is sometimes possible to consult a specialist on the group who keeps an up-to-date card catalog from reprints sent to him by various authors, but even such a catalog is seldom more than fragmentary until the next issue of the *Zoological Record* is out. (And this assumes that we already know the group in which the name was proposed, which may not be the case.)

Groups at the species level. Nomenclators for the names of species are almost nonexistent. There are none that cover all animals for the entire period from 1758 on, and all lists of species are to some extent lists of genera. For specific names published before 1800, there is one nomenclator; for specific names published between 1800 and 1850, there is a second:

[1758–1800] Sherborn, C. D. 1902. *Index Animalium* . . . , Sectio prima. . . . Cantabrigiae. 1195 pp.
[1801–1850] Sherborn, C. D. 1922–1933. *Index Animalium* . . . , Sectio secunda. . . . London. 28 pts., 7,056 pp.
Poche, F. 1938. Supplement zu C. D. Sherborn Index Animalium. *Festschr. 60. Geburtstage E. Strand,* vol. 5, pp. 477–615.

A gap of three years separates Sherborn's nomenclator from the first volume of the *Zoological Record,* but part of this is filled in by the early volumes. However, the first dozen or so of the *Records* are much less complete than later volumes, with many minor papers omitted and much detail lacking. There is no way to fill this gap except by extensive search in the literature of that period, or consultation of the catalogs of others who have so searched.

The most complete annual listing of new names is the *Zoological Record.* Any omissions from one volume are listed in later volumes as discovered. There is no guarantee of completeness, however, and for a complete study of the species it is necessary to search the literature separately, especially all the papers referred to in the revisionary works cited in the *Zoological Record.* There are no indexes to specific names, and generic transfers, which change half of the specific names, are listed only in later volumes and then only when they are clearly cited as new in the publication.

Aside from the nomenclators and the *Zoological Record,* species can be searched out only in the special literature of the group involved. The best sources here are major catalogs, worldwide lists of species and groups. The total number of worldwide catalogs published is not large, if we restrict ourselves to groups of ordinal rank or higher. They are almost restricted to the larger groups such as the Insecta.

Claassen, P. W. 1940. A Catalog of the Plecoptera of the World. *Cornell Univ. Agr. Exp. Sta. Memoir* 232, pp. 1–235.

von Dalla Torre, K. W. (also as C. G. Dalla Torre). 1892–1902. *Catalogus Hymenopterorum.* Leipzig. 10 volumes.

Ferris, G. F. 1916. Catalog and Hosts of the Anoplura. *Proc. California Acad. Sci.,* vol. 6, No. 6, pp. 129–213.

Gemminger, M. and Harold, E. 1868–1876. *Catalogus Coleopterorum.* . . . Monachii. 12 volumes.

Günther, A. C. L. G. 1859–1870. *Catalogue of the Fishes in the British Museum.* London, British Museum (Natural History), 7 volumes.

Hedicke, H. 1935– . *Hymenopterorum Catalogus.* 's-Gravenhage; Dr. W. Junk.

Hopkins, G. H. C., and Clay, T. 1952. *A Check List of the Genera and Species of Mallophaga.* London; British Museum (Natural History).

Horvath, G. et al. 1927– . *General Catalogue of the Hemiptera.* Northampton, Massachusetts; Smith College.

Kertész, K. 1902–1910. *Catalogus Dipterorum.* Budapest; Museum Nationale Hungaricum. 7 volumes.

Kirby, W. F. 1904–1910. *A Synonymic Catalog of Orthoptera.* London; British Museum (Natural History). 3 parts.

Kirkaldy, G. W. 1909– . *Catalogue of the Hemiptera (Heteroptera).* Berlin; Felix Dames.

Lethierry, L. F. and Severin, G. 1893–1896. *Catalogue Général des Hémiptères.* Brussels; Musée Royal d'Histoire Naturelle de Belgique. 3 volumes.

Peters, J. L. 1931– . *Checklist of the Birds of the World.* Cambridge, Massachusetts; Harvard University Press. 6 volumes and continuation in progress.

Quenstedt, W. 1913–1935. *Fossilium Catalogus.* Berlin; W. Junk. 70 parts.

Schenkling, S. 1910–1941. *Coleopterorum Catalogus.* Berlin, W. Junk, 171 parts, 31 volumes. Also supplements.

Snyder, T. E. 1949. Catalogue of the Termites of the World. *Smithsonian Misc. Coll.,* vol. 112, pp. 1–490.

Strand, E. 1911–1935. *Lepidopterorum Catalogus.* Berlin; W. Junk. 69 parts.

Through the name of the author. If the identity of the author and the group of animals on which he published are known, some of the following steps can be bypassed.

Directories. When the identity of the author and his field of specialization are not known, the most general sources must be consulted
before one can go to the specialized literature in which his publications will be found. First among these are directories. Frequently
listed here are such very general reference works as *Who's Who,*
American Men of Science, and *Who Knows and What,* with the counterparts of these in other parts of the world. In the situations encountered by taxonomists, these books are almost never of much help;
they are exclusively current, whereas the literature of taxonomy is
largely historical; and they are always nationally or regionally limited,
whereas an entirely unknown author can be from any part of the
world. There are no directories of all taxonomists of all groups for
the entire period from 1758 on.

One *world directory of taxonomists* now living or recently deceased
covers those working on all groups of animals:

Directory of Zoological Taxonomists of the World, by R. E. Blackwelder
    and R. M. Blackwelder. Carbondale, Illinois; Southern Illinois Univer-
    sity Press. 1961.

Several *directories of zoologists of the world* have been published,
but the requirements for listing have generally excluded some taxonomists. The information given on taxonomic specialties is usually scanty.
Examples are:

Index des Zoologistes. Paris; Union Internationale des Sciences Biologiques.
    429 pp. 1953. (Alphabetically, by country, and by groups of interest.)
Liste des Paléontologists du Monde. Directory of the Paleontologists. Paris;
    Union Paleontologique Internationale. 237 pp. 1961.
Minerva: Jahrbuch der Gelehrten Welt. 1891–1938. (The staffs of univer-
    sities, libraries, museums, and learned societies throughout the world.)
The World of Learning 1961–62, 12th ed. London; Europa Publications.
    1359 pp. 1962.
*Zoologisches Adressbuch. Namen und Adressen der Lebenden Zoologen,
    Anatomen, Physiologen und Zoopalaeontologen.* . . . Berlin; R. Fried-
    lander und Sohn. 740 pp. 1895.

Several regional directories have been published. The older ones
are very difficult to obtain. Sometimes there is a considerable amount
of information on taxonomic interests.

Guias de Naturalistas Sudamericanos, by E. Martinez Fontes and J. J.
    Parodiz. Buenos Aires; by the authors. 138 pp. 1949.
Systematic Botanists and Zoologists Working in Africa. London; Commis-
    sion for Technical Co-operation in Africa South of the Sahara
    (C.C.T.A.). 35 pp. 1954.

A number of directories covering the zoologists or *taxonomists of one country* have been published:

List of Japanese Zoologists Partaking Systematic Zoology, by T. Uchida. *Journ. Fac. Sci. Hokkaido Univ.*, ser. VI, (Zool.), vol. 12, no. 1–2, suppl. pp. 1–33. 1954.

Directory of Entomologists in Canada. Montreal; Tenth International Congress of Entomology. 29 pp. 1956.

The Naturalists' Directory, by M. E. Cassino. Boxford, Massachusetts; M. E. Cassino. 38th ed. 1958. (1st ed. 1878).

List of Japanese Palaeontologists, ed. by Nat. Comm. for Palaeontology. Tokyo; Science Council of Japan. 15 pp. 1958.

Various organizations and individuals have assembled lists of the *specialists in certain fields.* Many of these are mimeographed and difficult to obtain in libraries, but specialists often have personal copies. For example:

List of Members. Society of European Nematologists. 31 pp. mimeographed. 1960.

List of Members. Society of Systematic Zoology. (1950, In *News Letter No. 2*; 1951, suppl. to *News Letter No. 5*; 1959, Carbondale, Illinois.)

Directory of Stanford Mineral Scientists. 1962 ed. 265 pp.

List of American Writers on Recent Conchology. With the Titles of Their Memoirs and Dates of Publication, by G. W. Tryon, Jr. New York; Ballière Brothers. 68 pp. 1861.

Arachnologists and Curators of Arachnid Collections, by H. W. Levi. 4 pp. mimeographed. 1958. (Does not include acarologists.)

List of Members. The Lepidopterists' Society. August 1964 (and earlier editions), by J. C. Downey.

Second Preliminary List of the World's Rotifer Students, by J. J. Gallagher. 10 pp. mimeographed. 1959.

Membership List of the Paleontological Society. 1957.

1958 Directory of Conchologists. Los Angeles; John Q. Burch, 50 pp. mimeographed. 1958.

Entomologen Adressbuch, by A. Hoffmann. Wien, 3rd ed. 1930.

Acarologists of the World. A Directory, by J. M. Brennan. Hamilton, Montana; Rocky Mountain Laboratory. 41 pp. 1959.

List of Copepodologists, by R. U. Gooding. Boston. 39 pp. mimeographed. July 1964.

An International Directory of Oceanographers, 4th ed., comp. by R. C. Vetter. Washington, D.C.; National Academy of Sciences/National Research Council. 1964. (Lists many taxonomists)

Directory of Hydrobiological Laboratories and Personnel in North America, ed. by R. W. Hiatt. Honolulu, Hawaii; University of Hawaii Press. 1954.

A few bibliographies of large size, on broad aspects including many groups of animals, automatically contain extensive *lists of the publications* of many taxonomists:

Bibliography and Catalogue of the Fossil Vertebrata of North America, by O. P. Hay. *United States Geol. Surv. Bull. No.* 179, 868 pp. 1902. (Up to 1900.)

Bibliography of Fossil Vertebrates 1928–1933, by C. L. Camp and V. L. Vanderhoof. *Geol. Soc. America, Spec. Pap.* 27, 503 pp. 1940. *Same* 1934–1938, by C. L. Camp et al. *Spec. Pap.* 42, 663 pp. 1942. *Same* 1939–1943, by C. L. Camp et al. *Geol. Soc. America Memoir* 37, 371 pp. 1949. *Same* 1944–1948, by C. L. Camp et al. *Memoir* 57, 465 pp. 1953. *Same* 1949–1953, by C. L. Camp et al. *Memoir* 84, 532 pp. 1961. (And continuations)

Bibliography of Fossil Vertebrates Exclusive of North America, 1509–1927, by A. S. Romer et al. *Geol. Soc. America Memoir* 87, 2 volumes. 1962.

Index-Catalogue of Medical and Veterinary Zoology—Authors 1902–1958. *United States Dept. Agric. Bull.* 39.

Marine Borers. An Annotated Bibliography, by W. F. Clapp and R. Kenk. Washington, D.C.; Office of Naval Research, Dept. of Navy. 1136 pp. 1963.

Second Bibliography and Catalogue . . . , by O. P. Hay. *Carnegie Inst. Washington, Publ. No.* 390 (2 volumes), 2003 pp. 1929, 1930.

Personal bibliographies. Finding desired publications through the name of the author may involve identification of the author himself, as indicated above, but it certainly will include finding the appropriate publications of that author. The most complete source of help is a bibliography of his works. Many workers have published or distributed lists of their own publications, and frequently a list of publications is published with a biographical sketch after the death of a taxonomist. These lists are intended to help fellow workers and successors to know and obtain the papers of the taxonomist. (Every taxonomist should maintain an accurate and complete list of his publications; such a list should be exhaustive and should include all types of publications.)

It is usually not easy to find bibliographies of individuals. They are sometimes cited in other bibliographies, such as:

A World Bibliography of Bibliographies, by T. Besterman. London; T. Besterman. 2nd ed., 3 vols., 4,111 pp. 1947–1949.

Bibliographic Index. A Cumulative Bibliography of Bibliographies. New York; H. W. Wilson. 1945– . (Arranged by subject)

Bibliography of Biographies of Entomologists, by M. M. Carpenter. *American Midl. Nat.,* vol. 33, pp. 1–116. 1945. Supplement. *American Midl. Nat.,* vol. 50, pp. 257–348. 1953.

Checklist of the Coleopterous Insects of Mexico, Central America, the West Indies, and South America, part 6 (bibliography pp. 927–1388) by R. E. Blackwelder. 1957. *United States Nat. Mus. Bull.* 185. (Cites lists of publications for each author for which published lists were known)

There is one *journal* that contains indexes to several types of bibliographic materials:

Journal of the Society for the Bibliography of Natural History. 1, 1936–   . (Four volumes published by 1964)

In order to illustrate the titles and places of publication, a selection of personal bibliographies is given here.

(Anonymous) 1911. Bibliography of Cárlos Emilio Porter. *Rev. Chilena Hist. Nat.*, vol. 15.

Blackwelder, R. E. 1950. Bibliography of Herbert S. Barber. *Coleopt. Bull.*, vol. 4, pp. 55–59.

———. 1942. The Entomological Work of Adalbert Fenyes. *Pan-Pacific Ent.*, vol. 18, pp. 17–22.

Buchanan, L. L. 1935. Thomas Lincoln Casey and the Casey Collection of Coleoptera. *Smithsonian Misc. Coll.*, vol. 94, no. 3, pp. 1–15.

Korschefsky, R. 1939. Dr. Walther Horn. *Ent. Blätter*, vol. 35, pp. 177–184.

Although complete bibliographies of the work of individual taxonomists are available in published form for only a small proportion of such workers, there are partial or serial listings of many sorts. Some of the works of all authors are to be found in one or more of these. These works can be divided very roughly into two groups on the basis of their temporal coverage—those exclusively historical, covering only publications of the far past, and those which extend up into recent times. Of the former, most that can be usefully cited are of large size and broad coverage:

Bibliotheca Historico-naturalis, by W. Engelmann. Leipzig, 786 pp. 1846. (Covers all zoological literature from 1700 to 1846. Arranged by groups. Continued in the following two works.)

Bibliotheca Zoologica . . . , by J. V. Carus and W. Engelmann. 2 vols. (pp. 951–2144) 1861. (Covers years 1846–1860.)

Bibliotheca Zoologica II . . . , by O. Taschenberg. Vols. 1–8, 6,620 pp. 1887–1923. (Covers years 1861–1880.)

Catalogue of the Books, Manuscripts, Maps and Drawings in the British Museum (Natural History). London; British Museum (Natural History). 5 vols. and 3 suppl. vols. 1903–1940.

Catalogue of Scientific Papers (1800–1863). London; Royal Society of London. vols. 1–6. 1867–1872.

Catalogue of Scientific Papers (1864–1873). London; Royal Society of London. vols. 7–8. 1877–1879.

Catalogue of Scientific Papers (1874–1883). London; Royal Society of London. vols. 9–11. 1891–1896.

Catalogue of Scientific Papers (1800–1883) Supplementary Volume. London; Royal Society of London. vol. 12. 1902.

Royal Society Catalogue of Scientific Papers. (See preceding items. Covers more than 1500 periodicals from 1800–1900. Preceded by *Repertorium Commentationen,* which covers papers up to 1800.)

Bibliographia Zoologiae et Geologiae. A General Catalogue of All Books, Tracts, and Memoirs on Zoology and Geology, by L. Agassiz. London; The Ray Society. 4 vols. 1848–1854.

List of Geological Literature Added to the Geological Society's Library During the Year (1934). London; Geological Society. vols. 1–37. 1895–1934. (Continued as next item.)

Bibliography and Index of Geology Exclusive of North America. New York; Geological Society of America. vol. 1– . 1933–

Bibliography of North American Geology (1959), by R. R. King et al. Washington; *United States Geol. Surv. Bull.* (e.g., 1,145). 1940–1959.

A Bibliography of American Natural History. The Pioneer Century, 1767–1865, by M. Meisel. New York; Premier Publ. Co. 3 vols. 1924–1929.

Some of the preceding are continuing series. In addition to these, there are works that attempt to bring the literature up-to-date, such as:

Zoological Record. (Authors alphabetically by phyla.)

Archiv für Naturgeschichte. Leipzig. 1832– . (Authors alphabetically by phyla.)

Concilium Bibliographicum, 1896–1940. Zürich. (3 × 5 cards)

Bibliographica Zoologica. Leipzig. 42 vols. 1896–1932. (Covers references on the *Concilium Bibliographicum* cards; authors arranged alphabetically under each subject number.)

Zoologischer Anzeiger. Leipzig. 1878– . (With cumulative indexes.)

For some of the groups of animals, there are major lists arranged by authors. As authors usually publish only in one major field, these are generally useful references. (See also the paleontology and parasitology works cited above.)

Bibliography of North American Conchology, Previous to the Year 1860, by W. G. Binney. *Smithsonian Misc. Coll.,* (pt. 1) vol. 142, 650 pp. 1863; (pt. 2) vol. 174, 298 pp. 1864.

Bibliotheca Entomologica, by H. A. Hagen. 1862–1863. (Covers through 1861)

Index Litteraturae Entomologicus, by W. Horn and S. Schenkling. Berlin-Dahlem. 4 pts., 1,426 pp. 1928–1929. (Covers through 1863)

Bibliography of Australian Entomology 1775–1930 with Biographical Notes on Authors and Collectors, by A. Musgrave. Sydney; Royal Zoological Society of New South Wales. 380 pp. 1932.

Bibliography of Fishes, by B. Dean. New York; American Museum of
Natural History. 3 vols. 1916–1923.
A Bibliography of Birds . . . 4 parts. *Field Mus. Nat. Hist., Zool. Ser.,*
vol. 25. 1939–1946.

In a very few cases entire volumes have been devoted to listing the
works of one author and those written about him:

A *Catalogue of the Works of Linnaeus,* by B. H. Soulsby. London; British
Museum (Natural History). 246 pp. 1933.

A regional bibliography covering, from 1758 to some recent date, the
literature on a large group of animals, would be a possible source for
citations of major works:

Bibliographic Index of Permian Invertebrates, by C. C. Branson. *Geol. Soc.
America Mem.* 26. 1,049 pp. 1948.
Indicis Generum Malacozoorum Primordia . . . , by A. N. Herrmannsen.
Cassellis. vol. 1, pp. 1–637. 1846–47; vol. 2, pp. 1–717. 1847–1849.
Supplementa et Corrigenda, pp. 1–140. 1852.
Bibliography of Australian Entomology, 1775–1930, by A. Musgrave. (See
above)
An *Introduction to the Literature of Vertebrate Zoology* . . . , by C. A.
Wood. London; Oxford University Press. 643 pp. 1931.
A Bibliography of Birds. . . . (See above)
Bibliography of Japanese Palaeontology and Related Sciences—1941–1950,
by R. Endo. *Palaeontol. Soc. Japan, Spec. Pap.* No. 1, 71 pp. 1951.

Through a general subject. The principal taxonomic subjects other
than species or groups would be the activities of taxonomic study—
methods, collections, institutions, nomenclature, etc. Although these
do not involve taxonomic data as such, they are sometimes essential
to the location of certain types of information pertinent to taxonomy.

Methods and equipment. Although these subjects are seldom dealt
with in detail, a variety of books and papers are available, of which
the following are examples:

United States National Museum. 1911. Directions for Collecting and Pre-
serving Specimens. *United States Nat. Mus. Bull.* 39, parts A to S.
Many authors.
National Museum of Canada. 1948. Methods of Collecting and Preserving
Vertebrate Animals. *Nat. Mus. Canada Bull.* 69 (Biol. Series No. 18).
British Museum (Natural History). 1909–1965. *Instructions for Collec-
tors* . . . London; British Museum (Natural History), pts. 1–13. (Var-
ious authors and editions)
Zimmer, Carl. 1928. Zoologische Musealtechnik. In *Methodik der Wissen-
schaftlichen Biologie* . . . , ed. by T. Péterfi. 2 vols. Berlin; Julius
Springer.

A *Manual of Entomological Equipment,* by A. Peterson. Ann Arbor, Michigan; Edwards Bros. Part I, 1934; Part II, 1937.

Collecting, Preserving and Studying Insects, by H. Oldroyd. London; Hutchinson and Co. 327 pp. 1958.

Burt, W. H. 1957. *Mammals of the Great Lakes Region.* Ann Arbor, Michigan; University of Michigan Press. 246 pp. (Collecting and preparing specimens, pp. 160–178.)

Cockrum, E. L. 1962. *Laboratory and Field Manual for Introduction to Mammalogy,* 2nd ed. New York; Ronald Press. 121 pp.

Hall, E. R. 1955. *Handbook of Mammals of Kansas.* Lawrence, Kansas; Museum of Natural History. 303 pp. (Suggestions for collecting, pp. 256–287.)

Taxidermy and Zoological Collecting, by W. T. Hornaday, New York; Chas. Scribner's Sons. 362 pp. 1891.

A *Field Collector's Manual in Natural History,* prepared by members of the staff of the Smithsonian Institution, Washington, D.C.; Smithsonian Institution (Publication 3766). 118 pp.

Collections. Taxonomists are frequently interested in the present location of particular collections. A few general works list these locations.

Where is the —————— Collection? An account of the various natural history collections which have come under the notice of the compiler, by C. D. Sherborn. Cambridge; Cambridge University Press. 149 pp. 1940.

Über Entomologische Sammlungen, Entomologen und Entomo-Museologie, by W. Horn and I. Kahle. 3 parts, 536 pp. 1935–1937.

(See also individual biographical sketches of individuals, listed above, where such information is often given.)

Expeditions. A few general works will give a start toward finding out about expeditions. Many exploring expeditions are known by the name of the ship used (for example, *Challenger, Galathea,* or *Vega*) and can be found listed under these names.

A Reference Guide to the Literature of Travel, by E. G. Cox. *University of Washington Publications in Language and Literature,* vols. 9, 10, 12. 1935, 1938, 1949.

Scientific Expeditions, by E. Terek. New York; Queens Borough Public Library. 176 pp. 1952.

Tanner, Z. L. 1897. Deep Sea Exploration: A General Description of the Steamer Albatross, Her Appliances and Methods. *Bull. United States Fish Commission,* vol. XVI for 1896, pp. 257–424.

Townsend, C. H. 1901. Dredging and Other Records of the United States Fish Commission Steamer Albatross, with Bibliography Relative to the

Work of the Vessel. *Rep. Commissioner for 1900, United States Comm. Fish and Fisheries,* Part XXVI, pp. 387–562.

Localities. The localities published are not always readily identified. There are some books, in addition to atlases and gazetteers, that may help in these cases.

Dictionnaire des Bureaux de Poste, Cinquième Éd. Berne; Union Postale Universelle. 2 vols. 813 pp. 1951. (The post offices of the world listed alphabetically.)
United States Official Postal Guide July 1953. Part I. Washington, D.C.; Government Printing Office. 769 pp. 1953. (Post offices alphabetically for all of U.S.; by states; and by counties.)
Webster's Geographical Dictionary, Rev. ed. Springfield, Massachusetts; G. C. Merriam. 1962.
Selander, R. B. and Vaurie, P. A Gazetteer to Accompany the "Insecta" Volumes of the "Biologia Centrali-Americana." *American Mus. Nov.,* No. 2,099, 70 pp. 1962.
Directory of British Fossiliferous Localities. London; Palaeontographical Society. 268 pp. 1954.

Institutions with taxonomic interests. The American organizations now sponsoring taxonomic work are listed in Chapter 3. For identification of institutions throughout the world, there are a few general reference works:

Minerva: Jahrbuch der Gelehrten Welt. 1891–1938.
The World of Learning 1961–62, 12th ed. London; Europa Publications, Ltd. 1,359 pp. 1962.
Directory of Natural History and Other Field Study Societies in Great Britain. London; British Association for the Advancement of Science. 217 pp. 1959.
A *Bibliography of American Natural History. The Pioneer Century, 1767–1865,* by M. Meisel. New York; Premier Publ. Co. 3 vols. 1924–1929.

Periodicals. It is not infrequently necessary to identify an unknown periodical, especially when the title is abbreviated. The following lists of periodicals will sometimes help:

Bibliographa Zoologiae et Geologiae, by L. Agassiz. vol. 1, pp. 1–85 (periodicals). London; Royal Society of London. 1848.
Catalogue of Scientific Serials . . . 1633–1876, by S. H. Scudder. *Libr. Harvard Univ., Spec. Publ.* No. 1, 358 pp. 1879.
A *Catalogue of Scientific and Technical Periodicals,* 2nd ed., 1665–1895, by H. C. Bolton. Washington, D.C.; Smithsonian Institution. 1,247 pp. 1897.

List of Journals. International Catalogue of Scientific Literature. London; Royal Society of London. 312 pp. 1903.

Union List of Serials in Libraries of the United States and Canada, ed. by W. Gregory, 2nd ed. New York; H. W. Wilson Co. 3,065 pp. 1943.

Union List of Technical Periodicals, by E. G. Bowerman. 3rd ed. New York; Special Libraries Association. 285 pp. 1947.

List of Scientific and Learned Periodicals in the Netherlands. 1953.

A List of Abbreviations of the Titles of Biological Journals. Selected by permission from the "World List of Scientific Periodicals." Issued by the Biological Council. London; H. K. Lewis and Co. 2nd ed. 31 pp. 1954.

New Serial Titles. 1950–1960. Supplement to the Union List of Serials Third Edition. Washington, D.C.; Library of Congress. 2 vols. 1961. (Also monthly supplements and annual accumulative volumes.)

World List of Scientific Periodicals. 1900–1960. 4th ed. Washington, D.C.; Butterworth's. vol. 1, A-E. 1963. (18,907 entries.)

A few lists of periodicals cover those relating to a single group of animals:

Zoological Record. (Some volumes contain lists of the periodicals which are cataloged.)

A List of Printed Malacological Periodicals, by W. SS. Van Benthem Jutting and C. O. Van Regteren Altena. *Basteria,* vol. 22, pp. 10–15. 1958.

An Annotated Checklist of the More Important Entomological Periodicals, by E. S. Claassen. *Ann. Ent. Soc. America,* vol. 38, pp. 403–411. 1945.

List of Serial Publications on Entomology. In Entomological Nomenclature and Literature, by W. J. Chamberlin. 3rd ed. Dubuque, Iowa; Wm. C. Brown Co. 1952. pp. 76–94.

List of Serials, *In* Checklist of the Coleopterous Insects of Mexico . . . , by R. E. Blackwelder. *United States Nat. Mus. Bull.* 185, Part 6, pp. 1345–1388. 1957.

Books. Identification of books by title or author can be attempted from a variety of sources. There are current lists of new books, lists of books in various libraries, book dealers' catalogs, and card catalog lists of rare books. This is an aspect with which any librarian can be helpful.

Catalogue of the Books . . . in the British Museum (Natural History). London. 5 vols. and 3 suppl. vols.

A Catalog of Books Represented by Library of Congress Printed Cards Issued to July 31, 1942. Ann Arbor, Michigan; Edwards Bros. 167 vols. 1942–1946. *Supplement Cards Issued August 1, 1942–December 31, 1947.* Ann Arbor, Michigan; J. W. Edwards. 42 vols. 1948.

The Library of Congress Author Catalog. A Cumulative List of Works Represented by Library of Congress Printed Cards. 1948–1952. Ann Arbor, Michigan; J. W. Edwards. 24 vols. 1953.

The National Union Catalog. A cumulative author list representing Library of Congress printed cards and titles reported by other American libraries. 1953–1957. Ann Arbor, Michigan; J. W. Edwards. 26 vols. 1958. *Same.* 1958–1962. New York; Rowman and Littlefield, 50 vols. 1963. *Same.* 1952–1955. Ann Arbor, Michigan; J. W. Edwards. 30 vols. 1961. *Same.* 1963. Washington, D.C.; Library of Congress. 5 vols. 1964. (And continuations)

A Bibliography of American Natural History. The Pioneer Century. 1767– 1865, by M. Meisel. New York; Premier Publ. Co. 3 vols. 1924–1929.

Catalogue of the Printed Books in the Library of the British Museum. London; Wm. Clowes and Sons. 58 vols. 1881–1900, and 10 suppl. vols. 1900–1905. (Reprinted by Edwards Bros., Ann Arbor, 1946–1950.)

British Museum General Catalogue of Printed Books. Photolithographic Edition to 1955. London; British Museum. 252 vols. 1931– .

Catalogue Général des Livres Imprimés de la Bibliothèque Nationale. Auteurs. Paris; Imprimerie Nationale. 189 vols. 1897– .

Nomenclature. Although there are thousands of papers which deal with some aspect of nomenclature, a few works are cited here in which there is substantial coverage of the entire subject, usually with further bibliographic references:

International Rules of Zoological Nomenclature. *Proc. Biol. Soc. Washington,* vol. 39, pp. 75–104. 1926.

International Code of Zoological Nomenclature . . . , ed. by N. R. Stoll et al. London; International Trust for Zoological Nomenclature. 176 pp. 1961.

International Code of Zoological Nomenclature . . . , ed. by N. R. Stoll et al., Rev. anonymously. London; International Trust for Zoological Nomenclature. 176 pp. 1964.

Copenhagen Decisions on Zoological Nomenclature . . . , ed. by F. Hemming. London; International Trust for Zoological Nomenclature. 135 pp. December 31, 1953.

Bulletin of Zoological Nomenclature. London; International Trust for Zoological Nomenclature, vol. 1, 1943– .

Ferris, G. F. 1928. *The Principles of Systematic Entomology.* Stanford, California; Stanford University Press. 169 pp. (Stanford Univ. Publ., Univ. Series, Biol. Sci., vol. 5, no. 3, pp. 101–269.)

Follett, W. I. 1955. *An Unofficial Interpretation of the International Rules of Zoological Nomenclature. . . .* San Francisco, California; by the author. 99 pp.

Keen, A. M., and Muller, S. W. 1948. *Schenk and McMaster's Procedure in Taxonomy. . . .* Stanford, California; Stanford University Press. 92 pp.

Keen, A. M., and Muller, S. W. 1956. *Schenk and McMaster's Procedure in Taxonomy. . . .* Stanford, California; Stanford University Press. 119 pp.

Mayr, E., Linsley, E. G., and Usinger, R. L. 1953. *Methods and Principles of Systematic Zoology*. New York; McGraw-Hill Book Co. 328 pp.

Richter, R. 1948. *Einführung in die Zoologische Nomenklatur durch Erläuterung der Internationalen Regeln*. Frankfurt am Main; W. Kramer. 252 pp.

Savory, T. 1962. *Naming the living world*. . . . New York; John Wiley and Sons. 128 pp.

Schenk, E. T., and McMasters, J. H. 1936. *Procedure in Taxonomy*. . . . Stanford, California; Stanford University Press. 72 pp. (For later editions see Keen and Muller, 1948 and 1956, below.)

Interpreting the literature

When the relevant literature has been found, there is still a wide range of problems involved in understanding what the author was trying to record. These problems involve the actual contents of the book—the arrangement of materials, the intent of the author and his plan of presentation, the various forms of language that may have been employed, the interpretation of the biological data presented, and the conversion of bibliographical references into other useful works on the subject. Failure to resolve these problems in each case may result in failure to obtain from the book all that it is capable of giving.

The physical book. Nearly all books, and some papers in serial publications, consist of several parts, even though they are bound into one unit. In taxonomic work it is especially important to identify all the parts of a book, as species may be referred to in a key, a tabular appendix, or an index, as well as in the main text. Such secondary citations may be bibliographically and synonymically more important than the primary citation.

This is the time to note whether the book or paper is part of a continuing series, what the complete title is, whether there is an explanatory preface, what data are included in tables or other forms as appendices, whether there is an addendum or corrigendum at the back, and the nature of the indexes. Taxonomically, any part of the book can be important, and failure to examine it carefully can produce inadequate results. The following are some examples of the need to know the physical nature of a work:

Title page. In the so-called Disciples Edition of Cuvier's *Le Règne Animal* . . . , the full title reads

"*Le Règne Animal distribué d'après son organisation, pour servir de base à l'histoire naturelle des animaux, et d'introduction a l'anatomie comparée, par Georges Cuvier. Edition accompagnée de planches gravées, representant les types de tous les genres, les caractères distinctifs des divers groupes et les modifications de structure sur lesquelles repose cette classification; par une reunion de disciples de Cuvier, MM. Audouin, Blanchard, Desbayes, Alcide D'Orbigny, Doyere, Dugès, Duvernay, Laurillard, Milne Edwards, Roulin et Valenciennes."*

This statement that the types of the genera are indicated by illustration is not repeated anywhere in the text. If it is missed on the title page, this entire aspect and significance of the book would be overlooked. In fact, the significance of this title *was* overlooked for nearly a century.

Issuance in parts. Many books, which are now known only in bound form as a single item, were originally published in two or more parts or signatures. Often the title page shows no indication of this fact. For example, Erichson's *Genera et Species Staphylinorum* . . . always appears as a single work and is always dated 1840. Like almost all books, it was printed in "signatures" of 16 or 32 pages, which may be shown by small letters or numerals at the bottom of the first page of each one. Sometimes the signatures are numbered serially, and sometimes they are dated. These signatures show us that the first 400 pages of this book were actually published in 1839 and only pages 401–954 in 1840.

Even when it is well known that publication occurred in sections, it may be difficult to tell the dates of each section. For example, in 1893–1908 Heyne and Taschenberg published a large work entitled *Die Exotische Käfer in Vort und Bild.* It consisted of brief text and colored plates of many beetles, and it was published in 17 parts, only a few of which carried title pages or were dated. The plates are generally bound at the back of each volume and are not dated. From a variety of sources, it has been shown that the work was issued in small fascicles consisting of a few pages of text and usually four plates. It has been possible to date every section of the text and every plate, but this information does not appear in the book directly. The story of the detective work used to discover these dates is told in:

Blackwelder, R. E. 1949. An Adventure in Biblio-Chronology. *Journ. Washington Acad. Sci.*, vol. 39, pp. 301–305.

Many of the older works which were issued in parts have been recorded by bibliographers. The most active of these in zoological works were C. Davies Sherborn and Francis J. Griffin, both Englishmen. Their studies, and those of most other bibliographers, are published or listed in a special periodical:

Journal of the Society for the Bibliography of Natural History. London; vol. 1, 1936– .

In addition to earlier publication in parts there are of course several forms of obvious and intentional separation into parts. A long paper may be split between two or more issues of a journal, a large book may be published as two separate volumes, a supplement may be published after a lapse of time. All such fragmentation must be kept track of, and the taxonomist must be alert to detect it.

Reprintings and new editions. Especially with older works, a student is likely to assume that when he obtains a copy his problems are over. However, many books have been reissued in later editions, either with no change or with changes that may amount to complete rewriting. Sometimes there is little evidence of the existence of several editions. For example, the *Nomenclatoris Zoologici Index Universalis* of Agassiz, listed earlier, is well known as a commonly available quarto volume. Only exceptionally does one see the very small sized edition published two years later. They are not identical in content, and page numbers are entirely different.

Not infrequently a set will consist of volumes which were issued in differing editions. For example, the usual set of Kirby and Spence, *An Introduction to Entomology* . . . , consists of four volumes, bound as a uniform set but actually representing several editions:

    vol. 1, ed. 4, 1822
    vol. 2, ed. 3, 1823
    vol. 3, ed. 1, 1826
    vol. 4, ed. 1, 1826

If the work were of taxonomic importance, it would be necessary to use a set consisting of first editions of all volumes:

    vol. 1, ed. 1, 1815
    vol. 2, ed. 1, 1817
    vol. 3, ed. 1, 1826

Failure to cite the edition of all such works may result in real confusion, as pages may be different, text may be altered, or even authorship may be changed.

Although new editions are commonly shown to be such by notations on the title page or elsewhere, reprintings are frequently not shown in any way on older books. This makes their detection difficult. In taxonomic work, it is necessary to be meticulous in referring to the various issues, and especially to study the first issue before making nomenclatural decisions.

One other type of multiple publication can be troublesome. At times an author has submitted a paper for publication in a journal and then circulated advance copies before the journal issue appears. These copies may even be printed from the same type as used in the journal. These advance copies are called preprints. A series of examples is described in:

Blackwelder, R. E. 1952. Preprints of Proc. U.S. National Museum, 1890–1897. *Syst. Zool.*, vol. 1, pp. 86–89.

Although such preprints may be ruled unacceptable in nomenclature, they are present in libraries and must be taken into account.

There have been times when authors have obtained publication of an article in two or more journals or in a journal and separately as a book. An example is:

Mulsant, E., and Rey, C. 1872. *Histoire Naturelle des Coleopteres de France: Brevipennes* (*Aleochariens*). Paris, 321 pp. (Bolitocharaires)

Mulsant, E., and Rey, C. 1872. Tribu des Brevipennes: Famille des Aleochariens: Huitieme Branche: Bolitocharaires. *Ann. Soc. Linn. Lyon*, ser. 2, vol. 19, pp. 81–413.

Mulsant, E., and Rey, C. 1873. Tribu des Brevipennes: Famille des Aleochariens: Branche des Bolitocharaires. *Mem. Acad. Sci. Lyon*, vol. 19, pp. 73–123. (Pp. 89–104 are by error numbered 233–248.)

The first two are nearly identical, with principally the page numbers changed. The third set is in new type on a larger page and looks very different. Each of these must be examined and its date determined. Each is relevant in taxonomic work.

Authorship. One further problem remains to be noted. Obviously every contribution has been prepared by some person or persons. This "author" is usually indicated, but many exceptions occur. Some of the possibilities are:

1. One author, indicated as such.

2. Two or more authors, clearly indicated as co-authors. The one listed first is usually called the senior author, although he may not be senior in years and he may not have done much of the work.

3. One or more authors are indicated and some collaborators, illustrators, editors, etc., also noted. These latter are generally not cited in bibliographic references.

4. The original author may be cited but the current revision credited to a revisor. For example:

Schenk, E. T., and McMasters, J. H. 1956. *Procedure in Taxonomy*, 3rd ed., revised by A. M. Keen and S. W. Muller. Stanford, California; Stanford University Press.

Evans, C. L. 1956. *Starling's Principles of Human Physiology*, 12th ed. Philadelphia; Lea and Febiger.

5. There may have been no nominal author because the work was prepared by a committee or group. In some instances the group may be identified. In other cases the actual author may become known from other sources.

List of Members. Society of Systematic Zoology, Carbondale, Illinois. 1959. (Published anonymously by the Society but actually prepared by its secretary, who at that time was R. E. Blackwelder.)

Guide to General Zoology Laboratory, compiled and edited by The Staff, Dept. of Zoology, Kansas State College, Manhattan, Kansas. Minneapolis, Minnesota; Burgess Pub. Co. 1957. (There is nothing to show who was on the staff at that time.)

Such works are sometimes listed as anonymous, sometimes under the name of the sponsoring organization, sometimes under the name of the committee, and sometimes under the parenthetical name of the actual author if known.

6. An author is cited, but the work was actually produced in whole by another person. Cases of this are rare, but an example is:

Schönherr, C. J. 1817. *Appendix ad Synonymia Insectorum* . . . , 266 pp. (Actually by J. W. Dalman.)

7. Two or more authors are cited, but the senior author actually took no part in the writing. For example:

Mulsant, E., and Rey, C. 1839–86. *Histoire Naturelle des Coleopteres de France.* (In some early volumes Mulsant was the sole author, in some the work appears to be joint, and in some of the final volumes all the work was done by Rey after the death of Mulsant. These should be cited as Mulsant and Rey, Mulsant *in* Mulsant and Rey, or Rey *in* Mulsant and Rey, depending on the real authorship.)

8. Anonymously, with no indication of authorship. Such papers are not now acceptable in nomenclatural taxonomy, but if the author can be determined from other sources, as is usually the case, his name should be used in parenthesis.

9. Anonymously, with the author remaining unknown. These are cited simply as Anonymous or Anon.

General. The foregoing are some of the principal difficulties encountered in the physical arrangement, source, and dating of zoological works. Books have been published with a series of page numbers missing; page numbers have been duplicated by mistake; there have been typographical errors on title pages; the title given at the beginning of a work may be different from the one on the title page or jacket; a volume marked "volume 1" may actually be the only one ever published, and so on. Only constant care, inquiry, and meticulous citation will surmount these mechanical difficulties and so prevent inadequate conclusions.

The author's intent. The author's intent is not always clearly indicated, but, if there is no direct statement, a few moments spent in studying the contents, appendices, and indexes will often be rewarding. A Preface is the usual place for the author's comments about his book, but they may occur in a Foreword or even in the Introduction.

All books are written under limitations of space. Most reference books cover only stated aspects of the field. Frequently there is a limitation in time to the coverage. Many reference books include notations on certain side aspects of the data, frequently in coded form, which is comprehensible only after explanation.

During the early part of the nineteenth century, it was common to use very long titles, which gave a great deal of information about the coverage. These are now frequently abbreviated or shortened, and this explanation is lost or at least suppressed. In modern times, it has become customary to use short general titles. When a subtitle is added, this will usually not be quoted.

Sometimes the intent of the author can be made out by a careful study of the text. Such was the case with the works of Fabricus in the late eighteenth century, who described many species of insects, mostly with short descriptions. It was noticed by a careful Swedish worker that there was one longer description in each genus, and that the species so described was evidently considered to be more important than the others. It is clear that Fabricus was treating this species as a kind of type, what we now call the genotype or type species of the genus.

There can be no substitute for a clear understanding of the purpose of a book, the author's intentions, the details of arrangement, and the facts of publication if the user is to obtain all the aid which the book can give, but this is especially true of reference books. Much available information is missed by careless examination.

Language forms

In using literature, the taxonomist not only runs into various languages of the world, of which at least fifteen are now used for works of taxonomic interest, but also finds various forms of language in use in each one—abbreviations, contractions, typographical variations, spelling forms, foreign phrases, and differing translations or transliterations. Some of these can be profitably discussed.

Languages. Zoological literature can be published in any language of the world, no matter how obscure. Fortunately, the vast majority of works have been published in one of the Western European languages using principally what we call the English alphabet, more appropriately called the neo-Latin alphabet—the Latin alphabet plus *j*, *k*, *w*, and *y* (of which *k* was known to the Romans, as kappa, but seldom used). These languages are principally: English, French, German, Italian, Spanish, Portuguese, Dutch, Danish, and Swedish. Some of these use also a few letter forms made by adding diacritic marks to the basic letters (see next section).

Languages with completely different alphabets are numerous. In Russian and Japanese, there is a large literature, mostly on the faunas of those countries. Lesser numbers of works have appeared in Egyptian, Arabic, Hebrew, Turkish, and Chinese.

Although taxonomic data can appear in any language, if descriptions of new species are to be acceptable under the International Code, they must be written or summarized in English, French, German, Italian, or Latin. This rule has tended to keep the important taxonomic literature in these five languages. In recent years, zoologists in the Netherlands, Scandinavia, and Japan, especially, have published most of their major taxonomic papers in English, which is now used in close to a majority of all works. Although few American taxonomists are fluent in all five of these nomenclaturally acceptable languages, familiarity with technical German and French is still necessary for serious monographic work, and Latin must be familiar as a vocabulary through the technical terms of science.

Diacritic marks. The special letter forms mentioned before are formed by adding diacritic marks to the Roman letters. These marks serve to distinguish two forms of the same letter, for ease of recognition, for pronunciation, for accent, or to indicate a hidden diphthong or double vowel. There is literally no end to these marks, as they can be coined at will and for a variety of purposes. Many of them are reproduced with difficulty in the printing types of other countries, but their omission may substantially alter the meaning.

Some of these marks do not change the letter but merely indicate stress; some alter the pronunciation considerably; and some result in adding a vowel sound to a consonant, or a second consonantal sound to a consonant. No complete list would be possible, but the following includes most of those commonly found in European languages (and in latinized Japanese):

English–ö
German–ö, ä, ü
Swedish–ö, å, ä
Danish–Æ, ø
Spanish–ñ, ç, é, á, í, ó, ú
Portuguese–é, ã, õ, á, â, è, ò, ç, ì, ó
Italian–à, è, é, í, ò, ú
French–é, è, ê, à, â, ô, î, ï
Czechoslovakian–š, é, ě, ř, í, ů, ú, ý, ž, á, č, ď, ň, ó
Polish–Ł
Rumanian–ă, ş, ţ, î
Japanese (latinized [1])–ô

Such marks were formerly to be retained whenever they entered formal zoological nomenclature. The 1961 Code outlaws them in all names, new and old. This rule will itself produce many changes in spelling of old names. This attempt to get rid of a troublesome typographical problem will probably also result in the widespread misspelling of many names which normally employ such special letters. The German name Müller is simply wrong if spelled Muller, although the form Mueller may be considered an acceptable substitute. In the case of Łwow, a city in Poland, Lwow would be meaningless and no alternative native form is available. Similarly, č in Czech is not at all the same as c, as is also the case with the Rumanian ţ and t.

Ligatures. In printing with type, it is often found appropriate to combine two or more letters on a single base, making a unit of them. Examples that are nearly universal even today are ﬀ, ﬁ, ﬂ, ﬄ. Many others have been used in some fonts of type, and, among these, two have been almost universally employed from Roman days until recent decades: œ and æ (and their capitals Œ and Æ).

In modern American usage these two ligatures, which represent diphthongs or double vowels, are commonly spelled as ae and oe separately or replaced by e alone.

> hæmoglobin is spelled hemoglobin
> Hominidæ is spelled Hominidae
> alumnæ is spelled alumnae
> œsophagus is spelled esophagus (or rarely oesophagus)

There are two serious objections to this modern practice of "simplifying" these spellings. First, it is not applied uniformly. Besides the divergence in treatment shown above, there are a few that are never or seldom changed. For example, coelom is now seldom spelled cœlom

---

[1] Used principally in proper names, such as Yôrô.

but almost never celom. This would be a case similar to alumnæ, except that this root is commonly combined with hemo- to form (the blood cavity) hemocoel. If hæm- is to be changed to hem-, then surely in the same word -cœl should be changed to -cel. This has scarcely ever been done. All such spelling variations must be taken into account when using alphabetical indexes or nomenclators. Second, stems using e in Latin are often entirely different in meaning from similar ones using œ or æ. Thus, pædogenesis is reproduction by immature forms, whereas pedogenesis would refer to feet. The fact that in English writing the long vowel of ped- (child) and short vowel of ped- (foot) are not distinguishable leaves an unnecessary uncertainty over the meaning of all words compounded with ped-, of which there are scores in biology and medicine. Because some of the confusing terms are directly involved in taxonomic activities, the diphthong spellings are used throughout this book and are recommended to all zoologists, even if ligatures are not deemed necessary.

Abbreviations. There is no end to the contracted or curtailed forms that have been used to save space when writing words, particularly long technical terms. Many of these are considered standard or acceptable at the time of their use and cause no difficulty to the immediate audience. We are so used to abbreviations such as Ltd., Co., mm., n. sp., that we forget they may be less familiar a century hence, when our writings may still be in use. It is also noteworthy that even such a well-known abbreviation as U.S.A. can currently refer either to the United States of America or to the Union of South Africa.

Dictionaries in most languages explain the common abbreviations in that language. The technical abbreviations can sometimes be recognized from their spelling and the context, but these will not always serve. Many large works have adopted special lists of abbreviations and explain them in an alphabetical list or glossary. Such a list should always be searched for, wherever abbreviations are encountered.

Lists of standard abbreviations for geographical regions have been published in many countries. Abbreviations of the names of authors of zoological names are common, although they are officially said to be sources of possible confusion. A few formal glossaries published as separate works include many of the abbreviations in their particular field, but there is no good one for taxonomy. No very extensive list of abbreviations in general is available; it would be an impractical project. Several brief lists are available for reference, but their purpose is generally not to show which abbreviations are reasonable and acceptable but only to define certain ones in common use. These are:

Schenk, E. T., and McMasters, J. H. 1936. *Procedure in Taxonomy*. Stanford, California; Stanford University Press. (pp. 23–25)

Keen, A. M., and Muller, S. W. 1956. *Procedure in Taxonomy*, 3rd ed. Stanford, California; Stanford University Press. (pp. 27–29)

Smith, R. C. 1958. *Guide to the Literature of the Zoological Sciences*, 5th ed. Minneapolis, Minnesota; Burgess Pub. Co. (pp. 147–148)

A few of the abbreviations particularly relevant to taxonomy are listed below.

aff.	*affinis*, having affinity with but not identical with
auct.	*auctorum*, of authors
ca.	*circa*, about (with reference to dates)
cf.	*confer*, to be compared to; compare
e.g.	*exempli gratia*, for example
emend.	emended, emendation
et al.	*et alia*, and others
gen. nov.	*genus novum*, new genus
ibid.	*ibidem*, in the same reference (book or article)
i.e.	*id est*, that is
in litt.	*in litteris*, in correspondence
loc. cit.	*loco citato*, in the place previously cited (publication *and* page)
n.n.	*nomen nudum* or *nomen novum*
nob.	*nobis*, to us (referring to author of a name)
nom. dub.	*nomen dubium*, name of doubtful application
nom. nov.	*nomen novum*, new name
nom. nud.	*nomen nudum* (pl. *nomina nuda*), name without validation
nov.	*novus*, new
op. cit.	*opere citato*, in the work or article previously cited for this writer (no page cited)
part.	*partim*, part
p.p.	*pro parte*, in part
q.v.	*quod vide*, which see
s.l.	*sensu lato*, in the broad sense
sp.	species (singular)
sp. indet.	undetermined species
sp. nov.	*species novum*, new species (singular)
spp.	species (plural)
ssp.	subspecies (singular)
sspp.	subspecies (plural)
ssp. nov.	new subspecies
s. str.	*sensu stricto*, in the strict sense
subsp.	subspecies
viz.	*videlicet*, namely

Translation. Because the literature of taxonomy is in so many languages, translation is routine. It usually takes the form of on-the-spot reading of descriptions in whatever language they occur, as the taxonomist may well be called upon to integrate works in five or more languages into his own monographic work. This descriptive material is generally much easier to read than ordinary text, partly because of the high percentage of technical terms that are readily understood or learned. Not so frequently, formal translations are prepared (and perhaps published) for the use of taxonomists. Unfortunately, there is no easy way to find such translations, except by being alert when studying the literature and using libraries.

At the present time, there are increasing numbers of books being translated from Russian. Many of these are of taxonomic interest, but their ultimate value in taxonomy has not yet become apparent. In using any translation, a taxonomist must remember that he is not using the original. Even minute alterations in the translation may alter the text in ways that affect taxonomy, especially nomenclature.

Transliteration. Whereas translation is the rendering into a second language what is expressed by the words of the first language, transliteration is the spelling of the words of a language using a certain alphabet into the letters of a different alphabet. Thus, Greek words are transliterated into the Roman alphabet (Greek χαιτη = Roman chaite). There are several systems for transliteration of non-Roman alphabets, giving rise to varying spellings in the second language. For example:

χαιτη becomes chaite or chaete
красный becomes krasnaya or krasnaia

Transliteration is important in taxonomy in two ways. It can change the spelling of the names of authors (such as Russian), making them less readily identified in bibliographies, and it has repercussions in zoological nomenclature, where the method of transliterating Greek to Roman is now specified in the rules of nomenclature.

πιρρότης becomes *Pyrrhotes*, not *Pyrrotes*
χιτών becomes *Chiton*, not *Kiton*

Latin phrases. In the early days of taxonomy, all workers used Latin as their language of publication. As this practice dwindled, many persons, particularly English-speaking ones, continued to use Latin words and phrases under various pretexts. Although the number of Latin phrases in taxonomic use is dwindling, they are frequent in the older literature. They may still confuse the uninitiated, who must translate them to obtain the intended connotation. Although in general literature, there is no end to these, only a few are commonly found in taxonomic literature.

auctorum	of authors
ex nomine	by or under that name
ex parte	in part
fide	on the authority of, or with reference to publication, to a cited published statement
incertae sedis	of uncertain position in a list or scheme
lapsus calami	a slip of the pen (pl., *lapsus calamorum*)
mihi	of me, as *X-us albus* mihi, a species named by me
nec	not

nomen conservandum: a name that has been conserved (preserved) by the ICZN, although it is not otherwise acceptable

nomen dubium	a name not certainly applicable to any known taxon

nomen inquirendum: a name of doubtful assignment

nomen novum	a new name expressly proposed as replacement for an earlier name and valid only if the latter is preoccupied
nomen nudum	a name that fails to satisfy the technical rules for availability then in force (pl., *nomina nuda*)
nomen oblitum	a name that has remained unused in the primary zoological literature for more than fifty years
nomen triviale	the second part of the name of a species or the third part of the name of a subspecies. (Not recognized in 1961 Code, where "specific" name is used instead.)
pro parte	in part
sensu lato	in the broad sense
sensu stricto	in the strict sense
seu	either, or
sic	thus (to signalize exact transcription)

species indeterminata: a species not identifiable from the original publication

species inquirenda: a doubtfully classified species

species novum	a new species (pl., *species nova*)
teste	according to (verbal, not written, testimony)
vere	the true
vide	see

Interpretation of the contents

The main purpose of the use of technical literature is to obtain the information recorded therein. In taxonomy this is primarily the classified comparative data about the kinds of organisms. Taxonomic literature can be used to answer the questions: What kind is it? What are its features? Where does it occur? And in general: What is known about its kind?

In asking what kind it is, the taxonomist expects to find out that it is biologically part of a certain species of which other examples are known; he expects to find out the name now used for this species and also the other names that have previously been applied; and then he expects these names to lead him to all else that is known about this kind and similar kinds. This is the basic function of taxonomy.

Identification. The practical aspects of identification are dealt with in Chapter 5. It remains only to reiterate that this most basic recovery process is beset with many pitfalls. The forms of identification aids (such as keys) are highly artificial and can lead to serious error. A large percentage of identifications are made from the literature, but these are always at the mercy of the accuracy of the publishing author as well as the correctness of the reading by the user. Here as elsewhere, it is of vast importance to know a great deal about the author —how broad his background, how infrequent his serious failures, how accurate his use of words, how deep his search for distinguishing features, and how complete his search of previous literature. Good identifications can seldom be made from a work with which the identifier is not thoroughly familiar or which he only casually consults.

The late Thomas L. Casey was a careful student of the taxonomy of beetles. His work has been criticized by some other workers for being based on minute details that were individual variation rather than species characteristics. A person familiar with his best work and his manner of studying would know that such criticism applies properly only to that part of his work on beetles of large size, one-half inch or longer, and not to the minute forms which he also studied extensively. The explanation is that he used the highest powers of his microscope on everything he examined, thus seeing minute detail on the larger specimens. A knowledge of this practice and its effects will enable a later taxonomist to use Casey's works with less danger of misinterpretation. It will enable him to put confidence in certain works and be wary of the conclusions in certain other ones. This is the proper way to judge the works of a taxonomist; it will result in the maximum benefit to science from his labors, as the minutiae may prove to be of great value themselves.

Good identification requires that one obtain from every book the largest possible amount of information; everything-that-the-author-knew would be the ideal. But there are grave dangers in trying to go beyond the data, to guess about something that is not recorded.

Synonymy. Identification serves first of all to apply a name to the taxon represented. If the identification is to genus, a generic name will be applied, which means that the specimen belongs to some unidentified or unspecified species of the genus. (This is often written as *Drosophila* sp., for example.) If the identification is to species, the name will be a specific name (a binomial or binomen, such as *Musca domestica*). If the identification is to subspecies, there will be a subspecific name (a trinomen, such as *Pica pica hudsonica*). Unfortunately, a very large number of animals have, for various reasons, received more than one name. To recover all the data about a taxon, one must obtain what is recorded under *all* the names that have ever been applied to it, not merely what is under the currently used name.

Whenever two or more names have been applied to any taxon, *all* of them are *synonyms* (different names for the same thing). In spite of this indisputable fact, taxonomists commonly refer to the *rejected* names alone as the synonymy or the synonyms. Because in many cases no one can be absolutely sure of the identification of any specimen except the original one, the holotype, as even a direct offspring of the type (if sexually produced) may be a hybrid, the "correct" name for any species is subject to change for any of a variety of reasons. Therefore, the taxonomist must keep track of all the names that have been used, and he must seek out these synonyms to obtain the data they represent.

Synonymy is a subject closely affected by the rules of nomenclature. Much of this must be left for the chapters on generic and specific names. Here it is necessary to discuss only the interpretation of the forms used for the presentation of synonymy. Some sample synonymies are explained in detail, but, for brevity, the bibliographic citation that would normally accompany each name is omitted. (Notice the differences in punctuation that are acceptable.)

(I)　　　　　*Philonthus discoideus* (Gravenhorst)

Synonymy:

Staphylinus discoideus Gravenhorst, 1802	(1)
Staphylinus suturalis Marsham, 1802	(2)
Staphylinus testaceus Gravenhorst, 1806 (Misidentification of S. testaceus Paykull, 1789)	(3)
Quedius suturalis (Marsham) Stephens, 1829 (Not Kiesenwetter, 1845; not Thomson, 1867)	(4)
Quedius lepidulus Stephens, 1832	(5)
Staphylinus conformis Boisduval and Lacordaire, 1835	(6)
Philonthus discoideus (Gravenhorst) Nordmann, 1836	(7)
Philonthus suturalis (Marsham) Nordmann, 1836	(8)
Philonthus ruficornis Melsheimer, 1846 (Not Motschulsky, 1860; not Hochhuth, 1860; not Broun, 1880)	(9)

Philonthus conformis (Boisduval and Lacordaire) Ganglbauer, (10)
1895

Philonthus rufipennis Gerhardt, 1910 (As aberration) (Not (11)
Gravenhorst, 1802; not Solier, 1849)

Philonthus lepidulus (Stephens) Bernhauer and Schubert, 1914 (12)
(Not LeConte, 1863)

Philonthus testaceus (Gravenhorst), Bernhauer and Schubert, (13)
1914

Philonthus discoidens Cameron, 1933 (Misspelling) (14)

Philonthus gerhardtianus Scheerpeltz, 1933 (New name for (15)
rufipennis)

The heading is the currently accepted name for the species, the one accepted by the writer of this synonymy. The parentheses around the name of the author show that the species was first described in some genus other than *Philonthus.*

The first name in the synonymy, in this case, is the original publication of the Gravenhorst name in its original genus (1). The history of this name can be traced by following it in later lines, thus: In 1836 it was transferred to the genus *Philonthus* (7), and in 1933 it was misspelled *P. discoidens* (14).

In 1802 a second writer published another name for what later proved to be the same species, *S. suturalis* (2). In 1829 this name was transferred to *Quedius* (4); and in 1836 it was transferred to *Philonthus* (8).

The next name applied to this species was *S. testaceus* by Gravenhorst in 1806, who thought he had the species which Paykull had previously described under this name (3). This was an error, and there really is no such name as *S. testaceus* Gravenhorst. It is carried in the synonymy in this manner to give the citation to this publication about the species. The *S. testaceus* of Gravenhorst has been referred to by later writers, one of whom transferred it to *Philonthus* (13).

Later writers in other parts of the world believed they had discovered distinct species, which they named *Quedius lepidulus* (5), *Staphylinus conformis* (6), and *Philonthus ruficornis* (9). *Q. lepidulus* was also transferred to *Philonthus* (12), as was *S. conformis* (10).

In 1910 a supposed aberration or variant of this species was described and named *Philonthus rufipennis* (11). Inasmuch as names for aberrations are not accepted under the rules, this name is unacceptable. However, it was itself preceded by two other uses of the same name (Gravenhorst, 1802, and Solier, 1849) and is thus a junior homonym as well as a junior synonym. In 1933 Scheerpeltz recognized the homonymy but not the synonymy; he therefore unnecessarily renamed this form *P. gerhardtianus* (15).

There are thus eight basic names involved in this synonymy, with six additional generic combinations and one misspelling.

The synonymy in this form gives the chronology of the names of this species. The generic transfers of each of the names is also shown. Much other information about the names is available too. For example, if *P. discoideus* were sometime proven to be an unavailable name, the next available name, which would have to be adopted, is *P. suturalis* (Marsham, 1802) (2) and (8). And the next one after that would be *P. lepidulus* (Stephens) (5) and (12).

In item (4) there are listed two later uses of the same name, *Q. suturalis,* at much later dates. Under some rules these later uses might be considered homonyms, but under other rules they would not. This would be of no concern in any case in the synonymy of *P. discoideus.* The later uses are cited here to prevent these uses from being confused with the *Q. suturalis* Marsham, with consequent mixture of inapplicable data. Another example of this occurs in item (11).

The double authority in items (4), (7), (8), (10), (12), and (13) appears to be comparable to the double citations in botany, where the original author is cited in parenthesis after a generic transfer, and the author of the transferred name is listed after the first without parentheses. In these items a double authority is not intended, the second name merely being part of the bibliographic reference to where the transfer was made. Frequently in zoology, a comma is placed between the two authors' names; for example, *Quedius suturalis* (Marsham), Stephens, 1829, p. 278.

Item (12) illustrates another problem of homonymy. Long before *Q. lepidulus* Stephens (5) was transferred to *Philonthus* (12), a different species of *Philonthus* was described by LeConte as *Philonthus lepidulus.* This latter name was acceptable from 1863 until 1914, when it became a junior homonym and would, under some rules, have to be replaced. This piece of information is added to the synonymy simply to prevent confusion between these two species, both of which have at some time been called *Philonthus lepidulus.* (It is this problem that makes it so important to cite the author of the species in every case.)

Item (15) illustrates the danger of renaming a supposed homonym without first investigating its zoological position to see if it is a needed name. *P. rufipennis* Gerhardt is an unacceptable name, so no replacement is required even though it is a junior homonym.

This entire synonymy could have been arranged to emphasize the history of the individual names, rather than the history of the species, thus:

(II)
Staphylinus discoideus Gravenhorst (1802)	(16)
Philonthus discoideus (Gravenhorst), Nordmann (1836)	(17)
P. discoidens Cameron, 1933 (Misspelling)	(18)
S. suturalis Marsham (1802)	(19)
Quedius suturalis (Marsham), Stephens (1829)	(20)
P. suturalis (Marsham), Nordmann (1836)	(21)
Q. lepidulus Stephens (1832)	(22)
P. lepidulus (Stephens), Bernhauer and Schubert (1914)	(23)
(etc.)	

Commonly the accepted name of the species is not repeated in the synonymy, thus:

(III)
***Philonthus discoideus* (Gravenhorst)**	(24)
Gravenhorst, 1802, p. 38 (*Staphylinus*)	(25)
P. suturalis (Marsham), 1802, p. 509 (*Staphylinus*)	(26)
P. lepidulus (Stephens), 1832, p. 278 (*Quedius*)	(27)

Here the first reference (25) is to the original publication of *P. discoideus*, and the parenthetical name *Staphylinus* shows in what genus it was placed by Gravenhorst at that time. This parenthetical name thus explains or completes the parenthesis around the author's name in line (24).

Other features of species synonymies are illustrated by the following:

(IV)
***Crematogaster (Crematogaster) laeviuscula* Mayr**	(28)

Synonymy:
Crematogaster laeviuscula Mayr, 1870	(29)
Oecodoma (Atta) bicolor Buckley, 1867	(30)
Cr. clara Mayr, 1870	(31)
Crematogaster laeviuscula Dalla Torre, 1893	(32)
Crematogaster lineolata subsp. *laeviuscula* Emery, 1895	(33)
Crematogaster lineolata subsp. *laeviuscula* var. *clara*, Emery, 1895	(34)
Crematogaster laeviuscula Wheeler, 1919	(35)
Crematogaster lineolata subsp. *laeviuscula* Emery, 1922	(36)

This synonymy contains several confusing items that are not decipherable without prior knowledge of the customs of the author. For example, it is clear that this species is being placed in the subgenus *Crematogaster* because this name is in parenthesis in the heading (28). It would appear from item (29) that the species was originally described without mention of a subgenus, but in the cited text this original reference is quoted as *Crematogaster* (*Acrocoelia*) *laeviuscula* Mayr.

The listing of an older name, *Oecodoma* (*Atta*) *bicolor* (30), is not explained. If it is actually older, it must be accepted as the correct name, unless it is a junior homonym, which is not indicated.

The abbreviation *Cr.* in item (31) can only be an error or a quotation. If it is the author's purpose to quote each of these citations directly from the originals, that fact should be made clear. Other synonymies in this book vary so greatly that the inconsistencies appear to be due to carelessness.

Item (33) shows the name *C. laeviuscula* when its taxon is reduced to subspecific status. Item (34) shows it with a variety added. *C. clara* was first described as a distinct species (31), then reduced to varietal status (34), and must be treated under the 1961 Code as a synonym, because varieties are not recognized within subspecies, so far as nomenclature is concerned.

In item (34), the comma after "var. *clara*" could have been intended to show that Emery was not its original author. The same device should then have been used in (32), (33), and (36). Its absence in these makes its presence in (34) appear to be an error.

(V)   ***Canis latrans lestes* Merriam**   (37)
Synonymy:
    Canis lestes Merriam, 1897   (38)
    Canis latrans lestes, Nelson and Goldman, 1932   (39)
    Canis estor Merriam, part, 1897   (40)

This example illustrates the intentional use of the comma to show that the following persons were not the original authors (39). Item (40) shows another sort of synonym, because only part of *Canis estor* is the same as *Canis lestes*. *Canis estor* may still be an acceptable name for some other species or subspecies, but some of the original specimens were misidentified. There is some possibility of misconstruing this listing, because it appears at first that *C. estor* is a synonym of *C. lestes*. The key to the different nature of this item is of course the word "part," which is sometimes expressed as in part, *ex parte, pro parte,* or *p.p.*

The use of the comma in this example is fairly well standardized in modern works (although the Code does not approve of the omission of the original author's name), but in older works the comma is sometimes encountered between the name of the animal and the name of either original or subsequent author, or it may be replaced with a period, as:

A. *Staphylinus variegatus*, Dejean, 1821   (41)
    Emus variegatus, Fauvel, 1877   (42)

B. *Noctua marginata*. Fabricius   (43)
    Heliothis marginata. Ochsenheimer   (44)

The synonymy of generic names is a much more complicated matter than that of specific names, in spite of the obvious fact that only one name (a uninominal one) is involved instead of two names (a binomen). Much of this will have to be left until the problems of genotypy are discussed in Chapter 25. The more obvious things to be read in generic synonymy are cited here:

(VI)   *Quedius* Stephens
Synonymy:
    Quedius Stephens, 1829a, 1829b, 1832   (51)
    Raphirus Stephens, 1829a, 1829b, 1832 (Subgenus)   (52)
    Aemulus Gistel, 1834   (53)
    Sauridus Mulsant and Rey, 1876 (Same as *Raphirus*)   (54)
    Ediquus Mulsant and Rey, 1876 (Subgenus)   (55)
    Ediquus Reitter, 1887 (Not Mulsant and Rey, 1876)   (56)
    Quaedius Ragusa, 1893 (Misspelling)   (57)
    Farus Blackwelder, 1952 (New name for *Ediquus*   (58)
        Reitter) (Subgenus)

The original publication of this name was not unambiguous. In three successive works, Stephens proposed this name as new. It is necessary to recognize the exact dates and to record the fact that the supposed first publication in 1832 was not the earliest one. Even if the earlier publications are later proven to be invalid, they must be cited to refer to included information; but in that event the genus would date from the first acceptable publication, whenever that was.

At the same time (in all three works) another name was proposed (52) for certain related species. This name *Raphirus* included species that are now believed to be in part of *Quedius* but noticeably different from the species which Stephens put under that name. This second group is therefore included in *Quedius* but as a subgenus named *Raphirus*. At the generic level *Raphirus* is a synonym of *Quedius*, a partial synonym. At the subgeneric level it is distinct, a subgenus.

It is part of the synonymy of *Quedius,* as above, but it can also be listed separately from the other synonyms under a separate heading: Subgenera.

Aemulus is simply an unnecessary name given to *Quedius* by a man who didn't like the latter name. Such an excuse is not acceptable as a reason for changing names, so *Aemulus* becomes a useless synonym.

Sauridus is found to be the same as *Raphirus,* so it is a synonym of a subgenus. At the generic level it is thus a partial synonym of *Quedius* also.

Ediquus Mulsant and Rey was proposed for part of the genus *Quedius* and is now accepted as a subgenus of the latter. But eleven years later the same name was proposed independently by another author for yet another part of *Quedius* (56). This second *Ediquus* is also a subgenus, but that name cannot be used because it is a homonym, preoccupied by *Ediquus* Mulsant and Rey. It did not receive another name until 1952, when Blackwelder renamed it *Farus* (58). *Farus,* with synonym *Ediquus* Reitter, is a subgenus of *Quedius,* and thus a partial synonym of it also.

(VII)   ***Crematogaster*** Lund

Synonymy:

Crematogaster Lund, 1831	(59)
Acroceolia Mayr, 1852	(60)
Crematogaster Fitch, 1854	(61)

In several other places in this work the second name (60) is spelled *Acrocoelia,* which can be shown to be correct. So *Acroceolia* is a misspelling here. The second reference to *Crematogaster* should not have been credited to Fitch, who merely referred to the Lund name. It should be written thus:

Crematogaster Lund, Fitch (1854)	(62)

(VIII)   *Notemigonus* Rafinesque

Synonymy:

Abramis Cuvier, 1817	(63)
Notemigonus Rafinesque, 1819	(64)

According to the synonymy given, the oldest name is *Abramis.* No reason is given for rejecting the name in favor of *Notemigonus,* so the use of *Notemigonus* is not justified by the synonymy. (In other works this combining of *Abramis* and *Notemigonus* is not accepted, but the synonymy given here can only mean that its author thought at that particular time that the names were synonyms.)

(IX)   Genus *Chloroperla* Newman
Synonymy:
    1836. *Chloroperla* Newman   (65)
    1841. *Isopteryx* Pictet   (66)
    1862. *Isopteryx* Hagen   (67)
    1892. *Isopteryx* Banks   (68)
    1907. *Chloroperla* Banks   (69)
    1909. *Chloroperla* Enderlein   (70)

The chronological arrangement is emphasized in this synonymy. After the listing of the original reference to *Chloroperla* (65) and to *Isopteryx* (66), the later citations refer merely to uses of these names by later authors. They should be written *Isopteryx* Pictet, Hagen (1862) (67), *Chloroperla* Newman, Enderlein (1909) (70), and so on.

Bibliographic references. These are the devices used to lead the reader to source material, reference works, and further reading on a given subject. Most references are abbreviated to some extent, and nearly all are cryptic. The welcome exceptions are those that are well classified and annotated, so that the reader is left in no doubt about the nature of the contents. Much information on bibliographies is given elsewhere in this text. We are here concerned only with the means of extracting from the references the information necessary to enable us to procure the book or paper for further examination.

References to taxonomic literature occur in a variety of situations, such as (1) footnotes, (2) synonymies, (3) informal lists of works cited, and (4) formal bibliographies. Although the forms of these vary somewhat, the problems remain much the same for the user. The following are samples taken at random from a variety of books:

Merriam, N. Amer. Fauna, 2:20, October 30, 1889   (1)
Linsdale, Amer. Midland Nat., 19:180, January, 1938   (2)

The book containing these has a list of literature cited, in which the Linsdale item is given in full. Under C. H. Merriam there are four papers listed but not the one given here, so the reader will have to identify N. Amer. Fauna, find out the rest of the name of Merriam, and discover whether 2:20 means volume 2, page 20 or something else.

Haeckel 1896, Festschr. Gegenbaur, 34.   (3)

The context of this reference shows it to be concerned with echinoderms. The date will very likely show what type of reference to look in to identify it further. As Festschriften are usually not multi-volume works, the 34 probably refers to a page.

Logan, 1890, Pac. Rev., vol. 7, p. 40　　　　　　　　　　　　　　　(4)

Here either the author or the serial may lead to identification of the work. The context may help.

T. D. A. Cockerell, "Fossil Mammals at the Colorado Museum of　　(5)
　　Natural History," *Scientific Monthly*, vol. 17, no. 3 (September
　　1923), pp. 271–77.

Here nothing is left to the reader except where to obtain a copy of the work. This is an excellent reference, giving all appropriate information and not using unreasonable abbreviations. (The page reference would have been clearer as 271–277.)

D. L. Frizzell, "Terminology of Types," *American Midland*　　　　(6)
　　Naturalist, vol. 14 (1933), no. 6, pp. 637–68.

This is a footnote from the same work as (5). In comparison with (2), its superiority is evident. However, it takes two lines of print instead of one, which can be a serious drawback if it is referred to more than once.

Many editors prefer, and may insist upon, abbreviation of part of the reference. For example:

Jacobson, M., et al. *Science* 132:1011–1012, 1960　　　　　　　　(7)

Where there are no complications, this can readily be translated into vol. 132, pp. 1011–1012. If the periodical was issued in several simultaneous series, in sequential series with duplicating volume numbers, or in parts with duplicating page numbers, this simple arrangement would be hopelessly clumsy. A fictitious example is:

Ann. Nevada Acad. Knowl., A:(2):1:(3):1–3.

There is no way to guess that "A" stands for subject series A (Agriculture), "(2)" for second series of volumes, "1" for volume one in this series, and "(3)" for third part of the volume. Only the page numbers are fairly obvious.

Nothing but experience with books and bibliographies will circumvent these difficulties. Punctuation and explanatory words are intended to aid the reader; their omission adds to his problems. In reality they take so little space that the saving by their deletion can scarcely be measured and probably saves nothing at all in costs.

Summary

The recovery of published data about any kind of animal or any group begins as a simple problem in the use of the library. Standard reference works may give a start in obtaining the relevant publications. The librarian can be of great help.

Before long, the technicalities of zoological nomenclature and of synonymy may complicate the problem to the point where only a considerable experience with the literature of that group will enable the taxonomist to discover all the relevant works. Nothing less than all will suffice for high-quality taxonomic studies.

Many specialized reference works are available. Some are not well known and require careful search. It is not uncommon for valuable reference material to be published under misleading titles or buried in other works. There is thus no substitute for continuous detailed study of reference material as well as special works on the animals themselves.

Books have been published with so many unusual features, in so many editions, in such a variety of parts, in so many languages and language forms, that only constant alertness will avoid misinterpretations, errors or omissions that can seriously affect taxonomic work.

Other aspects of recorded taxonomic data are discussed in Chapter 7, Recording the Data; Chapter 15, Descriptive Taxonomy; and Chapter 16, The Publication of Data.

CHAPTER 15

Descriptive Taxonomy

Whether taxonomy be defined broadly or narrowly, there are several activities included in its scope. In the broadest sense, taxonomy includes all the procedures of systematizing. In the usual restricted sense, taxonomy includes the segregation, description, and naming of species and genera, the cataloging and identification of specimens, and the publication of the data, descriptions, and names. The term is thus still too broad to be the subject of a chapter in this book, and it is therefore dealt with in several chapters on the various aspects of taxonomy.

The describing of species and genera has occupied a large part of the time of taxonomists from before Linnaeus. It is the principal means of making known the kinds of animals and their distinctive features. New species are described to make their existence known to other students, and species already known may be redescribed at later dates when more information becomes available about them. Every revisionary and monographic study describes or redescribes the genera and species included. Description, in one form or other, is one of the essential steps in the taxonomic system.

Description is not the very first step taken by a taxonomist. The steps may vary according to the circumstances under which he becomes interested in a certain group of animals. Oftentimes, the sequence of events is as follows:

When a group of animals interests a student so much that he decides to do taxonomic work upon it, his first inclination is to collect, to amass examples of the available kinds, to preserve them in the accustomed way, and to use this collection in his studies.

Next he will try to identify these species, to establish their identity with previously known species. If they are previously known species, this will supply him with the scientific name for each one. If any

279

prove to be unknown species, they will stand in his collection as "new species."

This identifying will acquaint him with some of the recent literature on the group, or a museum where there is an identified collection. He will now become interested in all the literature of the group, back to 1758 if necessary. He may well take to collecting this literature almost as interestedly as he does the specimens themselves.

The complexity of the literature and the multiplicity of Latin names therein will very likely start him cataloging. This may take the form of cards which represent the detailed facts from the literature, to be rearranged into whatever pattern serves to help the taxonomist understand the various species and all that is known about them. A good job of cataloging requires an adequate search for pertinent literature and an adequate analysis of that literature.

By now he should have such a command of previous work that he can make certain of his identifications, but he will want to check these further by comparing them with the opinions of other people. The literature may contain many such opinions, but he will want to study specimens in museums and obtain named specimens from other sources, such as by exchange, purchase, or borrowing. If he has made the most of all opportunities so far, he is now in a position to begin a serious study of this group.

By study, at this point, we do not mean taxonomy. Now is the time for him to study the named species before him—those specimens available, the descriptive data in publications, *and* the natural history of the group in every aspect that is known or can be observed. He *must* learn how the sexes differ, the forms assumed in the life history, the seasonal or environmental forms produced, and the castes that occur. He *must* know how the animals live, with what other organisms they are associated, how they perform the various functions of living, and how they reproduce their kind. He *must* find out how individuals of one species vary and how the species are similar to and different from each other.

The word student was used above for this beginning taxonomist. It is a most appropriate word, because his work so far has been largely study—of specimens, of literature, of life histories, and of variations. The studying does not end here but goes on through all the taxonomic work he will do. The most successful and respected taxonomists can still appropriately be called students of the animals on which they specialize. The study never ceases. It should include the investigation of new ways to compare the animals, new methods of extracting information from the specimens, and new methods of collecting, preserving, and examining them. It should seek to apply new discoveries made in other aspects of biology to his own work. It should always strive to increase the general background knowledge and to broaden the perspective with which the specimens or species are examined.

If the student now has the background knowledge, has assembled an adequate collection and an adequate library, if he has learned the characters (or features) by means of which this group can be classified, and if he is prepared to avoid the special pitfalls of this group as well as those common to many groups, he is now ready to undertake descriptive taxonomy—the public discrimination and description of the species.

Some people would say that he must also learn what a species is and how one can be recognized in practice. Inasmuch as there has never been effective agreement upon either of these points, it must be concluded that he will have to rely on his knowledge of past work, of modern concepts, and of the nature of the animals concerned.

The actual description is the simplest part of this sequence. By observation of specimens, measurement, tabulation of data, statistical analysis, and so on, he will determine the features of each species and record them in words, figures, drawings, or photographs. His description may be composite or it may be taken entirely from one specimen.

If the species described has a name already, this will serve to connect the new description with the older literature and labeled specimens. If the species does not have a name (a new species), the taxonomist is in for an even more rigorous series of steps—this time ones forced upon him by the opinion of world taxonomists. He must propose a name for this new species—one acceptable to the taxonomic world as expressed in the rules of zoological nomenclature then in force. The steps involve the selection, spelling, and publication of the new name; they need not be itemized here. The application of any acceptable scientific name to a species serves at least two purposes. It provides a label for referring to the species, and it shows the opinion of the namer as to the genus to which this particular species belongs.

A description, or a series of descriptions, may be useful to a student in his further studies, but to be part of science in fact they must be communicated to others. This is accomplished principally by publication, which involves both printing and distribution. (The forms and requirements of publication are described in later sections.)

In a sense the work of the taxonomist ends here. To be sure, he does not stop studying the animals. He does not stop identifying, naming, or classifying. But most of these steps are now repetitions of steps already described. The exact order given here is not always followed, and at every step there may be deviations, skips, or additional steps.

These ten steps have really covered only the study and description of one species. A much larger task of the taxonomist is the revision of groups of species. This involves the same steps in even more comparative manner. It also involves more bibliographic work and the preparation of keys for identification of the groups and species. Here descriptive taxonomy merges into classification.

Among these eleven steps, there are a number of problems, some obvious and some hidden, but all posing difficulties for the beginning student or even for the expert. Some are discussed below, in order to make clear the nature of these problems and suggest some of the things that individual taxonomists may do to avoid the difficulties.

Observation of features

All taxonomy and, indeed, all biology begins with observation of the features of individual organisms. These features are sometimes referred to as characters. Some of the implications of this latter word make it unsuitable in a discussion of generalities, because the "characters" used in taxonomy must not be restricted to any one type of feature. All of the aspects mentioned in Chapter 10, Comparative Data, and any other comparative data ever discovered, must be included here.

Observation can be accomplished with the aid of any of the usual senses, but by far the largest number are observed with the aid of the eyes. Many others are detected by instruments with results that are again visual. Even the results of experiments must be observed. In most ordinary taxonomy individuals, either living or dead, are directly observed visually. The features deemed to be worthy of description are then recorded as a written description or graphically in pictures.

At this point it is of primary importance to observe everything— the usual taxonomic features, any unusual features of possible taxonomic value, and any features of the biology of the animals. Not everything that is observed is worthy of recording, or at least of publishing. The important thing is to *see* all possible features so that *the most appropriate ones* can be used and *no necessary ones* are missed. What is worthy of recording is an almost entirely subjective thing, varying from group to group. In a general way, it is proper to record the type of information found in earlier work to be useful in the particular group *and* such additional and perhaps unusual features as the observer thinks may prove useful in the future. The indiscriminate recording of all features has sometimes been advocated, but it defeats its own purpose. The volume of data alone would obscure the effective items; the person best qualified to judge the usefulness of the different data in taxonomy would abrogate his role of making this distinction; and the literature of descriptive taxonomy, already voluminous, would become completely overpowering.

It should be pointed out, however, that the use of only the customary procedures for observing, and the recording of only the customary data, will never make more than a minimum contribution to knowledge. It is always important to add to the useful techniques, to record new types of data that can lead to superior discriminations, and to clarify earlier work by more exacting study of the customary features.

Preparation of material. The observation of the features of the specimens generally requires some preparation of the specimens for study. This may be the skinning of mammals, the pickling of fishes, or the pinning of insects, but it is oftentimes also the preparation of microscope slides, the spreading of insect wings, the relaxation of insects for dissection of genitalia, the cleaning away of the matrix rock from fossils, the staining of translucent features, and so on. Effective preparation is essential to good taxonomic work, especially at the descriptive level.

> "The important thing in the preparation of material for study is not the following of any particular set method, but rather the treatment of material in such a way that the structures which it is necessary to employ in systematic work (that is, those structures, regardless of size or obscurity, which the study of comparative morphology shows to be significant for such purposes) shall be rendered clearly visible" (J. C. Chamberlin, 1931, p. 20).

If one reads here for the word structures, all features that can be examined, he will have a most appropriate rule for all preparation of specimens for descriptive work. It was Chamberlin's belief that the searching out of new methods of observation was of critical importance. Only through gradual improvement in methods can taxonomy improve its results. One example may be appropriate here. A few years ago Dr. Adam Bøving, beetle larva specialist at the U.S. Department of Agriculture, was studying the tiny almost colorless grubs of the family Anobiidae. He laboriously dissected to examine musculature, because external features were insufficient for his purposes. Then a non-biologist equipped his microscope with polarizing lenses, and he found that he could see with the polarized light most of the surface muscles without dissection, and in other cases muscle fibers and bands were much more clearly discernible. Thereby a real advance was made in the study of these animals as well as a better classification of them.

The series or sample. There have been many statements in recent years about the importance of samples and sampling techniques in taxonomy. In classification, where species are being grouped and distinguished, samples of the species, or representatives of some sort, must be used. In descriptive taxonomy, it is seldom that sampling is necessary or desirable, especially in isolated descriptions at the species level.

The "series" of the taxonomist is not a sample in the statistical sense. It simply consists of all the available specimens. These are usually a few individuals, collected at isolated times and places, and perhaps a few larger collections or lots each containing a number of specimens taken together at one time and place. If the species is common, these collections will themselves sometimes be samples of what was present at that place. Rarely it will be a carefully selected statistically unbiased sample, but more often it will simply be the specimens obtainable at that moment in that place, biased by the collector's interests, by the method of collection, and by other factors.

The use of such a mixed series, specimens from many places taken at different times by different people using different collecting methods, can have disadvantages, but frequently the advantages outweigh them. The most effective feature of such use is that it tends to eliminate the effects of the bias of any individual sample. In many cases it is superior statistically to any single intentional "sampling."

In the description of a new species, the series of specimens before the taxonomist usually consists of a few specimens taken at one time in one place, either because this is the only time the species has been seen or because it exists nowhere else. The series available will vary from this minimum up to an extensive collection from many places, available because the new species is being isolated from a previously known but composite species, in which the mixture of species was undetected. In the former case, it may be that little will be known about the species beyond what the few specimens show. In the latter case, there may be a great deal of information about the new species, including what the specimens show but also including its biology, its bionomics, and its relation to human welfare.

Those who assume that taxonomy is nowadays concerned with populations and not types will also feel that there must be a large enough series to make possible the drawing of inferences as to the total variation in the population from which the specimens came, but in most groups of animals the majority of new species are still described from a very few specimens—*solely because that is all that are available*. Even in revisionary work, it is generally true that many of the included species are so rare as to be available in very limited numbers, often in no single series of statistically useful size. Formal, studied inference as to the variation in a real but unseen population is almost never undertaken in a direct sense in taxonomy. It may be more important in population dynamics and evolutionary studies.

Taxonomic characters

A great deal of taxonomic work is based on observation, comparison, tabulation, and analysis of features of individual animals. These features may be positive or negative, involving presence or absence of structures, qualitative or quantitative, and structural, functional, behavioral, or developmental. The features are aspects of the comparative data described in an earlier chapter. In any kind of data each animal has certain features which can be isolated, described, and dealt with. These features are the pool from which the characters used in taxonomy and classification are drawn.

A taxonomic character has been defined as any attribute of an organism by which it differs from an organism belonging to a different taxonomic group. An important point is made here that taxonomic characters are possessed only by organisms, that is, individuals. In the sense of structural and physiological features, groups such as species and genera do not have characters. The features which we ascribe to such groups (such as generic characters) are merely the features which all included comparable individuals have in common.

Few taxonomic characters are shared by all members of a species. Immature specimens, damaged specimens, the wrong sex, may not show the features of the ones chosen for description. A character is thus a feature shown by all appropriate specimens at appropriate times. Few taxonomic characters are absolutely restricted to the species at hand. They do not have to be exclusive to be characters. They will, however, be useful characters only if they occur in a pattern that enables the taxonomist to use them.

Characters that are completely distinctive in one species or group are called *diagnostic characters.* They are the features which have been particularly chosen because they are sufficient to distinguish the individuals or taxa concerned from all other such individuals or taxa. They are always taxonomic characters, but not all taxonomic characters would serve well as diagnostic characters. A description will usually list many characteristics that are at least partially distinctive, but only the most distinctive and stable of these would be singled out for a diagnosis.

In some discussions of taxonomic characters it is assumed that evolutionary history is involved in the features used in taxonomy. Although such features certainly did have an evolutionary history, the history is never known as a fact. It is not the phylogenetic nature of the features that makes them useful in taxonomy but their comparative nature—their stability, distribution, and clear-cut delimitation. The extended discussion under this heading in Mayr, Linsley and Usinger (1953, pp. 105–124) deals primarily with the kinds of taxonomic data (see Chapter 10, Comparative Data) and the phylogenetic aspects of characters.

Selection of characters. The means actually used in making the choice of what characters to use in classifying species has been discussed briefly in Chapter 12 (The Practice of Classification). Similar methods are used in the selection of characters for discrimination of species.

It is not hard to see what is required of the characters—that they be universal in the appropriate part of the species and that they be distinctive for this species. It is quite a different matter to state how they may be selected to produce these results. There is no question that the selection is largely subjective, but this does not mean that it need be arbitrary or based on guesses. It will be based on the accumulated experience of the taxonomist, especially his knowledge of the literature. The latter will recall to him what characters have previously been used, and which of these have proven to be effective and which have not. His knowledge of the animals and their life will enable him to judge what new characters are suitable and what new techniques may yield helpful new characters.

The taxonomist will thus observe many characters of the animals. He will intuitively or statistically evaluate the effectiveness of each character with reference to several things: (1) its stability within the species or group, (2) its distinctiveness in separating one group from the others, (3) its availability on the specimens which will have to be classified or identified, and (4) its comparability with the descriptions and classifications already published on this group. None of these factors can be neglected with safety. Preliminary surveys of the features of the specimens may be helpful and so may statistical or graphic analysis of variability.

Good characters. Obviously selection is aimed at finding the most effective characters. Some taxonomists may believe that this means the ones that show the real genetic relationships between the species.

Others may think it means the maximum correlation of attributes. Still others may think it means merely that the results give the most satisfaction in identifying, classifying, and other taxonomic work. Whatever the views on this, the characters thought to be most satisfactory are called *good characters.* They are simply the ones that produce the desired results.

There are, however, a few attributes that can be ascribed to good characters in general: (1) They are not subject to wide variation among the known specimens, (2) they do not show a high intrinsic genetic variability, (3) they are not readily modified by the environment, (4) they are consistently expressed, (5) they are available in the specimens which must be used, (6) they are visible with reasonable procedures, and (7) they can be effectively recorded. Of these the ones representing the inherent nature of the animals, (1)–(4), are the most important, but failure to consider the others can seriously affect subsequent use of the characters.

Measurements. An integral part of most description is the citation of pertinent measurements. These may be the dimensions of the whole animal, the dimensions of certain parts, ratios of any two dimensions, body or organ weights, or other numerical or mensural data. In most groups of animals, some measurements are considered to be desirable or necessary. In some cases, elaborate measurements and ratios may be necessary. There is, however, no reason to use either measurements or ratios unless these add to the usefulness of the description.

Five criteria of good numerical observations are listed by Simpson, Roe, and Lewontin (1960, p. 20) thus: "they should be logical, related to a definite problem, adequate, well delimited, and comparable and standardized."

Statistical characters. It is customary to consider statistical analysis to be a major part of taxonomy. It certainly occupies an increasingly important place in many branches of biology. However, in the description of taxa, which means the description of the features of individuals or of groups of like individuals, there is little place for statistics as such. Numerical data will often be useful, but it is not usually possible to deal statistically with the features of one specimen or with any features held in common by a group of individuals. The place where statistics becomes useful is in the study of populations. In spite of some published statements that taxonomy deals basically with populations, it is easy to see that almost all descriptive taxonomy deals solely with individuals and with groups of individuals which hold the pertinent features in common.

The statistical characters used extensively in the study of micro-organisms are numerals representing growth rates, reactions of cultures to experimental treatment, population or culture features, etc. These are true statistical characters of populations. They do not represent features of individuals. Although at some levels animal taxonomy can employ such characters under special circumstances, they are not usually available. They are not among the features that are herein described as comparative, because taxonomy does not usually compare populations.

Many characters of animals can be converted into numbers, ratios, proportions, etc. These are sometimes supposed to be statistical in nature. However, real taxonomy does not normally result from statistical manipulation of such figures, no matter how useful they may be elsewhere in biology. Some taxonomic uses of statistics are discussed in Chapter 10, Comparative Data.

Description

The word description is applied to two slightly different things. In dealing with species, the published statement of its features is called a description. This may be long or short, detailed or general, verbal or pictorial, but its purpose is to convey to others the features of this species (or specimen) that are deemed to be of interest in further study. Some workers think that all observable details should be described, and they sometimes produce descriptions that occupy many pages apiece. Other workers feel that economy and even usefulness demand that they restrict themselves to what seem to them to be the useful characters. No one has ever produced a really complete description, even in the cases in which an entire book has been devoted to one species, and taxonomic practice generally expects a carefully selective description that will permit detailed comparison with the descriptions of the other species in the genus.

In reality there is no such thing as a description of a species. The species is merely the sum of the individuals. The actual features described are therefore the features of the individuals available. No two specimens are ever exactly alike, so a description may include the range of variation of each character, analysis of the statistical frequencies, and comparison of these with the geographical distribution of the species so far as known.

There is another purpose for describing; according to convention, a new species can be named only if it is described, but the nature or extent of the description is not specified. It is best to fulfil this legal requirement by preparing the best possible (that is, the most useful) description, so that both purposes are served at one time. The legal aspects will be discussed in Chapter 26, The Names of Species.

The taxonomic purpose of the description is to enable subsequent workers to recognize the group described. There have been attempts to restrict the word description to a particular aspect of recording data—the general features as opposed to the comparative features that distinguish the taxon from other taxa. This seems to yield no useful distinction, because any adjective is descriptive and no one ever knows for sure what characters do in fact distinguish all forms of the two species effectively in the light of later discoveries.

No one would attempt to write a description without prior experience. In determining that a species is new, or assembling material for a revisionary study, a general familiarity with the characters and descriptive problems in this group must be gained before starting on the work of writing the new description. This advance knowledge of descriptive characters has both good and bad aspects. It is helpful to have all descriptions in a group comparable in style, characters, and descriptive terminology, so that they can be compared. But it is unfortunate if following the old format prevents the taxonomist from seeing and recording new kinds of characters not previously utilized.

The word *description* is often taken to mean all the material published together about that species. This may include any or all of the following, depending in part on whether or not the species is new:

The name of the species
The author of the name
The date of publication of the name
N. sp. or New species (if appropriate)
Reference to illustrations
Synonymy
Diagnosis (a condensed statement of distinctive characters)
Description (a more detailed exposition of characters)
Measurements
Type locality
Distinguishing features (in case a formal diagnosis is not given)
Specimens examined
    Collector
    Location (in what collection)
Other localities represented by the specimens
Types and their location (in what museum)
    Collector
Geologic occurrence

Discussion of variability among the specimens
Remarks on any of above, on habitat, habits, or other
List of material examined

In the case of a genus, the description may and should include most of the following items:

Name of the genus
Author of the name
Date of publication of the name
N. gen. or New genus (if appropriate)
Synonymy
    Subgenera
Genotype, name of species
    Manner of fixation
    Date of fixation
    Name of person designating
    Synonymy of this species
    Original genus
Genotypes of all synonyms and subgenera
Distinguishing characters of the genus
Diagnosis (a condensed statement of distinctive characters)
Key
Notes on relationships
Included species (in addition to the genotype)
Notes on history of genus
Source of material

A formal style is desirable in describing species or groups, but the style should not interfere with the presentation of relevant information. A final paragraph of Remarks can always serve to include any information that doesn't fit in the formal arrangement. It should never be assumed that *any* information will be of no value, as later workers may recognize a value not obvious to the describer. Of course, there are practical limits to the descriptive material.

In the description itself there should be a definite sequence of features, preferably the same ones used by other workers. In insects, for example, this might be color, general appearance, head, antennae, mouthparts, eyes, thorax, wings, legs, abdomen, and genitalia. This sequence makes it much easier to compare descriptions and to find out if a certain feature is described. Measurements and counts should be used whenever appropriate. Illustrations may be more useful than words, but poor figures may be worse than none. In many groups photographs are of little value in description. They cannot compare with careful drawings of diagnostic characters or features, in most cases, although they may be effective in showing posture, habitat, or natural appearance.

Some individual points worth considering are the following:

1. Uniformity of style in presentation.
2. Comparison directly with similar species.
3. Record absence of features as well as presence.
4. Excessively long descriptions may obscure the important points.
5. Very brief descriptions may omit needed characters.
6. Telegraphic, concise wording, however, does save space.
7. Italicize key words, or otherwise emphasize them.
8. Where analyses of variations were made, include summaries.
9. Try to describe so that the species can be understood by a person who does *not* have specimens at hand, or by an inexperienced person.
10. Try to improve on previous descriptions in the group, in accuracy, usefulness, breadth, and depth.

Graphic descriptions. Verbal descriptions are not the only means of recording the features of specimens or populations. The use of graphic representations is increasing in importance and in some groups is of greater utility than words. Some technical journals have rules requiring the illustration of all new species. The rules of nomenclature have always accepted a "figure" of a specimen in lieu of a verbal description, although, regrettably, the new Code is not clear on this point. There are, however, illustrations that leave much to be desired, or are simply downright useless. There is a tendency to believe that "the camera cannot lie"; but photographs *can* distort, *can* overemphasize, and *can* reproduce so poorly as to be ineffective. There are many taxonomists who believe that photographs are seldom as useful as good drawings.

The case for the use of drawings is well made in the valuable but out-of-print book *The Principles of Systematic Entomology,* by G. F. Ferris. In the chapter entitled Entomological Drafting, which is equally applicable to animals of other kinds, Ferris makes these comments:

"No purely verbal description, with the hampering limitations that are inevitably associated with it, can thus form a satisfactory substitute for the (described) object. It is possible to describe a simple geometrical figure in terms of words, but as these figures become more complex the difficulty is enormously increased. How much more difficult will it be then to describe an insect with its complex spatial relations of parts, its structures of varying form, its variety of curves, its various axes, its different planes, its wealth of detail! The task is one for which words are an unsuitable tool. *There remains but one other way by which this ideal may be approached, and that is by means of some sort of pictorial or graphic presentation."*

Sometimes the illustrating of taxonomic material is referred to as art-work. The implication that art has a major place in scientific illustration is unfortunate; accuracy of representation is the chief criterion by which such illustrations should be judged. There have been some highly skilled and artistically talented illustrators, but art must be a secondary consideration.

The one essential rule in all taxonomic drawing is given by Ferris thus: *"Never cease studying the object until the drawing is finished."*

Ferris also gives some specific instructions for making drawings. Others will be found in Mayr, Linsley, Usinger (1953), where there is also a short bibliography of books on this subject. (See also works cited on p. 319.)

Although in some specialties there is a strong feeling about the inadequacies of photographs as a part of descriptions, they have been effectively used in other instances. It is sometimes both possible and effective to give photographs to show the general appearance, even when details must be shown by drawings. There are of course other uses of photographs where drawings will not serve. These are in showing the habitats in which the animals live and other non-taxonomic features. Use of photographs in publications may introduce difficulties or extra expense and must be planned in advance. In straight description of specimens, diagrams are inappropriate. In the representation of life histories or in the presentation of statistical data on populations, they are essential. They should be used whenever they will contribute to an understanding of the species.

Types in descriptive taxonomy

The relation of type specimens and the ideas of typification are discussed in Chapter 11, Species and Subspecies. In descriptive taxonomy, types are of considerable direct importance, and their employment is outlined here.

The use of types has grown with the development of taxonomy. By the end of the nineteenth century, they were being used by many taxonomists. Some of the early codes of nomenclature contained some reference to them, but, although some abuses arose, the *International Rules of Zoological Nomenclature* (1905) made no mention of the types of species. The nomenclatural use of types is unequivocally required by the current Code, which regulates their employment. This does not, however, make it necessary to use types in any other way. Nevertheless, taxonomists do, with very few exceptions, select certain specimens as the basis for their descriptions, and these specimens are universally known as types.

When a describer has only a single specimen of a new species, or when he chooses to set aside a single specimen, it is the *holotype* (original single type). If he has other specimens which he considers identical, he may designate these *paratypes* (subsidiary original types). If one of the paratypes is of the opposite sex to the holotype, it may be called an *allotype* (first paratype of the other sex). If the original describer of a species does not set aside a holotype, his entire series is considered to be a type series; each specimen is a *syntype* (formerly called cotype). Later on, another taxonomist may select one of the syntypes to be a *lectotype* (single type selected subsequently). The remainder of the paratypes are then called *paralectotypes* (subsequent paratypes).

If a holotype is definitely lost or destroyed, another of the original syntypes is selected to take its place (it also becomes a lectotype). This could be chosen from what are called cotypes, syntypes, or paratypes. If the holotype and all other type specimens are lost or destroyed, some other specimen can be designated as a *neotype* (a new holotype) to take its place, under very rigid regulations.

The types are used as standards. There can be no question that a type (at least holotype, lectotype, or neotype) belongs to the species it typifies. By comparison with other specimens, the taxonomist determines if they too, in his opinion, belong to that species.

Although other specimens than the holotype may be used in the description, the type is always in complete agreement with all its details (except where some items were intentionally included to show variation). The description thus acts as a verbal picture of the type (and perhaps other specimens) for the use of persons who cannot examine the type itself. The paratypes are thought to be unquestionably conspecific with the type (belonging to the same species), and they are often sent to other museums where they can act as stand-ins for the holotype. Thus, the descriptions and the paratypes serve to spread knowledge of the features of the type without risking the unique specimen by frequent shipment to every interested student.

If any question arises as to exactly what the species is, especially if it is found that the original series of specimens (holotype and paratypes) consists of more than one species, the holotype is the court of last appeal. It is always a representative of its species, even if all the paratypes prove to be something else that was originally confused with it.

A good description includes two distinct things. It cites the features of the holotype in detail, and it cites any variation in these features shown by the other specimens, as well as any features shown only by them. It should be made clear which is which. The first is a description of the holotype; the second should be a description of the entire species as nearly as it can be put into words at that time. Both aspects are important.

It has sometimes been assumed that taxonomists use the type specimen alone when describing a species, that the variation shown by other specimens is not considered, and that all "duplicates" or other specimens are disposed of or traded to other collections. These statements contain just enough truth to make it difficult to deny them, although they misrepresent the real situation. There have been taxonomists who described only one specimen; there have been cases of exchange of most duplicates; there have been cases in which the variation of other specimens was not recorded. But none of these has ever been considered good practice. Work based on such procedures probably would not be entirely worthless, but it would be less valuable than it should be.

In order to select a type specimen to describe, even the persons who may have restricted their descriptions to a single type must have examined all the available specimens. The basis of the choice may have been sex, size, condition, source, ownership, or other, but there was a choice, based on some examination. If there was little or no diversity shown by the series of specimens (one or many, as the case may be), there is small need to dwell at length on this aspect. If the series shows considerable diversity, the taxonomist will know that there is a chance that his "species" will eventually be found to consist of several. If he describes the "species" from all the specimens, including all the variation, it will be difficult to separate the description into two later on. He should therefore concentrate on one specimen, so that his entire description will always apply accurately to that one (the holotype).

Theoretically, the describer should describe the species. In practice, he can generally do so only in well-known groups where a new species is found as a segregate of a known but composite species. Otherwise, the "species" is not known but only a few specimens of it. (This is not usually even a statistical sample.)

There have been cases, especially in paleontology, in which the specimens other than the type (holotype) were arranged in sets to show the variation and these sets distributed to other collections. This is an excellent way to show the diversity in the original series. In some cases, these specimens have been labeled as paratypes. These paratypes thus do *not* necessarily show the features of the holotype. This use of paratypes is condemned by most taxonomists, who believe that a specimen labeled as paratype should agree with the holotype in all pertinent features. These are the taxonomists who deny the claim of Simpson (1961) that types can serve only one purpose, that of an anchor for the name. These taxonomists know that types (and similar paratypes and other compared specimens) are constantly being used to show what the original species was like—as standards for comparison, and as a last resort in identification—as well as for the anchor for one or more names.

No taxonomist believes that the type will tell him what the species is, because the species consists of other sexes, other developmental stages, other seasonal forms, other phases, castes, positions, and so on. But he knows that the type will tell him *one* condition that indisputably does belong to that species, one point around which he can rally the diversity, the life forms, of the species. There is every reason for treating the holotype as a specimen of unique value and use. Paratypes that agree closely with the type in every taxonomic feature are also valuable taxonomically and can spread the understanding of the original concept of the species to other collections and students. *Any* specimen that has been carefully compared with the holotype will have extra taxonomic importance because of that fact. It should be carefully labeled to show all the relevant facts of the comparison (with what it was compared, by whom, when, and to what extent).

There are scores of other terms that refer to specimens and include the root *-type*. Many of them have real value in taxonomy as specimens of particular sorts or histories. The terms, however, are usually less informative than a descriptive phrase. They should be used with great care, if at all, because their meanings are not always inherent in the term and may not be widely understood. Among these are *hypotype* (a described or figured specimen), *topotype* (a specimen from the same locality as the type), *metatype* and *homoeotype* (a specimen compared with the type and believed to agree with its features), and *plastotype* (a cast of a type).

Type-locality. Because a species (or a subspecies) had a definite geographic range and its types were collected at some spot in that range, the place of collection can be of considerable importance in taxonomy. The place at which the holotype was taken is called the type-locality. In any description of a new species, this locality should be recorded as accurately as possible, using well-known points of reference and exact distances.

This appears to be a simple concept, easy to put into practice. There are confusing features, however. Some writers believe that the type-locality is the published locality rather than the actual place where the specimen was found. If the species was originally described from America septentrionalis, the type-locality would be North America. These writers then find it necessary to restrict the type-locality, to limit it to some reasonable part of the continent. This concept of type-locality, which is in reality the published-locality, leads to endless difficulties. It has not been formally accepted in the rules of nomenclature, although there is mention of restriction of type-locality. There seems to be no good reason for not accepting the simple straightforward concept and definition that the type-locality is the place from which the type was collected.

It is seldom that a type-locality is published in the greatest possible detail. In addition to incompleteness, there have been cases of erroneous citation. Any later taxonomist can correct the statement of locality, make it more definite and explicit, or separate out the relevant part, just as he may correct and supplement the description of the species. When citing the locality of the type of a new species, it is important to record the relevant detail but not to obscure this with pointless data or ineffective minutiae.

Descriptive publications

Diagnosis. As suggested above, diagnosis is the statement of the features in which a species (or other taxon) is completely unique. These may not be the most important taxonomic characters, but they are the distinctive ones, the most immediately useful ones. Diagnosis is not really distinct from description, but it is always useful to have a separate statement of the distinguishing features. Description must give both the features which make possible the grouping of taxa and those which serve to distinguish taxa. Diagnosis denotes merely the latter function.

In nomenclature, diagnosis has an additional function. In order to publish a name for a new species, it is necessary to give some statement of the distinguishing features. This statement is in principle a diagnosis. (Unfortunately, in many cases, the actual diagnosis is so sketchy that it serves little purpose except as an attempt to fulfil the requirement of the rule.)

Revisions and monographs. Revisionary work in general is largely descriptive. Revisions and monographs usually include redescription of all previously known species as well as all new ones. The description may be on a more-detailed scale than the original, or it may be somewhat summary. Monographs tend to be more elaborate and complete than other revisions. They also generally cover a larger group of organisms.

There is nothing unique about revisions or monographs among descriptive publications. Besides the descriptions, they generally include most of the following: keys, complete synonymies, detailed distributions, records of locations and identity of types and other specimens, analysis of previous publications, and summaries of available knowledge of non-taxonomic sorts.

Keys, classifications, synonymies, and bibliographies are aspects of revisionary works which can be published alone. Their usefulness is great even when separated from the descriptive material. It is not always possible to do monographic work and, in lieu of this most complete work, any of its parts will be welcomed.

Keys. Taxonomic keys are devices for permitting the identification of taxa. The key may enable identification of species, genera, orders, or any other taxa. It may key out the species in a genus, or the genera in a family. It may key out several levels in succession. One recent key, to the phylum Pogonophora, keys out all the orders, families, and genera in the phylum. Separate keys to each genus then key out the species.

There have been a variety of styles of keys used, but those that have proven to be practical are all rather similar. Whatever real differences occur are usually the result of attempts to make the key serve other purposes than just identification. These other purposes may be classification (arrangement), sequence of primitive to specialized, phylogenetic speculation, and so on. Where these functions are appropriate for presentation, they should be separate from the key, both to prevent interfering with the function of identification and to make more effective presentation of the other data.

As long as the one purpose of the key is to aid in identification, there is little to choose between the minor forms of successful keys. The requirements of a good key are:

1. It must be workable, which means that
    (a) its arrangement must be self-explanatory,

(b) it must be simple,

(c) its couplets must be distinctive, and

(d) its length must be reasonable.

2. It must be possible to work it in reverse.

3. It must use characters that are simple, clear, and direct.

4. It must be illustrated, if necessary for clarity.

5. It must be composed *only* of couplets that consist of mutually exclusive statements, with the first one positive, the second negative, or both positive if qualitative.

6. It must show clearly its limitations as to sex, developmental stage, age, and any other pertinent condition.

7. It must deal with all appropriate taxa, including any that cannot be keyed out.

There is really only one major type of key that is widely practical. An example is given here, with notes on possible modifications.

Key to the Suborders of Coleoptera

1. Hind coxae immovably fused to metasternum, completely dividing first visible abdominal sternite (fig. 33) . . . . Adephaga

   Hind coxae not fused to metasternum, not dividing first visible abdominal sternite 2

2(1). Pronotum with notopleural sutures 3

   Pronotum without notopleural sutures . . . . . Polyphaga

3(2). Antennae filiform (fig. 8); wings with apex spirally rolled in repose . . . . Archostemata

   Antennae clavate to capitate (figs. 15 and 16); wings with apex folded in repose, never rolled . . . . Myxophaga

This key illustrates the capacity for reverse use, by means of the parenthetical numbers after the couplet numbers. The couplets consist of clear-cut characters, generally with positive condition first. Where necessary more than one character is used in a couplet. Appropriate characters are illustrated. The author has not been concerned with the order of the groups keyed out, as he deals with them later in the order Archostemata, Adephaga, Myxophaga, Polyphaga. (Slightly modified from R. H. Arnett, Jr., *The Beetles of the United States (A Manual for Identification)*.)

Other types of keys are described in the works cited below.

Chamberlin, W. J. 1952. *Entomological Nomenclature and Literature*, 3rd ed. Dubuque, Iowa; Wm. C. Brown Co., pp. 12–14.

Ferris, G. F. 1928. *The Principles of Systematic Entomology*. Stanford, California; Stanford University Press, pp. 105–107.

Mayr, Linsley, and Usinger. 1953. *Methods and Principles of Systematic Zoology*. New York; McGraw-Hill Book Co., pp. 162–168.

Metcalf, Z. P. 1954. The Construction of Keys. *Syst. Zool.,* vol. 3, pp. 38–45.

Voss, E. G. 1952. The History of Keys and Phylogenetic Trees in Systematic Biology. *Journ. Sci. Lab. Denison Univ.,* vol. 43, Art. 1, pp. 1–25.

Some taxonomists have found the making of keys to be difficult or laborious. This would no doubt be true if one set out to construct a key to an unfamiliar group of animals. The making of keys should not be undertaken by anyone who is not thoroughly familiar with the taxonomy of the group. An experienced person will know so much about the features of the group, its comparative zoology and its biology, that the making of the key should not be difficult at all at the genus or species level. There are exceptions, however, where certain groups prove difficult to key out.

As one progresses up the hierarchic scale, the making of keys becomes more difficult because so much diversity is involved. There are Arthropoda without jointed legs, without an exoskeleton, without tracheae, and so on. *Most* arthropods will key out readily, but some will upset the simple key characters. A key at this level is of little use if it works only for the obvious forms—those that could be placed without reference to the key. It is never sufficient to key out the common kinds alone and forget the exceptions. The user of the key will never know whether he has one of the exceptions or not. As pointed out in an earlier chapter, identification is a tricky business at best, and in many groups the keys are suitable only for use by specialists.

There have been times when it was claimed that keys, like classifications, should employ characters of great biological significance. This is entirely unnecessary. The important things are that the character be readily detectable and clear-cut in its alternative appearances and that these distinguish the groups already recognized by the taxonomist.

Classifications. Although keys are very common in taxonomy, formal classifications are rare, except as incidental adjuncts to other works. In some cases, arrangement is mistaken for classifying, but serial arrangement doesn't necessarily involve union into classes, which is the essence of classification.

The purpose of zoological classification is to present the classifier's views on the grouping and subgrouping of the animals involved. These views may involve belief in the phylogenetic origin of the groups, or intent that the grouping show new correlations of the features, or merely a summary of existing knowledge of the organisms.

Such classifications are useful as reference works, for teaching the

Hyman, 1940	Copeland, 1956	Honigberg et al., 1964
Phylum Protozoa	Phylum Phaeophyta	Phylum Protozoa
SP. Plasmodroma	Cl. Heterokonta	SP. Sarcomastigophora
Cl. Flagellata	Phylum Pyrrhophyta	SpC. Mastigophora
Cl. Rhizopoda	Cl. Mastigophora	Cl. Phytamastigophorea
Cl. Sporozoa	Phylum Protoplasta	Cl. Zoomastigophorea
SC. Telosporidia	Cl. Zoomastigoda	SpC. Opalinata
SC. Cnidosporidia	Cl. Mycetozoa	SpC. Sarcodina
SC. Sarcosporidia	Cl. Rhizopoda	Cl. Rhizopodea
SC. Haplosporidia	Cl. Heliozoa	SC. Lobosia
SP. Ciliophora	Cl. Sarcodina	SC. Filosia
Cl. Ciliata	Phylum Fungilli	SC. Granuloreticulosia
SC. Protociliata	Cl. Sporozoa	SC. Mycetozoia
SC. Euciliata	Cl. Neosporidia	SC. Labyrinthulida
Cl. Suctoria	Phylum Ciliophora	Cl. Piroplasmea
	Cl. Infusoria	Cl. Actinopodea
	Cl. Tentaculifera	SC. Radiolaria
		SC. Acantharia
		SC. Heliozoia
		SC. Proteomyxidia
		SP. Sporozoa
		Cl. Telosporea
		SC. Gregarinia
		SC. Coccidia
		Cl. Toxoplasmea
		Cl. Haplosporea
		SP. Cnidospora
		Cl. Myxosporidea
		Cl. Microsporidia
		SP. Ciliophora
		Cl. Ciliata
		SC. Holotrichia
		SC. Peritrichia
		SC. Suctoria
		SC. Spirotrichia

Figure 15-1

diversity of animals, as evidence of relations that will help solve practical problems, and so on. Much of our classification has become so well known that it is taken for granted in ordinary daily life.

There have been few really new classifications published in the past half-century. The reason for this is that most of the real groupings have been recognized, and all that can be done now is to redefine them or make changes in detail. Several supposedly quite distinct schemes differ chiefly in the hierarchic level of groups or in the names selected. Of course, shifts of position of individual taxa, splitting of taxa, union of taxa, and recognition of new levels are common in revisionary work.

For example, three supposedly rather different classifications of Protozoa are shown in parallel columns in Fig. 15-1.

At first glance there is little similarity between these three classifications, but if some of the unfamiliar names are replaced by synonyms and the nominal levels of the taxa are not indicated, these three schemes appear as follows:

Hyman 1940	Copeland, 1956	Honigberg et al., 1964
Protozoa	(pt. of king. Mychota)	Protozoa
Plasmodroma	Phaeophyta	Sarcomastigophora
Mastigophora	Mastigophora (pt.)	Mastigophora
		Opalinida
	Protoplasta	
Sarcodina	Sarcodina	Sarcodina
		(+ Sporozoa pt.)
Sporozoa	Sporozoa	Sporozoa (pt.)
		Sporozoa (pt.)
Ciliophora	Ciliophora	Ciliophora
Ciliata	Ciliata	Ciliata
(Opalinida)	(Opalinida)	(+ Suctoria)
Suctoria	Suctoria	

There are still differences between these three, but they are in reality only slight, consisting of difference of opinion as to whether two groups should be united or not at a higher level and whether an aberrant group belongs closest to one or another of the major groups. Compared to the five-class scheme given in many textbooks (Mastigophora, Sarcodina, Sporozoa, Ciliata, and Suctoria), Hyman shows two major groups: $1 + 2 + 3$, $4 + 5$, Copeland shows four major groups: 1, 2, 3, $4 + 5$. Honigberg et al. show four quite different major groups: $1 + 2$, 3a, 3b, $4 + 5$, with small parts of 3 and 4 transferred to $1 + 2$.

Classifications are the means of presenting the groupings discovered in the diversity of animals. Here the hierarchic nature of grouping becomes evident, with groups, supergroups, and subgroups. Because the taxonomist believes that the correlations that give the groupings, and thus the groups themselves, are the result of evolutionary processes, he looks at classifications to see whether they contain any evidence of the phylogeny of the groups. There usually is no direct evidence, but even indirect evidence can lead us to interesting speculation on the ancestry. It is thus from classification that most studies of phylogeny spring. There may also be feedback from the hypothetical phylogeny that will lead him to consider a change in the classification. At best, then, the classifications may be modified in detail because of phylogenetic speculation, but they are never based on phylogeny in the first place.

Phylogenetic trees. The diagrams used to record the supposed ancestry of a taxon are generally based on the prior classification of the group. This speculation is not a part of taxonomy as such, and the interested reader is referred to the extensive literature of Biosystematics, The New Systematics, and Evolution.

Synonymies. All revisionary and monographic studies must be based on the detailed study of past taxonomic work. The record of this study is ordinarily presented in the monograph in the form of a complete synonymy, which includes all such bibliographic data. A synonymy is a list of the synonyms (names) that have been applied to the taxon under consideration, including the name that is therein accepted as the correct name.

In a formal sense, the synonyms in zoological nomenclature include only such names as are acceptably published—what the 1961 Code calls "available." In actual practice, a synonymy generally includes more than just a list of the "acceptable" names and their bibliographic references. It will also include (1) any names that have been printed (but not acceptably) and may thus be mistaken for real zoological names, (2) expressions which are not names but have been attached to the taxon in question, (3) misidentifications, which are uses of wrong names for a species, and (4) bibliographic histories of all of these.

It is a commonplace error to speak of "the synonymy" and "the synonyms" of a name, meaning only its rejected synonyms. This does no harm so long as it does not obscure the fact that *all the names applied to a taxon are synonyms,* from which one is determined under the Code to be the currently acceptable one.

In arranging a list of synonyms, there are a variety of forms, arrangements, and special devices. If the list is very short, the arrangement will not be a matter of great concern, as the list can be rearranged mentally on the spot. If the list is long, it will have to be arranged for a specific purpose, to show (1) the chronological history of each name, (2) the chronology of all of them together, or (3) the bibliographic history of each name.

The most complete synonymy will quote exactly every citation of each name in the literature, with bibliographic reference and annotations about its status. For example:

X-us albus (Jones)

Y-us albus Jones, 1820, *Journ. Nat. Hist.,* vol. 14, p. 27.   (1)
Y-us albus Jones, Smith, *Journ. Nat. Hist.,* vol. 17, p. 166 (1823)   (2)
Y-us (X-us) albus Jones, Doe, *Proc. Miami Acad. Sci.,* vol. 7,   (3)
     p. 109 (1920)
X-us albus (Jones), Roe, *Microsc. Journ.,* vol. 107, p. 286 (1927)   (4)
X-us albus (Jones), Moe, *Cat. Exidae of World,* p. 428 (1935)   (5)

In this example, it is assumed that these five are the only times that this species has been referred to in print under any name. (Such a short list is unlikely for such an old name, but might occur for recently described species, especially from relatively unstudied regions of the earth.) Such complete synonymies are not common, however, because of the large amount of space required to print them. Monographers are usually forced to adopt other styles and abbreviated arrangements. For example:

B-us niger (Moss)   (6)

A-us niger Moss, 1852, p. 29; Peters, 1894, p. 66; Daly, 1914,   (7)
    p. 267; Doane, 1946, p. 144; Black, 1960, p. 422.

B-us ebonus Dart, 1924, p. 66; Doane, 1946, p. 144; Black,   (8)
    1960, p. 422.

B-us niger (Moss), Doane, 1946, p. 144; Black, 1960, p. 422.   (9)

Here, the names are in chronological order, the references are chronological only under each name; and the references are greatly abbreviated so that they must be identified in an accompanying bibliography.

In the case of widespread species, especially variable ones to which many names have been applied, it is often practicable to list only the original reference to each name or name form. For example:

Staphylinus maxillosus Linnaeus

Staphylinus maxillosus Linnaeus, 1758	(1)
Staphylinus anonymus Sulzer, 1761	(2)
Staphylinus tertius Schaeffer, 1766 (not as a name)	(3)
Staphylinus balteatus Degeer, 1774	(4)
Staphylinus fasciatus Fuessly, 1775	(5)
Staphylinus nebulosus Fourcroy, 1785	(6)
Staphylinus maxillosus (Linn., 1758) Brahm, 1790 (typ. err.)	(7)
Staphylinus villosus Gravenhorst, 1802	(8)
Creophilus maxillosus (Linn., 1758) Samouelle, 1819	(9)
Emus nebulosus (Fourcroy, 1785) Mannerheim, 1830	(10)
Creophilus ciliaris (Curtis, 1829, n.n.) Stephens, 1832	(11)
Emus maxillosus (Linn., 1758) Dejean, 1833	(12)
Creophilus fasciatus Laporte, 1835 (not Fuessly, 1775)	(13)
Creophilus villosus (Gravenhorst, 1802) Laporte, 1835	(14)
Emus villosus (Gravenhorst, 1802) Dejean, 1837	(15)
Staphylinus arcticus Erichson, 1839	(16)
Staphylinus cinerarius Erichson, 1839	(17)
Creophilus cinerarius (Erichson, 1839) Erichson, 1839	(18)
Creophilus arcticus (Erichson, 1839) Erichson, 1839	(19)
Staphylinus ciliaris (Stephens, 1832) Erichson, 1841	(20)
Staphylinus bicinctus Mannerheim, 1834	(21)

Staphylinus maxillosus **Linnaeus** (*Continued*)

Emus mandibularis (Dejean, 1833, n.n.) Mannerheim, 1843 (stillb.)	(22)
Creophylus villosus (Gravenhorst, 1802) Duval, 1856 (typ. err.)	(23)
Creophilus orientalis Motschulsky, 1857	(24)
Creophilus maxillaris Thomson, 1859 (lapsus)	(25)
Creophilus fulvago Motschulsky, 1860	(26)
Creophilus bicinctus (Mannerheim, 1843) Gemm. & Har. 1868	(27)
Creophilus anonymus (Sulzer, 1761) Gemm. & Har. 1868	(28)
Creophilus balteatus (Degeer, 1774) Gemm. & Har. 1868	(29)
Creophilus nebulosus (Fourcroy, 1785) Gemm. & Har., 1868	(30)
Creophilus fasciatus (Fuessly, 1775) Gemm. & Har., 1868	(31)
Creophilus tertius (Schaeffer, 1766) Bertolini, 1872	(32)
Creophilus subfasciatus Sharp, 1874 (var.)	(33)
Creophilus medialis Sharp, 1874 (var.)	(34)
Creophilus imbecillus Sharp, 1874 (var.)	(35)
Emus cinerarius (Erichson, 1839) Fauvel, 1878	(36)
Creophilus pulchella Meier, 1899 (var.)	(37)
Creophilus canariensis Bernhauer, 1908 (subsp.)	(38)
Creophilus ciliaroides Hatch, 1938 (aberr.)	(39)
Staphylinus orientalis (Motschulsky, 1857) Blkwr. 1943	(40)
Staphylinus pulchella (Meier, 1899) Blkwr., 1952	(41)
Staphylinus fulvago (Motschulsky, 1860) Blkwr., 1952	(42)
Staphylinus subfasciatus (Sharp, 1874) Blkwr., 1952	(43)
Staphylinus medialis (Sharp, 1874) Blkwr., 1952	(44)
Staphylinus imbecillus (Sharp, 1874) Blkwr., 1952	(45)
Staphylinus fasciatus (Laporte, 1835) Blkwr., 1952	(46)
Staphylinus canariensis (Bernhauer, 1908) Blkwr., 1952	(47)
Staphylinus ciliaroides (Hatch, 1938) Blkwr., 1952	(48)
Staphylinus mandibularis (Mannerheim, 1843) Blkwr., 1952	(49)
Staphylinus maxillosus Linn., 1758 (Blackwelder, 1952)	(50)

Many things are illustrated by this example, besides the abbreviated style. Some of these points are:

1. The species is now in the original genus (1, 50) even though it was on intermediate occasions transferred to other genera (9, 12).

2. The widespread and variable species received several names in different parts of the world (2, 3, 4, 5, 6, 8, 11, 16, 17, 20, 21, 24, 26). These are now believed to be straight synonyms (subjective).

3. One synonym (22) was published in synonymy and is therefore a stillborn synonym (objective).

4. Typographical errors or lapsus calamorum occur in the generic names (7, 23) and also in the specific epithet (25).

5. Homonymy occurs (5, 13), which is somewhat unusual within one species. The 1961 Code holds that S. *fasciatus* Fuessly (1775) and S. *fasciatus* (Laporte, 1835) are not homonyms because they apply to the same species; the Laporte name is to be considered merely as a reference to the Fuessly name. This is a procedure that could lead only to synonymic confusion, because the names were, in fact, completely distinct and are, in fact, homonymous.

6. One unacceptable name (3) appears. It has been ruled that the word *tertius* in this case is a number, not a name. Because the word has been previously cited in synonymy, it is best to continue, as it may have acquired standing in some way in subsequent works. Listing it helps to avoid misunderstandings, if it is adequately annotated.

7. Forms named as varieties are now thought to be identical with the typical form (33, 34, 35, 37). An aberration has been similarly suppressed (39), as well as a subspecies (38).

8. The minimum of bibliographic reference is given, including only the original work or the work in which the transfer to another genus is made.

9. The names were originally proposed in three different genera; that is, in three genera still believed to be distinct. The present placement (50) in the original genus (1) is the result of new evidence about the type species of that genus, so that the *Staphylinus* of 1952 (50) is not the same zoological genus as that of 1758 (1).

10. Some published references to names are not acceptable under the Code. To fail to cite them would hide the fact that they have been noted and found to be of unacceptable status. Frequently they later meet the requirements and become acceptable as of that later date. Listing the earlier reference (22), with annotation, records the situation.

11. This synonymy is complete in these features: (a) it lists all the specific epithets that have been applied; (b) it lists all the generic combinations of all these epithets; (c) it lists all the original references, for the names and for the generic transfers (thus, for all the binomina); and (d) it lists all known misspellings of the epithets (but not all those of the generic names in other applications).

The purpose of the synonymy is to convey information about the names that have been applied, their status at the time of publication, any circumstances affecting their proposal, history, applicability, and so on. It is therefore highly desirable to annotate all synonymies fully. Much bibliographic and taxonomic study can be wasted if the resulting synonymy fails to show all the information obtained and all the taxonomic conclusions reached.

Specifically, it will generally be assumed that any unannotated synonym is a subjective synonym in the opinion of the author of the synonymy, and that there are no special circumstances or technicalities involved.

Bibliographies. In taxonomic works, a bibliography is usually a device to permit abbreviation and condensation of synonymies. It thus serves as a list of literature cited. In much taxonomic work, other than revisionary or descriptive, the listing of pertinent works may assume greater proportions and importance. The extent of the taxonomic literature and the relevance of the works of all authors over a two-hundred-year period make the need for bibliographic aids very great.

A variety of bibliographic works are cited in Chapter 14, The Use of Literature. They may be consulted as samples of style, coverage, purpose, and so on. The present book contains several types of bibliographies, from brief lists in the text to a substantial bibliography at the end of the text. Some books useful for the preparation of bibliographies will be found listed there.

Naming the taxa

The description of individuals or groups is of little use unless each is identified with a name. The application of the names is a subject of such technicality that it is not directly covered by any chapter in this book except those on the conventions of zoological nomenclature (Part VI). The naming of taxa, presumably new or previously unnamed ones, is so closely governed by the rules of nomenclature that a taxonomist cannot properly use a name at all without reference to the rules.

Professional taxonomists have come to look upon descriptions and names as responsibilities of the author. If he publishes them, he has a responsibility to all future workers to make them accurate, effective, and properly integrated into previous work. This responsibility includes being right in his opinion that the group is new, correctly placing it in the classification, making sure the relevant features are all recorded, and ensuring that all the information is made available to others, not buried in an unknown publication.

On the other hand, it is frequently impossible to be absolutely sure that a species is new or that it is certainly the same as some older species. In this circumstance, professionals frequently publish their data under a new name, knowing that it may eventually have to be "sunk" into synonymy. This is often preferable to running the risk of associating new data with an old name only to find out later that the supposed connection was nonexistent. It is easier to record a synonym than to erase data which have been erroneously associated with a name.

Just as there is no special honor attached to the publication of a new name, there is also no dishonor in having a name reduced to synonymy. Some workers in the past have been insulted by such an act by a colleague, but there is no justification for such a reaction. The science grows continually, by addition and correction, and no one can expect to know all that will be known to his successors.

The names of species and groups are involved in problems of spelling, citation, publication, synonymy, homonymy, types, and others, all discussed in great detail in Part VI of this book. They cannot be proposed or used effectively without a detailed knowledge of the rules by which they are governed.

Summary

Descriptive taxonomy refers to the principal scientific activity of most taxonomists, the describing of the features of the specimens and groups which they discover. Without the descriptions, and the publication of them, each taxonomist would be working alone, with no way to use the discoveries of others and no way to communicate his discoveries to them.

In order to describe, it is first necessary to observe. Among the things observed, it is then necessary to choose those relevant to the study being made. In taxonomy, the features chosen become the taxonomic characters.

Although descriptions may refer to only one individual or taxon, the basic purpose of describing is to make possible comparisons, of similar things to group them, or of dissimilar things to distinguish them. The characters are the features selected to be compared with the corresponding features of other individuals or groups.

In addition to formal descriptions, descriptive taxonomy deals with types, with keys and classifications, and with synonymies—the record of the names applied to the taxon.

CHAPTER 16

The Publication of Data

Information accumulated by an individual is not really part of science until it is communicated to other scientists—made part of the general knowledge of the subject. For this purpose, communication is possible only by words, or symbols representing words, and by pictures. Words combine into statements by means of which concepts are described. The statements may be vocal or written. Vocal statements are seldom heard by more than a few of the directly interested persons in the world, even at large meetings, but many types of scientific contributions have been initially presented in this manner. Similarly, written communications (letters, information sheets, and drafts of possible publications) seldom reach more than a few people and have little effect on science, particularly taxonomy.

This leaves as the only effective means of communication the direct publication of the knowledge. The word publication could involve, besides printing, such practices as distribution of recordings or microfilm, but in science it seldom does. In taxonomy, all forms of reproduction other than printing are specifically outlawed or recommended against if there are any nomenclatural considerations. Nearly all dissemination of taxonomic knowledge is through formal publication by printing, in books, technical periodicals, or separate pamphlets.

General problems of publication

Rather diverse problems face the taxonomist preparing to publish taxonomic data and conclusions. In order to undertake a taxonomic publication a zoologist needs rather extensive taxonomic experience. Just how extensive depends on the type of publication. Recording new localities or new facts about a well-known species would require only a minimum of background. Describing as new a species formerly mixed with another species, or transferring a species to a new genus,

308

would require very substantial knowledge of the technical aspects of both taxonomy and nomenclature.

What to publish. There have been many published comments on the abuses of publication, suggesting that there are limits to what should be published. There is no possibility of setting standards, however, and the individual must follow his own conscience. He should be careful with his facts, reasonable in his interpretations, considerate of his colleagues, and critical of his own efforts; he should publish everything that contributes usefully to the knowledge of animals; and he should try to publish in the most useful form possible.

Single large publications are often more useful than a series of small ones, yet frequently it is difficult to obtain the publication of a large work, no matter how good. Some projects that result in diverse data are best published in separate papers in different journals. It is thus often possible to justify either a single large paper or a series of small ones. Many factors may be relevant to the decision.

Some employers judge a taxonomist's qualifications or activities by the number of his papers. Few scientists believe this to be a good criterion, but, if the situation exists, the individual cannot afford to ignore it. No real monograph should ever be broken up into short papers, but any peripheral aspects may profitably be reported separately, whereupon they actually become more available than in a large work. There are no limits to the things that may be worth publishing in taxonomy. A single taxonomist might publish examples of all the following, over a period of fifty years or so:

New records of species from new localities
Revision of a genus or larger group
Description of an unusual structure
Monographic study of a group
Checklist or catalog
Notes on synonymy and records
Description of a new species or genus
Study of genera in a larger group
Classification of phyla, classes, orders, or families
Essay on taxonomic trends or ideas
Essay on theory of taxonomy
Description of a previously unknown larval form
Notes on a new method of study, collecting, or storage
Itinerary of an expedition for collecting
Study of the nomenclature of a group
Comparative study of a group
Biography or bibliography of a taxonomist

Preliminary notes for a new classification
Regional or faunal studies
Essay on relation of taxonomy to some other field
Critique of new proposal
Studies on nomenclature rules and problems
Summaries or tables for reference
Index to literature
Studies on dates of taxonomic books
Systems for citing or filing literature
Book reviews

These by no means exhaust the possibilities. Taxonomists are interested in pertinent aspects of most other zoological sciences and appreciate summaries of these for their special benefit by other taxonomists. They are interested in learning where to find collections, who is working on a particular group, in what museums there are noteworthy collections, the history of taxonomy, how taxonomy directly benefits other fields, how animals can be identified by non-taxonomists, how to teach taxonomy, and many other subjects. Diversity in the publications of a taxonomist adds spice to his efforts, broadens his outlook, makes more of his ideas available to others, and increases his understanding of taxonomic problems.

Where to publish. It seems useful to comment briefly on taxonomic works published in inappropriate journals. Recently a zoological student discovered a population of animals that seemed to represent a new species. He was without any direct training or experience, and his advisors were not taxonomists. He wrote a paper describing the species as new and submitted it to an appropriate taxonomic journal, which turned it down because it failed to meet taxonomic standards. It was thereupon submitted to a game management journal, which published it because it had some practical aspects. The paper, as published, showed the inexperience of its author in taxonomic matters. It is now in a journal that carries little such material and is not well known to workers on this group of animals.

Several conclusions can be drawn from this: (1) Taxonomic experience is necessary for the publication of new species and other technical taxonomic material; (2) a taxonomic paper should be published only in a journal devoted largely to such material; and (3) a rejected manuscript should be rewritten until it is acceptable to the most appropriate journal; anything else shows willingness to publish under lower standards.

It is a sad addition to the above story that the description therein fails to conform to several of the requirements of the rules of nomenclature and will probably be held unacceptable. In this condition, it not only fails to fulfil the desired end of describing, naming, and making known a newly discovered species, but it places the author on record in the literature of this species for all time as an inadequate taxonomist, because the reason for rejection of the oldest name will always have to be cited.

It must be admitted that journals sometimes reject manuscripts because of lack of space or other reasons besides inadequacy of the manuscript. However, there is no part of the animal kingdom so isolated that there is only one journal appropriate for taxonomic papers dealing with it. Numerous alternatives are available, before resorting to a non-taxonomic journal.

There are cogent reasons for publishing taxonomic papers only in taxonomic journals. First, most taxonomists subscribe to the chief journals in their field and thus obtain the paper immediately. Second, the taxonomic journals are the ones most likely to be indexed in the *Zoological Record*, often by special arrangement between the editors. Third, the taxonomic journals are more likely to be available in the centers where taxonomic work is being done and thus available to those who must come there to see collections, literature, and specialists. Fourth, more respect is engendered by a good paper in an appropriate journal than by the same paper in an equally professional and prestigious journal whose subject of specialization is not taxonomic. Fifth, the *1961 Code of Nomenclature* has instructions to editors of taxonomic papers, which are very unlikely to be known to the editors of non-taxonomic journals.

Authorship of taxonomic papers. Occasionally, in the general monthly *Science,* there have been series of letters discussing the problems raised by multiple authorship of articles. In some fields, short articles are not infrequently signed by five or more authors. It may be that the aspect of obtaining "credit" for a share of the work is important. It may even be sufficient justification. But such a practice lays a real burden on librarians, bibliographers, and all who need to cite these papers. The most common solution is to cite such a paper as "Mothy et al." (Mothy and others.) This effectively reduces the credits of most of the authors back to zero!

In taxonomy, the problems are not merely those of the bibliographer. The name (or names) of the new species will be cited whenever that species is referred to, very likely for centuries to come. A name such as *Pseudopentarthrum subcylindricum* Champion is clumsy enough as is, but to have as authors, for example, Boisduval and Lacordaire would extend the name over fifty letters; while to extend it to three or more authors imposes a burden on the user too great to be accepted, so that drastic abbreviation or curtailment will be practiced.

Although there are a great many monographic papers signed by two authors, and double authorities for names are thus common throughout the animal kingdom, it can be doubted that two men actually collaborated in the preparation of each description. If they did not, the species should be credited to the one who prepared the description. It looks clumsy to find a species cited as *X-us albus* Bonhomme, in Malheuse and Bonhomme, 1892, but in citing the species only Bonhomme need be given. The longer form is a reference to the publication.

Probably the best solution when two people collaborate, and especially if there are more than two, is for one to publish the new species separately and then to collaborate on the rest of the study, if appropriate. There are many sorts of taxonomic papers which do not involve description of new species or groups. Co-authorship is often advantageous for these, but even here it is well to avoid multi-authorship whenever possible.

Ethics in publication. In many of the professions, there are codes of ethics that have the support of organizations and individuals in the field. Taxonomy has gradually built up such standards for certain practices, and it has also adopted a very elaborate code of behavior in one particular subject—zoological nomenclature.

Beyond this technical aspect, discussed later, there are some things of which people need to be reminded that involve the giving of credit for prior study or data, the accumulation of a collection, the borrowing of specimens, the exchange of specimens, the suppression of hard-to-explain data, the use of unfounded criticism, deliberate pre-empting of a field of study, use of research funds improperly, and all aspects involved in contacts with other taxonomists. Not all of these are directly involved in publication, but most of them may be reflected in the published results. Specifically, professional ethics in taxonomic publication involves: (1) integrity, (2) the giving of credit for the help of others, (3) courtesy and forebearance in language, and (4) humility.

Integrity. Integrity requires scrupulous adherence to fact, presentation of all sides of a situation, leaving room for difference of opinion, and doing all the things that one would normally be expected to have done to establish facts. It does not require conformance to fads or popular schemes; in fact, it requires the taxonomist to follow his own conscience even when it challenges the ideas accepted by others.

The giving of credit for the help of others. No scientist or scholar produces his contribution to knowledge without help both from the past and from contemporary colleagues. In taxonomy most "new" work is a reworking of previous data, often with only a relatively small addition of fresh data and no really new ideas. Every person publishing in taxonomy should have a proper respect for the previous work, even though he is now in a position to improve on it. In a short time his work, also, will be improved on by his successors.

It is customary for this respect to take the form of direct acknowledgment. In taxonomic material itself, the references to previous work and the name of the author of each species fulfill this requirement. In the general aspects of books and papers there are a few more direct acknowledgments that may be appropriate. For example:

1. Cite the name of the collector of specimens used.
2. Acknowledge the specimens lent or donated.
3. Identify all unpublished data provided by others.
4. Acknowledge permission to use previously published material.
5. Give credit to persons preparing photos, drawings, tabulations, and so on if their contribution was more than clerical.
6. Acknowledge assistance of advisors, manuscript critics, financial aid, facilities made available by institutions, and so on.
7. Acknowledge permission to quote from unpublished letters or manuscripts.

All of these admonitions must be treated with sense. It is possible to carry acknowledgment to ridiculous lengths. When material is brought together from many sources, the only reasonable acknowledgment may be in the bibliographic citations. People whose business it is to help in the publication, but who have no other connection with taxonomy, need not be cited except in special cases. The overloading of a paper with references to every person who ever wrote on the subject may serve only to hide the contribution of the writer.

Courtesy and forebearance in language. There are admonitions to be found against the use of emotional phraseology, indulging in controversy, and personal attacks on other scientists. All of these admonitions will bear repetition, but they also cry out for a more understanding statement. Although personal attacks are never justified in taxonomic papers, emotion is not necessarily bad and may be worth communicating. Controversy, furthermore, is the source of much clarification of ideas; the admonition should be rather not to take offense at argument or criticism, but to study it for its possible contribution.

Humility. Humility is not always highly regarded these days. What is intended here is the avoidance of making claims for the importance of conclusions. The facts will speak for themselves, regardless of whether the author claims "a transcendent hypothesis" or only the discovery of a "new" fact. The "new" fact is new to him and to the literature he happened to see. It may not be new to the reader. The claim of newness does not increase the importance of the discovery, if it is one, and it doesn't soften the criticism if it turns out not to be.

Much use of the first person in writing shows lack of humility. In formal writing, the first person is usually restricted to the Preface of books and to some essays. Whenever necessary, exceptions to this rule are allowable.

Never refer to the value of your own work. You are not a disinterested judge and no qualified colleague will pay any attention to your rating of yourself. Do not be misled into thinking you can shift the blame to the editor or the advertising department or someone else who speaks well of you.

A problem related to ethics is the attitude of a taxonomist to the new species which he describes. There have been persons who sought professional status or other recognition because of the number of new species they had named. Whatever the motives of individuals may have been in the past, professional taxonomists do not now rate another taxonomist by the number of species he has named. Rather, they ask about the effectiveness of his revisions, classifications, and monographs. They look upon the publication of a new species not as an honor but as a responsibility—a responsibility to other taxonomists to have made this species recognizable, to have determined correctly that it is previously undescribed, and to make all information about it available to all other workers.

If there ever was a cult of new-species worshipers, it is fast dying out, even though there are still thousands of new species in museums waiting to be described.

Technical problems of publication

There are many aspects of taxonomic publication that have special requirements. Some of these are discussed briefly below.

Descriptions. Many of the problems of publishing descriptions of animals or taxa are cited in Chapter 15, Descriptive Taxonomy. Only one further point requires emphasis here: in taxonomy description serves two purposes. First, it records and conveys to the reader the data about the thing described. Second, it establishes part of the legal basis for a new name. For the latter, the describer must know and take into account the implications of the rules of nomenclature and the procedures which have come to be accepted as standard in that part of taxonomy.

Works that are relevant to this subject or have direct suggestions on preparation of descriptions include:

Ferris, G. F. 1928. *The Principles of Systematic Entomology.* Stanford, California; Stanford University Press.

Keen, A. M., and Muller, S. W. 1956. *Procedure in Taxonomy.* Stanford, California; Stanford University Press.

Mayr, Linsley, and Usinger. 1953. *Methods and Principles of Systematic Zoology.* New York; McGraw-Hill Book Co.

Rensch, B. 1934. *Kurze Anweizung für zoologisch-systematische Studien.* Leipzig; Akademische Verlagsgesellschaft M.B.H.

International Code of Zoological Nomenclature. . . . London; International Trust for Zoological Nomenclature. 1961.

Keys. Publication problems with keys involve principally the style of key most useful to the person identifying specimens. Keys do not need to use characters of great biological significance. They are often most effective with simple characters that happen to have clear-cut distinctions that can be readily used. Keys are not classifications; there should not be any attempt to combine the two, as this generally does violence to both functions. Some forms of keys waste a great deal of space on the page by indenting each couplet farther than the one above it. In a very short key this will not be serious, but in long keys it will probably be unreasonably expensive.

In general, a key of several hundred couplets will be less useful than the same couplets arranged as a series of keys. For example, to key out the 124 families in Arnett's beetle book cited on p. 298 would have taken 123 couplets. A mistake anywhere along the line, when using the key, might appear only at the end, when the identification proved to be wrong. Going clear back to the beginning to check all couplets could be unnecessarily time-consuming. Instead Dr. Arnett arranged the key to the order as separate keys to the four suborders and to ten groups within one of these. The user can thus verify the suborder before proceeding to key out the group, and he can then verify the group before keying out the family. One of the keys is 47 couplets long, one has 27 couplets, and the rest have fewer than this.

Classifications. The technical problems of publishing classifications are those of obtaining a style which permits communication of a maximum amount of information. The major problem is the tendency of authors to try to incorporate into a classification what they believe to be the phylogeny of the groups. This would be appropriate if the phylogeny were an independent source of data. It is not, because all ideas of phylogeny are based on comparative studies of the taxonomic features of the animals. The classification should use this comparative data directly; it should be little influenced by speculation on phylogeny based on these same data. There will, of course, always be some feed-back from phylogenetic studies to the taxonomic data on which it was based.

The conclusion here is that if phylogeny has been studied, it should be presented separately from the classification. The latter is a listing of groups, in which the sequence is a minor consideration and the history of no direct interest.

(It is not intended here to imply that there is no connection between classification and phylogeny. There is an inescapable connection in that they are both based on comparative data. Furthermore, every classification after the first one may be influenced by the phylogenies derived in large part from the earlier classification. It is principally the fact that a classification lends itself mechanically to phylogenetic conclusions that makes it seem that the two are connected, whereas the real connection is indirect, through the comparative data. The taxa are probably of evolutionary origin, so we tend to think of them as evolutionary facts, whereas they are comparative facts which very likely justify evolutionary conclusions.)

Synonymies. The tracing of the various names applied to each genus or species and the forms of all these names is tedious and time-consuming, but it is absolutely necessary to effective taxonomic work. The publication problems are chiefly the reactions of editors, who find the cost of the tabular material high and the reader appeal low. As a result synonymies are seldom published except in monographic works, where subsidized publication may be stretched to include them.

Bibliographies. On a small scale, bibliographies accompany many types of publications. They are more likely to be selective than exhaustive. Large bibliographies are difficult and expensive to publish, but they are of unending benefit to specialist, librarian, and general zoologist alike.

Nomenclatural aspects. First of all, the publication of new zoological names is deeply affected by certain provisions of the rules of nomenclature. The 1961 Code specifies that any material affecting nomenclature must be published in conformance with the following stipulations:

(1) It "must be reproduced in ink on paper by some method that assures numerous identical copies." This means that microfilms, microcards, and

other photographic reproductions are not acceptable. It also means that mimeographing and hectographing (Ditto or spirit duplicating) are not to be used.

(2) It "must be issued for the purpose of scientific, public, permanent record." This means that it cannot be intended only for those at a particular meeting, nor for a limited group of colleagues or students. It also means that distribution of a few copies of proof sheets is neither acceptable nor sufficient.

(3) It "must be obtainable by purchase or free distribution." This is to ensure the availability of copies to the zoological world at large.

(4) The document must not be anonymous, although before 1953 such anonymous publications were usually held to be acceptable.

(5) "Mere deposit of a document in a library" does not satisfy the distribution requirements, whether or not it is suitably printed. This rules out dissertations or theses which are made in too few copies for general distribution.

If nomenclature is affected by the proposed publication, the author should conform to these requirements exactly. They are not unreasonable. While they do not really ensure satisfactory methods of printing and distribution, they are helpful in eliminating unsatisfactory methods.

Languages. The description of new taxa, to be acceptable under the rules of nomenclature, should be written in English, French, German, Italian, or Latin. This is an attempt to make all new publication available to all workers, who cannot possibly deal with all the many languages of the earth. In the past only local works have been published in other languages, but important works have been published in more recent years in Japanese and Russian.

So far as the zoologists represented in the International Congresses of Zoology are concerned, these five languages are the acceptable ones for all papers of nomenclatural interest. Most taxonomic papers by Dutch, Danish, and Swedish writers, for example, are published in English. Many papers in the past have had descriptive matter in Latin, regardless of the language of the text, and, of course, in botany a Latin diagnosis is always required.

Historical documentation. Because nomenclatural aspects are usually prominent in taxonomic studies, it is necessary to know the entire history of every taxon discussed, especially the name-history. It is therefore frequently necessary to include chronological summaries, often in the form of synonymies. A high degree of accuracy must be maintained in these items in every publication.

Spelling of names. In nomenclature, all erroneous spellings become a matter of record and are studied, as they *may* produce real problems later on. Spelling variations should be avoided, even at the cost of extra proofreading, final checks back to original sources, and letter-for-letter rechecking at every opportunity.

Preparation of papers

The remainder of this section will be devoted principally to listing some of the books that will aid a taxonomist in planning, writing, and obtaining publication of technical papers.

Manuals of style. Most serial publications, in which nearly all taxonomic papers are published, require certain special forms of presentation. These may affect title, abstracts, tables, illustrations and their legends, and bibliographies. The manuscript must be prepared with these style requirements in mind, or it will have to be done over.

Many American biological journals have now adopted as their guide the following recent style manual:

Style Manual for Biological Journals. Washington, D.C.; American Institute of Biological Sciences. 1960.

Other style manuals and papers on particular aspects of style include the following:

A Manual of Style Containing Typographical and Other Rules for Authors, Printers, and Publishers recommended by the University of Chicago Press, Together with Specimens of Type, 11th ed. Chicago; University of Chicago Press. 1949.
Style Manual. Washington, D.C.; Government Printing Office. 1945.
Blackwelder, R. E. 1957. Journals and Abbreviations, pp. 1344–1388. *In* Checklist of the Coleopterous Insects of Mexico, Central America, the West Indies, and South America. *U.S. Nat. Mus. Bull.* 185 (part 6).
Sumney, G., Jr. *American Punctuation.* New York; Ronald Press. 1949.

Terms. The proper use and spelling of technical terms can be facilitated by reference to special dictionaries and glossaries, such as:

Brown, R. W. 1954. *Composition of Scientific Words. A Manual of Methods and a Lexicon of Materials for the Practice of Logotechnics.* Washington, D.C.; by the author. (Available only from the Smithsonian Institution, Washington, D.C.)
Jaeger, E. C. 1950. *A Source-Book of Biological Names and Terms,* 2nd ed. Springfield, Illinois; Chas. C Thomas.
Melander, A. L. 1940. *Source Book of Biological Terms.* New York; Department of Biology, The City College.
Nybakken, O. E. 1959. *Greek and Latin in Scientific Terminology.* Ames, Iowa. Iowa State College Press.
Woods, R. S. 1944. *The Naturalist's Lexicon.* Pasadena, California; Abbey Garden Press.

Technical writing manuals. There are many books and papers on writing as a means of expression. These include a variety of helpful information of more general nature than the style manuals. For example:

Brennecke, E., Jr., and Clark, D. L. 1942. *Magazine Article Writing.* New York; Macmillan.

Crouch, W. G., and Zetler, R. L. 1954. *A Guide to Technical Writing,* 2nd ed. New York; Ronald Press.

Emberger, M. R., and Hall, M. R. 1955. *Scientific Writing.* New York; Harcourt, Brace.

Mills, H. G., and Walter, J. H. 1954. *Technical Writing.* New York; Rinehart.

Perrin, P. G. 1950. *Writer's Guide and Index to English,* rev. ed. Chicago; Scott, Foresman.

Trelease, Sam. F. 1958. *How to Write Scientific and Technical Papers.* Baltimore; Williams and Wilkins.

The use of words. Many books are available to help the writer find the correct word to express his ideas and to help him avoid errors of usage and of grammar. For example:

Bierce, Ambrose. 1909. *Write It Right. A Little Blacklist of Literary Faults.* New York; Walter Neale.

Fowler, H. W. 1965. *A Dictionary of Modern English Usage.* 2nd ed., rev. by Sir Ernest Gowers. Oxford; Oxford University Press.

Kierzek, J. M. 1954. *The Macmillan Handbook of English,* 3rd ed. New York; Macmillan.

Mathews, William. 1907. *Words: Their Use and Abuse.* Chicago; Scott, Foresman.

Roget, P. M. 1933. *Thesaurus of English Words and Phrases . . . ,* enl. and rev. ed. New York; Grossett and Dunlap.

Roget, P. M. 1931. *Thesaurus of the English Language in Dictionary Form,* rev. by C. O. S. Mawson. Garden City, N.Y.; Garden City Publishing Co.

Soule, R. 1938. *A Dictionary of English Synonyms,* rev. and enl. by A. D. Sheffield. Boston; Little, Brown.

Woolley, E. C., Scott, F. W., and Berdahl, E. T. 1944. *College Handbook of Composition,* 4th ed. Boston; D. C. Heath and Co.

Preparation of illustrations. What to illustrate and how to prepare the illustrations for publication are discussed in numerous books and papers, such as:

Cannon, H. G. 1936. *A Method of Illustration for Zoological Papers.* Association of British Zoologists.

Dreisbach, R. R. 1952. Preparing and Photographing Slides of Insect Genitalia. *Syst. Zool.,* vol. 1, pp. 134–136.

Ferris, G. F. 1928. *The Principles of Systematic Entomology.* Stanford, California; Stanford University Press.

Hanna, G. D. 1931. Illustrating Fossils. *Journ. Paleontology,* vol. 5, pp. 49–68.

Olsen, T. H., and Morrow, J. E. 1959. Guide for Preparing Figures. *Bull. Bingham Oceanogr. Coll.,* vol. 17, pp. 147–153.

Papp, C. S. 1963. *An Introduction to Scientific Illustration.* Riverside, California; Chas. E. Papp.

Reeside, J. C., Jr. 1930. The Preparation of Paleontological Illustrations. *Journ. Paleontology,* vol. 4, pp. 299–308.

Ridgway, J. L. 1938. *Scientific Illustration.* Stanford, California; Stanford University Press.

Zweifel, F. W. 1961. *A Handbook of Biological Illustration.* Chicago; University of Chicago Press.

Common errors. There are some common errors that may be pointed out individually because of their special applicability to taxonomic papers.

Names in apposition in titles. When a species is referred to by both common name and Latin name in a title, the names are separated by a comma. The names are in apposition—they refer to exactly the same thing.

"Notes on the housefly, *Musca domestica*."   (1)

Here, "the housefly" is not merely a loose expression for some indoor insect but a definite name for the species *Musca domestica*. Since "the housefly" and "*Musca domestica*" refer to exactly the same thing, they are said to be in apposition.

"Notes on the fly *Musca domestica*."   (2)

Here, "the fly" is not equivalent to the species named and is therefore not separated by a comma. The latter expression (2) is equivalent to saying "the particular kind of fly which is called *Musca domestica*." The former expression (1) is equivalent to saying "the housefly (also known as *Musca domestica*)."

These two forms can be distinguished only with knowledge of what "housefly" and "fly" denote. The word "the" is restrictive, but it is not equal to the task of restricting "fly" to any one species, as it can with "housefly."

"Notes on a fly, *Musca domestica*."   (3)

Here, the comma is used even though "fly" is indefinite, because "a fly" is in apposition to the Latin name. The Latin name could be omitted without making the title entirely meaningless. In the second form above, omission of the Latin name would leave only "Notes on the fly." This is taxonomically meaningless, because there is no such thing as "the fly."

The article "a" in this situation always denotes a particular thing, although ordinarily it is permissive; therefore in such cases it involves apposition and a comma.

The rule must be that if the two expressions are fully synonymous (if they refer to exactly the same thing), either expression could be omitted without making the entire statement meaningless; they are in apposition and must be separated by a comma. If not, the comma must be omitted.

Actual examples of an erroneous and of a correct use of the comma are the following:

"The locust, *Zonocerus variegatus* L."   (4)

(The comma should have been omitted, because that species is not "the locust" but only "a locust.")

"The morphology . . . of the green peach aphid, *Myxus persicae* (5) (Sulzer)."

(The comma is correct because the two expressions are fully synonymous, in apposition; either one could be omitted.)

Use of abbreviations. There is a great temptation to abbreviate words or names in certain circumstances. Although it is sometimes permissible to do so, it is best to avoid this wherever possible.

Where the Latin name of a species is repeated several times in a page or section, it is permissible and sometimes actually helpful to abbreviate the generic name after the first use of it.

Musca domestica . . . *M. domestica* . . . *M. domestica*   (6)

It must be noted, however, that if more than one such genus is abbreviated in a paper, there may be two with the same initial.

In descriptions or technical material, abbreviations may be troublesome to the reader. Even so-called standard abbreviations may not be familiar to students in other parts of the world. Much taxonomic work is eventually worldwide in interest. It is therefore best to write out words and terms wherever possible, even in charts and tables.

In bibliographies most journals and publishers will permit or insist upon abbreviation of the scientific periodicals cited. The author must conform to these requirements, but it is always better to err on the side of writing out too much rather than too little. Most systems would abbreviate *Revue d'Entomologie* as *Rev. Ent.* However, there are also journals entitled *Revue Entomologique* and *Revista de Entomologia*.

Both of these would also be abbreviated as *Rev. Ent.* To be explicit, these must all be written out. It is a mistake to assume that there is no duplication of an abbreviation, unless one has at least checked the *Union List of Serials.*

An excellent but little-known system of abbreviating names of journals has been in use for many years by the Smithsonian Institution and the U.S. National Museum. The Institution's publication of this system is now long out of print and not widely available, but the system is described in:

Blackwelder, R. E. 1949. Citing literature in the Coleopterists' Bulletin. *Coleopt. Bull.,* vol. 3, pp. 55–59.

In essence, this system recommends: (1) write out all titles consisting of a single word; (2) use all the principal words of the full title; (3) never insert words not in the title, except in parenthesis; (4) write out in full all words of one syllable and all proper nouns; (5) abbreviate other words by stopping before the second vowel; (6) if any confusion is anticipated, write out the word or add syllables; and (7) use the exact wording of the full title and retain its order. Exceptions are noted for compound germanic words and for all cases where confusion between titles may occur.

A very complete list of abbreviations of entomological (and zoological) journals under this system (including about 800 journals) is given in *U.S. Nat. Mus. Bull.* 185 (part 6), pp. 1345–1388. (See Blackwelder, 1957, in the bibliography.)

Use of generic name alone. It is common in non-taxonomic papers to find references to the structure or behavior or some feature of a genus. For example:

Fig. 145. B. Longitudinal section through the proboscis of *Gyrocotyle.*   (7)

It is obvious that no genus has a proboscis and that it is impossible to draw a picture of a genus. It must be assumed in such cases that the author means "in some unspecified species of *Gyrocotyle.*" This would be clearer if it were written ". . . of *Gyrocotyle* sp.," meaning "of one species of *Gyrocotyle,*" or, if all the species share this feature, as ". . . of *Gyrocotyle* spp.," meaning "of the species of *Gyrocotyle.*"

Where various animals are being discussed, and there is only one species per genus, it does no great harm to refer to them by the generic name alone. The generic name assumes something of the nature of a common name, and some writers do not italicize it in such a situation. For example:

Necator americana is the original American hookworm. . . . Necator   (8)
  is a worm of historic sociological significance.

It would seem to be a pomposity to repeat the scientific name in such
instances. It is only necessary to be careful that there is no possibility
of confusion.

The first subspecies. When in a previously known species a population is found that is recognized to be a subspecies, there is a tendency
to record the facts in such a manner as this:

X-us albus David (1866). Black, length 7–9 inches. Range: New England
  to Virginia.
X-us albus carolinensis n. subsp. Tinged with gray, length 6–8 inches.
  Range: Coast of South Carolina.

The new population has a few features that fall partly outside the
range of the previously known specimens, and it comes from a peripheral locality.

Assuming that zoologically the new population *does* represent a new
subspecies, the above descriptions are misleading. The *species* occurs
from New England to South Carolina (or from New England to Virginia, and in South Carolina). Its specimens are black or grayish black,
6–9 inches in length. Only the nominotypical (or nominate) subspecies
X-us albus albus fits the first description above. The correct complete
description would be:

A. *X-us albus* David (1866). Black sometimes tinged with gray, length 6–9
    inches. Range: New England to South Carolina.
  1. *X-us albus* subsp. *albus* s. str. Black, length 7–9 inches. Range: New
      England to Virginia.
  2. *X-us albus* subsp. *carolinensis* nov. Black tinged with gray, length
      6–8 inches. Range: Coast of South Carolina.

The rule should be that every species consists of at least one subspecies; if there is only one recognized, it is not mentioned; if there
are two or more, the nominate one must be distinguished, and any
description given under the specific name *must* cover *all* the subspecies.

Quotation of foreign languages. In quoting from works in foreign
languages, and especially in citing foreign titles in bibliographies, it
not infrequently happens that alterations are made in the orthography
(spelling). The names of authors, titles of papers, and names of journals should be quoted in the exact form of the original, so far as physically possible, and any deviation should be shown as such.

In German, all nouns are capitalized; in French, accent marks are universally used; in many languages, capitalizing of particles in personal names is a matter of individual preference and cannot be changed by others. For example:

W. C. van Heurn	J. C. Von Bloeker
V. S. Van der Goot	H. von Boetticher
J. De La Paz	A. M. da Costa Lima
J. R. de la Torre Bueno	E. A. Da Rosa
M. A. V. D'Andretta	J. de Beaumont
J. d'Aguilar	L. A. P. DeConinck

Errors in the quotation of names and titles may be the fault of the author, the editor, or the printer. In any case, it is the responsibility of the author to see that the final form is correct.

Over-use of technical terms. All science uses technical terminology and finds that accuracy in communication depends upon accuracy of definition and use. It is, however, possible to over-do the use of such terms. The following example was quoted by J. R. de la Torre Bueno in an editorial entitled "Heavy, Heavy Science." It was originally published in the *New York Sun*.

"It would appear from what evidence is available that the act of oviposition is immediately stimulated by the crepuscular diminution in the intensity of illumination and the rise in relative humidity as the diurnal temperature decreases."

Translation:

"Egg-laying seems to be stimulated by twilight and the dampness of evening."

In addition to unnecessary use of technical terms in ordinary writing, there is a tendency for taxonomists to coin terms for use in their descriptive work. Sometimes these are necessary and reasonably employed. At other times they are synonyms of equally acceptable terms or not even needed at all.

New terms should be proposed only after thorough search has shown that no suitable word is available, and then only with assurance that it is correctly formed. Dictionaries are full of useful but forgotten words that can be effectively used. A useful book is:

R. W. Brown. 1954. *Composition of Scientific Words.* Published by the author. (Obtainable from Smithsonian Institution, Washington, D.C.)

Inadequate titles. In addition to misusing commas, many faults can be found in the titles of papers and books. They may be too short or too long, indefinite or unnecessarily detailed, ambiguous or so specific as to be intelligible only to a specialist. It would be difficult to illustrate all of these.

Titles that are too short and too indefinite include:

Nouveautés diverses.
Remarques en passant.
Geänderte Namen.
New neotropical myrmecophiles.
Miscellaneous notes and new species.

Titles that are too long and include unnecessary matter are:

Esploracion cientifica practicada par órden del Supremo Gobierno i segun las instrucciones del doctor don R. A. Philippi, par don Carlos Juliet, ayudante de la Comision esploradora del mar i costas de Chilóe i Llanquihue, a bordo del "Covadonga."

Illustrations of exotic entomology, containing upwards of six hundred and fifty figures and descriptions of foreign insects, interspersed with remarks and reflections on their nature and properties. [With subtitle] A new edition brought down to the present state of the science, with the systematic characters of each species, synonyms, indexes, and other additional matter.

A title that is specific but ambiguous as to the locality of the specimens is:

Eine neue Odacantha (Ins., Col.: Carabidae) des Senckenberg-Museums.

Titles intelligible only to a specialist on the (unnamed) group are:

Stilpnastus nov. gen.
Addenda au genre *Petalium.*
La larve du genre *Scirtes.*
On the genus *Opoleon,* Gorh.
Synopsis of the species of the tribe Lebiini.

A title should always be carefully selected with at least the following considerations in mind:

1. Reasonable brevity is desirable.
2. The necessary information must be communicated.
3. Unnecessary information should be omitted.
4. Both general and specific information should be given.
5. The title should aid in indexing the article by subject.
6. It should not include punctuation, abbreviations, or local words not intelligible everywhere in the world.

The distribution of publications

Taxonomic publications are normally distributed by two rather different means. In the first place, papers published in serials (magazines, journals, reports, occasional papers, or other named series) are automatically distributed to the mailing list of that serial, whether by subscription, exchange, or gift. If published in or as a separate book, the sale of the book will automatically provide this primary distribution.

An author has little to do with this aspect, except to consider the extent and selection of distribution in deciding where to publish his paper. In the case of commercially published books, this primary distribution will probably be the only type of distribution.

In the second place, some serials provide the author with extra copies of his paper removed from the volume. These copies are called reprints, author's extras, or separates. These terms are often used loosely, but they do have definite meanings that can be readily distinguished.

Author's extras are pages removed unchanged from extra copies of the publication. They frequently contain parts of other papers and may not show the place and date of publication. This fact causes them to be sometimes regarded as a nuisance, but they are the most accurate type of separate copy so far as exact duplication of the serial paper is concerned. (The bibliographic reference can and should be added later.)

Separates (or separata) are copies printed from the same type as the original but with all extraneous matter eliminated. If the article begins in the center of a page in the serial, it will begin in the same place in the separate. It may have a line of type added to show the bibliographic reference of the original, but it is otherwise identical to the original (in completeness, arrangement, and pagination).

Reprints are, strictly speaking, copies rearranged to fit most satisfactorily on a page and thus not printed at the same time as the original but from substantially the same type. Pages may be renumbered, the text arranged differently on the page, new type set for the title, etc. These copies give the least assurance of duplication of the original and are definitely disadvantageous in taxonomic papers. The word reprint is often used loosely to include all three types of copies.

In working from separated copies of articles, especially older ones, it is very necessary to recognize those which differ in any material way from the original. It is often desirable to compare a copy with the serial, and then to note upon it that it has been so compared and found to be (or not to be) in agreement with it in all pertinent ways.

The purpose of these extra copies is direct distribution by the author to interested colleagues. Whether the copies are supplied free (still a common practice in taxonomic journals) or are purchased by the author (usually at cost), they are mailed out to specialists on the particular group to serve as convenient personal copies in place of the inconvenient library set of the periodical.

The mailing of "reprints" is a kind of professional courtesy. It frequently results in an informal exchange of papers, to the benefit of both parties. The mailing may be as soon as possible after publication, or it may be periodic, containing several papers at a time.

The purchase and distribution of separates is a good practice for several reasons.

1. It makes papers available in the most convenient form to those in the field most likely to profit from them.
2. It serves as a notice that the writer is still active and is interested in other people's papers on this subject.
3. It makes the paper available to persons who do not have ready access to the complete serial.
4. It serves as a record for the writer's employer that he is producing research, and aids in the maintenance of files of his work.
5. It enables a teacher to use copies of his papers (and those of others also) in his teaching, as sources of data, as examples of methods, and so on.

Dealers in secondhand books. In a few branches of taxonomy there are dealers specializing in books of technical nature. Some of these sell also papers of smaller size and even reprints. Because the normal mark-up on such secondhand items is high, often 100% or more, some taxonomists have looked upon these dealers as unprincipled profiteers. It is doubtful if any of them ever made more than a moderate living. A more realistic view is that they are really indispensable and should be encouraged to continue their service to the science. Such dealers provide places where reprints and monographs can be obtained by those who need them. They are an unexcelled means of distributing these papers. A person who is willing to buy a paper is probably one who will use it. What better way is there to ensure that copies will get to the persons who really need them? Some taxonomists make a practice of giving or selling a few reprints of their papers to secondhand-book dealers. The practice has much to commend it.

Summary

Publication is the essential activity that makes knowledge available to others. Although science is theoretically possible without publication, a scientific effort which dies with its discoverer cannot claim either the support or the plaudits of mankind. Taxonomy is a science that traditionally publishes freely and voluminously. It continues to use its publications longer than most sciences. For some purposes the oldest publications are often of most importance.

The problems of publication in taxonomy are the same as in other fields, with one notable exception. The international conventions embodied in the rules of nomenclature add many special problems, for which the solutions lie only in the stipulations of the rules. These nomenclatural aspects of publication are cited in Part VI, Zoological Nomenclature, but some general problems relating to publications are illustrated in Chapter 14, The Use of Literature, as well as in this chapter.

PART V

Theoretical Taxonomy

"If we agree that taxonomy is not a purely practical exercise planned to attach individuals to certain names, then we must I think admit, with the majority of taxonomists, that it has a deeper purpose which may be defined as the definition of the things of nature and their unification into a system of intelligible relations." (Thompson, 1955)

"Systematics is that branch of biological science which is engaged in unraveling the mystery of evolution by the only method by which this can be done, namely by a comprehensive study of relationships between living things, based on all available data." (Petrunkevitch, 1952)

"In general, it is more correct to speak of a phylogenetic background for taxonomy than of a phylogenetic basis. And we must constantly beware of arguing in a circle and giving independent existential value to the phylogenetic groupings which we have merely deduced from the distribution of characters and structural plans in existing groups." (Huxley, 1942)

"Just as most people believed in witches in the Middle Ages, so may agreement in taxonomy result from the copying of mistakes." (Davis and Heywood, 1963)

Chapter 17. Taxonomy as a Science 331
    The science of taxonomy 331
    The broader taxonomy 333
    The New Systematics 334
    Classification vs. population studies 339
    After *The New Systematics* 340
    The old and the new in taxonomy 341
    The opposing view 346
    The taxonomy of today 348

Chapter 18. The Nature of Classification and of Species 351
    Classifying species and groups 351
        Confusion of terms 351
            Concept 351
            Groups 353
            Categories 353
        Groups and categories 354
        Reality and objectivity of categories 358
        Phylogeny as basis of classification 360
    Species 364
        The nature of species 365
        Outbreeding data 369
        The "species problem" 371
        The biological species concept 372

CHAPTER 17

Taxonomy as a Science

The matters dealt with under this heading are in large part the controversial aspects of systematics. Various interpretations have been advanced, but logical arguments have not always been presented for them. It is necessary to discuss some of these things only because there have been illogical but authoritative pronouncements repeatedly made about them, using two well-known rhetorical devices for persuading the unwary reader. These are the repetition of catch phrases and the use of derogatory labels.

Among these controversial matters is the scientific standing of taxonomy itself, the relation between taxonomy and phylogeny, the so-called "biological species concept," The New Systematics, the nature of natural classification, and the reality and objectivity of categories. It is not possible to discuss these adequately here, but it is necessary to present some of the arguments for the assumptions made in this book as to the status of each of these problems. (Some of these items are discussed in Chapter 18, The Nature of Classification and of Species.)

It must be emphasized that this book is not intended to cover all of systematics. An attempt is made to deal with all aspects of taxonomy, but the other branches of systematics are more appropriately discussed under other headings. It is not intended to cover Evolution, Speciation, Phylogeny, Population Dynamics, or Genetics. Much of what is sometimes called Biosystematics is omitted as belonging to one of the above fields. What has been called The New Systematics by American evolutionists is discussed only to the extent of showing that it has had little effect on taxonomy.

The science of taxonomy

There is literally no end to definitions of the word "science." Whether or not any particular study is a science depends entirely on the definition adopted. It is pointless to carry on a discussion of such a matter, but it is possible to arrive at a better understanding of a field of knowledge by examining its basis, its methods, and its results.

There are several things to be said about science that are pertinent to systematics, or more especially to taxonomy. First of all, science is knowledge and it is the process which makes knowledge. The knowledge is organized, and therefore a science is a system of organized knowledge. We can scarcely go wrong with the statement of Lenzen (1955) that: "The problem of empirical science is the acquisition and systematization of knowledge concerning the things and phenomena experienced in observation."

Inasmuch as the principal business of taxonomy is the discovery of the comparative facts of the kinds of organisms, and the principal business of classification is to provide a system in which these facts can be integrated, the part of systematics which includes both of these would appear to fit Lenzen's statement exactly.

The field of taxonomy has been criticized as not being a science. This conclusion is wrong, even though there is some evidence to support it. The amount of data to be collected and organized in taxonomy is so tremendous, and taxonomists have always been so far from completion of the work, that many of them have never accomplished much of the organizing but spend their lives accumulating the data. Some of them have even been relatively untrained in a scientific background and have operated much like stamp collectors or amateur naturalists.

This type of work is, of course, not typical of taxonomy as a whole, even though it may have been much in evidence. But even this work, as long as it produces new data even in small quantities, is entitled to the same recognition in science as any other activity that produces facts singly. There is no justification for declaring any experimental work to be unscientific merely because the experimenter published only the results of the experiment, leaving it to an Einstein or a Schrödinger to work out the implications and propose an explanation.

In taxonomy the conclusions drawn from the data are the assignments to species, genera, and phyla. These are always tentative assignments and therefore in the nature of hypotheses. All aspects of classification involve organization of knowledge, and the data of taxonomy are the basis of the classification. Even some taxonomists sometimes forget that their work is at the time a part of science only if it is fitted into the organized knowledge. As it was put by W. R. Thompson (1937), "We must not expect to constitute a science of purely individual phenomena. The idea involves contradiction." And ". . . the science of the laboratory and museum deals with material properties and their temporal or spacial concatenations."

In the end we must agree with Popper (1959) that "what is to be called a 'science' and who is to be called a 'scientist' must always remain a matter of convention or decision."

The broader taxonomy

From the very beginning of modern biological classification, from the time of Linnaeus, animals were classified principally on the basis of visible structural features. Various methods were devised for assessing the relative usefulness of different structures for this purpose, and occasionally other sorts of attributes were also employed. It was found necessary to preserve specimens for later comparison, as verbal descriptions and even pictures were not always adequate. The only attributes that can be readily preserved are the structural ones, and it became nearly universal to rely on such structural characters in taxonomy. When other attributes seemed to offer additional material for comparisons, it was usually found that these were correlated with structural features. This strengthened the taxonomists' view that structure is an effective key to most inherent attributes.

After the publication of Darwin's works, it was expected that taxonomy would be greatly changed by the new ideas. Again, in the decade after 1938, the publications on The New Systematics led to expectation of another revolution in the basis of taxonomy. Long after the publication of Darwin's *Origin of Species,* astonishment was expressed at the fact that the classifications of the taxonomists were not much affected by the evolutionary ideas. More than twenty years after The New Systematics was announced, there is a great reticence to admit that the classifications of the taxonomists have again remained unchanged. Were the same factors responsible for these two unexpected developments? The answer has been overlooked by the evolutionists of both periods, and a different revolution in taxonomy and classification has also been overlooked by them. It occurred in the two decades before The New Systematics.

In the third and fourth decades of the present century an important trend was started in the study of the largest "groups" of taxonomic subjects, the insects and the invertebrate fossils, and was felt in other groups as well. As these two include over three-quarters of all known animals, the trend was of substantial importance. Unfortunately, it has seldom been referred to, because it was not immediately recognized as a successful trend and was pushed from the limelight by later developments.

Beginning in the 1920's, an increasing number of professors taught that taxonomy and classification should not be based on a few key characters but on all available information of whatever sort. This did not mean that equal weight was to be given to every feature, but that comparative and analytical methods should determine the usefulness of each fact. Not only was a much wider range of individual features to be used, but also data from new methods of study. Here was one of the major factors in the early growth of biometry, now grown into the vast field of biostatistics. Here was acceptance of biological features on a wider scale than before. Here was wider recognition of the animal origin of fossils and the necessity of studying both the living forms *and* the fossil record to understand either one. This movement gained considerable momentum simply because it gave better results in taxonomy. Its earliest devotees did not happen to be widely influential or in positions of authority. Their ideas produced superior monographs, helped to solve longstanding problems in "difficult" groups, and proved themselves against prejudice. The spread of these ideas seems to have continued in some aspects under the guise of the more publicized trends of the following decades, and it is likely that additional impetus was given by the discussions of the new ideas.

The New Systematics

In 1940 there was published in England a collection of essays and reports entitled *The New Systematics*. In it, several authors tried to assess the taxonomic work of the past and to suggest improved attitudes and methods for the future. It was not claimed that there was any New Systematics in existence then or even that the old systematics was inadequate. The keynote was set by the editor, Julian S. Huxley, in the first two sentences, thus: "To hope for the new systematics is to imply no disrespect for the old. It has been largely the rapid progress made by classical taxonomy itself that has necessitated the introduction of new methods of analysis, new approaches to synthesis."

At the same time the editor did introduce several concepts new to taxonomy, new in the sense that taxonomists had not accepted responsibility for them. Chief among these was the need to discover the mechanisms of evolution. There is good reason to believe that the data of taxonomy will be important in solving the problems of evolution, but solving the problems of evolution is not the function of taxonomy. The people who study the processes of evolution will be partly drawn from the ranks of the taxonomists, where their taxonomic training will be invaluable, but when working on evolutionary mechanisms they will be evolutionists or whatever they choose to call themselves. The real taxonomists will still be busy with the overwhelming job of recording and systematizing the data on the myriads of kinds of animals.

Subsequent books that took up the expression New Systematics were unanimous in applying the expression to the latter problem—the origin of the groups used by taxonomy in its classifications, particularly those called species. This was largely a new field. It investigated mechanisms of speciation, rates, variability in populations, the genetic control of these, and other aspects of genetics and evolution. It made substantial advances in these fields, setting up what is in effect a new branch of biology that could well be called variation or the science of organic change.

Not content with developing a new field for investigation, the students of this changed use of the expression New Systematics sought to make the new discoveries the basis for major changes in the approach to classification of organisms. If this attempt had been based on sound arguments and valid premises, it might have resulted in a revolution in taxonomy. Such a revolution has actually been claimed by the proponents, but there is much evidence that it was restricted almost entirely to the few who gave up the study of taxonomy for the study of speciation.

The New Systematics did not claim to inaugurate a new era. Even the title was the subject of apology by the editor, who admitted it might better have been called *Modern Problems in Systematics.* Before The New Systematics could be born, "the mass of new facts and ideas which the last two or three decades have hurled at us must be digested, correlated, and synthesized." The editor might have gone further and included the mass of facts from the previous one hundred and eighty years, because taxonomy must always add all new facts to the sum of those previously discovered, keeping them all systematized and available.

Within a few years, however, there were published other books that took up the idea of a New Systematics that would explain evolution. In these it was reported that there had been a revolution in animal taxonomy, that The New Systematics had already displaced the old except in a few relatively unknown groups, and that the study of species formation and the factors of evolution had become the principal task of "modern" taxonomists.

The idea that a revolution has recently occurred in taxonomy has received wide publicity among biologists. There have been few attempts to determine whether there is any truth to the claim. No one can deny that there have been evolutionary studies by persons with taxonomic backgrounds, and there have also been ecological studies by such persons. This does not make both ecology and evolution part of taxonomy. There may have been classifications that were widely different from those of thirty years ago, but if so they have been kept well hidden. There seems to be no serious doubt that most taxonomic work and most classification has been almost completely unaffected by the so-called great change.

One of the principal conclusions of *The New Systematics* (the book) was that taxonomy would do well to pay more attention to facts from the fields of cytology, genetics, ecology, physiology, and behavior. It did not state that taxonomy should *start* to give attention to these fields, because it recognized that attention was already being given. It admitted the dual responsibility, that in each case the two fields must work together to increase knowledge, that, for example, genetics must make its data available in such a way that they can be useful to taxonomists just as taxonomy must assemble all its data into classifications that will be useful to geneticists. There was nothing new in this to those who had been teaching for twenty years that taxonomists must use all available data. These men also taught that "the systematist may and should employ any means that are available in order to arrive at a knowledge of the biological facts, whether these means be found in morphological, anatomical, physiological, experimental, genetical, or even chemical studies" (Ferris, 1928, p. 33). A corollary of this is the stated need to study the importance or validity of each type of data for whatever purpose it is to be used.

There have been few voices raised since 1940 in support of these ideas. The two outstanding ones are of botanists. Heslop-Harrison (1956, p. 107) clearly expresses both the validity of orthodox taxonomy and the need for broader horizons. "Many taxonomists," he says, ". . . hold to the legitimacy of the aims of orthodox taxonomy, and acknowledge the practical value of the existing taxonomic structure. . . . At the same time, the implications of the new work must be taken into account in bringing up to date taxonomic procedure and in framing, defining, and describing the units of orthodox taxonomy." Among this new work he cites population structure, ecology, geographical variation, cytology, and genetics. It should be noted that Heslop-Harrison is not suggesting that taxonomists do the work on population structure, or ecology, or genetics. He claims only that taxonomists need to take into account the implications of work done by students of these fields. This is exactly in line with the view of Ferris, but it is not the viewpoint of The New Systematics in its later form.

Keck (1957, p. 47) accepts a more direct goal of taxonomy to reflect phylogeny but recommends the same expansion of viewpoint to utilize more data from other fields than just "comparative morphology and geographic distribution." The taxonomist should seek assistance also from genetics, anatomy, cytology, paleobotany, embryology, ecology, physiology, etc. He also cites as the tools of modern taxonomy of plants: field studies, pollen studies, parasitism, biochemistry, and cytogenetics.

All of this applies equally to animal taxonomy. It can be pointed out, however, that all these remarks stop short of the obvious generalization that the attributes or data useful to taxonomists in classifying organisms are all *comparative in nature*. At the time of recording they may not be stated as comparisons, but their value to taxonomy lies principally in eventual comparison with the corresponding data from other individuals or groups. This generalization would not be of great moment if it did not lead to a further generalization, one that has been inherent in the teaching and taxonomic work of men like Ferris, Robson, and Turrill. Just as all taxonomic data must be fundamentally comparative, *so all data which are comparative* are of significance to taxonomy. There is little basis for doubt that for most groups of animals comparative structure will remain of prime interest, but all other comparative data must be considered whenever available, and their bearing on taxonomic problems analyzed.

If any justification is needed for the continued use of comparative structure as the prime factor in classification, it will be furnished by the continuing demonstration that data from genetics, ecology, parasitism, physiology, and behavior are usually and almost inevitably represented in structure at some level, so that they can be used indirectly in classification in the form of the correlated structure. Of course, the correlation must be established in every possible case. Furthermore, since its presence strengthens the comparison and its absence weakens it, the correlation must be reported. Likewise, the nature and extent of the correlation must be analyzed. The final effect of most of these experimental or non-morphological attributes is thus to adjust and strengthen the morphological system.

There is no limit to the advantages to taxonomy afforded by this all-encompassing approach. It absorbs all comparative data of whatever nature and uses it to correct or bolster the classification previously erected. It can be extended indefinitely and always leads on to a closer agreement between the classification and the nature of the organisms. It can utilize any comparative data provided by the study of speciation or other evolution. But the study of these fields is no more a part of taxonomy than is the cytogenetics and biochemistry that provide other data for its use.

It is often overlooked that taxonomy makes no claim to the discovery of all the data it employs. The bacteriologists Lamanna and Mallette (1953, p. 11) clearly state the situation, thus: "Whether there is conscious realization of the fact or not, all persons scientifically investigating organisms are making a contribution to the data of taxonomy. In a very real sense the best taxonomy is a synthesis of all knowledge of biology." Realization of this situation has been slow, partly because of the gradualness of the accumulation and acceptance of the data from new fields and partly because of the obfuscation produced by recent verbose over-emphasis on one of the fields.

The expression "comparative zoology" is not often seen in publications, but it represents the entire legitimate field of modern animal taxonomy, just as it did earlier for classical taxonomy. In a major book on problems of systematics, entirely neglected by all the more recent writers on problems of species and taxonomy, written by Robson in 1928, there is this appropriate statement of the source of taxonomic data: "Every living organism exhibits a large number of attributes to which we give the name 'characters.' Such 'characters' include every structure and property of the animal or plant, whether they be organs, cytological structure, physiological activities, habits or ecological relationships." The use of gross structural characters as keys to the differences of other nature is justified by Robson because of correlations demonstrated to exist. "If we thus are compelled to regard the sum total of an organism's attributes—metabolic, structural, habitudinal and reproductive—as the expression of its fundamental biochemical and biophysical constitution, it follows that the differences which we recognize as specific at the structural level must be likewise founded on more deeply seated differences."

It is now evident that in many fields there has been an increase in the acceptance of a broader base for taxonomy. There is more willingness to consider new techniques and new types of data. There is probably more open-mindedness about some aspects of taxonomy. There is certainly more interest in taxonomic data on the part of biochemists and geneticists, who see the need for making their data available to taxonomy.

There appears to be no basis for the claim that there has recently been a revolution in systematics. There *has been* an increase in interest in many aspects of systematic theory and practice. The expression The New Systematics has been so consistently and persistently misused as a supposed switch to evolutionary interests that it is no longer of much use in its original meaning. Nevertheless, there is continued improvement in training of taxonomists, in the breadth of data used in taxonomy, and in the use of taxonomic data in other fields.

Classification vs. population studies

Several recent books [1] have stated or implied that taxonomy, or classification, must be based on phylogeny. With this as a premise, it is argued that study of the evolutionary origin of species and other groups is not only part of taxonomy but the most important part. Based on this assumption, taxonomy becomes principally a means of recording data about populations, and much of the later New Systematics deals only with this aspect. In these same books, species are defined as populations. Classification is said to group populations rather than individuals. Statistical methods are advocated which are applicable to populations but not to either species or individuals. The entire purpose, the methods, and the justification implied in these books are different from those of classical taxonomy.

This view has been persuasively proclaimed. It has appeal because much of modern biology is unquestionably concerned with populations. It is probably the fault of taxonomists that the relationship of classifying and identifying to populations is not recognized to be almost non-existent, because it is only the individuals and groups of individuals that can be grouped and categorized.

Before a taxonomist can use knowledge of populations, he must complete some classification at the species level and above. If he takes an interest in populations, it can only be after he has classified the group and identified the species. He can then distinguish subspecies or populations after the species are at least partly known. This will usually be the result of an evolutionary interest, not a taxonomic one. The well-trained taxonomist is likely to be the best-qualified zoologist for evolutionary studies, so it is not uncommon for him to do part of his research in this distinct field. His interest in populations is usually not taxonomic but evolutionary.

A claim has been made for the unifying effect of the Darwinian theory, as it attempts to account for the diversity of animals. Many zoologists look upon Darwin's work as the source of concepts which give meaning to classification. These same persons sometimes marvel that, as we can now see, the theory had practically no effect upon taxonomy. It is impossible to tell by a man's taxonomic work whether he believed in evolution or even knew about it. Classifications have

[1] For example, Sprague, T. A. 1940. *In* The New Systematics. Mayr, E. 1942. Systematics and the Origin of Species. . . . Simpson, G. G. 1945. Classification of Mammals. . . . Alexander, G. 1956. General Biology.

been practically unaffected by Darwin's theories or the later developments. This was actually inevitable, because taxonomy was and still is the study of the groups found among animals; it is not the study of how the groups came to be. Any knowledge of this latter subject will be of great interest to taxonomists and will add to the data available to them, but this knowledge of mechanisms is not the goal of the study of taxonomy.

In this book it is assumed that taxonomy is the study of individuals first, of groups of individuals second (including species), and of groups of groups third. It is agreed that the individuals are the product of heredity. It is agreed that the species are the result of evolution (speciation), and it is admitted that the features held in common by the individuals are in part the result of common ancestry. All this is as important to him as the fact that the individuals are composed of systems of carbon-chain molecules organized into cells and organs. But neither the phylogeny nor the physiology is the immediate business of taxonomy. It has more than enough to do in its proper classical role of making known the kinds of animals and the attributes in which they are alike or unlike.

After *The New Systematics*

Although it was the first book to suggest the study of new attitudes and aspects of systematics, *The New Systematics* seems to be at the same time the last to recognize the nature of taxonomy. Books that followed in quick succession adopted the deprecatory epithet of The Old Systematics; they labeled the previous work as typological, nondimensional, meaningless, and inadequate. By labeling The New Systematics as new, modern, objective, biological, and multidimensional it was implied that in all these ways the new approaches were better than the old. This would have been of little consequence if it had not been for the fact that these succeeding books failed to note the implications of the original essays and proceeded to change the emphasis from improvement in taxonomic method, philosophy, and breadth to concentration on the means by which evolution had produced the kinds taxonomists study. For example, one of the early books stated that the systematist who studies the factors of evolution wants to find out how species originate, how they are related, and what this relationship means. This is probably true, but the implication that the working taxonomist is among those who have time to study the factors of evolution is certainly unjustified in the broad view.

This new subject, of interest to taxonomists, to be sure, is as distinct from taxonomy as genetics or ecology. Its data will have to be recorded and systematized by taxonomists, but this does not make the field a part of taxonomy.

The original New Systematics had primarily the same implications for taxonomy as had the trend of the previous two decades described in a previous paragraph—to increase the breadth, soundness, accuracy, and usefulness of the classifications of organisms. This worthy goal was soon forgotten in the changed emphasis placed on the expression The New Systematics. Perhaps this was inevitable, because one of the book's authors stated that such a thing as a New Systematics was impossible; another said that its major premises were unsound; and the editor's conclusions cited nothing in the way of radical new ideas or approaches but merely increases and improvements in programs and methods already being pursued. Subsequent works used the expression for the quite different aspect of studying the origin of the groups to be classified. This complete change in the meaning of what has become a popular catch-phrase makes it difficult to discuss the original ideas or assess their impact on taxonomy.

The old and the new in taxonomy

One of the major features of recent articles by some of the zoological New Systematists, if not itself a trend in systematics, is the use of derogatory labels on the taxonomic work of those who still classify organisms on the basis of comparative data. The tone of this disparagement has become very caustic at times, using authority, ridicule, sarcasm, repetition, and other rhetorical devices in lieu of facts. Not all the persons who have been quoted as supporting these views really do so, but effective rebuttal is seldom made.

The New Systematics was ushered in by Huxley and his collaborators with a recounting of the needs of taxonomy, to correct its failures and to bring it abreast of the times, but without any claim that it had failed completely in its purpose. It was in fact the very successes of taxonomy in making known nearly a million kinds of animals that necessitated the attempted re-evaluation and led to the suggestions for the future. Turrill (1940, p. 47) remarked: "On the whole, taxonomists have every reason to be proud of the work they have accomplished since the time of Linnaeus by the use of descriptive and comparative morphological methods." Calman (1940, p. 455) wrote that while new species are brought in almost every day: "What is very remarkable and significant, however, in this constant influx of novelties, is the rarity of the unexpected. . . . Seldom, very seldom indeed, do we come across a species for which there is not a place waiting in the accepted classification."

Reference is frequently made to the "difficult" groups, to *Rubus, Taraxacum, Crataegus,* and *Salix* in plants and to *Peromyscus, Cynips, Daphnia,* and *Artemia* among animals. The word confusion is frequently applied, and it would be possible for a non-specialist to get the idea that all taxonomists are confused and all taxonomy incomplete and inadequate. To be sure, there are difficult genera in nearly all groups of animals. There are also some whole groups that are more difficult than the rest. Which ones might be cited as examples depends largely on what level of the classification is involved and on what we are thinking of as difficulties. Of genera we may cite *Aimophila* (in the birds) and *Graphognathus* (in the weevils); of families, Muscidae (in the flies); of orders, Charadriiformes (in the birds); of subclasses, Branchiopoda (in the crustaceans); and of classes, Turbellaria (in the flatworms).

In spite of these examples, and all the others that can be cited, any implication that the major work of taxonomy is meaningless, or confused, or incapable of producing the results for which it was designed, is unjustified. Such an implication, when coupled, as it invariably is, with the use of derogatory labels, does no credit to the persons who claim that taxonomists must change their viewpoint and their approach to classification.

Of the supposedly derogatory labels pinned onto classical taxonomy, the one most evident was Mayr's 1942 (page 6) reference to "The Old Systematics" as contrasted with "The New Systematics." This was not the first time these expressions had been used (Hubbs, 1934, p. 116; Huxley, 1940, p. 1; Huxley, 1942, p. 111), but it was the first time that full definitions were added that increased the potential effect of the word "old" and added other deprecations. These definitions are quoted here in full, with bracketed numbers referring to the subsequent discussion paragraphs.

"*The old systematics* is characterized by the central position of the species. No work, or very little, is done on infraspecific categories (subspecies) [1]. A purely morphological species definition is employed [2]. Many species are known from only single or at best a very few specimens [3]; the individual is therefore the basic taxonomic unit [4]. There is great interest in purely technical questions of nomenclature and 'types' [5]. The major problems are those of a cataloguer or bibliographer, rather than those of a biologist [6].

"*The new systematics* may be characterized as follows: The importance of the species as such is reduced, since most of the actual work is done with subdivisions of the species, such as subspecies and populations [1]. The purely morphological species definition has been replaced by a biological one, which takes ecological, geographical, genetic, and other factors into consideration [2]. The material available for generic revisions frequently amounts to many hundreds or even thousands of specimens, a number sufficient to permit a detailed study of the extent of individual variation [3]. The population or rather an adequate sample of it, the 'series' of the museum worker, has become the basic taxonomic unit [4]. The choosing of the correct name for the analyzed taxonomic unit no longer occupies the central position of all systematic work and is less often subject to argument between fellow workers [5]."

In what amount to formal definitions of the old and new systematics, we can only take these statements literally. A comparison of the two definitions discloses some strange contrasts.

[1]. It is simply not true that little work on subspecies is done in the classical taxonomy. The whole idea of subspecies was developed there. In appropriate groups a very satisfactory start had been made in determining the nature of the variation patterns in species. It is only in such groups that any work of this sort has been done within The New Systematics, because in most classes of animals so little data are yet available that virtually nothing can be done with infraspecific variation or species structure.

[2]. The expression "a purely morphological species definition" is definitely a derogatory label. Even in Linnaeus' time more than structure was used, as shown by consistent recording *and consideration* of locality. In this century it has been realized that many other features are correlated with structure, and structure has often been used as a guide to these other features—the easiest way to take them into account. Furthermore, whenever occasion demands, other data are used: hosts, parasites, life history, habits, physiology, breeding capabilities, ecological preferences, genetics, stratigraphic position, and others. There has not been a major monograph in many years that can justly be called "purely morphological," and the best ones have been far from that, even without any overt acceptance of The New Systematics.

[3]. It would be very interesting to know how it is that the classical taxonomist in 1930 had only a single specimen or a very few to work with, in spite of all efforts to assemble as many as possible, whereas in 1950 he would have been able to obtain the whole population or at least an adequate sample of it. This must be a blessing reserved for

those who give the password "New Systematics." The truth is that the vast majority of animal species are still known from the few specimens obtained. Where specimens were plentiful, the older workers frequently gloried in the possession of a large series of some particular species. There were some series studied as early as 1920 that had thousands of specimens from all parts of the known range. And every specimen was studied, as well as the range of variation. It is hard to see how a New Systematist can work with populations or series without ever looking at the individuals that compose them.

[4]. There seems to be some disagreement as to the basic unit used in systematics, as discussed elsewhere. Some taxonomists have always believed that they were classifying individuals first and then groups of individuals. It is hard to know what others think, but all taxonomic work familiar to the writer assumes individuals as the start—specimens are always present. A taxonomist may assume that these specimens before him are part of a breeding population, but he usually does not see the population and has little data on it except what is derived from inferences on the individuals before him. The first level of classification, the groups of individuals, has usually been the one called the species level. It differs from higher levels only in that the members of its groups are individuals rather than groups of individuals. This is not true if we employ subgroups within the species, as most speciationists and some taxonomists do. In practice, then, the level species is not different in nature from the level genus or the level order. The claim that the species is reduced in importance in The New Systematics because of interest in subspecies confirms that the *species level* is no different from other levels. It can be made more or less important by changes in interest of workers.

[5]. There has always been interest in nomenclature. There will always be as long as zoologists hold it a necessity that they have distinctive names for the different kinds of organisms. There have doubtless been persons who acted as if the giving of names—their validation—was the goal of their work, but this does not justify the implication that all taxonomists considered the naming as anything but the labeling of the groups they recorded so that others might recognize them and use the recorded data. Without names, there could be no classification. Without types, there could be no fixity of names. The New Systematics has done nothing at all to change this situation. It will do nothing in the future to change it so long as it is agreed to refer to the groups of animals by a formal system of names. Naturally, the nomenclature gradually comes to require less attention as a group becomes well known. Thus the larger burden of nomenclature falls on

the pioneers in each group. It is churlish to imply that this essential phase of the work is unimportant, and it is fatuous to imply that pioneering taxonomy is finished. There are still groups relatively untouched.

It seems obvious that this point of view is centered about the work on vertebrates, particularly that on birds. No doubt the taxonomic work at the nomenclatural level has been very nearly completed in this group, but the birds are no more than a hundredth part of the known animals, and a specialist's acquaintance with some of the invertebrate groups would give substantially different views on species, taxonomy, nomenclature, and other aspects of comparative zoology.

[6]. Because some taxonomists have done more than their share of cataloging and bibliography, some other taxonomists are in position to parasitize this work. With catalogs and bibliographies at hand, they can go on to other types of work, oftentimes giving no credit or thought to the labors that made possible their new outlooks. It would be interesting to see how, without catalogs or bibliographic aids, a New Systematist would deal with a homogeneous subfamily of ten thousand species. Up to now such workers have been able to work in smaller and well-cataloged groups.

It is hardly necessary to point out the illogical nature of all such sweeping derogatory statements. They are not warranted by the facts but are confused with false implications. They would ordinarily be classed as propaganda—appearing to be what they are not and convincing by deceit if at all. They serve no useful purpose in science.

The reference in [2] to "morphological" came in some later works to be replaced by "typological." This also results in a derogatory implication when it is contrasted with the supposedly more scientific "biological."

Of far more importance than this derogatory implication, however, is the faulty thinking which makes possible its application to taxonomy and classification rather than to nomenclature. Typological, as applied to taxonomy, appears to refer to the study of types as representatives of the species. There have no doubt been a few taxonomists who put so much weight on types that they seemed to build their whole knowledge from them, but this is an unreasonable conclusion in general. Even the most determined users of types had to study the available specimens before selecting the types; they made the selection on the basis of prior study of this series and of other species; they almost always knew something about the variation represented by the specimens. The better workers, and nearly all workers in recent decades, used the types as name-bearers, as hitching-posts to settle points of name-application.

The value of the type specimens for the purpose of name-bearing was so great that it provided basis for the misunderstanding so widespread among non-taxonomists that the types were alone considered in taxonomy. This is far from the truth. The same idea in more sophisticated form is the basis for the derogatory use of "typological" for a taxonomic approach. It would be quite impossible to prove that there is any such thing as a typological taxonomy of substantial proportions, just as it would be impossible to prove that there is none at all. The implication is unreasonable, however, unless evidence is presented to show that these are more than exceptional occurrences.

The basis of taxonomy in general has never been types. It has usually been the specimens available to the taxonomist, one or many. Some have chosen their types to be middle-of-the-road in the variation pattern; some have chosen the types because of sex, condition, source, ownership, or other circumstance. Some have chosen many types for each species, having a variety of purposes in mind. But the only direct purpose served by most types is to show to what the name is to be applied. This gives us a type-founded nomenclature; it does not give us a typological taxonomy.

The opposing view

So much discussion of the ideas of a few persons on what taxonomy should concern itself with might lead to a belief that the latter-day interpretations of The New Systematics are the only ones held or put into print. They are certainly not the only ones held, but the aggressive and illogical attacks that have met some publications have deterred many from expressing their views in print. The correspondence of the writer as an officer of the Society of Systematic Zoology for sixteen years shows that many taxonomists are troubled by the implications of the new ideas and many are definitely not in agreement with them. It is often overlooked that there have been some who expressed these views in print.

First of all, there are those with philosophical and logical training who see the fallacies of the claims of objectivity, reality, greater importance, more naturalness, phylogenetic basis, and so on. Unfortunately, none of these have effectively argued their case. Not that their presentation has been unsound, but it has not reached and interested enough taxonomists to be effective. First among these was Gilmour in *The New Systematics* itself (1940). His views of the nature of natural and phylogenetic classifications were accepted by the editor, J. S. Huxley (as well as T. H. Huxley, by quotation), and are substantially those argued in the next chapter.

These views of Gilmour were misconstrued by one later writer, who uses the appropriate term "phenotypic species concept" but misquotes Gilmour in several ways, implying that his conclusions lead to the fall of all science. This is not justified. That writer's views were based on an assumption made early in his paper but not labeled as such: that systematists accepted a new labor, after the evolution theories appeared, of arranging the groups of animals so as to reflect the actual course of evolution. With this assumption, some of the conclusions would follow logically. There is evidence that ideas of this sort were talked about in the post-Darwinian days, but it is hard to find any classifications of that period that were changed because of these ideas. The *working* taxonomist did not and could not take them into direct account. Several evolutionists have admitted that there was no great change in the classifications, and one searches in vain for examples of classifications that were changed.

Next there are Woodger (1937, 1952) and Gregg (1950, 1954), supported in the abstract by all the non-biological authors of books on logic. The evidence in these books, bearing on "natural" classification and on the nature of categories and groups, is overwhelming. Their thesis is that rigorous use of language is essential to clarity, and in the case of these words dual meanings cause much of the difficulty. These biological philosophers believe that the ordinary language of science, merely an extension of the language of the everyday world, is not sufficiently precise to serve the purposes of a complex field such as taxonomy and the methodology of taxonomy. Even if taxonomists are not prepared to adopt their symbolic language to obtain the maximum rigor, they must make every possible effort to obtain clarity and mutual understanding in their language. Too often, apparent differences of opinion can be shown to be merely differences in the use of words.

It is not necessary to go to philosophers to get backing for the importance of rigor in language. In the introduction to Roget's *Thesaurus* there appears the following very pertinent passage:

"It is of the utmost consequence that strict accuracy should regulate our use of language, and that every one should acquire the power and the habit of expressing his thoughts with perspicuity and correctness. Few, indeed, can appreciate the real extent and importance of that influence which language has always exercised on human affairs, or can be aware how often these are determined by causes much slighter than are apparent to a superficial observer. False logic, disguised under specious phraseology, too often gains the assent of the unthinking multitude, disseminating far and wide the seeds of prejudice and error. Truisms pass current, and wear the semblance of profound wisdom, when dressed up in the tinsel garb of antithetical phrases, or set off by an imposing pomp of paradox. By a confused jargon of involved and mystical sentences, the imagination is easily inveigled into a transcendental region of clouds, and the understanding beguiled into the belief that it is acquiring knowledge and approaching truth. A misapplied or misapprehended term is sufficient to give rise to fierce and interminable disputes; a misnomer has turned the tide of popular opinion; a verbal sophism has decided a party question; an artful watchword, thrown among combustible materials, has kindled the flame of deadly warfare, and changed the destiny of an empire."

Taxonomists have insisted on exact and clear terminology in description of organisms, in the terms used for structures and situations, and in the names used for the groups. It is strange they have not insisted on similar clarity in the expressions used for the ideas involved in discussing the basis of taxonomy, the concepts, theories, problems, and other aspects of methodology. As Gregg puts it, taxonomists are frequently guilty of ambiguity of reference as well as ambiguity of meaning.

A number of other writers have examined one or more aspects of The New Systematics and denied the validity of the so-called modern view of it. Many more have ignored the claims entirely and simply continued to use the idea of species distinguished by characteristics that are primarily structural simply because structure is the most easily investigated aspect of most organisms. Virtually all monographers have tacitly followed this method.

It would be a serious error to leave this subject without reference to W. R. Thompson. His book and papers listed in the bibliography contribute substantially to these discussions, with major emphasis on logical arguments and justification of assumptions. His writings abound in worthwhile clarifications, viewpoints, and critiques.

The taxonomy of today

If the statements of the New Systematists were correct, taxonomy today would be very different from the taxonomy of a few years ago.

Most work would now be done on subdivisions of species, since the latter would be mostly known; structural features would be largely replaced by ecological, genetic, and geographical data; for generic revisions hundreds or even thousands of specimens would be available, instead of only one or a few; the population would have become the basic taxonomic unit; and names would be no longer time-consuming.

Many taxonomists will recognize that these so-called changes are actually descriptive of the current work in only a very small segment of the animal kingdom—the higher vertebrates and a genus or two among other groups. Elsewhere, what little data of new types are available are welcome but are insignificant compared to the comparative data that has always been the basis of the taxonomy. In most groups of animals the supposed great change in practice simply has not occurred. Most taxonomists who continue to do taxonomic work are still doing it in almost exactly the same way as before. Many of them do have a better understanding of broad biological concepts and theories and improved methods, but no one has given them any better way to classify animals or any reason to expect a better way to appear.

It is a little strange that while the New Systematists claim a great change in taxonomy, they also see great confusion in taxonomy. It does not seem to occur to them that the supposed change and the confusion could be cause and effect instead of effect and cause. It is possible to deny that there has been any widespread change, but it would be difficult to deny that there is confusion. Fortunately, the working taxonomist has often been untouched by the confusion, which has seemed to trouble mainly the New Systematists. Actually, it is exceedingly difficult to find a piece of taxonomic work that is different in basic approach from the best work of a few decades ago. For the second time in a hundred years, taxonomy has been left substantially unchanged by a new concept heralded as cataclysmic.

A few biologists have recently attempted to force taxonomists to abandon the methods and concepts that have been the basis of its successes. They have insisted that the taxonomists *must* study the origins of the kinds of animals they have been studying. This so-called "modern" approach is firmly entrenched in the sense that it is being actively proclaimed, and giving results that justify the labor. These results are not part of taxonomy, although they should interest taxonomists; they are a part of a new science of speciation. Recent publications show that only a few taxonomists have really accepted the views of the speciationists *as they apply to taxonomy* to the extent of weighing the evidence presented. Certainly many taxonomists have been passive or completely aloof. Others have been actively opposed to certain implications, and this group is larger than is usually realized.

Those taxonomists who have accepted the recent statements have seldom given any evidence that they have themselves examined the arguments, the logic, and the conclusions and are willing to say that they actively agree in the implications. There is no reason to suppose that the casual quoter of the new ideas sees all their implications and accepts them all. It is certain that the implications have not all been brought out for him to examine at leisure, and there is good evidence for believing that a few important implications have not been faced even by the proponents.

The preceding paragraphs are believed to show that the application of speciation ideas to taxonomy has not been universally accepted or even widely practiced, that the arguments purporting to show that these ideas must be adopted by taxonomists are far from conclusive and in fact often illogical and unfounded, that the purpose of taxonomy is to classify animals and our knowledge about them rather than to illuminate their evolution, and that current use of language falls far short of adequacy for discussion of the methodology of taxonomy.

It is specifically not intended to deny evolution or that species have had histories (phylogenies), or even that the phylogenies are somehow related to the present nature of the species; nor the importance of the study of speciation, populations, genetics, evolution, or any other biological subject; nor the desirability of finding and using data from all fields of biology in our classifications.

There is nothing but language and acceptances [2] standing in the way of recognition of the real goals and uses of classification. If taxonomists can make the language adequately rigorous and free from rhetorical obfuscations, it should be possible for them to deal with the acceptances on the basis of fact and logic. If one refuses to consider semantics and epistemology in discussions of methodology, one can expect no end to the "problems" of taxonomy.

[2] An acceptance is a system of statements which cannot be shown to be true but which is helpful in satisfying some need. Although not proved, it is believed in so intensely as to become dogma and to generate vigorous verbal acclaim. Its falsity can be shown only by careful analysis of all premises and their logical consequences. (See also Chapter 18, under Phylogeny as basis of classification.)

CHAPTER 18

The Nature of Classification and of Species

Classification is grouping. The groups are known as taxa. A group of groups (also a taxon) is more inclusive than one of its included groups and is said to be at a higher or more-inclusive level. The level is the category. A system of levels or categories is a hierarchy.

The groups themselves, the taxa, are the principal result of classifying. With more than a million kinds of organisms known, it is necessary to group them into classes to be able to deal with so many. This grouping into classes is really the only purpose of classification, although the taxa will be useful in a variety of ways.

The first level of grouping involves specimens. It is known that many individuals belong to a single kind, having many features in common and being able to interbreed freely. These kinds are known by the term species. Species are of importance in other fields than taxonomy, and the use of these groups may be different in the other fields. It therefore happens that there is a difference of opinion as to what constitutes a species, either in practice or in theory. Misunderstandings have arisen and continue to cause difficulties between fields.

The higher levels of grouping are not so much subject to controversy. The theoretical views may not correspond closely to practice, but reasonable use of these groupings is possible.

The practical aspects of classifying have been discussed in Chapter 12, The Practice of Classification. Some theoretical aspects are discussed below.

Classifying species and groups

Confusion of terms. Before the ideas of grouping and categorizing can be usefully discussed, it is necessary to define and distinguish four words that must be employed in this connection. These words are concepts, groups, categories, and species.

A *concept* is an idea in the mind of a person. It results from sense impressions and imagination. It is the mental picture "seen" at the time and remembered. It exists only in the mind of the one individual, for there is no direct way to communicate concepts. By converting the concept into a statement, verbal or pictorial, one can induce another person to form a concept from the statement. In the case of concrete concepts, those representing objects, it is sometimes possible to compare two concepts by comparing the statements about them with the objects and reach a degree of certainty that the two concepts represent the same thing. In the case of abstract concepts, those representing classes of objects, hypothetical objects, and transient phenomena, there is no way to compare the concepts of two individuals, no way to be sure that they are the same. The nearest approach to assurance is by defining the terms used in the statement by reference to other concept-statements previously agreed upon. All language is based on such definition.

It is unlikely that any two individuals have exactly the same concept or idea which they associate with any given term. Their ideas are always colored by their past experience. It is therefore ineffectual to speak of a concept held by many persons, such as "the biological species concept." We can only study the statements made about the concept, and it is possible for many persons to express agreement with the statements. Unless each statement is completely free of ambiguities, however, there can be no assurance that the individuals are agreeing to the same thing, because they in turn agree only to the new concept they have formed from the statement, and any ambiguities will lead to different ideas of what the statement means. This use of the term concept in place of such a term as definition serves to obscure the nature and source of the information or ideas under discussion and to prevent logical treatment of the statements.

Inasmuch as concepts can be known to others only through the statements made about them, the statement becomes the thing that is important whenever concepts are involved in science, because there can be no science without communication. It must therefore be understood that when the word *concept* is used in a general sense, the actual object of reference is the *statement* derived from the concept. Only the statement can be examined; only the statement can be judged as to accuracy, truth, or agreement with other statements.

An obvious misuse of the term concept occurs in the recent statement that certain concepts are *valid* and that two concepts are not *invalidated* by the existence of an intermediate concept. This shows lack of understanding of what concepts are. There is no such thing as an invalid concept. A statement about the concept may be contrary to fact, but every concept exists in some mind and cannot be described as valid or invalid, true or false, philosophical or practical. When the concept is described in words, the description can be labeled as inadequate, if one wishes to do so.

Groups, by which we mean groups of objects or groups of groups, are not at all like concepts. They may arise from concepts, from sense impressions, but they can be entirely definite if the features setting them apart are clear-cut. The groups may be of any size; they may be founded on intrinsic features or extrinsic decisions. Groups of objects will be equally definite to all persons, but groups of groups will be recognizable only if their definitions are unambiguous.

For example, the words which have been assembled into this sentence form a group that is entirely definite. Any person can recognize the group and its limits. On the other hand, if one suggests two groups of dogs, the large dogs and the small dogs, no one will be certain where to draw the line between the two. The former group (the sentence) is objective, the latter ones are subjective, simply because of the nature of the distinctions.

Categories are levels in a hierarchy, to which are assigned groups. They are almost exactly comparable to a set of shelves. Any given group under consideration may be placed on any one of the shelves (categories). The place of the shelf or level in the hierarchy is definite, the group may or may not be definite, and the placing of it on the shelf is entirely definite and unequivocal.

The shelf (level or category) is part of a series, the hierarchy, which is originally indefinite in the sense that it is the result of choice, but once the choices have been made and the levels determined, the hierarchy becomes entirely rigid and definable.

In order to use the categories there must be a name for each one, and it is accepted procedure to call them species, genus, family, etc. But these names are also applied to the groups placed in the levels. A group placed in the family level is called a family. This leads readily to misunderstanding when it is not clear whether the word family refers to a particular family group or to the level at which such family groups are placed.

The species level seems to hold the most interest for many taxonomists. Many have attempted to define the group which should be placed at the level called species, but these attempts have not met with obvious success. This seems to have been the result of failure to consider the nature of the various things represented by the word species.

These four terms are frequently used without definition and with meanings that seem to vary with the circumstances. They are very often ambiguous. In any discussion of the confusion of language in taxonomy, it is therefore necessary to distinguish clearly what meaning is being attached to them. They will be used here with the following meanings, and the usages of certain other writers will be contrasted with these: *Concept,* the idea produced in the mind of an individual by sense impressions and imagination; *group,* the limited set of objects or sets segregated on any basis by any individual; *category,* one of the levels in the hierarchy, to which the groups may be assigned; *species,* (1) one of the groups, the one placed in the category called the species level (a species group); (2) the category or level at which the species groups are placed (the species level).

Groups and categories. The entire concept of classification is based upon the fact that objects or ideas can be assembled into groups, the groups into more-inclusive groups, and the more-inclusive groups into still-more-inclusive groups. When there are many groups on several levels involved, groups, larger groups, and still larger groups, we may recognize a similarity in level between certain groups, so that some of the groupings are placed in the level of larger groups and some in the level of still larger groups. This is the process known as categorizing; the different levels are each a category. The categories are higher or lower depending solely upon whether they are made up of objects, or groups of objects, or groups of groups. There is no real connection between the groups and the categories.

Categorizing is one of the basic methods of thinking. Bruner et al. (1956, p. 10) point out that judgment, memory, problem-solving, inventive thinking, esthetics, perception, and concept formation all involve categorizing. The development of formal categories is tantamount to science. But there is no separate existence to any category; each is purely an invention of the mind intended to give order and accessibility to the objects of classification. The objects classified may be real or conceptual. One can classify anything that can be represented by a substantive (a noun, or a gerund). The groups, whether of objects or of concepts, can always be given reality or objectivity by definition. The categories cannot have reality or objectivity.

Among biological taxonomists there has been constant use of the term category. Examination of these uses shows that in many, if not most, cases the word is misleadingly applied. Reference is sometimes made to the level in a scheme but more often to the groups that are put at this level. For example, among recent writers we find these remarks:

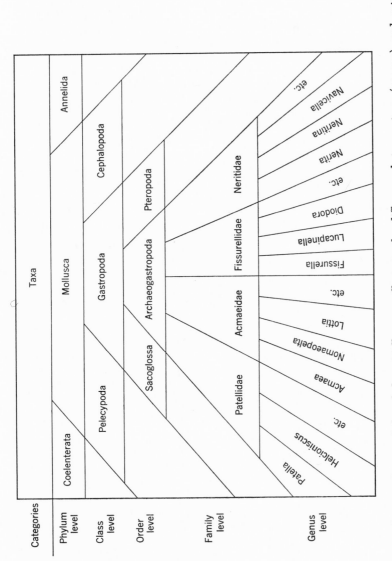

Figure 18-1. A classification of part of the phylum Mollusca to illustrate the difference between taxa (groups) and categories (levels) of the hierarchy. The arrangement of categories is fixed by convention; the placing of any group at a certain level is a matter of individual opinion, as Sacoglossa, for example, may be thought to be better placed at either a higher or a lower level.

(1) "Systematic categories are . . . based on an enumeration and evaluation of morphological resemblances." The author is referring here to objects or groups, which *can* resemble each other, not to categories which cannot.

(2) "Some of the major categories are useful, as, for example, the fusulinids and the graptolites." These, of course, are animal groups, not categories.

(3) "The discovery of that category of animals that we call the mammals." Obviously, the mammals are a group, not a category.

(4) "The evolution of higher taxonomic categories. . . ." It is impossible for a *category* to evolve, but few biologists would fail to believe that *groups* have evolved in some way.

(5) "The higher systematic categories are formed by uniting lower categories that share certain characters." Categories are not made up of subordinate categories, which cannot be united; they are simply individual levels. The group which is placed at any given level is made up of subordinate groups (or of objects). The latter have characters (or attributes), but the categories have only the relative attribute of being higher or lower in the system than some other category. This same book also says that "a family may be defined as a systematic category . . . which is separated from other families by a decided gap." A family, in actuality, is a group of genera which has been placed at the level in the hierarchy which is called the category family. And again, ". . . the taxonomic characters are a consequence of the categories." Obviously they are a consequence of the grouping, if we wish to make such a statement.

(6) "Because of the ambiguity 'Phyllopoda' is best not used in systematic nomenclature to designate a definite taxonomic category." This name represents a group of animals, in the opinion of someone, but no category is represented by such a name. The category would be Order, or Suborder, or some other, according to the terminology applied to categories in zoological classification. Again this writer says, ". . . there are many major taxonomic categories that are confined to salt water. . . ." Of course, no category from phylum to sub-subspecies can be considered to be restricted to salt water, whereas there are many large *groups* that are so confined.

(7) "Genera are grouped into somewhat larger categories, however, and these in turn into still larger groups." Here the false usage is practically admitted, since category is equated with group. Nevertheless, there are no such things as larger categories, only levels that are higher in the system. The groups put at the higher levels usually will be larger than those in the lower levels, but not necessarily so.

(8) A paleontologist writes: "The recognition of species, genera, and higher systematic categories depends on real or assumed discontinuity of the distribution of certain definable morphologic features." Again, of course, groups are meant, as categories have no foundation in morphological or other features but only in the system we adopt for arranging the groups on the basis of *their* features.

(9) "There is extremely little disagreement in well-worked taxonomic groups as to the limit of the species. . . . Such agreement is utterly lacking as regards the higher categories. . . ." Categories do not have limits at all, and it is doubtful if there is any noticeable disagreement about categories, whether order is higher or lower than class, or whether the level between class and family shall be called phalanx or order. The real disagreement applies only to the groups to be placed in each level. A few years later the same author writes: "They (sibling species) are found in nearly all systematic categories, although they seem to be decidedly more frequent in some than in others." Sibling species are particular groups of organisms. By definition they are placed in the level we call the species level. These same groups exist in the genera to which the species are assigned and in the orders to which the genera are assigned. Obviously they exist in every group of whatever size that includes them. Each one exists at each higher level, so that there are exactly as many sibling species at the phylum level as there are at the specific or generic levels. They cannot be more frequent in one category than in another, although they may be more frequent in one group of animals.

Such examples could be extended indefinitely, for it is a rare biology book that has escaped this error. There are doubtless persons who would claim that such a large number of writers cannot be wrong, yet this is clearly the case, because these writers have failed to distinguish between two things that are different in nature. It would be possible to go on using the word category as a synonym of group, but this would only necessitate use of a new term for the level in the scheme.

There are a few taxonomists who have clearly distinguished between categories and groups as used in zoological classification. One is Simpson (1945, p. 19), who makes the distinction explicitly and then uses the words in their proper meaning. He states: "The framework of classification is to equate . . . groups of animals with the categories of the hierarchy. . . ." Also: "All categories above the species have in common that they may include groups discontinuous . . . between themselves." Although he does not directly discuss the distinction between categories and groups, it is clear that the categorical hierarchy is an arbitrary arrangement and that the groups, as natural as they can be made, are assigned to the categories by the taxonomists.

Another writer who makes this distinction very clearly is Gregg (1950, p. 419). In discussing the loose usage of such terms as "objective reality" and "exist," Gregg has occasion to emphasize the difference between membership in a class (group) and the relation of a part to a whole. He points out that an organism can be a member of several taxonomic groups but cannot be a member of any taxonomic category. For example, a fruit fly may be a member of the groups *Drosophila,* Diptera, and Arthropoda. On the other hand, the group Drosophilidae not only can be a member of other groups, such as Diptera, Insecta, and Animalia, but can also be a member of *one* taxonomic category, that of Family. This category family is *not* included in any other category; it is simply a level chosen and named at an arbitrary place in a series of levels.

Things are not classified into categories but into groups. It makes no difference what things are being classified. When a taxonomist sets out to classify groups, he does not classify them into categories either but into more-inclusive groups. Only when he has groups, more-inclusive groups, and still-more-inclusive groups can he make use of the hierarchical levels which are called categories. He does this by assigning each group to some level, the more-inclusive groups to the higher levels. It also makes no difference what basis he uses for the grouping—the hierarchical system is not dependent on the nature of the groups.

In most of the examples cited above, it is evident now that the writers use the expression "the category genus" (for example) as equivalent to "the groups placed in the categorical level called the genus level." If this were always clear to the casual reader, there might be no serious objection to the looseness of the phraseology. However, it is sometimes claimed that categories have certain attributes at one level and not at another. If this claim is made the basis for further argumentation, it becomes of real importance to know that the groups but not the categories can be so described. When coupled with several other terminological obfuscations, this becomes a real hindrance to communication.

Reality and objectivity of categories. The misapplication of a word may not be a serious thing in ordinary language, although there is bound to be misunderstanding. Perhaps no harm would result even in science if no conclusions were drawn from the statement. In the case of the term category, however, there has been continuing obfuscation of a very important problem because of the failure to distinguish between categories and groups. This problem is the profound difference of opinion that appears to exist over the question of the reality of the category species and the other categories.

There has been widespread reference to the greater objectivity or reality of the "category" species. The only justification that is possible for these statements is the fact that there is a possibility that species *groups* can be more objectively delimited than other groups. There is ample evidence that in practice most species are in fact no more objective than genera or classes—they are all based on gaps in the variation of evolving animals. But the known mechanisms of evolution seem to justify the belief that species in certain groups of animals can be based on the attribute of interbreeding capacities, which is not available at lower or higher levels in the hierarchy. It has sometimes been assumed that, because this is possible in certain groups, it must be generally true. The fallacy of this is shown by Sonneborn (discussed in the next section).

Although a category, which is a concept, can have no reality, the groups placed at that level may have reality regardless of what the level is, from subspecies to kingdom. If any group is defined upon attributes which are real and exclusive, it will have real existence if it consists of actual individual objects with material existence. This of course does not mean that all groups of whatever size or placed at whatever level are equally real or objective. They vary according to their individual basis as defined. It may well be that *individual* species can be more objectively defined than *individual* orders, but this does not mean that the *concept* species is more objective than the *concept* order, or that the *category* species is more objective than the *category* order.

Dobzhansky and others have stated that species are more natural than the groups in other levels. This may be true, depending on what is meant by natural. It is probably not possible to become aware of the whole nature of a species, although many facets of its nature can be known. But one can delimit the various species in such a way that the resulting groups conform to the nature-so-far-as-known most completely. Dobzhansky's statement was in the form that "the category of species is more natural than other categories . . ." and this can have significance only if he means "the groups placed in the category species" and "the groups placed in other categories."

The man who has given the closest study to the use of language in taxonomy is unquestionably Woodger. His conclusion (1952, p. 21) on the reality of species and absence of reality at other levels is summed up thus: "[There is] no justification for distinguishing between species and other taxonomic sets. As abstract entities they are, so to speak, all in the same boat; they sink or swim together." But if one is speaking of evolutionary species and genera, "there is again no justification for distinguishing species as real from genera as unreal. . . . Only if we regard species exclusively in the evolutionary sense and genera exclusively in the taxonomic sense, can we say that species are real and genera unreal, and then only if we wish to deny the existence of abstract entities. But this is clearly an unsatisfactory mode of comparison. The taxonomic system and the evolutionary phylogenetic scheme are quite different things doing quite different jobs and only confusion will result from identifying or mixing them."

The maintenance of a clear distinction between groups and categories would eliminate the arguments over "reality" of species. It would leave open the question of whether the groups at one level can be made more objective than those at other levels, so that the speciationists could argue logically the case for species. Up to now the argument has been clouded by the misuse of the word category and also by the indistinguishability of the two meanings of the word species—the speciationists' concepts of the individual groups of interbreeding individuals, and the taxonomists' concepts of groups of individuals isolated from other such groups by gaps in the variation of attributes.

Phylogeny as basis of classification. It has many times been stated that it is the aim of taxonomists to detect evolution at work, and of classification to reflect phylogeny, which is the history of the evolutionary changes. These statements have great appeal and will find general assent on the part of many modern biologists. Nevertheless, no one has taken the trouble to present a justification for the statements, except to say that the apparent arguments against these ideals are not really pertinent. Why they are not pertinent is always left to the imagination of the reader.

The claim of a recognizable phylogenetic basis for taxonomy and classification is left without justification apparently because no real argument is possible. No single fact of phylogeny is definitely known for any species in nature. All are based on assumptions of varying validity. In order to base a classification on phylogeny, one would have to know the pertinent facts of the phylogeny of every species to be included in the classification, and there is no likelihood that this level of knowledge will ever be attained. Mayr et al. (1953, p. 42) deny that this is so: "Since it is the avowed aim of a modern classification to reflect phylogeny, one might assume that classifications could not be attempted until phylogenies are clearly and unequivocally established. This is not the case." There follows as the only proof of this denial, "Many of our existing classifications are actually pragmatic and based on the degree of similarity, regardless of whether they reflect blood relationship or not. Such a system may occasionally be more useful than a strictly phylogenetic system." This seems to be an argument against his claim and certainly does not support it at all, as there is no proof that any of these pragmatic classifications actually do reflect phylogeny. The examples which follow this argument compare classifications labeled as "practical" with ones labeled as "phylogenetic" and lead to the conclusion that the former is not more useful than the latter. It seems to have escaped notice that these classifications are all based on *the facts known at the time the classification was made,* and that the so-called phylogenetic classifications cited are different only because their authors had available additional information, not because their basis is truly phylogenetic.

It is not quite correct to say that there have been no arguments given to support these claims of phylogenetic basis. Mayr gives two, as follows: "(1) it [the phylogenetic system] is the only known system that has a sound theoretical basis . . . ; and (2) it has the practical advantage of combining forms . . . that have the greatest number of characters in common." These statements are not only inconclusive but irrelevant. No usable basis whatever is available for the phylogenetic system other than *inferences from the same data* that are used *directly* in the so-called practical system. The phylogenetic system not only does not have a "sound theoretical basis" but would always be identical with one of the possible "practical" classifications of the same material, and so never has a distinct existence.

As to the practical advantage claimed, this is exactly the advantage demonstrated by Gilmour many years before for the practical classification. In fact, this is the whole basis for practical classification. The maximum correlation of attributes enables us to make the largest number of inferences about the things classified. It is true that many classifiers have not had access to all possible data on attributes and have furthermore failed to use as many correlations as they might, but these failures do not in any way affect the theoretical potential of this system. Neither is there any distinct basis shown on which a phylogenetic system *might be based,* nor is there anything shown that might helpfully be obtained from such a system which cannot be obtained from what these writers call the practical system.

Here is a major case of the obfuscation that can result from deprecatory labels. Since all biologists must be concerned with evolution, and since evolution inevitably produces phylogenies, any system ostensibly based on phylogeny has great appeal. The usual name given to these is natural classification. All other systems are then labeled artificial or practical, usually with the implication that they are unscientific. It is not clear what meaning is herein given to the word "natural," unless it means reflecting nature or the true nature of the subjects. If a natural classification could be based on proven phylogenies, then it probably would reflect nature substantially. It surely would reflect the *phylogenetic* nature of its subjects. It would not necessarily reflect the ontogenetic nature of these subjects, their bio-chemical nature, their psychological nature, or any other of their natures. Such a system would in fact be based on correlation of phylogenetic attributes.

The true alternative to this is not an artificial system, however that term is used, but a system frankly based on *all available attributes* of whatever nature. No classification has been proposed that utilized all possible attributes, but at any stage in the development of the classification of a group, the best system was the one that used knowledge of the largest number of attributes. This is also a natural system; it is based on the nature of the subjects, all the natures that are known. It happens to be a useful natural system at the present time, because it uses available data; the phylogenetic system is not at the present time a useful natural system because *independent* phylogenetic data are not available for its basis.

It is possible to have an unnatural system, one not related to any aspect of the nature of the subjects. For example, one can classify objects on the basis of who collected or owned them. This is actually done in some museums by keeping separate the collections of various individuals or expeditions. These groupings are related to something other than the nature of the subjects. But classifications of this sort have never figured in the arguments over natural and artificial systems. There simply have been no unnatural systems, just as there have been no phylogenetic systems.

Simpson (1945) gives the most reasonable and complete discussion of phylogenetic classification of any recent writer. He denies categorically that a classification does or can express phylogeny, and he describes the "primary purpose of a classification" as "simply to provide a convenient, practical means by which zoologists may know what they are talking about and others may find out." Many of his readers seem to have overlooked these forthright statements, preferring to take note of a later sentence that seems to contradict the first: "The basis of this system is phylogenetic, as has been strongly emphasized here, and this means that the groups to be recognized in classification should be as nearly as possible valid phylogenetic entities and that the criteria of definition are to have phylogenetic implications. . . ."

These same readers fail to note the implication of still another statement: "Phylogeny must itself be determined before classification can be based on it." This is the very thing denied by Mayr et al., but it is the most important point that can be made. As a paleontologist, working with sequences of forms that are interpreted as giving data on phylogeny, Simpson apparently believes that there *are* phylogenies known, although he admits that none is perfectly known and universally accepted in detail. In fact, for it to be possible to claim that there are phylogenies known, one would have to define a phylogeny as the *supposed* or *postulated* history of the species. If it is defined as the actual history, surely there are none that could qualify.

The use of a hypothetical phylogeny as basis for classification leaves us with the likelihood that every student will suppose a different phylogeny and thus produce a different classification. Agreeing to the facts of structure and occurrence of fossils does not necessarily lead to agreement on the phylogeny, as witness the repeated changes in the "phylogenies" of such animals as horses. On the other hand, in a system based on correlation of attributes, agreement on the appropriate attributes very nearly insures similar results in classification.

Grant (1957, p. 58) claims that: "A system of classification of the biological species must be judged, not on the basis of convenience, but according to whether it represents accurately or inaccurately the realities of nature." There may be argument over what realities are to be represented, but it is hard to picture a classification that substantially represents the realities in nature without being highly practical as well. If it really represents nature, it will permit us to draw important generalizations from the data, and this is the purpose of all classification, as pointed out by Gilmour.

The idea that classification must show the phylogenetic relationships of animals is the one referred to by Pearl (1922), when he denies the common view that the evolution theory *must* have an effect on taxonomy. It simply has not had the expected effect, although many aspects of evolutionary study have contributed to the advance of taxonomy and have contributed some data of importance in classifications. But the fact that we concede that evolution has taken place and that therefore at least some animals are related by descent to some others does not force us to use genealogical relationships as the basis of our classifications. And although this has frequently been stated to be a logical necessity, which it is not, there have been almost no cases in which a classification has actually been proposed on the basis of a pre-established phylogeny. Even the outstanding recent apparent exception to this (Michener and Sokal, 1957, p. 130) proves to be based primarily on comparative data previously assembled and analyzed, with both the phylogeny and the classification based on these comparative data. This is, in fact, an inevitable situation in the present state of our knowledge of phylogeny.

It was truly said of classification that "its value lies solely in the aid which it can give in the understanding and the interpreting of the facts which it reveals." And one of its points of aid is in supplying inferences and data on which we can base the speculations about phylogeny. It was Ferris (1928, p. 22) who wrote this, and he recognized that the more data there are available for the classification, the more aid it will give us in various applications. The aim thus should be to make the classification "display in the most advantageous and most nearly correct manner the facts that have been discovered." This will include all facts of whatever nature, even phylogenetic.

The classifications now in use are all "natural" in the only useful meaning of that word. They can be made more and more useful, and will be, by incorporation of new data whenever such become sufficiently available. This includes even phylogenetic data, but at present there are no useful data of this sort, because the correlations obtainable from the phylogenetic inferences that are obtained from other data have already been utilized in the system under the guise of structure, distribution, ecology, physiology, etc. Any attempt to distinguish between phylogenetic and artificial systems will do nothing but confuse the true nature of natural classification that reflects the maximum amount of all the natures of the subjects. Such confusion would delay the improvement of the useful natural classifications, still based largely on structure but increasingly on data from all other fields of biology.

It is greatly to be hoped that taxonomists will not be so attached to the "acceptance" represented by the phylogenetic basis of classification that they will fail to recognize its true position as just one of the possible natural systems, one at present completely unattainable, and one not yet demonstrated to be even theoretically superior to any other. Taxonomists need not fear that they will lose anything by recognizing this state of affairs. They will actually gain several important things: (1) freedom from the confusion of believing in and aiming at an unattainable goal; (2) a practically as well as theoretically justifiable basis for their important scientific work; and (3) a start toward the recognition that rigor in language is a factor of tremendous importance in a field as complicated as taxonomy. Language should be our servant. We must not allow it to be our master, overpowering us by means of the notions attached uncritically to words.

Species

There have been a large number of publications in recent years dealing with the subject of species. The newer branches of zoology have had to deal with the kinds of the taxonomists, but their chief interest is in other problems than how to distinguish the kinds that exist. The ecologist may want to know how kinds affect each other and what this effect means to the biota as a whole; the geneticist may want to know how the kinds reproduce themselves and maintain their identity, and how they can change; the evolutionist may want to know how the changes are accepted or rejected, how one kind changes into something different, and how the known kinds came to exist; and the phylogenist may want to know the course of the changes, the ancestry of each kind and group. All of these are worthy biological questions. Each of these scientists looks at the kinds from the viewpoint of his questions. He sees in the idea of distinct kinds what his questions lead him to look for. The result is a series of different viewpoints about the basis and nature of the assemblages we call kinds.

Nearly all recent discussions of the word species, or the concepts behind the word, relate to its use in these modern fields. Two things have been clearly brought out—that the usages are related in the basic nature of living things and that the needs of the various fields of study are so diverse that there is no imminent likelihood that the several concepts can be directly correlated on the level of the definition of terms.

The taxonomist is still very largely concerned with kinds. He is interested in what is discovered by the geneticist about mechanisms of change, and so on, and in the speculations of the phylogenists and evolutionists as to the course of the successive changes, but as a taxonomist he works exclusively with the distinguishing of the kinds and the grouping of them. He deals with animals, not genes; with life histories, not phylogenies; with the relative stability of today, not the probable changes in prehistoric time.

Some of the evolutionists have made strenuous efforts to unite the taxonomic and the phylogenetic concepts of kinds. Perhaps it would be more accurate to say that they have tried to force taxonomists to use their evolutionary kind in taxonomy. This has been a wasted effort, because taxonomy continues to distinguish and group kinds by their attributes, using the evolutionary ideas only to temper their judgment. There is at present no possibility that taxonomists can use directly the theoretical bases of kinds that are so helpful to the geneticist and the evolutionist, because he has a tremendous system, based entirely on direct features of the organisms, that cannot be mixed with the new concepts.

The nature of species. In *The New Systematics* (1940, p. 269) Hogben remarks that "we need not prolong a barren controversy about the various definitions of species." Later writers apparently disagree as to whether the subject is barren, because much additional material has been written. If agreement on a definition of species, or on what the basis of species is to be, is taken as the criterion, the more recent discussions have been as barren as the old. It would seem that this was inevitable, because the arguers have never settled the preliminary problem of what it is the argument is designed to settle. A major symposium (1957) on The Species Problem did not even attempt to define the problem or the issues.

No one can hope to get anywhere in a discussion of any subject unless he takes the trouble to distinguish between the multiple meanings of key words. "Species" is such a word; its meanings are several and rather distinct, including these: (1) "Species" without any article would usually be plural and mean the various populations of organisms which are recognized by some means. These populations are groups of individuals. (2) "A species" would refer to one of the populations or groups. (3) "The species" would sometimes be singular but collective and denote the general concept covering all the groups which are known as species. (It would also sometimes be plural and have the same meaning as (1).) (4) "The level of species" or "the category species" refers to the hierarchical level chosen for these populations or groups.

The first of these refers to group of individuals, groups which may have reality, objectivity, importance, or a definite relationship to nature, if so defined and distinguished. The second refers to one of the same groups and thus has the same reality and basis. The third as a plural is the same as the first, but as a singular it refers only to the idea of groups which fit the prescribed conception. There is nothing objective or real about this idea; it has no relationship to the nature of organisms or to the nature of the groups. (It doubtless does have a relationship to the mind and to the sense-impressions which gave rise to it.) The fourth is a man-made thing, or more properly a man-selected position. It has a definite relationship to the other categories in the hierarchy, but it bears no relationship to nature. It is not objective, natural, or significant in itself. As part of the hierarchy it has the significance given to it by the act of placing it between two other categories. There are other meanings than these of the word species. Apparently, however, most meanings will fall into one of the two types illustrated by (1) and (2) or by (3) and (4).

If one now goes back to examine some of the statements about species, to see how this distinction applies to them, some interesting things appear. There will be the constant misuse of the word category for group, which will have to be taken into account. In 1941, one writer stated that "the category of species is more 'natural' than other categories used by systematists." If he really meant categories, his statement would be wrong, because the categories are all exactly alike in their nature and are all completely artificial. If he means the actual groups or populations, as the context shows he does, then he is correct if the particular species (plural) referred to are ones which are de-limited on natural grounds. There is no question that the concepts of breeding populations and gene pools sometimes enable one to fix the groups at one level more clearly than at any other so far as these features are concerned, and the groups which correspond with such concepts may then be said to be more natural, objective, and sig-nificant.

In 1942, it was stated that "species are real and objective units, because the delimitation of each species is definite. . . ." Here the individual groups are obviously intended, and the statement is correct if the qualification about the definite delimitation is true. In 1953, the same author wrote, ". . . the species occupies a unique position in the taxonomic hierarchy." Here there can be no question that ref-erence is made to the level known as the species level, the level to which species are assigned. The basis for the statement is given thus: "Essentially there are three kinds of categories: 1. The species. 2. Groups of populations within species. . . . 3. Groupings of spe-cies. . . ." This is the same as saying that the species level is unique because it occurs between the next lower level and the next higher level. In this case, of course, the imputation of greater reality or significance to one level in the hierarchy is unjustified. The level has no reality. This can be ascribed only to the groups placed in it.

In 1953, another zoologist wrote, "The species, with its included subdivisions, is a different sort of group from those above it." Here the references to subdivisions and group make it seem that "the spe-cies" refers to "a species." If so, he is right if the species in question is based on some natural phenomena that are different from those used for the groups at higher levels. On the other hand, the category spe-cies cannot have subdivisions; it consists of groups that consist of subgroups clear down to individuals, but the level is fixed and in-divisible.

In another example, in 1957, it is stated that "an objectively defined species concept is available to take its place" (the place of the typo-logical species concept). This apparent imputation of objectivity to a concept is false by definition. Apparently what may have been meant and what might have been true is that there is new interest in objectively defined individual groups that can be placed at the level in the hierarchy called the species level. This writer claims that the species of the typological concept can be defined only subjectively. The only thing objective about the new species concept is the delimitation of the groups placed in it. But on this basis the groups of the older concept are on exactly the same foundation as those of the new concept, *provided* that the same basis is used in delimiting the groups. If the newer groupings are based on more clearly definable features, then the new groups may be more objective to that extent, regardless of whether they are species or families.

Part of the confusion over species is due to the fact that some persons believe that they are classifying actual animals into groups having like attributes, whereas others believe that they are classifying species produced by evolution. The taxonomist is called upon to record the structures, behavior, distribution, and other relations of groups of individuals. He finds that correlations between these things are useful and meaningful. The phylogenist is called upon to interpret the probable evolution of groups. He is little concerned with individuals and their attributes but more with the ranges of characteristics and the factors that cause groups to remain separate. Each of these specialists has called certain of his groups species, but they may have in common only this name. The concept which the taxonomist calls by the name species is based on data gathered from individual organisms by means of his senses and used for the business of taxonomy, which is discovery and systematization of data. The concept which the phylogenist calls by the name species is based on data drawn from populations by means of inferences and used for the business of phylogeny, which is the deduction of the past history of the group.

Nothing whatever is gained by denying the existence of these two very different concepts. No solution has been attained by the alternative scheme of claiming the existence of many other concepts of species, which turn out to be mere variants of these. No end to the arguments over species concepts can be reached until it is recognized that at the present time there is no way to correlate the species composed of individuals having attributes with the species composed of populations supposedly phylogenetically related. To accept the one is not to deny the other. They both exist; they both serve useful purposes; but they are not necessarily correlated. At the present time zoologists can deal with each separately, but cannot yet combine them into a single concept in practice.

This leaves two things represented by a single term. Misunderstanding and confusion are inevitable. Taxonomy may again be forced to surrender and abandon an established term (species) to a new group of zoologists who use it for a more recent concept, as has happened also in the case of the term genotype. Such an unreasonable eventuality might actually be necessary in order to counter the illogical arguments of the speciationists that the species are not properly understood by the scientists who discovered and distinguished a million of them.

Sonneborn recognizes this difficulty (see below), and so do others, but their voices are lost in the chorus that considers semantics a part of philosophy and therefore of no value. A little attention to semantics would prevent a substantial waste of time, energy, and ink by making clear the lack of a common ground for argument.

Outbreeding data. It has been argued above that there is no such thing as a *biological species concept* or a *typological species concept*. There cannot be, in a general sense, because it is not possible to assure that the same concept is held in the minds of many people. These terms should be replaced by less ambiguous ones. Concept can be replaced with definition, which is what is intended anyway, but this still leaves two ambiguous words in each term. The only kind of species that can be defined is the group of individuals, either those sharing certain attributes or those that can interbreed. So species can be replaced with group or species group. Biological refers to data drawn from the fields of biology. There have simply been no species that were not based on biological considerations. Typological denotes a type-centered approach which simply does not exist in practice.

What is left is a *species-group definition,* which may be based on few or many characters, features from one field or all possible fields, features related to ancestry, structure, or function. There is simply no other way in which the species of classification can be defined.

One can define the species level in the hierarchy by its assigned position. Species can be defined as interbreeding populations in nature if their breeding capabilities are known. Taxonomists can define species as groups of specimens agreeing in pertinent attributes. Phylogenists can define in very indefinite terms phylogenetic species, but they cannot define individual species of this sort. No one can define any general species concept supposed to be held by more than one person.

In discussions of the so-called biological species definition, it is often found necessary to note that in most forms this definition can be applied only in sexual animals. Since sexuality is widespread among animals, it is assumed that the biological species can be recognized throughout the kingdom, except for a few exceptional groups. A closer look at this idea reveals that if the interbreeding of sexual animals is to be taken as the criterion of species, there must be available either one of two things: (1) knowledge of the breeding capacities of all the forms whose species status is to be established, or (2) substantial evidence that inferences can safely be drawn from species in which it is known to ones in which it is not. There must be either direct information or reasonable inferences for all species.

The number of times in which actual interbreeding capabilities are known between two populations is so small compared to those in which it is unknown, that the percentage is virtually zero. This leaves inferences from these few cases as the only evidence that could be used in actual determinations of species status. There is some evidence to show that inferences can reasonably be made over fairly large groups. There is also evidence to show that in most phyla of animals there are unexpected exceptions which make it impossible to draw such inferences safely. These exceptions include not only the many forms of asexual reproduction, but also the ones in which parthenogenesis and self-fertilization occur.

A list of the dozens of groups of animals in which asexual processes occur would show them to be too numerous to be passed off as "exceptions" or "in the minority." Sonneborn (1957) has estimated that it will eventually turn out that bisexual forms do not outnumber the asexual and unisexual ones. This view will appear extreme to some zoologists, especially to those principally familiar with vertebrate groups. Nevertheless, the list should serve to show that asexual individuals are sufficiently widespread that they *must* be taken into account.

All of this is not the main point of interest to taxonomists. Whether species are sexual or asexual, inbreeding or outbreeding, is not the whole picture in taxonomy. If taxonomists are to make use of a species concept that is based on outbreeding, they must be able to determine that outbreeding *can* occur. They must be able to do this in at least some cases in every group and in all cases *if the groups are to be defined on this basis.* This is the crux of the matter: Can they determine, in all forms that they wish to classify, that crossbreeding can or cannot occur between the individuals?

The groups in which they *cannot* do so include at least the following: All those in which binary fission is the rule; all those in which sexes do not exist; all those in which parthenogenesis occurs; all those in which there is obligate self-fertilization; all those in which budding, gemmulation, sporulation, or dichotomous autotomy occur; all those in which we find it impossible to perform breeding experiments; and all those known to us only by fossil remains. There is no question that these include a majority of the materials on which taxonomists work. In fact, they probably include more than 90% of all organisms. This is the reason why the breeding population can never be the actual unit of classification, why the ability to produce fertile offspring cannot be the actual basis for classification among animals in general.

Furthermore, classification, if it is to take account of all available data, must deal with individuals who do not breed or at any given moment are not part of a breeding group. Here are those castes of social insects which have no reproductive function, the non-breeding stages of forms with polymorphic life histories, and the individuals which simply do not breed because of lack of opportunity, malformation, malfunction, or other factors. These are all living animals; they cannot be omitted from our classifications.

The argument that these have no effect on the future of the species, no place in evolution, is true, but it has no bearing on the need to recognize their existence and place them appropriately in the classification. The natural system based on all known attributes is quite capable of dealing with these forms. The phylogenetic system refuses to deal with them, usually claiming that individuals are not the units classified. When species or breeding populations are the objects classified, these "exceptional" individuals are not troublesome because they are either ignored or blanketed into the population on the basis of non-breeding attributes. Here the choice is between two procedures: first, sidestepping the problem by ignoring many individuals along with the data that pertain to them; and, second, recognizing that a system based on individuals incorporates all forms of them along with their important contribution of data. Taxonomy has everything to gain and nothing to lose in frankly accepting the system which it has in fact always used.

The "species problem." The species of the speciationist's biological species concept are based on what he calls a "biological" definition. Taxonomists are not supposed to ask the meaning here of the word biological but to accept the fact that it is more important, meaningful, and useful in the study of evolution than something called typological or morphological. The biological species is based on breeding populations, barriers to the spread of genes, and mechanisms for the control of mutations. The evolutionists seem to be convinced that this concept is useful to them, in fact indispensable.

The taxonomist on the other hand rarely finds himself in possession of data on breeding populations, barriers, or genetic mechanisms in sufficient quantity to be directly useful in classifying animals. There can be no question that he does not have these data in sufficient quantity to make them *the* basis of a classification on a large scale. Because of this, the taxonomist is forced to use other data. Since the classifications upon which he must build are all based on such other data, he finds this no hardship at all. As a result he uses a concept of species based on correlation of attributes, principally structural ones. He is convinced that such a concept is essential to his work of classification.

Speciationists emphasize the evolutionary importance of species. The importance of species to the taxonomist is also great, in some opinions far greater. Keck (1957, p. 51) gives the main argument for keeping the concept for taxonomy: "From the utilitarian point of view the retention of the taxonomic concept of species is extremely important, for its replacement by a new category on a different basis would play havoc over a big segment of biological literature." He further warns against continued employment of species as a basic evolutionary unit in place of breeding population or some other unit.

It is evident that the classical method of taxonomy, the use of species based on comparative data, has produced the present-day classification which proves itself every day in practice. The so-called "modern" method of the speciationists has produced much valuable data about evolutionary mechanisms and pathways, but it has produced no species or classifications useful for the recording and systematization of knowledge.

The biological species concept. It is not appropriate for anyone to use the expression "biological species concept" in reference to a particular way of defining the term species. The use of this expression suggests that a definition is used that involves aspects of the biology of organisms not used in some other definition that is not called biological. This distinction does not exist.

Everyone seems to agree that there are species in nature, or at least that there are now gaps in the continuum produced by evolution, and gaps between the end products of the branches even when the branches are unbroken. Taxonomists have studied or recorded a tremendous variety of features of organisms, all of them biological features and including all known types of biological features. It is impossible for there to be anything more appropriately called "biological."

PART VI
Zoological Nomenclature

"We cannot dispense with names. They are an essential part of the equipment with which we must work . . . any one who is properly to be called a systematist must consider it a part of his equipment for his profession to know something of the commonly accepted rules and the procedure that governs the application of names." (Ferris, 1928)

Chapter 19. Rules of Nomenclature   379
    History   380
    The individual's freedom   387
    The complexity of the rules   388
    New features in the 1961 Code   390
    Unsatisfactory aspects of the 1961 Code   398
    New features of the 1964 amended
      code   403
    Treatment of the rules in this book   404

Chapter 20. The Nature of Scientific Names   407
    Names of taxa in the higher
      categories   408
    Names of taxa in the family group   408
    Names of genera and subgenera   408
    The names of species   422
    The infraspecific categories   431

Chapter 21. Names of Taxa in the Higher Categories   435
    The names of the taxa   436
        Historical background   436
        Types of taxa in higher categories   437
        Uniform endings   438
        Terminology   439
    Comments on names of taxa in each category   444

Chapter 22. Names of the Family-group Taxa   449
    The superfamily taxa   451
    The family taxa   453
    The subfamily taxa   469
    The supertribe category, its taxa, and
      their names   470
    The tribe taxa   471
    The subtribe category, its taxa, and
      their names   473

Chapter 23. The Names of Genera   474
    The genus taxa   474
    Collective groups   495

Chapter 24. The Names of Subgenera   497
    The subgenus taxa   497
    The names of sections   501

Chapter 25. Genotypy and the Types of
        Genera   502
    The concept of genotypy   502
    Terms   509
    Genotype determination   512
    Recommendations of the Code   533

Chapter 26. The Names of Species   535
    The species taxa   535
    The "work" of an animal   565
    Names of hybrids   567

Chapter 27. The Names of Subspecies   568
    The subspecies taxa   568
    Categories below the subspecies   575
    Names of hybrids between subspecies   576

Chapter 28. Epithets   577
    The sources and grammar of epithets   579
    Orthography of epithets   583
    Emendation   587

Chapter 29. Types of Species and
        Subspecies   589
    Kinds of species types   591
        Primary types   592
        Secondary types   600
        Tertiary types   600
        Other types or specimens   601
        Type-localities   601
    Subspecific types   602
    Ownership and preservation   602

Preface to Part VI

It is the purpose of this Part to provide background for the present rules, to outline the problems of zoological names, to spell out the rules now applicable to these problems, to add considerations that are not covered by the 1961 Code, and to discuss the rules and their application to a wide variety of examples and cases.

This Part is based primarily upon the writer's own experience with taxonomy and nomenclature and his analysis of the new codes and their implications. The experience was largely shared with other American nomenclaturists.

The analysis of the codes was prepared for inclusion in this textbook. Its preparation led to the conviction that the present situation in nomenclature is far more complex than generally realized and that the 1961 and 1964 Codes contain more confusing and controversial provisions than a quick study can reveal.

This Part therefore consists in large measure of a strong critique of the recent codes. So many points are criticized that there may appear to be rejection of the rules *in toto*. This is not the conclusion of the writer, although rejection of certain provisions or implications in the codes appears to be inescapable.

To confirm that this critique was necessary and the details correctly dealt with, the manuscript was reviewed by W. I. Follett of the California Academy of Sciences. His agreement on all major criticisms greatly strengthens the account, and his correction of details has added accuracy to many statements. This collaboration is gratefully acknowledged.

CHAPTER 19

Rules of Nomenclature

Zoological nomenclature is a complex and specialized subject. Names are given to animal groups, and used to refer to them, under rules which are neither simple nor unchanging. The current rules are now available in published form for study or consultation by taxonomists. It would be inappropriate to duplicate those rules here in their original arrangement, even if space were available.

The present rules are the result of a long history or evolution, during which there was much discussion of the various solutions and special cases. The rules cannot be fully understood without some knowledge of this history and of the arguments presented for or against the many items.

This introductory chapter gives the writer's views of the relevant part of the historical background, summaries of the new features of the two recent codes, a list of the features which seem destined to prove unsatisfactory to the working taxonomist, and a demonstration of the sad fact that the recent codes have greatly increased the complexity of nomenclature and confused both terminology and the relation between taxonomy and nomenclature. In later chapters it will also appear that the code is not arranged in a manner to be useful to the student or the working taxonomist; it is a nomenclaturist's code.

Although these faults, or supposed faults, affect a large part of the code, it is not concluded here that the code should be abandoned. It should be followed whenever possible. It should be studied in detail, and its weak spots identified. Its rulings should, however, be disregarded wherever they interfere with taxonomy, are unworkably complex, or fail to recognize the decisions correctly taken under previous codes.

The present code exists in two editions, indistinguishable externally, as follows:

379

International Code of Zoological Nomenclature adopted by the XV International Congress of Zoology, by Editorial Committee, N. R. Stoll chairman. London: International Trust for Zoological Nomenclature. 1961. (Effective November 6, 1961) (Title also in French)

International Code of Zoological Nomenclature. . . . 1964. (As above except for changes in pagination. Revision anonymous. Effective May 1964. Changes to four Articles and in many subsidiary details)

The 1964 Code is a slightly amended version of the 1961 Code. The 1961 Code is thus the definitive revision of the rules. It will be referred to exclusively in the following discussions, except in the few items in which the 1964 Code differs. A person possessing only the 1961 version can readily annotate it with the changes listed later in this chapter. A person possessing only the 1964 edition will find all references to Articles apply to his copy, even when referred to the 1961 edition, with the deletions noted in every case.

History

The use of rules in zoological nomenclature has a history extending back at least a hundred years. The only part of this history which concerns us much now occurred in the first two-thirds of the twentieth century—from about 1900 to 1964.

In 1901, at the 5th International Zoological Congress, there was adopted a code of rules called the *International Rules of Zoological Nomenclature* (or more exactly *Règles Internationales de la Nomenclature zoologique*). A few changes were made by subsequent congresses, but most of the imperfections found in the *Règles* were taken care of by issuance of interpretative opinions, prepared by a special Commission on Zoological Nomenclature set up by the Congress. Until about 1948 these *Règles* were presumably followed by all taxonomists. (In reality a rather large number of taxonomists failed to follow certain rules to which they had personal objections.)

In 1948, at the 13th International Congress of Zoology, in Paris, a large number of proposals were considered for revision of the *Règles.* There was inadequate preparation for this study, and it was found impossible to arrive at acceptable conclusions on many points. By the time of the next Congress, there had been no progress of note on these revisions.

After the 13th Congress, the proceedings were published in volume 4 of the *Bulletin of Zoological Nomenclature.* The amended rules were to be effective as soon as published, but separate publication never occurred. In the meantime the *Bulletin* announced that "zoologists are advised . . . to guide themselves in their work by reference to the decisions in regard thereto recorded in the Official Record. . . ." It has not been claimed that these were binding on zoologists, and, in fact, even the later codes ignore the Paris actions.

In 1953, at the 14th International Zoological Congress, a colloquium or study group of about forty zoologists, including many nomenclaturists, studied aspects of the *Règles* and worked out solutions or compromises to many problems. These conclusions were published in 1953 as *Copenhagen Decisions on Zoological Nomenclature* . . . , the first recent publication of the International Commission on Zoological Nomenclature to be sold at a price within the reach of the average zoologist.

Both the Commission and zoologists in general were admonished to follow the amended rules cited therein even before publication of the complete new code. This admonition has been taken by many persons to have established an interim set of rules on December 31, 1953, which remained in effect until publication of the 1961 Code. The 1961 Code does not recognize the Copenhagen Decisions directly, but many of the provisions of the latter are herein cited as having been in effect from 1954 until 1961.

In 1958, at the 15th International Congress of Zoology, a second colloquium, averaging a hundred members, made several basic decisions on the coverage of the developing new code and then authorized a special committee to prepare and publish it, based on the general decisions and a long list of particular decisions on individual problems.

The new code was published in 1961, the English title being *International Code of Zoological Nomenclature adopted by the XV International Congress of Zoology*. It is published in both French and English throughout, with both texts being considered substantive.

Little reaction to this 1961 Code has been published. The arrangement, organization, and indexing are good from the point of view of nomenclaturists. The coverage of many problems is analytical, logical, and clear. Other problems are treated in what seem to some users to be illogical and unreasonable manner. The code is extremely complex in operation, there being at least five basic dates at which differing rules apply. There is a real chance that the code is so complex it will not be followed exactly even by those who believe it to be proper to do so—that it will defeat its own purpose.

The 1961 Code is not the end of the story. Not only will there continue to be Opinions issued on specific cases in which the application of the code is not clear or on cases resulting in instability or undesirable change, and Declarations in effect amending the code itself, but

there have already been one revision of the code (1964) and preparations for another (presumably 1968 or 1969). With this start it is
likely that every Congress will feel free to make substantive changes,
quite possibly spurred by national preferences of the host country.
The zoologists have not adopted a representation system for voting
on rules, as has been done by the botanists. It is unlikely that stability of *rules* can be attained under the present system, and it is
unlikely that wide adherence can be attained by the Commission
and others to a code that is forever changing.

The Hemming era. A person who did not take part in the activities of nomenclature between 1940 and 1960 might well be puzzled
by some of the occurrences and the undercurrent of continued dissatisfaction. Participants in these activities have so far failed to record
the unpleasant aspects of this period, perhaps because of a desire
to avoid personalities or criticisms of individuals. The following notes
are intended to show that there has been ample cause for dissatisfaction.

During the first three decades of the existence of the International
Commission, it acted on many problems. It was slow, because its
work was done by correspondence, and some members took little part
in its deliberations. It accomplished what it did under the leadership
of the parasitologist Charles Wardell Stiles, of the U.S. Public Health
Service in Washington, D.C. Upon the retirement of Stiles in 1936,
the Commission lapsed into inactivity, largely as a consequence of
the worldwide upheaval caused by World War II.

At the Sixth International Congress of Entomology at Madrid in
1935, much interest in nomenclature appeared. Discussion was
led by Francis Hemming, an amateur lepidopterist from England.
By the time of the next zoological congress, in Paris in 1948, Hemming had been made secretary to the Commission, succeeding Stiles,
and had started a new series of Opinions, as well as plans for a new
code.

At Paris in 1948, the members of the Section on Nomenclature of
the 13th International Congress of Zoology spent much time discussing a series of changes in the rules, amounting to a revision of the
entire code. Although this action was not illegal, it had not been
adequately announced in advance and few American nomenclaturists
were present. The resulting new rules were to be drafted by Hemming (and a group of unnamed "jurists") and presented to the next
Congress.

As the 14th Congress (at Copenhagen in 1953) neared, Hemming
had not yet been able to formulate the new rules. He thereupon

called for a Colloquium of nomenclaturists to meet for a week before the Congress to make decisions on many points. This Colloquium was attended by eighteen nomenclaturists from America, sixteen from Britain, and seventeen from the rest of the world.

Some outstanding nomenclaturists were unable to attend, and one of the best was not invited by Hemming for political reasons. This was Rudolf Richter of Frankfurt-am-Main, one of the few commissioners to contribute substantially to the literature of nomenclature. Richter's membership on the Commission was illegally canceled by Hemming, because he was suspected of having been a Nazi collaborator. This was a poor beginning for an international agreement of scientists.

The Colloquium found compromises for some disputed questions and failed to solve others. One problem, that of imposing some time limit on the Law of Priority (variously called a Law of Prescription, Principle of Conservation, or Law of Recency), proved to be a stalemate. When the Colloquium reported to the Section on Nomenclature for the final vote, this one item was singled out and scheduled for voting at a pre-announced hour. As this time neared, after the Section had been at work for two hours, over forty persons entered the room, loudly voted for the limitation, and then left without taking part in any other of the deliberations. These forty were all young Germans who had been urged to attend for the purpose of passing this one disputed provision. A rule adopted in such a manner could not attain the respect of zoologists in general. This highhanded action could not have taken place without the acquiescence of Hemming and the chairman of the Section on Nomenclature.

During this Congress several new members were appointed to the Commission. Taxonomic organizations throughout the world had been invited to recommend candidates. Several American nomenclaturists were so nominated, Mr. C. W. Sabrosky of the U.S. Department of Agriculture being by far the most often mentioned. Mr. Hemming passed over all the nominations and appointed an American not nominated by any organization and not previously known as either interested or qualified in nomenclature. Because of the absence of nomination, he was designated a commissioner-at-large, a new title. (Several American nomenclaturists, including Sabrosky, were appointed at the time of the next Congress, after Hemming's retirement.)

These unfortunate actions were only part of the events which brought the Commission and its secretary into the regrettable position of inspiring little confidence among the nomenclaturists of several countries. The most damaging feature, however, was something that has become a veritable byword—the verbosity of Hemming in all his writings.

The Copenhagen Decisions were published in one hundred and thirty-five pages of size eight and a half by eleven inches. The words therein printed could readily have been published in half as many pages. The really relevant information, omitting all the pomposity, repetition, ritual, and "officialese," could easily have been published in one-third of the condensed pages or one-sixth of the actual total. The *Bulletin of Zoological Nomenclature* was of similar style. The resulting price was so high that almost no individual nomenclaturists could afford it. Even a manuscript submitted by an outside author would be padded up to several times its proper size by headings, official authorizations, and, frequently, lengthy replies by Hemming.

The Introduction to the 1961 Code states of Hemming that "his capability and zeal in organizing and conducting the Paris discussions and the Copenhagen Colloquium, and in preparing for the London Colloquium, eventuated in the comprehensive exchange of ideas on zoological nomenclature that brought to culmination an instrument embodying the majority views of interested zoologists throughout the world." That much is to Hemming's credit. It appears likely, however, that history will see even greater influence of *these* features: (1) his attempts to involve "jurists" in the drafting of the rules, (2) his attempts to involve departments of state of each country in selection of the Commission, (3) his extreme verbosity and obfuscation, (4) the excessively high cost of his publications, (5) his intransigence on points of nomenclature on which he held personal views, (6) his refusal to allow any other organization to assist in the international field of taxonomy, and (7) his refusal during one period to permit even a nomenclaturist to express views on a particular case unless in his own field.

It is quite possible that zoology would be better off with no major revision than with one brought about in this manner. The rivalries and antagonisms aroused along the way will damage international cooperation in nomenclature for years to come.

Authority. Since 1901 the rules of zoological nomenclature have been the responsibility of the International Congress of Zoology. This is a loose organization which meets every five years but has no permanent membership. The authority of the Congress consists solely in public opinion and the willingness of zoologists to follow its lead in trying to attain uniformity, stability, and simplicity in nomenclature. Not all zoologists have followed the Congress rulings in the past, but it has come to be more and more common to recognize that there is at the present time no better system and that conformance is definitely advantageous in many respects.

In the paragraph of the 1961 Code quoted above, the editors imply that the code embodies "the majority views of interested zoologists throughout the world." As this is likely to be understood, it is a fantastic overstatement. According to the same Introduction, only about two hundred persons participated even briefly in the London Colloquium and about three hundred communications had been received from interested zoologists. Considering the overlap between correspondents, participants, Commissioners, and consultants, it is doubtful if over two hundred taxonomists took any real part in formulation of the new code. Compared to the more than five thousand active taxonomists in the world, this is a strange "majority."

In matter of plain fact, there is no way for most taxonomists to be heard effectively. The rules have always been made and interpreted by a handful of taxonomists. If they have done their work well, it is not because of the influence or control of a "majority" of taxonomists, and it is not with their approval in any formal sense, because they have no direct opportunity for expressing approval or disapproval. They have frequently not even known what was being discussed.

The Commission has greatly exaggerated its authority in recent years, from *advising* to *ruling*. European zoologists did at one time express dissatisfaction with the International Rules by failing to follow them. There is some chance that the same fate awaits the 1961 Code in America.

The International Commission. The commission which prepared the first International Rules for the 5th Congress was thereafter made permanent, as the International Commission on Zoological Nomenclature. The authority of the Commission is nebulous, but the successive congresses have ratified its rulings. It has interpreted the *Règles*, given opinions on specific problems and cases, and proposed changes in the *Règles* for consideration by the congresses.

In 1913 the Congress gave the Commission authority to suspend the *Règles* in certain circumstances. This was intended to prevent changes of well-known names for inadequate reasons produced by strict application of the *Règles*. Many rulings have been made under these powers, and lists of the illegal but officially accepted names have been published.

The original Commission (1895) had five members. In 1898 the membership was increased to fifteen, and it was made a permanent commission, recommending to the Congresses its own membership. At the Paris Congress in 1948 the limitation on the number of members was removed, and the Commission has since had up to twenty-eight members. There is no system of national, regional, or specialty representation.

The actions of the Commission, both as to its membership and as to its nomenclatural rulings, must be approved by the next Congress before becoming fully effective. This approval has in recent years been automatic, there sometimes being no discussion of the issues themselves. By appointment of Alternate Commissioners to fill vacancies at meetings, the Secretary has exerted a great influence on the Commission, amounting at times to personal control.

Among the commissioners have been some of the best taxonomists and nomenclaturists in zoology. Unfortunately, some others have been largely inactive or honorary, some have been prevented from taking part by political considerations, and some have been appointed in spite of lack of experience in nomenclature. At one time it was proposed (by Hemming) to have the members of the Commission nominated by the national academies of science through their respective departments of state. Fortunately this proposal was summarily rejected.

At present the Commission lists its duties as these:

"(1) to consider for a period of at least one year in advance of a Congress (or for such less time as the Commission may agree) any proposal for a change in the Code;

(2) to submit to the Congresses recommendations for the clarification or modification of the Code;

(3) to render between successive Congresses Declarations (i.e., provisional amendments to the Code) embodying such recommendations;

(4) to render Opinions and Directions on questions of zoological nomenclature that do not involve changes in the Code;

(5) to compile the Official Lists of accepted, and the Official Indexes of rejected, names and works in zoology;

(6) to submit reports to the Congresses on its work; and

(7) to discharge such other duties as the Congresses may determine."

The Commission states its own powers thus (Article 78):

"The Commission has the power, when an application is referred to it by any zoologist, to interpret the provisions of the Code and to apply such interpretation to any question of zoological nomenclature. . . . If a case before the Commission involves a situation that is not properly or completely covered by the

Code, the Commission is to issue a Declaration (a provisional amendment to the Code) and to propose to the next succeeding Congress adoption of this amendment in the manner prescribed in Article 87. . . . If the case in question involves the application of the Code to a particular situation relating to an individual name, act, or publication, the Commission is to render a decision, termed an Opinion, and either (1) to state how the Code is to be applied or interpreted; or (2) acting in the interests of stability and universality, to exempt, under its plenary powers [Art. 79], the particular case from the application of the Code, and to state the course to be followed. . . . Opinions have force immediately upon publication of the ruling of the Commission. . . . All decisions are to be rigidly construed and no conclusions other than those expressly specified are to be drawn from them. . . . The Commission is empowered to suspend, on due notice as prescribed by its Constitution, the application of any provisions of the Code except those in the present and the next succeeding Chapter [XVII—The International Commission on Zoological Nomenclature, and XVIII—Regulations Governing This Code], if such application to a particular case would in its judgment disturb stability or universality or cause confusion. For the purpose of preventing such disturbance and of promoting a stable and universally accepted nomenclature, it may, under these plenary powers, annul or validate any name, type-designation, or other published nomenclatural act, or any publication, and validate or establish replacements."

The International Trust. Sometime before 1947 the financial needs of the Commission and its secretariat led to the formation of an International Trust for Zoological Nomenclature, incorporated in England. This was an unofficial organization consisting of certain individuals associated with Mr. Hemming. The activities of the Trust have been viewed with suspicion by some taxonomists, but it has been the nominal publisher of the recent Opinions of the International Commission, its Declarations, the *Bulletin of Zoological Nomenclature*, and the 1961 and 1964 Codes.

The individual's freedom

The relationship of individuals to the *Règles* and to the 1961 Code has seldom been publicly discussed. The original rules were for the guidance of zoologists; they presumably would be most effective if universally followed. The rules were usually couched in terms of

recommending the best solutions to problems. It was gradually forgotten that the rules were not binding on zoologists—that they were
not actually laws. The Opinions published by the Commission, instead of being advisory, came to be treated as supplementary rulings,
supposedly binding on all workers. The 1961 Code continues this
trend.

There have always been taxonomists who believed that certain
rulings were inadequate, illogical, unworkable, or ill-advised. This
will certainly be true of the 1961 Code as well. There have also been
taxonomists who believed that certain rulings would act to stifle or
channel taxonomic work, to the detriment of the study of the animals.
Some of these taxonomists have refused to follow these rules, in order
to preserve their ability to discover and present the facts of the
diversity of animals. These men are often among the most active
and effective taxonomists.

Although universal acceptance would insure the maximum benefit
in the future from any set of rules, it is not necessary for an individual to act against his conscience in pursuing his scientific work,
merely because an organization of a few taxonomists has recommended a certain rule which affects that work. It is advisable that
every taxonomist or zoologist should follow the rules explicitly until
he has made a thorough study of them, discussed his problem and
proposed solutions for it, and is certain that adhering to the particular
rule would interfere with his recording the facts of animal diversity.

The complexity of the rules

There have always been taxonomists who felt that the *Règles* were
unduly detailed, even though there were many points of theory and
practice not covered. It is unquestionably true that the accumulation
of interpretative Opinions made the rules much harder to use. By
1948, there had been Opinions interpreting previous Opinions, as
well as Opinions contradicting previous Opinions. Whatever the
complexities of the *Règles* may have been, they are simplicity itself
compared with the body of rules now governing zoological nomenclature.

On the surface it would seem that a unified code, superseding all
previous rules and all the interpretative opinions, would be a simpli-
fication. To those who were familiar with the verbiage and obfuscations of Francis Hemming during the period from 1948 to 1960, the
statements in the new code will seem to be clarity at its best. Both
these judgments are based on misunderstanding of the present situation.

To put the situation in its worst light may not be entirely fair, but it will emphasize the forgotten factors that do still operate. On this basis we must recognize that:

1. After 1960 all decisions must be based on the new code, with its extremely complex provisions.
2. Decisions taken between 1901 and 1953 under the *Règles* cannot all be upset by the new code, so it is necessary to know the old rules also to judge these actions.
3. Decisions taken between 1953 and 1961 under the Copenhagen Decisions also cannot be upset by the new code, so these Decisions must be studied.
4. No decision under any of these codes is of any value unless it is determined that there has been no contrary action by the Commission under its Plenary Powers. There are now eight separate printed Official Lists of the various exceptions made before 1958. All subsequent issues of the *Bulletin of Zoological Nomenclature* must be searched for supplements to these.
5. A new edition of the 1961 Code was issued in 1964, incorporating changes in at least four rules.
6. None of these twelve works is adequately indexed from the point of view of the taxonomist who is not an experienced nomenclaturist.
7. Further revisions are already being prepared to form another edition after the next Congress.
8. In at least thirty places, individual rules apply only between specified dates, making it necessary to follow the history of a name through successive forms of the rule.
9. The extent of mixture of taxonomy into the supposedly strictly nomenclatural rules is much more than the amount necessary for regulation of names. This does interfere with taxonomic work in several respects.
10. Any decision taken in good faith by any taxonomist now stands in greater danger of being reversed by the Commission, because of the established practice of acting in response to individual applications.

For an individual to find out for certain all the rules which apply to a given problem is much more difficult than ever before. And when the necessary search is made and a solution agreed upon, it is no longer possible to feel that the solution will not be upset by any reversal of some ruling in the future.

Any person doubting the actual complexity of the 1961 Code should examine Articles 21 and 22, on the Date of Publication. If one thinks that the index will serve a person looking for specific answers, he should try to find the four ways in which the Code permits the publication (or establishment) of a new species name.

There were obviously substantial efforts made to organize the provisions of the code, to explain them and define words, to bring similar things together. If one wishes to argue the principles underlying homonymy, this organization is helpful, but if one is trying to find the answer to a problem of homonymy of species names, he will find the section cluttered with references to family name homonymy and stipulations that may apply to all levels or not. Nomenclaturists need a certain arrangement in arguing the appropriateness of underlying ideas, but the taxonomist user of the code needs a very different sort of organization to help answer his questions.

The remainder of this book is one such different arrangement. When the code has been in use for a longer time, it may be possible to simplify it in other ways.

New features of the 1961 Code

A detailed comparison of the new code with the old would yield little of value to the student who is attempting to learn and follow the current rules. Some of the most controversial things in the 1961 Code are new there, and some of these are among the things most likely to cause difficulties in use of the code. Included in the latter are these:

1. Recent demands for some limitation on the application of the so-called Law of Priority, in cases of resurrection of long-forgotten names, were embodied in proposals using the terms "Principle of Conservation," "Law of Prescription," or "Law of Recency." The 1961 Code adopted such a limitation, stipulating (in Article 23) that "A name that has remained unused as a senior synonym in the primary zoological literature for more than fifty years is to be considered a forgotten name (nomen oblitum)" and referred to the Commission for possible placement on the Official Index of Rejected Names, or otherwise ruled upon.

2. When type specimens are destroyed, it becomes necessary to replace them with new specimens. Rules for the proposal and employment of neotypes are included.

3. Some ambiguity existed previously in the rules on homonymy of specific names. After 1960, secondary homonyms are to be revived if the senior homonym is removed from the genus [Article 59]. Of course, if a later author puts the two species back together, the secondary homonym is again rejected. Thus, two authors may correctly use different names for a single species.

4. The neo-Latin alphabet used for zoological names is spelled out explicitly. All diacritic marks, hyphens, and apostrophes are outlawed, requiring respelling of all older names incorporating such signs. (One exceptional use of the hyphen is still permitted—Article 26.)

5. The term parataxa has been used for "assemblages" based on diverse fossil fragments rather than on the animals themselves. It was expected that

parataxa might be recognized sometime in the future, but the 1961 Code explicitly rejects them from the arena of nomenclature.

6. Much more detailed rules are included on such subjects as requirements of publication, dates of publication, family names, availability of names, rejection of names, errors and emendations, source and spelling of names, the type concept at all levels, and genotype fixation.

7. The operation of the International Commission on Zoological Nomenclature (ICZN) is expressly described under the following headings: Status, Duties, Exercise of Powers (including publication of Declarations, Opinions, Directions, and Official Lists and Indexes), Plenary Powers, Status of Case under Consideration, Exemption, and Constitution and Bylaws.

8. Extensive appendixes are included to provide collateral information that may be of use to taxonomists. These include transliteration of Greek words into Latin; latinization of proper names; the formation of names from various types of words; determination of gender; compounding classical words; the stems, terminations, and family-name forms of Latin and Greek derivatives; and general recommendations to taxonomists and others.

9. The code creates a distinction between "primary zoological literature" and other literature, without defining them [Article 23]. It is necessary to make this distinction in order to apply the new rule of limitation on priority.

10. Numerals are no longer acceptable in hyphenated names, such as *4-maculatus,* but must be written out in the Latin form, as *quadrimaculatus.* This rule is retroactive and will thus cause many changes of names.

11. In attempting to provide for the importance of the date of publication of names, rules are given embodying provisions, for fixing and citing the dates, that are extremely detailed, unnecessarily complex, in part impractical, and also incomplete [Articles 21 and 22].

12. Detailed rules are given for determining the gender of generic names. These seem to add up to assigning the apparent Latin gender in most cases. An extensive appendix gives the gender of many words.

13. Family names must be changed if the name of their type genus is found to be a junior homonym [Article 39].

14. The date of a family name and its type genus that are replacements because of generic homonymy are to be changed to the earlier date of the homonymous name [Article 39]. (Deleted in 1964)

15. After 1960, in case of synonymy of family names, an earlier change is not to be upset (as required by Article 40) if it has been widely accepted in the meantime, but its date is to be changed to that of the rejected name [Article 40].

16. At all levels from family to species, the terms "typical (subgenus)" and "nominotypical (subgenus)" are to be replaced with "nominate (subgenus)," the one which contains the type of the subdivided higher taxon [Articles 37, 44, 47; Glossary].

17. The term "binomen" (pl. binomina) is introduced for the two-part name of the species, the combination, the scientific name of the species [Article 48; Glossary].

18. Two names of the same spelling, given independently to a single species, are not homonyms under Article 52.

19. For the first time it is made clear that two species names of identical spelling are not homonyms if they are in different genera that happen to be homonyms themselves [Articles 52 and 57(c)].

20. The Code abandons the variety of waiting periods specified in the Copenhagen Decisions.

21. Interpretations are given of what is a type series, which specimens are to be labeled, the responsibility of individuals and institutions possessing types, and so on [Article 72].

22. The terms *valid* and *available* are clearly defined for the first time. These terms are still inadequate to describe the status of names, as discussed in a later paragraph.

Instructions to non-taxonomists. A strange feature of the 1961 Code, not present in the previous codes, is the inclusion at various points of instructions to several types of non-taxonomists or even non-zoologists as to how they should perform their work when it involves use of zoological names. For example:

Editors are admonished not to divide a paper for publication into two parts so that descriptions are divided.

Publishers are admonished not to put into circulation advance copies of a forthcoming publication.

Editors are admonished to state the exact date of publication of each part of a serial.

Librarians are admonished not to remove covers before binding serials, in order to preserve information that may be thereon.

Printers are admonished to add to reprints or separata the exact place and date of publication.

Publishers are admonished to identify all preprints as such.

Secretaries of meetings are told not to include in their published reports the new names involved or anything else affecting nomenclature.

Reporters at such meetings are similarly admonished about their published reports.

All biologists are told that they should never use the word "genotype" in its only correct meaning, because the word is currently employed in genetics with a different meaning.

Every institution possessing types is admonished as to their care and labeling.

All editors and publishers of zoological papers are admonished to censor these papers to eliminate any breach of a specified Code of Ethics.

Every one of these things is desirable and would lessen the burden of future taxonomists. It seems a little naïve to suppose that so technical a publication as this will have any audience or following among these varied non-taxonomists. Nothing that man can do is likely to prevent other men from acting in such a way as to produce problems. It is of little use to tell those taxonomists who study this code what should be done by all these non-taxonomists.

General recommendations in the code. The 1961 Code includes more than one hundred and thirty recommendations as to how specified things should be done when there are no rules. Nearly half of these are attached to the Articles of the 1961 Code; the rest are principally in two lists in the Appendices. Not all of these appear here for the first time.

Some of the recommendations included with the Articles result from the belief that some things permitted by the Code would in reality be better not done. For example:

"Recommendation 2A. Names already in use outside the animal kingdom.—It is preferable not to propose for a genus of animals a name already in use for a genus outside the animal kingdom."

Some recommend against the use of procedures which the Commission had tried to prohibit but could not isolate effectively. For example:

"Recommendation 8A. Mimeographing and similar processes.— Zoologists are strongly urged not to use mimeographing, hectographing, or similar processes for a publication containing a new name or a statement affecting nomenclature."

Some instruct non-zoologists as to how to conduct their non-zoological activities that might have some effect on taxonomists. (See list under Instructions to non-taxonomists.)

Others recommend practices upon which the zoologists concerned were unable to reach agreement. The recommendation presumably serves to mollify the adherents and leaves the opponents a way out. For example:

"Recommendation 29A. Superfamilies and tribes.—It is recommended that the termination -*oidea* be adopted for the names of superfamilies and -*ini* for the names of tribes."

(The ending -*oidea* is thus widely used in insect names but elsewhere is often used for classes and phyla.)

Others recommend how a taxonomist shall make decisions on the nomenclatural matters left to his judgment. For example:

"Recommendation 69A. Preference for figured species.—In designating a type-species for a genus, a zoologist should give preference to a species that is adequately figured."

One recommendation requires the proposer of a new name to know or determine that the particular combination of letters will not suggest an objectionable meaning *in any language.*

"Appendix D.9. A zoologist should not propose a name that, when spoken, suggests a bizarre, comical, or otherwise objectionable meaning."

Many of the recommendations are wholly reasonable and generally acceptable, although there is some question of the appropriateness of such non-rules in the Code. In too many cases the recommendations are tacitly based on the qualification "other things being equal"; inasmuch as they never are equal, the taxonomist will frequently have to determine his action by the circumstances instead of by the reasonable but irrelevant recommendation.

Use of terms in the 1961 Code. In any activity as technical and complicated as a code of laws, the rigorous use of terms is essential to communication and application. In drafting the 1961 Code it was found necessary to define some groups of terms to express the intended regulations concisely. Several of these terms were new ones or were defined in new ways. In the opinion of the writer, the efforts made in the 1961 Code to use defined terms in a consistent manner have proven to be inadequate. Substantial confusion has been produced. The usages and definitions of some of these things, as defined and prescribed in the Code, are discussed in the texts of the following chapters. Others are presented here.

Specific name. Used in the Code for the second word of the species. Compare *trivial name, subspecific name, name of the species, specific trivial name, specific epithet;* discussed in Chapter 28, Epithets.

Nominal genus or species. The one to which any given name applies. Discussed principally in Chapter 25, Genotypy and the Types of Genera.

Valid and available. A *valid* name is defined in the Code as: "The correct name for a given taxon; a taxon may have several available names, but only one of those names . . . is the valid name" [Glossary]. The *available* name is "One that satisfies the provisions of Chapter IV" as to publication, formation, and description. This usage

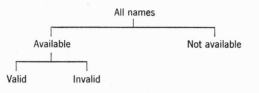

Figure 19-1. Categories of names, as used in the 1961 Code.

involves terms that have been employed in diverse ways. The attempt to standardize meanings may suffice in the Code, but it will surely cause confusion in discussions. It is incomplete in any case, as several other terms are used in the Code. It can be diagrammed as in Fig. 19-1.

The Code also uses the terms "preoccupied" and "published" but does not place them in this scheme.

This set of terms makes no provision for several sorts of names which must be distinguished. It groups as "Not available" (1) non-Latin names, including ones neither latinized nor treated as Latin, (2) unpublished Latin names, and (3) unacceptable inadvertent spellings, either original or subsequent. It cites as "available" names which cannot under any circumstances be employed—that are not in fact available for use (such as junior homonyms). It neglects entirely the nomina dubia, which would be "available" but are not the "valid" names of any known taxa. It groups as "invalid" (1) rejected synonyms, (2) either temporarily or permanently rejected homonyms, (3) unacceptable emendations, and (4) nomina dubia.

These distinctions are insufficient for sound nomenclatural work. The scheme in Fig. 19-2 is adapted from a paper by Blackwelder,

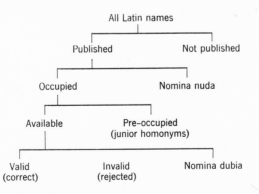

Figure 19-2. A more complete hierarchy of name categories.

Knight, and Smith in *Science* (1950), with slight changes in terms to correspond with the 1961 Code. It presents a more complete classification of these terms.

This scheme represents a hierarchy of five categories, in which the kinds of names are placed, just as in a biological classification, the "higher" groups being more inclusive. The categories may then be defined and named as follows, according to these conclusions: Any name that is printed and circulated is *published;* any published name that meets the publication requirements of the Code is *occupied* in zoological nomenclature (if it fails to meet the requirements it is an outlaw name; that is, unpublished, illegally published, or a *nomen nudum*); any occupied name that is not pre-occupied by an older name of the same spelling is *available* (if it is pre-occupied it is a junior homonym and is not available); the oldest available name is the *valid* (*correct*) name, unless it has been specifically set aside by the Commission under the Plenary Powers. (The correct name will, of course, vary with changes in our knowledge of subjective synonymy or discovery of unknown facts in the history of the names, such as homonymy and objective synonymy. An available name whose genus cannot be identified is a *nomen dubium.*

In this scheme, the only serious shortcoming is the term "valid." This has been much used in the same sense as "available," that any name is valid if it has been accepted into nomenclature—acceptably published, or validly published. There seems to be no good reason for not using the ordinary word "correct," which is exactly what is to be conveyed.

The opposite of "correct" would not be "incorrect," in speaking of names. In a sense all names for a taxon that are not the correct name for it are incorrect names, but more specifically and usefully they are *the rejected names.* They are rejected because (1) they are not published, or (2) they are nomina nuda, or (3) they are pre-occupied, or (4) they are not the ones selected. (Nomina dubia probably are the correct names for some species, but which ones is not known.)

The term "occupied" is not well known in such a connection. If one imagines that there is theoretically a niche for each possible combination of letters that could form a name under the Code, and that when a name is published it would occupy its particular niche, the concept of occupation may be clearer. Note also that the common use of the familiar term *pre-occupied* makes it easier to understand the corresponding but more inclusive term *occupied.*

Binomen. The 1961 Code, by its usage of "specific name" for the second half of the binominal name of a species, has left no convenient term for that dual name. It calls it variously "the name of the species," "species-group name," "the scientific name of the species," and "the binomen." None of these is effective in ordinary statements, as when genera and species are being contrasted, because it is not correct to say that generic names are uninominal and specific names are binominal. "Generic names" must be contrasted with "names of species," not with "specific names." Perhaps to circumvent this, the Code uses "binomen" to refer to the combined name, but this term is not useful to general zoologists who wish to point out the difference involved in the binominal system. One would scarcely say that "generic names" are uninominal and "binomina" are binominal.

In this book it has been found impractical to accept the Code's use of "specific name" as not being the same as "name of a species." It is therefore possible herein to refer to the binomen as the specific name (the name of the species). Binomen is then reserved for occasions when it is desired to emphasize the dual nature of the name of the species—the specific name.

Designation of types. The 1961 Code uses three terms to indicate the establishment of the types of genera (genotypes). These terms are: *designation* (by direct statement in original or subsequent work); *indication* (by one of three specified automatic methods) [Article 67(b)]; and *fixation* ("Used in this Code as a general term for the determination of a type-species, whether by designation (original or subsequent), or indication" [Glossary].)

These terms cover the situation adequately, but the use of "indication" is unfortunate and unnecessary. It lacks the finality needed for actions which are direct and automatic, and it adds no useful distinction. To say that a type is indicated by monotypy is neither so clear nor so definite as to say that it is fixed by monotypy.

The Code states that certain situations "indicate" the type. It would be much more expressive and accurate to say that the type is "automatically fixed." There are several situations resulting in automatic fixation, in both original and subsequent publications, although they are not all covered in the 1961 Code.

There are several special rules that provide for automatic "indication." These are more expressively described as "fixation by special rules."

The application of these terms to the actual situations and rules is discussed in Chapter 23, The Names of Genera, and in Chapter 25, Genotypy and the Types of Genera.

Type of genus or genotype. The strong recommendation in the 1961 Code against the use of the term *genotype* for what is therein called "the type species of the genus" is discussed in Chapter 25, Genotypy and the Types of Genera.

Nominate taxa. A nominate taxon is the subordinate taxon which contains the type of the subdivided "higher" taxon and bears the "same" name. In previous usage, this was called the typical subgenus or subspecies or the nominotypical one.

Nomen dubium, and so on. The 1961 Code uses four Latin expressions which have not always been used in the rules. They are (in Glossary):

1. "nomen dubium. A name not certainly applicable to any known taxon."
2. "nomen novum. A new name published to replace an earlier name and valid only if the latter is preoccupied. A *nomen novum* is a new name that is expressly proposed as a replacement name."
3. "nomen nudum. A name that, if published before 1931, fails to satisfy the conditions of Article 12 and 16, or, if published after 1930, fails to satisfy the conditions of Article 13a."
4. nomen oblitum. "A name that has remained unused as a senior synonym in the primary zoological literature for more than fifty years is to be considered a forgotten name *(nomen oblitum)*" [Article 23(b)].

Unsatisfactory aspects of the 1961 Code

The 1961 Code is a great improvement over other publications of the Commission in the last twenty years in many ways. In its rules, the compromises necessary to obtain adoption make it unlikely that any nomenclaturist will find all features acceptable or even workable. The members of the drafting committee did not approve of all decisions, but they drafted a code from the rulings made by the Congress and the Commission and produced an excellent synthesis of them.

Nevertheless, the 1961 Code (and the 1964 Code also) contains many things that seem to some to be defects, some serious and some merely annoying. The code cannot be used much without contact with these aspects, and it cannot be used effectively without recognizing their presence and making allowances for them. It is for this reason that some of the supposed weaknesses are listed here.

(1) Although the Codes seem to imply that they are complete and supersede all previous rules and Opinions, this is certainly not true. It is impossible in many cases to determine from the Codes alone whether actions taken in the past were correct. This can only be determined by reference to the *Règles,* the Copenhagen Decisions, and the Opinions. The wording of the

Codes seems to imply that all such things are covered, if not explicitly then by retroactive coverage. If this were really so, there would be thousands of generic name changes *produced* by these retroactive rules themselves. Zoologists will certainly not change names long accepted under the *Règles* because of a new rule under which they would now be unacceptable. Involved here are such things as names first published in synonymy and the types of generic names.

(2) The retroactive ruling on diacritic marks, hyphens, and so on, makes it impossible for taxonomists even to cite in synonymy the actual original spelling of a name which employed these signs. No matter how desirable the simplified spelling of them is thought to be for the future, the sweeping prohibition against their use even historically is unworkable, unnecessary, and ill-advised.

(3) The limitation on priority, insisted upon by a bare majority at several Congresses, is worded in such a manner as to be highly controversial in its application. As pointed out by W. I. Follett in the first major critique of the 1961 Code (New Precepts of Zoological Nomenclature, *A.I.B.S. Bulletin*, vol. 13, pp. 14–18; 1963), this rule opens the way to several subjective interpretations, to difference of resulting action, and to indecision on the correct procedures.

(4) The 1961 Code has gone into considerable detail in specifying how the date of publication is to be determined. This is a helpful feature of the Code. It is a shame that this constructive exposition is marred by two sets of Recommendations unworthy of the rules. The first of these forms part of the instructions to non-taxonomists cited above. The second is the recommendation of means to indicate how the date was determined, when citing it in print. The proposed scheme is not in itself unreasonable, but it is unworkable in that it includes no means to tell when the system is being used. If it happens not to be used in a given work, the very existence of the system will introduce confusion over the question of whether it was used or not. False conclusions could easily result.

(5) In introducing into the Code a new section regulating family names, long needed and sought after, the 1961 Code inserts an unfortunate subjective element in requiring zoologists to determine whether a name not ending in *-idae* or *-inae*, but possibly denoting a family-group taxon, is actually a latinized noun used as a name for a suprageneric taxon or merely a plural noun in a non-Latin language, denoting the members of a genus. For example, was "Staphyliniens" in a French work used as a family name or as a vernacular name for the species of *Staphylinus?*

(6) In attempting to ensure continuity of family names, with reasonable success, the 1961 Code requires the back-dating of certain replacement names. As described by Follett (cited above): "The antedating procedure specified (Article 39) in the case of homonymy of the name of the type-genus is bibliographically inaccurate and may result in the absurdity of imposing on a name a date earlier than that of the birth of its author."

(7) Although the problems of homonymy have never been fully covered by any code, the 1961 Code has taken a long step backward in requiring (after 1960) the restoration of all secondary homonyms that are moved out of the genus where the homonymy occurs. This permits, or forces, continual alternation of the rejected name and its replacement name, or simultaneous use of both names as the correct name for the species, in cases where there is difference of opinion as to the generic assignment of the species.

(8) In the rules for designation of genotype (called "type-species"), the Code has spelled out the details of several provisions. It left several subjective rulings, however. It also gives the impression of being fully retroactive in most items, although this would result in change of hundreds of names established under the rules previously in force.

(9) The complexity of any set of rules can be mitigated by careful arrangement, cross-referencing, annotation, and indexing. All of these devices are used extensively and well in the Code. In spite of this the over-all complexity of the present rules of nomenclature is excessive. (This is discussed and tabulated in previous paragraphs.)

(10) It is of course impossible to regulate names for things which themselves have no stability. We could not make useful rules if a given name could be used at any level in the hierarchy, or for a taxon among ones based on a variety of other concepts than their similarities. The Code claims to remain aloof from all the non-nomenclatural aspects, but it fails to do so. It introduces new terms such as "nominal species" for things that are neither names nor zoological species. In attempting to steer away from taxonomic problems, the Code has lost itself in a fog of these undefined terms.

(11) The new term for names, "specific name" instead of "trivial name," although a return to a common practice of the past, has been handled in such a way as to prevent the clarification of nomenclatural discussions. It is now held that the second part of the binomen is a name and can be a homonym. Two identical "names" of this sort are homonyms, however, only when (and usually while) they are placed together in a single genus. In reality, such *words* are homonyms wherever they are placed; but zoologically they can be homonyms only when they are combined with one generic name. Two "specific names" *albus* and *albus* are not homonyms if one is placed in the "genus" *X-us* and the other in the "genus" *Y-us*, even if these two "genera" are known to be the same zoologically. The homonymy does not exist in reality until the two *albus* words are placed under one genus name.

Other difficulties in use of these terms as defined in the Code are described in a later chapter.

(12) An astonishing feature of the 1961 Code is the series of eleven instructions to non-taxonomists. Although reasonable in purpose, they are naïve in conception and probably will be useless in practice. A single ad-

monition directed to taxonomists to attempt to have these good practices followed in the publication of their works would probably have had a much better chance of success. These admonitions are listed in an earlier paragraph.

(13) The large number of recommendations in the Code are cited in a preceding section. The nature of some of these makes them inappropriate in the Code, because they are mere sops to defeated viewpoints and desires.

(14) The term "valid" has had many definitions in nomenclature. It has been most commonly used (in the writer's experience) for a name that is validly (that is, properly) published. The use of this word in the Code in place of the simple word "correct" will be confusing for at least a generation, and has nothing to recommend it in any case. This and other terms of name status are discussed in a previous section.

(15) The implication that this Code was drafted with the help of and adopted with the approval of a majority of interested taxonomists is true only if "interested" is taken to mean those able to participate. The views of a majority on this or any other subject have never been known since the fraternity of taxonomists grew to maturity. Most taxonomists, even vitally interested ones, were unable to attend the Congresses; some were repulsed in their efforts to make their views known in correspondence with the Commission (including the present writer and some of his colleagues); and many were prevented by the cost of the Commission's publications from even finding out what was being proposed. (The small number of copies printed were available in only a few of the world's libraries.)

(16) There is repeated reference in the Code to categories. The word is neither defined nor indexed. Features are ascribed to categories which they cannot have and which belong to the taxa placed in them. They are said to be co-ordinate, which is impossible. Provisions (of Chapter XV) are said to apply to categories, whereas they all apply only to taxa and names of taxa. This confusion of terms clouds many statements and will help postpone the day when these concepts can be rigorously discussed.

New problems raised by the 1961 Code. For the first time, the 1961 Code has raised a series of new problems for taxonomists in general. Taxonomists are probably all convinced of the disastrous effects that would follow abandonment of internationally acceptable rules of nomenclature. No modern writer has suggested such abandonment. But the new problems and the general complexity may result in more lack of compliance than the previous less-complete rules did.

There will always be occasional failure to follow particular rules, either from ignorance of the rule, failure to recognize its applicability, or refusal to accept its implications. The latter may result from self-centered dissatisfaction with a restriction on personal freedom or from a carefully reasoned position that the rule is inadequate at that point. The present writer is among those who may refuse to follow an unreasonable rule after its unreasonableness has been demonstrated and argued, and after attempts have been made to "correct" the rule. That there will be other taxonomists in this group is already evident, inasmuch as suggestions for changes in the 1961 Code appeared less than a year after its publication.

The new problems include these: *The complexity of the code is now so great and so detailed that there is serious danger that taxonomists will not be able to use it effectively.* The style and indexing of the code give an illusion of simplicity that is soon dispelled in a serious attempt to find all the rulings relevant to a particular problem. The index, although an apparently analytical one of the kind useful to nomenclaturists, is largely useless to a non-nomenclaturist who is unfamiliar with the ideas under which the rules have been customarily arranged and indexed. For example, there are four very distinct ways in which a species can be given a name, all provided for in the Code. There is no place in the index where these are referred to together, so a person trying to find out how names can be established has no choice but to search the entire code item by item. (In this case the pertinent rules would be in Articles 2, 11, 46, and 70.)

Non-nomenclatural rulings have been introduced so frequently in the Code as to raise repeated problems of their own. Categories are not nomenclatural concepts, yet they are the subject of scores of rulings. (In spite of this, the term "category" is not in either the glossary or the index, as an entry of its own.) The entire code is suspended in mid-air between taxonomy and nomenclature by the constant use of the expressions "nominal species" and "nominal genus." These are neither taxa nor names. In the code, these nominal taxa have the types, not real genera and species, and not names. Yet these "nominal taxa" are nowhere *effectively* defined.

Name changes caused directly by the rules are for the first time a substantial item of the changes in the rules. Thousands of names already in use must now be changed in orthography because of the new ban on diacritic marks, apostrophes, hyphens, and numerals. The retroactive ruling on restoration of secondary homonyms will not only cause the change of specific names but will leave the door wide open to future changes and oscillations.

Numerous rules apply only to actions taken between specified years, giving cut-off points to many rulings. Although the "basic dates" are not new in the 1961 Code, only three were previously commonly recognized. The Code lists five such dates, and there are two more which must be dealt with. For some rules there are now five periods, during which different stipulations apply. Not all of these are listed in the rules themselves, making it difficult to keep track of the changes.

For example, of the nineteen possible ways of fixing the genotypes of genera and subgenera (as listed in Chapter 25), twelve apply to the entire period 1758–1964; fifteen apply to the period 1758–1930; sixteen apply to the period 1930–1953; eighteen apply to the period 1954–1960; and nineteen after 1960.

The 1961 Code is not final. The Code does not even suggest that it is intended to be a permanent instrument. The 1964 changes are already scheduled to be followed by other changes in 1968, and there is no reason to think that these will be the last. The 1964 amendments have in effect added a new deadline, and others may be added at each later revision.

It may seem that this is a normal process that might be expected to improve minor points of doubt in the rules. Actually, the likelihood of even minor changes tends to discourage exact conformance to the rules as well as careful study of the rules by students. Stability of reasonable rules is far more desirable than perfection and completeness at the expense of stability. (This is quite a different problem from that of stability of names.)

New features of the 1964 Amended Code

At the Washington Congress in 1963, several changes in the Code were proposed and approved. Some of these became part of the Code and were incorporated into a new edition published in 1964. Others of the changes required more careful drafting and were scheduled to be presented for approval at a subsequent Congress.

The 1964 edition, which is herein called the 1964 Code, is similar to the 1961 Code in the following ways:

1. The binding and cover title are identical.
2. The title page is identical except that a strip of paper with a new date has been pasted over the old date.
3. The Preface and Introduction are unchanged.
4. All Articles, Recommendations, Appendices, and Index are the same, except as noted below.

The 1964 edition differs from the 1961 Code in the following particulars:

1. Article numbers have been added to each page to facilitate reference, and page numbers have been shifted to accommodate this.

2. On the title page a strip of paper with a new date has been pasted over the old date.
3. To the Table of Contents has been added a reference to the (new) Preface to Second Edition.
4. Page numbers are changed in the Preface.
5. A Preface to Second Edition is added.
6. A List of Amendments is added.
7. Article 11(d) is amended. It formerly read: "A name first published as a synonym is not thereby made available." It now accepts such names if prior to 1961 they were treated as available names. (In other words, it is no longer fully retroactive.)
8. Article 31(a), on automatic correction of modern patronymics in the nominative case to the genitive case, is deleted. The rest of Article 31, on the endings of names based on modern personal names, is converted from a rule to a recommendation.
9. Article 39(a), on the choice of a replacement for a family-group name which must be changed because its type genus is a junior homonym, is entirely deleted.
10. Article 69(a)(i), which refers to the rule in Article 39(a), is deleted.
11. Cross-references and index entries are said to be amended accordingly.

Treatment of the rules in this book

In view of the tabulations already presented in this chapter, it is obviously useless to pretend that zoological nomenclature can be presented for study or reference by mere quotation of the provisions of the new code. Even the latest version is less than complete, not always workable, and often in need of further explanation and illustration. Perfection is scarcely to be attained by compromise, and not all the ill effects of past history have yet been eliminated.

It is the function of the following chapters to present the problems of zoological nomenclature, the past and present rules relating to those problems and the solutions for them, and a critique of each of those rules in the light of what is here believed to be a logical view of the basis and function of taxonomy.

It would be unsatisfactory for anyone to use these tabulations in place of the actual code. The code has a specific arrangement which brings together subjects such as spelling, dates, and types. This arrangement is not always suitable for student use, so the present chapters are arranged differently, to help answer some other questions whose answers are to be found only with difficulty in the code itself. But the exact statements of the code, in original context, can be determined only by direct reference to the code.

Quotation. In preparing the eleven chapters of this book on the names of taxa, it was intended to quote from the 1961 Code whenever possible. In a few sections this proved to be feasible and effective. In most others it was found that the appropriate rules could not effectively be directly quoted.

The absence of direct quotation in many sections is not in itself an indication of criticism or of supposed inadequacy of the Code. It is merely the result of a different arrangement of subject matter, so that many rules are partly irrelevant at any given point. In other cases parts of several rules need to be combined to cover a given point fully. This can best be done by indirect quotation.

Nominal taxa. One major terminological difference between the present discussions and the Code must be cited. The problem raised by the use of the adjective "nominal" has already been alluded to. Its implications are discussed in Chapter 25, Genotypy and the Types of Genera. There seems to be no question that this word is used or omitted in a haphazard manner. There can be little doubt that a "nominal genus" is sometimes the same as a zoological genus and sometimes the same as a generic name. In quotations the word must be retained, whether ambiguous or not. In the discussions of these Articles, the word is avoided, being replaced with "genus" or "generic name."

By using "nominal species" the Code seems to evade the issue of whether a type-species is type of a genus or of a name. In the present discussion it is assumed that the usage of the past, that every name has a type, is the only workable concept. It should be clear also that the type of one of the names, the "correct" or "valid" one, is also the temporary type of the zoological genus.

New code vs. old rules. Article 84 of the 1961 Code states that "all previous editions of the International Rules of Zoological Nomenclature are thereby superseded." This statement can be misleading if it is intended to imply that the previous *Règles* no longer have any use.

In some parts of the Code, two sets of provisions are included, to be applied before or after a rules change deadline, such as 1930/ 1931. This gives the impression that the Code actually includes all the provisions necessary to judge decisions taken in the past. This impression is heightened by Article 84, in which is added to the above quotation: "All amendments affecting the Code, adopted by the Congresses prior to the XV Congress, are no longer valid unless reaffirmed herein, and then only as here expressed." Article 86 states also: "The provisions of the Code apply to all zoological names and works, published after 1757, that affect zoological nomenclature."

If these implications are followed exactly, there will be innumerable changes of previous actions properly taken under the rules then in force. Stability of nomenclature cannot be attained by such means.

The 1961 Code assumes a starting point of 1960/1961 for many new rules. Some of these were first promulgated in the Copenhagen Decisions (December 31, 1953). Zoologists were instructed to follow the Copenhagen Decisions, so these particular rules actually date from the beginning of 1954. The Code makes no substantive reference to these decisions of the prior Congress. Although no thorough search has been made here, references to the 1953/1954 deadline formed by the Copenhagen Decisions are incorporated in many places in these discussions.

The Code also uses a deadline of 1950/1951, which is the date of promulgation of decisions taken at the 1948 Paris Congress and published in volume 4 of the *Bulletin of Zoological Nomenclature*. This date now applies only to the anonymous publication of names, although many other decisions were taken at that time and not put into effect until 1961.

In certain problems, particularly genotypy, it seems to be necessary to recognize that decisions made under the *Règles*, or under the Copenhagen Decisions, must now be judged under those rules, not under the 1961 Code. To this extent the older regulations are still in force.

CHAPTER 20

The Nature of Scientific Names

The system used for the naming of animals in zoology is generally described as a binominal system—a two-name system employing Latin names in two parts. Among the several million names applied to animals, probably less than half are such dual names, and the most commonly used group of animal names is not binominal.

The names of animal groups (taxa) are of several sorts, which can be listed thus:

1. Names of major groups; e.g., Echinodermata.
2. Names of family groups; e.g., Hominidae.
3. Names of genera; e.g., *Drosophila*.
4. Names of species; e.g., *Homo sapiens*.
5. Names of species with subgenus; e.g., *Aedes* (*Stegomyia*) *aegypti*.
6. Names of subspecies; e.g., *Microtus montanus nanus*.
7. Names of subspecies with subgenus; e.g., *Aedes* (*Ochlerotatus*) *canadensis mathesoni*.

Although these are the kinds of names recognized by the rules of nomenclature since 1960, workers in at least one group of animals still employ two others, below subspecies:

8. Names of varieties; e.g., *Myrmecocystus melliger semirufus testaceus*.
9. Names of varieties with subgenus; e.g., *Camponotus* (*Myrmentoma*) *caryae discolor cnemidatus*.

Of the above, (1) through (3) are uninominal names. Item (4) includes the actually binominal names. Items (5) and (6) consist of three parts each. Item (7) has four parts. The "illegal" names in (8) and (9) consist of four and five parts, respectively, except that they are sometimes written like (6) and (7) as three and four parts. The rules do not count the parts of the names in this way, because some of the parts are optional and can be omitted.

It is possible to argue that all the names in items (5) through (9)

407

are basically binominal in nature, with the additional parts added for a variety of other purposes than mere naming.

Thus, names of species are binominal but may be written with three parts. Names of subspecies are trinominal but may be written with four parts and for some purposes must be treated as if binominal. Names of segregates within a subspecies cannot be named under the rules but do actually exist in four or five parts. The names of genera and subgenera and groups at all higher levels are uninominal.

Names of taxa in the higher categories

The uninominal names of the taxa placed at the higher levels, the more-inclusive taxa, are comparatively simple and present few problems in ordinary use. There can, of course, be spelling problems, synonymy, and homonymy. At the lower levels these are regulated by the codes, but above the superfamily level there are at present no direct regulations.

These names are further described in Chapter 21, Names of Taxa in the Higher Categories.

Names of taxa in the family group

The families are frequently included with the preceding among what are loosely called "the higher categories" but more correctly the higher taxa or more-inclusive taxa. Because names of the family taxa *are* covered by regulations in the code, the present book deals with them separately.

The very existence of rules for these names increases the complexity of their employment, but basically they are simple uninominal names used by themselves to indicate particular groups of animals placed at any of the family levels.

Names of genera and subgenera

Generic and subgeneric names are of the same general nature. They are alike in most nomenclatural aspects, but they represent taxa of quite different sorts and are governed by two sets of nomenclatural rules that differ in several details.

The genus-group taxa are unquestionably more frequently referred to than all other taxa put together, because one is referred to whenever a species or subspecies is cited, as well as in their own right. The taxa are the genera and subgenera, of which the genera are groups of species, and the subgenera are of hybrid nature, sometimes subdivisions of a genus, sometimes groups of species, and always both at once in some manner.

Classification of genus-group names. The technical names of genera can be divided roughly into three groups. The first includes those that have not been acceptably published, the second includes all the acceptably published names, whether considered valid or not, and the third includes published names that are not accorded separate status under the rules. These are tabulated below and discussed in the following sections:

I. Names not accepted into formal nomenclature
    A. Unprinted names
        1. Manuscript names
        2. Museum labels
    B. Printed names
        3. Nomina nuda
        4. Anonymous names
        5. Conditional names
        6. Names of varieties or forms
        7. Names of hypothetical taxa
        8. Names given to terata
        9. Names of hybrid specimens
        10. Non-taxonomic names
        11. Names of parataxa
II. Names accepted into nomenclature
    C. Names currently adopted (valid, in 1961 Code)
        12. For genera
        13. For subgenera
        14. For sections
        15. For collective groups
    D. Names not currently adopted
        16. Junior homonyms
        17. Junior synonyms
        18. Emendations
III. Name forms not accorded separate status
    E. Misapplied names
        19. Misidentifications
    F. Errors
        20. Lapsus calamorum (slips-of-the-pens or -of-the-minds)
        21. Copyists' errors
        22. Printers' errors
        23. Misspellings

Although it would appear at first glance that only group II includes real zoological names of interest to nomenclaturists, all of group I can later be made into acceptable names. The name forms cited in group III(F) also can later be introduced into nomenclature, but such action is to be avoided because it results in possible confusion of dates, authors, priority, and application.

I. Names not accepted into formal nomenclature

1. Manuscript names are latinized names used by a taxonomist as generic names but not formally published. They may be intended for eventual publication or be used in correspondence or in public speech. They have no standing in nomenclature until they are acceptably published, at which time they become accepted into nomenclature as in group II. They can be a source of confusion if they are printed, for example in synonymy, by anyone.

2. Museum labels are a special case of manuscript names. They have no standing in nomenclature, but they sometimes are published carelessly and then become names with which the taxonomist must deal.

3. Nomina nuda are printed names which do not meet the requirements of the current set of rules and are therefore not accepted into nomenclature. In the *Règles* printed names were nomina nuda if they were (a) printed for private rather than public circulation; (b) printed without a description or a reference to one; or (c) printed after 1930 without designation of a genotype.

In the 1961 Code printed names are nomina nuda if they were (a) printed before 1931 without conforming to all provisions of Articles 11, 12, and 16; or (b) printed after 1930 without conforming to all provisions of Articles 11, 12, 13, and 16.

4, 5, 6. According to Article 17 of the 1961 Code, a name published anonymously after 1950 is unacceptable; a name published conditionally after 1960 is unacceptable; and a name expressly given after 1960 to a "variety" or "form" is unacceptable.

7, 8, 9, 10. According to Article 1 of the 1961 Code: "Names given to hypothetical concepts, to teratological specimens or hybrids as such, . . . or names proposed for other than taxonomic use, are excluded" from zoological nomenclature.

11. Parataxa are collateral groupings of individual fragments of animals that can sometimes be found in assemblages of several types of fragments. The assemblages may be assumed to represent species and seem to be identifiable to species and genus. The individual fragments (usually fossils) cannot be assigned to species or genus, but they can be utilized in stratigraphic correlation. In order to have means of referring to them, paleontologists have proposed to recognize the assemblages as individuals assignable to species as taxa and the fragments individually as parataxa. The 1961 Code does not per-

mit this within the framework of zoological nomenclature. Therefore, names given to all of these objects or assemblages are treated alike, as species, genera, or whatever taxa are appropriate, or they are used outside the framework of formal zoological nomenclature.

Parataxa are generally thought of as species. If they are given names as species, it would be possible to group these species into genera that would be indistinguishable from other genera. These need not be called parataxa. If the "generic" groupings are thought to be arbitrary, then they would also be classed as parataxa.

II. Names accepted into nomenclature

The names in classes 12 and 13 are the only generic names that are normally applied to animals in practice. Of course, some in any of the subgroups except (A) and (E) may be in use because their true status is not recognized, but without sanction of the rules.

For many purposes all names in classes 12 to 18 are treated alike under the rules. For example, they must meet the same publication requirements, they must all be Latin or treated as such, they can be rejected only because of stipulated reasons, and all except 15 require genotypes. Their genotypes are determined or fixed by the same methods, the explanation of which is the chief purpose of Chapter 25, Genotypy and the Types of Genera.

A name may belong in several of the categories at once, as 16, 17, and 18. An emended name that is a junior homonym may also be a junior synonym. It might turn out that it was based on a misidentification.

Generic and subgeneric names as outlined above are the names properly applied to genera and subgenera respectively. Article 43 of the 1961 Code states: "The categories in the genus-group are of co-ordinate status in nomenclature, that is, they are subject to the same rules and recommendations. . . ." Thus in determining priority, genotypes, and other nomenclatural matters, these two groups of names are in general treated as one.

When certain groups of species are listed as subgenera rather than as genera, however, a zoological factor has been introduced—the recognition of the zoological category (subgenus) of these groups. This is exactly similar to the assignment of a so-called group and its names to synonymy. Once this zoological factor has been introduced, the subgeneric names (and synonyms) assume a status quite different from that of generic names. For all strictly nomenclatural purposes, classes 12, 13, 14, 16, 17, and 18 are treated alike. Where zoological considerations have been admitted, class 12 differs from the others, which are similar in being all part of "the synonymy" (the rejected names, whether complete or partial synonyms).

Names currently adopted

12. Acceptable generic names are those that are properly published, not eliminated by homonymy, and applied each to one genus in preference to any of its junior synonyms (if any).

13. Acceptable subgeneric names are identical in status with generic names except that the group of species so designated is ranked as part only of a genus—as a subgenus. The subgeneric name must qualify in all the ways specified for generic names.

14. Names given to "sections" of genera are, under the 1961 Code, to be treated as subgeneric names and the sections themselves as subgenera.

15. Names for collective groups ("assemblages of identifiable species of which the generic positions are uncertain") are to be treated as generic names but require no genotypes.

Names not currently adopted

Homonyms (class 16) are identical names for different things. They must be further identified for priority purposes as senior and junior homonyms. Since identical names for different animals cannot be used under the Code, the younger or junior homonym must be replaced (with a junior synonym, if there is one, or with a new name—which is itself automatically a junior synonym also). Thus, all junior homonyms are or should be also synonyms. They are often the senior synonym but can never be used because of their homonymy.

Synonyms (class 17) are two names for the same thing. They may also be designated as senior and junior. Of far more importance, however, is the distinction of objective (nomenclatural, absolute, or isogenotypic) synonyms and subjective (zoological or temporary) synonyms. Unlike homonyms, many junior synonyms are the correct names for genera, because the senior synonyms cannot be used (since they are also junior homonyms).

Emendations are intentional changes in spelling of a name. They may be justified under Article 33 of the 1961 Code or unjustified. If justified, they replace the original spelling in all respects, amounting to the correction of the original error. If unjustified (class 18), they do not replace the original but are treated like entirely separate names. They are synonyms of the original spelling and objectively so (having the same genotype automatically). An unjustified emendation may replace the original if the latter is not usable (because of homonymy). The emendation is then merely one of the junior synonyms among which priority will dictate a selection.

III. Name forms not accorded separate status

Names in classes 19 to 23 do not have a separate status of their own. They are errors of some sort and are best ignored. That is to say, they should be corrected as soon as recognized and in most regards treated as if the error had never been made. Of course, in some outstanding cases, it is necessary for convenience to carry the erroneous spelling in synonymy like a synonym.

19. Misapplied names result from failure to recognize the true genotype and use it in determining the nature of the genus. This may occur through accepting the wrong species as genotype or through including in the genus species that are not congeneric with the genotype. In either case the genus as understood by the later worker may be different from that of its original proposer, and much confusion can result. It is necessary to correct these misapplications, usually by citing them in the synonymy of some other generic name. They do not have genotypes, and in fact have no real existence as names, although in some cases they may have met the requirements of the rules and be actually junior homonyms of the original name. If a misapplication of an old name were granted a separate status in nomenclature, it would be logically necessary to grant separate status to every use of every name. This is patently absurd, and nothing is gained by giving the misapplications the permanence of such acceptance into formal nomenclature.

20. A lapsus calami (plural, lapsus calamorum) is literally a slip-of-the-pen. In practice one may result from a temporary lapse of the mind, which permits a wrong name to pass uncorrected, or a wrong spelling. These are not typographical errors, since they are made by the author himself. For example, an entomologist familiar with ants once had occasion to refer to the little-known beetle genus *Campoporus*. He inadvertently wrote it as *Camponotus*, a well-known ant name. In a sense this error is a junior homonym of the real *Camponotus* and a junior synonym of *Campoporus*, but it is best not to accord it any such definite status. We may have to list it in synonymy to give a reference to the data published under that name, but we should identify it as not having a place in nomenclature.

Copyists' errors (21) and printers' errors (22) are inadvertent errors in spelling due to striking a wrong key, misreading a letter, or inserting a letter in the wrong place. They are to be corrected whenever recognized. If they occur at the beginning of a name, where they will

substantially affect its position in an alphabetical sequence, they will
need to be cited in synonymy.

> *Ocypus* Leach, 1819
>> *Acypus* of Erichson, 1839 (misspelling)
>
> *Goerius* Westwood, 1827
>> *Georius* of Wilson, 1836 (misspelling)
>> *Coarus* of Wu, 1937 (misspelling)

23. Misspellings are not clearly distinguished from the preceding
and result from several causes. Typographical errors are not uncommon, but not nearly all errors on the printed page are the fault of the
typesetter. They may result from ignorance or a lapse of the author,
from an illegible manuscript, or from misguided attempts of editors
or proofreaders to "correct" what appear to be errors. Like the lapsus
calami, the misspelling has no status of its own, although it sometimes
appears to be a junior synonym. In extreme cases it must be carried
in synonymy to avoid confusion, but it has no genotype.

The names of genera. Generic names are unique in zoological nomenclature in being an essential part of the name of a lower-rank
taxon, the species (and the subspecies). The name of this lower-rank
taxon cannot be written or spoken without inclusion of the generic
name. This means that the genus is indicated whenever a species is
cited by name, and thus every reference to a species indicates two
levels of its zoological classification.

The generic name in the binomen (the dual name of the species)
indicates more than just the group into which the species was placed.
The generic name also implies that the species referred to is believed
to be congeneric with (belong in the same genus as) the species
which is the type of that particular genus. It is unlikely that this is
often consciously reasoned, but, if it is not assumed or believed to be
true, the species cannot properly be referred to by that binomen.
This means that the type of the genus becomes of great importance
in the use of names for species. It is also essential in the use of names
for genera.

The names of subgenera. The subgenus is normally the only category other than the genus in the generic group. There are no levels
above the genus except those in the family group, as supergenus is
not used. No formal levels are normally recognized between the subgenus and the species, although sections of subgenera are sometimes
noted without formal names.

Subgeneric names do not share with generic names the position of
being a necessary part of the names of species. They approach it,

however, in a unique situation of their own in that they *may* be written as part of the name of the species if desired but can be omitted at any time without affecting that name. Their presence is so ephemeral that they are not even counted among the names making up the binomen.

When used, the subgeneric name adds two things to the citation of the species; first, that the species has been assigned to a section of the genus designated by that name and, second, that the species is assumed or believed to be consubgeneric with the species which is the type of that particular subgenus, just as for genera, above.

This means that the type of the subgenus also is of importance in the use of names for species, but this is an occasional, not a universal, phenomenon. The nomenclatural kinds of subgeneric names are shown with the generic names below. In most of the rest of this chapter, generic and subgeneric names are discussed as a single kind. Unless a separate statement is made for subgeneric names alone, all statements may be assumed to apply to names at both levels.

The category subgenus is an optional one, to be used or not at the discretion of the taxonomist. Even within a single genus, one worker can use subgenera while another ignores them. Zoologically, there is little confusion caused by this, but nomenclaturally all names in both levels must be dealt with as a single group of names by each of these workers.

Employment of subgeneric names. When the subgenus is recognized, its name is inserted between the generic and specific components of the specific name. It is always placed within parentheses [Article 6]. In this position it is understood to be optional and therefore to have no effect on the rule that the names of species are binominal. For example, *Aleochara* (*Ceranota*) *daltoni* (Stephens). This name is considered to be binominal, and (*Ceranota*) may be omitted at any time.

Neither generic nor subgeneric names are applied irrevocably to groups at one level. If a subgenus is believed to be actually a separate genus, its name becomes a generic name. If a genus is "sunk" as a subgenus of another genus, its name becomes a subgeneric name. The only distinction between the two is the optional matter of placing the subgenus in parenthesis as an additional part of the specific name.

Whenever subgenera are recognized in a genus, there must be at least two subgenera. One is the "typical" or nominate subgenus and has the same name (and genotype) as the genus. The other subgenus (or subgenera, if more than two) has a distinct subgeneric name. The generic description should include all the subgenera—the subgenera (including the typical one) each having a more restricted description in addition.

Generic synonymy. In citing the synonyms of a genus, all names applied to that genus or to any part of it are to be included. The synonyms are:

1. The accepted name for the genus.
2. The older names which are junior homonyms and so not usable.
3. The objective or isogenotypic synonyms of (1).
4. The subjective synonyms, proposed for species now believed to belong in this genus.
5. The accepted names for all subgenera.
6. The older names (for any of the subgenera) that are junior homonyms.
7. The objective synonyms of each subgeneric name.
8. The subjective synonyms of each subgeneric name.

For example:

A.

    Paederus Fabricius, 1775

        Geopaederus Gistel, 1848. (Objective) (A new name proposed for erroneous reasons)

        Poederomorphus Gautier, 1862. (Subjective) (Published as a new genus)

        Paederidus Mulsant & Rey, 1878. (Subjective) (Published as a new subgenus)

        Paederillus Casey, 1905. (Subjective) (Published as a new genus)

        Leucopaederus Casey, 1905. (Subjective) (Published as a new genus)

        Pseudopaederus Bernhauer, 1915. (Subgenus)

        Gnathopaederus Chapin, 1927, June 30. (Subgenus)

        Gnathopaederus Wendeler, 1927, November 1. (Junior homonym of *Gnathopaederus* Chapin) (= *Paederognathus*)

        Paederognathus Wendeler, 1928. (Originally as a subgenus)

        Neopaederus Blackwelder, 1939. (Subgenus)

This is the complete synonymy of the genus *Paederus,* but it does not show clearly the subgenera and their own synonymy. For this purpose a different arrangement is more effective, as follows:

B.

    Paederus Fabricius, 1775.

        Geopaederus Gistel, 1848. (Objective)

        Poederomorphus Gautier, 1862. (Subjective)

        Paederidus Mulsant & Rey, 1878. (Subjective) (A subgenus reduced to synonymy)

        Paederillus Casey, 1905. (Subjective)

> *Leucopaederus* Casey, 1905. (Subjective)
> *Gnathopaederus* Wendeler, 1927. (Not Chapin, 1927)
>   (= *Paederognathus*) (A subgenus reduced to synonymy)
> *Paederognathus* Wendeler, 1928. (Subjective) (A new name
>   for a junior homonym)
> Subg. *Pseudopaederus* Bernhauer, 1915.
> Subg. *Gnathopaederus* Chapin, 1927.
> Subg. *Neopaederus* Blackwelder, 1939.

In this case none of the currently accepted subgenera has received more than one name, so there are no junior synonyms listed under the subgeneric names. Wendeler renamed *Gnathopaederus* Wendeler because it was a junior homonym of *Gnathopaederus* Chapin, but this was not actually necessary, as it represented a supposed subgenus that was later rejected. The new name, *Paederognathus*, thus falls into the subjective synonymy of the generic name, while remaining an objective synonym of one of the other subjective generic synonyms.

The synonymy of the subgenera is illustrated in the following partial synonymy:

C.
> *Ischnopoda* Stephens, 1835.
>   *Acrotona* Thomson, 1859. (Subjective)
> Subg. *Aloconota* Thomson, 1858.
>   *Glossola* Fowler, 1888. (Subjective-objective)
>   *Taphrodota* Casey, 1906. (Subjective)
>   *Terasota* Casey, 1906. (Subjective)
> Subg. *Stethusa* Casey, 1910.
>   *Hypatheta* Fenyes, 1918. (Subjective)
>   *Athetalia* Casey, 1910. (Subjective)
> Subg. *Umbala* Blackwelder, 1952.
>   *Stictatheta* Cameron, 1939, August. (Not Cameron, 1939,
>   May) (Objective)
> Subg. *Stictatheta* Cameron, 1939, May.

Two items here are unusual. *Glossola* Fowler has as its genotype a species which is believed to be a synonym of the genotype of *Aloconota*. So long as this synonymy is accepted, the two names are objective synonyms, but if it is denied, they become subjective synonyms. In 1939 Cameron twice published the name *Stictatheta* for subgenera. The later one was renamed by a subsequent reviser, but the earlier one stands as a subgenus.

It may be emphasized that all subgeneric names are in effect subjective synonyms of the generic name. When arranged in strict chronological order, as in the first synonymy above, (A), the generic name is determined as the first available name on the list, whether it was originally or is now used for a genus or for a subgenus.

An additional minor question arises about the effect of the subgeneric name on the use of parentheses around the name of the author. The following examples illustrate the various situations that can arise.

1. *Staphylinus stercorarius* Olivier is transferred to *Platydracus* and becomes *Platydracus stercorarius* (Olivier).

2. *Staphylinus pubescens* Degeer is transferred to *Dinothenarus,* which is a subgenus of *Platydracus,* and becomes *Platydracus* (*Dinothenarus*) *pubescens* (Degeer).

3. *Litolathra suspecta* Casey is transferred to *Lathrobium,* of which *Litolathra* is a subjective synonym, and becomes *Lathrobium suspectum* (Casey).

4. *Lathrobiopsis texana* Casey is transferred to *Lathrobium,* but left in the subgenus *Lathrobiopsis,* and becomes *Lathrobium* (*Lathrobiopsis*) *suspectum* (Casey).

5. *Bledius* (*Blediodes*) *chinkiangensis* Bernhauer is changed to *Bledius* (*Hesperophilus*) *chinkiangensis* Bernhauer, because *Blediodes* is found to be a junior synonym of *Hesperophilus.*

6. *Coprophilus* (*Zonoptilus*) *longicollis* Cameron becomes *Coprophilus longicollis* Cameron, when *Zonoptilus* is reduced to synonymy of *Coprophilus* s.str.

7. *Paralispinus* (*Clavilispinus*) *siargaoanus* Bernhauer becomes *Clavilispinus siargaoanus* (Bernhauer), when the subgenus is recognized as a separate genus.

8. *Creophilus maxillosus* (Linnaeus), originally in *Staphylinus,* becomes *Staphylinus maxillosus* Linnaeus again when the name of the genus is changed back to *Staphylinus.*

9. *Medon* (*Charichirus*) *spectabilis* (Kraatz), originally in *Lithocharis,* becomes *Charichirus spectabilis* (Kraatz) when *Charichirus* is recognized as a distinct genus.

10. *Cryptobium* (*Biocrypta*) *magnolia* (Blatchley), originally in *Biocrypta,* becomes *Biocrypta magnolia* Blatchley again, when *Biocrypta* is again recognized as a genus.

The parentheses are used in the last form of items (1), (2), (3), (4), (7), and (9) because *the final generic name is different from the original generic name.* The parentheses are *not* used in the resulting name in items (5), (6), (8), and (10) because *the final generic name is the same as the original generic name.* The intermediate generic names and the subgeneric names have no bearing on the use of parentheses, unless as in (10) the one-time subgenus is not only the final genus but also the original genus.

Subgeneric vs. generic names. Generic and subgeneric names are the names properly applied to genera and subgenera respectively. Article 43 of the 1961 Code states: "The categories in the genus-group are of co-ordinate status in nomenclature, that is,' they are subject to the same rules and recommendations. . . ." Thus in determining priority, genotypes, and other nomenclatural matters, these two groups of names are in general treated as one.

In speaking of *strictly nomenclatural* problems, this statement in the Code is accurate and appropriate. In dealing with the names of genera and subgenera, however, few problems are strictly nomenclatural. The very fact that the names are used at two categorical levels is a zoological aspect which is all-pervading. Whenever this aspect is relevant, the two groups of names are likely to be at least slightly different in nature and to require different regulations.

For example, in listing the species in a certain genus, a writer chooses not to make use of the subgenera that have been proposed. In effect he deals with the entire genus at once (as he must, for example, in determining specific homonymy). If he desires to list the generic synonyms, he must include among them the subgenera, which for the purpose of that particular moment are equal to them in status. It is obvious that at this point the subgeneric names and the junior synonyms are of equal rank but are not on a plane with the generic name. The recognition of their zoological status through the category assigned to the concepts they represent makes it impossible to treat them as coordinate with the generic name.

Again, in citing the number of genera in a family or other higher group, a taxonomist counts only the true genera as he recognizes them, paying no attention to any subgenera. For this purpose the subgenera are on a lower level with which he is not at that moment concerned.

In short, in anything that involves recognition of the fact that a name applies to a subgenus and not to a genus, the subgeneric name has a status that is somewhat different from that of a generic name. This is not a contradiction of Article 6, since this is a zoological consideration, not a nomenclatural one. For example, the determination of genotypes is a strictly nomenclatural function, but it has no nomenclatural use. The fixation of a genotype will not fix the name of any zoological group until the zoological status of the group is worked out. Thus the purpose of the nomenclatural fixing of genotypes is the tying of names to zoological entities so that recognition of zoological identity (and sometimes also nomenclatural synonymy) can determine the correct name.

In considering all the names that have been applied to a particular genus (and its parts), they are all in a single category, according to Article 43, and are treated alike as a series of names. When the fact is stated that they apply to parts of one zoological genus, we can still say that they are all in one group—they are all synonyms. One of these synonyms will be the oldest available name for each subgenus, if they are recognized as such. Among the other names, however, we can see several kinds. There may be some objective synonyms of the generic name that can never be anything but objective synonyms. There may be some subjective synonyms, whose status depends on the judgment of each worker. Any subjective synonym is potentially a partial synonym, that is, corresponds only to part of the genus (a subgenus). By his treatment of the entities represented by these names, each writer distinguishes between the complete synonyms (synonyms of the genus) and the partial synonyms (subgenera and their synonyms).

Although nomenclaturally all these names belong in a single class, zoologically the synonyms of any generic name form a definite class distinct from the generic name, and require different treatment in certain non-nomenclatural details.

In ordinary taxonomy strictly nomenclatural use of names is uncommon. Most workers do not concern themselves with rechecking the validity of the publication of each name and the fixation of its genotype. They assume that these matters have been adequately dealt with by nomenclaturists. Thus, in normal use, generic and subgeneric names are always used with assumption of zoological status. We see this in revisionary work, in cataloging, and in synonymy. In all these, subgeneric names and synonyms are together classed apart from generic names.

Thus, according to this interpretation, the statement in Article 43 that generic and subgeneric names are coordinate from a nomenclatural standpoint is true in general but cannot be extended to cover situations in which the zoological status of the entities represented by the names is involved. As long as the names are dealt with purely as names, they are coordinate for most purposes. When they are used as names for entities in different zoological categories, they are not coordinate. In the latter case they must be treated in four groups— the names of genera, their synonyms, the names of subgenera, and their synonyms. The exceptions to Article 43 are listed below.

Names of genera and subgenera are *similar* in the following ways:

1. They are the names of taxa.
2. The taxa represented by the names have genotypes, fixed by methods that are mostly applicable to both.
3. The names can be proposed in the same manner.
4. They are published under the same requirements.
5. They are formed in the same way from the same source words and spelled under the same rules.
6. Their gender is determined by the same rules (but there is no use made of the gender of subgeneric names).
7. Original and subsequent spelling variations are subject to the same rules.
8. The author of the name is determined in the same manner.
9. The date of publication is determined in the same manner.
10. The author and date are cited in the same manner.
11. Both may have synonyms (but there are differences in the treatment of them).
12. Both may be homonyms and are then treated under the same rules.

Names of genera and subgenera are *different* in the following ways:

1. There are no categories subordinate to the subgenus level, so no names are transferred from below.
2. The genus is the basic category in the group and the one which cannot be dispensed with; the subgenus can, at any time, be reduced to the status of a rejected taxon with its name a synonym of the generic name.
3. The generic name must always be cited in the binomen; the subgeneric name may be omitted altogether, even if the writer does recognize the subgenus.
4. If they are used, the subgeneric names are cited in a different manner from generic names.
5. The genotypes of genera and subgenera are fixed by the same means, except that (a) the genotype of a subgenus cannot be fixed by subgeneric monotypy; and (b) the genotype of a subgenus cannot be fixed by elimination (of the only acceptable sort).
6. The subgeneric name is not involved in the agreement of the epithets and generic names in gender. It may have a gender different from the genus.
7. "The subgenus that contains the type-species of a subdivided genus bears the same name as the genus and is termed the 'nominate' subgenus" [Article 44(a)]; this nominate subgenus will change "when the invalid name of a subdivided genus is replaced by the name of a different subgenus" [Article 44(b)].
8. In case of synonymy, the generic name is selected from a combined list that includes the generic name, its synonyms, the names of all of

its subgenera, and all their synonyms. The subgeneric name is selected from the names applied only to that one subgenus. Thus, priority applies under partly different rules to the two sorts of names.

9. If a generic name and a subgeneric name are homonyms, and they were published simultaneously (either in the same work or in different works), the one proposed for the generic name takes precedence.

10. A generic name which must be rejected can be replaced with any of its subgeneric names, but a subgeneric name (other than that of the nominate subgenus) cannot be replaced with a generic name.

11. A genus can be divided into two separate genera or into two or more subgenera; a subgenus can only be divided into two subgenera.

12. Genera can be internally subdivided into subgenera; subgenera are not normally subdivided and never within the framework of the standard hierarchy.

It must be concluded that Article 43 of the 1961 Code, quoted in part above, is merely a statement of similarity in general treatment. The specific statement that the categories are subject to the same rules is completely irrelevant, and the implication that the names of the taxa at these categorical levels are governed by the same rules is a considerable exaggeration.

It would be a mistake to assume that generic and subgeneric names are "co-ordinate" or subject to the same rules. Inasmuch as they refer to taxa at different levels in the hierarchy, they cannot be identical in all respects.

The names of species

Unlike the names of genera and all more-inclusive taxa, the names of species consist of two parts. They combine the name of the genus with an additional epithet that signifies the particular species within that genus. This is the Linnaean binominal nomenclature, the two-name naming system.

For example: the genus of modern horses is *Equus*. Among its species are *Equus caballus* and *Equus asinus*. The words *caballus* and *asinus* standing alone have no meaning in taxonomy; they are not names of species or anything else. Only when they are part of a binominal combination are they meaningful taxonomically, and then it is the combination that is the name of the species.

Some writers have used the term "specific name" to refer to the second part of the binomen. This leads to considerable confusion. It is illogical to try to maintain a difference in meaning between the two adjective expressions "specific name" and "name of a species." The name of the species must by rule be binominal, so the specific name must also be binominal. One way to refer to the second part of the name is to call it a "trivial name." But here again, use of the noun "name," even with the adjective trivial, is misleading because the word is *not* a name, only part of a name. More distinctive is the expression, used by some other writers, *specific epithet,* the specific part of the binominal name.

Species-group categories. In the past the number of categories (levels) in the species group was indefinite. Linnaeus used two, species and variety. Subsequent workers used also subspecies, form, race, aberration, and others. Only subspecies was directly recognized by the *Règles,* although names were commonly given to taxa at some of the other levels as well.

The 1961 Code definitely limits the species group to the species and subspecies levels. All other possible levels below species are directly excluded from zoological nomenclature by Article 45. Definite criteria are given for determining whether names apply to subspecies or to infrasubspecific taxa, and therefore whether or not they are covered by the Code.

At the species and subspecies levels (that is, in the species and subspecies categories) are placed taxa of the rank of species and subspecies, respectively.

The species-group taxa. The species-group taxa are the species and the subspecies. The species are assemblages of individuals, variously understood. The subspecies are defined as geographical or stratigraphic segregates of the species, although in practice they have often been structural segregates as well.

The 1961 Code contains a special article [45(b)] which reads in its entirety: "Each taxon of the species-group is objectively defined only by reference to its type-specimen." The purpose or intent of this statement is an enigma. It appears to be intended to convey the fact that all taxa in the group are typified by a specimen rather than by a taxon at a lower level. However, a single taxon cannot be defined in practice by any one object. A definition of it must include limits of some sort. Objectivity in the definition may be obtained by use of a clear-cut break in the variation of each of the features used in the "definition," but not by reference to one specimen. The type specimen serves taxonomically as a reference point in the diversity of specimens. Nomenclaturally it serves as an anchor for the name.

No bearing of this article [45(b)] on zoological nomenclature has been noted. The implied bearing of types on taxonomic treatment of taxa (their definition) is inadequately described and not appropriate to the nomenclatural function of the Commission. In any case no such function of types is possible.

The names of species-group taxa. Many aspects of the establishment, use, and rejection of the names of species and subspecies were regulated in the *Règles*. Most of these regulations were little changed in the 1961 Code. The rules and their implications are considered in the Code under the following headings:

> Criteria of publication
> Criteria of availability
> Date of publication
> Validity
> Formation
> Emendation
> Authorship
> Homonymy
> Types

What is named. The binominal scientific names of zoological nomenclature apply to taxa of animals that occur in nature, either now existing or extinct. The taxon must be represented by actual individuals (specimens) or the "work" of individuals. Names are not given to hypothetical concepts, nor to recognized hybrids, nor to parts of species below the level of subspecies, nor to individual specimens believed to be unique in nature (terata, aberrations, and so on). Of course, some supposed species are represented by only one known specimen, which may be a hybrid or a teratum; it can be named as a representative of a supposed species, if its true nature is not known. Only names which are intended for taxonomic use are recognized in this system.

In general, only names of animals are recognized. However, if organisms first named as animals are transferred to another kingdom, the names continue to be recognized in zoological nomenclature. Organisms first named in another kingdom and transferred into the animal kingdom have their original names accepted into zoological nomenclature at the time of transfer but with original date.

Co-ordinate categories. The *Règles* stated (in Article 11) that "Specific and subspecific names are subject to the same rules and recommendations, and from a nomenclatural standpoint they are co-ordinate, that is, they are of the same value." This simple rule has been understood to refer to names of taxa at the levels of species and subspecies and to mean that in establishment, rejection, replacement, and orthography they were subject to the same rules. It was not usually assumed that they were subject to all the same rules and no different ones, because a subspecies is not the same as a species, and their names will be treated differently in some respects because of this difference.

The 1961 Code has seriously confused this simple picture by stating (in Article 46) that "The categories in the species-group are of co-ordinate status in nomenclature, that is they are subject to the same rules and recommendations. . . ." Categories do not enter into nomenclature at all. They contain taxa whose *names* enter into nomenclature. Two categories cannot be co-ordinate in any way, because categories exist only by virtue of being different levels or ranks. There are no useful rules in the Code applying to categories.

It would probably be assumed that these statements are intended to refer only to the *names of taxa* within these categories. This interpretation is not possible, because the quotation above is immediately followed by this elaboration: ". . . and a name established for a taxon in either category in the group, and based on a given type-specimen, is thereupon available with its original date and author for a taxon based on the same type-specimen in the other category." It is clear that neither categories nor taxa can be co-ordinate, whereas the names of the taxa in two categories *can* be subject to the same nomenclatural rules.

Names in the species group are co-ordinate in all respects as long as the category of the taxon is not involved. One can discuss validation, formation, priority, homonymy, and replacement without reference to the level of the taxa. For these strictly nomenclatural purposes the names are subject to the same rules, but whenever the category of the taxon is relevant, the names come under some rules that make them distinct.

Terms. A few terms are used with technical meanings in the following discussions, requiring definition.

Species are taxa, groups of individuals, assigned to the level called the species category.

Nominal species are defined in the 1961 Code as named species. It is not clear how "a named species" can be different from "a species." In the Code the "nominal species" are what taxonomists would usually call "names of species." (See discussion in Chapter 25, on Genotypy and the Types of Genera.)

Species inquirenda, doubtfully identified species needing further investigation.

Nomen dubium, a name not certainly applicable to any known taxon.

Nomen nudum, a name that fails to satisfy the minimum requirements of the rules in effect at the time of its publication.

Nomen oblitum, a name that has remained unused as a senior synonym in the primary zoological literature for more than fifty years; a forgotten name.

Nominate subspecies, the subspecies which contains the type specimen of the species and bears the same name as the species. (At one time called the typical or nominotypical subspecies.)

Classification of species-group names

The names of species-group taxa can be separated into three distinct groups, depending on availability and acceptability. The first includes those not accepted in formal zoological nomenclature, the second includes all the acceptably published names whether properly in use or not, and the third includes published name forms that are not accorded separate status under the rules.

I. Names not accepted into formal nomenclature
    A. Names not yet published (printed)
        1. Manuscript names
        2. Museum labels
    B. Names published but not acceptably
        3. Common names
        4. Names not meeting the minimum requirements of the rules
        5. Names of parataxa
        6. Names of forms, varieties, and so on
        7. Names of hybrids
        8. Names rejected by the Commission
II. Names accepted into nomenclature
    C. Names currently adopted
        9. For species
        10. For subspecies
    D. Names not currently adopted
        11. Junior synonyms
        12. Emendations (unjustified)
        13. Junior homonyms
III. Name forms not accorded separate status
    E. Errors and spellings later emended
        14. Lapsus calamorum
        15. Copyists' errors
        16. Printers' errors
        17. Misspellings
    F. Misapplied names
        18. Misidentifications

I. Names not accepted into formal nomenclature

1. Manuscript names are latinized names used by taxonomists for species or subspecies but not formally published. They are usually used in the same manner as published names, perhaps with expectation of eventual publication, but only in manuscripts, correspondence, or speech. They have no standing in nomenclature until they are acceptably published, at which time they enter into group II. Although they have no status in nomenclature, they can be a source of confusion, especially if they have been published (printed) by anyone, for example in synonymy.

2. Museum labels are a special case of manuscript names. They have no standing in nomenclature, but they are sometimes published carelessly and then become names with which the taxonomist must deal.

3. Common names that are published (printed) are either quoted vernacular names like dodo, Indian elephant, Japanese beetles, or brown hydra, or part of a formal system of non-Latin names paralleling zoological nomenclature. They have no status in nomenclature and do not generally affect the use of zoological names.

4. Legally unacceptable names are those which do not at that time meet the requirements of publication of the current rules and are therefore not accepted into nomenclature. In the *Règles* names were unacceptable if they were

(a) published before 1758;
(b) not binominal;
(c) printed without a description or a reference to one or an acceptable indication of its application (nomina nuda);
(d) after 1930, published without a differentiating description or a reference to one; or
(e) published in a work rejected by the Commission.

Under the 1961 Code, all of these conditions prevent acceptance of a name, and, in addition, names are now unacceptable if they were

(f) given to a teratological specimen as such;
(g) based on a hypothetical concept rather than an actual species or subspecies represented by specimens;
(h) after 1960, proposed for a hybrid as such;
(i) given to infrasubspecific forms as such;
(j) proposed for other than taxonomic use;
(k) published without meeting the publication requirements of Article 8;

(l) distributed solely in any microform, or merely as an explanation accompanying a photograph, or as proof sheets;

(m) merely in a manuscript deposited in a library;

(n) merely mentioned at a scientific meeting (and not reported in the minutes thereof);

(o) anonymously published (after 1950);

(p) so constructed that it cannot be treated as a Latin word (if this is possible);

(q) published after 1960 only in synonymy of an acceptable name;

(r) not a single word of more than one letter;

(s) formed of two or more words connected by a conjunction;

(t) proposed conditionally;

(u) not formed in accordance with Article 26 to 31, or capable of being emended appropriately.

5. Names of parataxa. Parataxa are collateral groupings of individual fragments of animals (usually fossils) that can sometimes be found in assemblages of several types of fragments. The assemblages may be assumed to represent individual animals and seem to be identifiable to species and genus. The individual fragments cannot be assigned to species or genus, but they can be utilized in stratigraphic correlation. In order to have a means of referring to them, paleontologists have proposed to recognize the assemblages as individuals assignable to taxa and the fragments individually as parataxa. The 1961 Code does not permit this within the framework of zoological nomenclature, so all such names of parataxa are unacceptable. The names already in use will stand when they are believed to represent species, but if they represent parataxa they become nomina nuda.

6. Names of forms, varieties, and all other infrasubspecific taxa were tacitly acceptable under the *Règles* or at least were widely proposed. Under the 1961 Code they are, after 1950, not accepted into nomenclature. All such names previously proposed will be outlawed unless they are now believed to represent species or subspecies.

7. Names of hybrids were provided for in the *Règles*, but in a formula rather than a name such as applied to a taxon. Hybrids are individuals, not taxa, and it was inconsistent of the *Règles* to include them in rules for the naming of taxa. The 1961 Code excludes all names of hybrids from zoological nomenclature, whether they are newly proposed or in the older literature.

8. Names rejected by the Commission under the Plenary Powers. These are

(a) names individually rejected by the Commission to prevent confusion or upsetting of well-known names; or

(b) names published in works which the Commission has ruled unavailable in nomenclature, usually because the author did not use binominal nomenclature. (Most such works are dated before 1800.)

II. Names accepted into nomenclature

The names in classes 9 and 10 are the only names for species-group taxa that are correctly used for animals. With classes 11 to 13, they form the total of all the names available for use for such taxa. Other names, such as those in classes 4 to 8 and 14 to 17, may be in use because their true status is not recognized, but, if so, it is without sanction of the rules and will be subject to change.

Names in the classes from 9 to 13 are the real zoological names. They have met the requirements of the rules and are available for use. They all have type specimens to determine to what taxon they will be applied. A name may belong to several of these classes simultaneously, as a junior synonym (11) may be the correct name for a species (8) if the senior synonym is a junior homonym. Or a name that is an emendation may also be a junior synonym, a junior homonym, or both.

Names in classes 9 to 13 are dealt with alike under many of the rules, those which deal with strictly nomenclatural matters. In other cases, whenever it makes a difference whether the name applies to a species or to a subspecies, there are differences between classes 9 and 10. These exceptions are listed under Names of Subspecies, below.

Names currently adopted

9. Acceptable names for species are those that are properly published and spelled, not eliminated by homonymy, correctly applied to a species taxon, and not preoccupied by another name that also fulfills these requirements.

10. Acceptable names for subspecies are those that are properly published and spelled, not eliminated by homonymy, correctly applied to a subspecies taxon, and not preoccupied by another name that also fulfills these requirements.

Names not currently adopted

11. Junior synonyms are acceptable names for a taxon that has another acceptable name of prior date. The term is also loosely applied to any rejected synonym, such as the oldest name when it is also a junior homonym.

12. Emendations that do not replace the original name (unjustified emendations) are separate names in their own right. They are simply junior synonyms. They may also be homonyms, senior or junior, of other names of the same spelling.

13. Junior homonyms. Homonyms, at the species levels, are identical names for different taxa. The names may be: (a) both names for species, (b) both names for subspecies, or (c) one name for a species and one for a subspecies. The homonymy involves the entire binomen; that is, *albus* and *albus* are not homonyms in the zoological sense, because they are not names. *X-us albus* Bell is a homonym of *X-us albus* Ring.

Among homonyms the first one published is the senior homonym and all later ones are junior homonyms. Normally, the senior homonym is the only one which can be used, but action by the Commission under the Plenary Powers can reverse this.

In specific homonymy there has been a problem over binomina which are identical but are not within the same zoological genus. For example: *Disticta* Wasmann, 1916, Coleoptera, is a junior generic homonym of *Disticta* Hampson, 1902, Lepidoptera. If each of these contains a species named *Disticta alba,* these two binomina are *not* specific homonyms, because they are not in the same genus but only in genera with identical names.

III. *Name forms not accorded separate status*

All names in this group are to be corrected whenever found. The original spelling thereafter has no standing in nomenclature and theoretically need not be cited in synonymy. In complete synonymies and catalogs of literature on a species, it is necessary to cite the misspellings in giving the reference to that use of the name. This can be done by citing the misspelling as a synonym or by citing it in parenthesis after the reference that contained it. For example:

A.

> *Creophilus maxillosus* (Linnaeus, 1758) (*Staphylinus*)
> *Creophilus maxillaris* Thomson, 1859 (lapsus)
> *Staphylinus maxilosus* (Linnaeus), Gmelin, 1793 (misspelling)

B.

> *Staphylinus maxillosus* Linnaeus, 1758, p. 421; Poda, 1761, p. 48; Fabricius, 1775, p. 265; Gmelin, 1793, p. 673 (*maxilosus*); Erichson, 1839, p. 432; Thomson, 1859, p. 23 (*maxillaris*); etc.

14. Lapsus calamorum (sing. lapsus calami). It is easy to write down the wrong name without realizing it, as *C. maxillaris,* above.

The only difference between these and the other errors listed is that these may look like real names—the error may not be at all obvious.[1]

15, 16. Copyists' errors and printers' errors are completely indistinguishable, in the final publication. They include what are often called typographical errors, but the error in what is set in type can originate with the author, a copyist, the editor, the typesetter, or the proofreader.

17. Misspellings are not easily distinguishable from other errors. The term is here reserved for unacceptable spellings that were deliberate—made through ignorance of the rules of formation—not accidental errors.

18. Misidentifications are mistaken uses of a name for a taxon to which it does not properly apply. They have some of the features of homonyms, and under some circumstances they are accepted into nomenclature as distinct names (see Article 70(b) of the 1961 Code). Misidentifications result in attaching to a species some data or specimens that do not belong there. It is important to remove such false data, and every adequate monograph will cite these misidentifications under the species to which they really belong, thus:

> X-us albus Linnaeus, 1758
> X-us pallidus of Mill, 1866, not Fabricius, 1800

There is no such species or name as X-us pallidus Mill, 1866; he merely referred to X-us pallidus Fabricius, 1800. The monograph will also take note under X-us pallidus Fabricius that the reference to that by Mill in 1866 refers to a different species.

The infraspecific categories

The categories or levels below the species level are sometimes referred to as the infraspecific categories. When we refer to the groups placed at these levels, we find that these terms become confusing, because, whereas we call the groups in one of the species levels "species," we do not call the groups at any infraspecific level "infraspecies."

From the earliest times one level below the species has been recognized, inasmuch as Linnaeus distinguished varieties in some of

---

[1] The question has been raised as to whether the plural of the Latin expression lapsus calami is lapsus calamorum or lapsus calami ("slips-of-the-pens" or "slips-of-the-pen"). When discussing such errors as a group, it is surely assumed that the slips are not all made by the same "pen" (author), and this would lead to lapsus calamorum. As a sort of collective expression, it would also be possible to use lapsus calami as plural.

his species. These varieties were simply specimens that differed from the others in characters that were not then considered to be of specific value. This concept was used in a rather loose manner for over a hundred years, and then it gradually became to be believed that the taxa in this level could be based on geographic variation and thereby be more useful for certain purposes. The new basis was accompanied by a change of term, and "subspecies" came to be used widely for these groups. Even today there is no objective basis for the groups distinguished, and "subspecies" is used in various ways by different taxonomists.

All of this did not prevent taxonomists from inventing and using other concepts of groups within the species. They believed that these were helpful in showing the facts of occurrence of the individuals. Here are found races, transition forms, aberrations, and so on.

Among possible infra-specific categories, the *Règles* provided for the use and naming of taxa at only one, the subspecies level, but they did make it possible to accept as subspecies all those groups within species which were, before 1961, called varieties or forms. So far as nomenclature was concerned, they were all treated as subspecies. The 1961 Code explicitly limits for the future the naming of segregates below the species level to subspecies, leaving the earlier usage as before, but the code provides means for determining whether any given taxon is to be accepted as a subspecies or not. The rules are these:

1. "Names given to . . . infrasubspecific forms as such . . . are excluded [from this Code]" [Article 1].
2. "Infrasubspecific forms are excluded from the species-group and the provisions of this Code do not apply to them . . ." [Article 45(c)].
3. "The original status of any name of a taxon of lower rank than species is determined as
   (i) subspecific, if the author, when originally establishing the name, either clearly stated it to apply to a subspecies or, before 1961, did not clearly state its rank . . . , and as
   (ii) subspecific, if the author when originally establishing the name, stated the taxon to be characteristic of a particular geographical area or geological horizon and did not expressly refer it to an infrasubspecific category; but as
   (iii) infrasubspecific, if the author, when originally establishing the name, either expressly referred the taxon to an infrasubspecific rank, or, after 1960, did not clearly state that it was a subspecies" [Article 45(d)].

4. "Before 1961, the use of either of the terms "variety" or "form" is not to be interpreted as an express statement of either subspecific or infrasubspecific rank" [Article 45(e)(i)].
5. "After 1960, a new name published as that of a "variety" or "form" is to be regarded as of infrasubspecific rank" [Article 45(e)(ii)].

The names of subspecies. The literature of some groups of animals contains many names that appear to be part of the Linnaean system but are not actually acceptable under the rules. For example, in the ant family Formicidae quadrinomials are almost universally employed. These are not acceptable names, but nothing is to be gained by upsetting such a subsidiary system merely for legalistic reasons. Erection of such schemes elsewhere is generally discouraged, and the names are not governed by the Code, except as myrmecologists agree among themselves to deal with them by analogy with subspecies names. They raise special problems of homonymy and synonymy.

The *Règles* and the 1961 Code both group subspecific names with the specific names, making them subject to the same rules and recommendations. One real difference is evident, although often neglected. Inasmuch as the subspecies is included in the species, just as the subgenus is included in the genus, the name of the subspecies is in a sense a part of the specific synonymy. At any time the subspecies can be abandoned, whereupon the subspecific name becomes a synonym of the specific name. This fact creates confusion because of the binominal nature of the specific name and the trinominal nature of the subspecific name. For example, *X-us albus pallescens* Smith is the name of a subspecies of *X-us albus* Jones. If the subspecies is shown to be only a sexual form (for example), *X-us albus pallescens* becomes a synonym of *X-us albus* and will have to be listed in the specific synonymy as:

> *X-us albus* Jones (1900)
> *X-us pallescens* Smith (1920)

Although the synonym may also be written in the trinominal form, it seems to be clear that the subspecific name is actually binominal in nature but is written as a trinominal to identify the species in which the subspecies is being placed.

This is not the same as the elevation of a subspecies to the rank of species. Here, as provided in the *Règles*, the subspecific name becomes a specific name. For example, *X-us niger brunnescens* Brown (1910) is removed from *X-us niger* Doe (1890) and becomes the name of a separate species: *X-us brunnescens* Brown (1910).

Both the *Règles* and the Code also stipulate the subspecific epithet be written immediately after the specific epithet, without the interposition of any mark of punctuation (as in the above examples).

The rules also class subspecific names with specific names for purposes of homonymy. A name in either category is to be rejected if the same name had earlier been used for some other species or subspecies of the same genus. Thus *X-us punctatus* Klein (1850) preoccupies *X-us rufescens punctatus* Fox (1900), and the latter must be replaced because it is a junior homonym. Similarly, *X-us albus glaber* Nein (1875) preoccupies *X-us glaber* Tisch (1920). For this purpose the subspecific name is considered to be a binomen. (It is assumed in these examples that the two species are in the same genus, not merely in two genera with identical names.)

CHAPTER 21

Names of Taxa in the Higher Categories

The categories

In the hierarchy of categories, taxa of differing inclusiveness are placed at the various levels (categories). The categories are higher or lower according to their relative positions. The taxa cannot properly be said to be higher or lower except with reference to the hierarchy. The higher categories are not phylum Protozoa and class Insecta, but phylum, class, infraorder, and so on.

Among the categories, what is higher or lower depends entirely upon the point of view. In the past it has often been considered that all categories above the genus category were "higher." With the advent of a complete set of rules for names of taxa in the family-group categories, it is not unreasonable to reserve the expression "higher categories" for those above the family-group levels.

The category groups used in this book are therefore the species group, the genus group, the family group, and the group of the higher categories.

The taxa

The taxa are the groups of animals, generally groups of species. At all levels above species any taxon is basically a group of species, but in ordinary usage only the so-called basic categories (genus, family, order, class, phylum, kingdom) are treated as such groups. The super-taxa at all levels are treated as groups of the basic taxa (a superclass as a group of classes) and the sub-taxa at all levels as subdivisions of the basic taxa (a suborder as a section of the order). These can be listed as follows:

Category	Consists of
1. Kingdom	Group of species
2. Subkingdom	Part of a kingdom
3. Superphylum	Group of phyla
4. Phylum	Group of species
5. Subphylum	Part of a phylum
6. Superclass	Group of classes

Category	Consists of
7. Class	Group of species
8. Subclass	Part of a class
9. Infraclass	Part of a subclass
10. Superorder	Group of orders
11. Order	Group of species
12. Suborder	Part of an order
13. Infraorder	Part of a suborder
14. Superfamily	Group of families
15. Family	Group of species
16. Subfamily	Part of a family
17. Supertribe	Group of tribes
18. Tribe	Part of a subfamily
19. Subtribe	Part of a tribe
20. Genus	Group of species
21. Subgenus	Part of a genus

In nomenclature these are divided into three groups: (1) those that have a species as type (genus and subgenus), (2) those that have a genus as type (superfamily to subtribe), and (3) those that have no type (kingdom to infraorder). This last group is the subject of the present chapter, the higher thirteen categories in the above list.

The names of the taxa

The taxa or groups of zoological classification receive Latin names in much the same way as the species do, except that the names are uninominal. The number of these names in the animal kingdom runs into the hundreds of thousands, but this is a comparatively small number beside the millions used for species and taxa at lower levels.

The names of the taxa placed in the categories above the family group are alike in all fundamental respects and in most details. In recent discussions, and in the Copenhagen Decisions, they were divided into two or three groups, the Order/Class Group (or Order Group and Class Group) and the Phylum Group. The nearly universal similarity in their nomenclatural aspects makes it unnecessary to deal with them in such separate groups, and the changing of level to which they are subject makes it undesirable to do so.

Historical background. In the International Rules, provision was made for regulating a few aspects of the names of family-group taxa and the names of all taxa at higher levels. Four stipulations applied to names of taxa at family-group levels, and two of these applied also to all names of taxa in categories above the family-group level.

At the 13th Congress, at Paris in 1948, proposals were made to incorporate in the *Règles* provisions for detailed regulation of family-group names. These were adopted by the 14th Congress, at Copenhagen in 1953, where it was proposed to extend these rules, with some important differences, to the names of taxa at the "order/class" and "phylum" levels. These proposed rules were outlined in the Copenhagen Decisions.

At the 15th Congress, at London in 1958, all provisions for names of taxa above the family group were rejected, and the 1961 Code not only includes no such provisions but expressly states: "This Code is concerned with such names in the family-, genus-, and species-groups."

There are thus *no rules* governing the names of the taxa covered by the present chapter. There are (1) several aspects generally agreed to by zoologists, (2) several that are the subject of proposal and counterproposal, (3) several that were covered by implication in the 1901–1953 *Règles*, (4) several proposed as rules in the Copenhagen Decisions (binding on taxonomists for seven years), and (5) some others that must follow from the general taxonomic system now in use. Altogether these provide a fairly complete system for names at these higher levels, but a system leaving much more freedom for choice than do those for names at the genus, species, or even family levels. The stipulations and recommendations listed below are thus an alloy of implied rules, previous official recommendations, and past practice.

General considerations. Several aspects of taxa and their names at these higher levels (above the family group) are omitted from the detailed provisions listed later in this chapter for simplicity or because there has been no agreement as yet on how they are to be handled. These include the types of the taxa, the possibility of uniform endings for the names, and the use of terms, all of which are discussed briefly here.

Types of taxa in higher categories. Names of the taxa at the phylum, class, and order levels, and of all the taxa placed at super or sub levels of these, or at intermediate levels, differ from all other names of taxa in zoology in not being based on any kind of type. Some writers have assumed that each level should be typified by one of the taxa at the next lower level, as a class would have as its type a particular included subclass. This at once proves to be an unworkable system, because use of any category, and especially the subordinate ones, is optional, and one taxonomist would take a subclass to be type of the class while the next taxonomist, not using subclass at all, would take an order or superorder as type. Only if there were a fixed number of categories could such a system work, and it is not appropriate for anyone or any group of people to decide what categories will someday be needed in every group.

The Copenhagen Decisions proposed a workable plan, to have each higher taxon typified by a genus in the same manner as family names. Each taxon would have a type genus, but the type genus would not be used in formation of the taxon name.

With the abandonment of the proposals of the Copenhagen Decisions, which were officially in effect from 1953 to 1961, no formal type exists for any higher-category taxon. However, this does not really leave the taxa untypified. A type is never needed unless a taxon is to be divided or two taxa are to be combined. A simple rule for each of these situations will generally result in the outcome acceptable to the largest number of taxonomists.

1. When a taxon is to be divided, the original name is to be retained for the larger of the resulting groups.
2. When two taxa are to be combined, if they are of substantially the same size, a new name for the combined taxon will avoid confusion, but if one taxon is much larger, its name should be retained for the combined taxon.

This amounts to having as the type of a taxon the subordinate taxon that is the largest at whatever level is involved. For example, the type of a class that is to be divided is the larger of its subgroups, whether these have received recognition as subclasses or as orders or have not been previously recognized at any formal level.

Uniform endings. Zoologists have found it to be useful to have all family names end in *-idae,* making each one immediately recognizable as the name of a family. There has been an understandable desire to have the level of the other taxa similarly shown, by distinctive endings.

The 1901–1953 *Règles,* as well as the 1961 Code, provide that subfamily names shall end in *-inae.* The 1961 Code also recommends the ending *-ini* for the names of tribes and the ending *-oidea* for the names of superfamilies. The Copenhagen Decisions refused to stipulate endings for names of family-group taxa other than families and subfamilies and recommended against the use of standardized endings for names of taxa at any level above the family group.

The advantages of uniform endings have impressed many people. Some of these have proposed that such endings be adopted now. For names in several relatively small groups, mostly vertebrates, endings at certain levels, usually ordinal, have been adopted by national organizations or groups of specialists. None of these has attained general international acceptance. No one system for all higher-level names has met with wide acceptance.

Aside from various difficulties of agreeing on the exact scheme, there are two over-riding difficulties that have led some taxonomists to reject all such schemes. First, there is no way to alter the thousands of usages of the unstandardized names in the literature. For generations to come students would have to learn both systems. Second, most of the proposed endings are not distinctive among the various names of animal groups. Hominidae with the ending *-idae* will very seldom be confused for the name of a genus, because genera only rarely end in *-idae* and seldom even in *-ae.* The same is true for *-inae, -ini,* and *-oidea* endings. But too many of the endings proposed for names of taxa at higher levels, such as *-ida* and *-ina* are commonly found as the ending of generic names. No system has yet been proposed with truly distinctive endings at all levels.

If such uniform endings had been adopted early in the development of zoological nomenclature, the system would be effective and useful. After two hundred years, it appears to be too late to attain those benefits by this device.

Terminology. It should perhaps be made clear in one place how various terms are being employed in this chapter.

The groups of animals are *taxa.* Each taxon is placed at some level in the *hierarchy.* This level is the *category.* A category can be higher or lower than some other one, so we may speak of a *higher category.* The taxa are not higher or lower but are placed in higher or lower categories (levels or ranks). The taxa are more- or less-inclusive (include more or fewer subgroups or species).

The categories have names, but these are terms and not names in zoological nomenclature. (They are kingdom, phylum, class, and so on.) The taxa also have names—the zoological names.

It is an error to state, "This animal belongs in the category Mammalia." Mammalia is the name of a taxon, not of a category. To refer to "the origin of higher categories" is not taxonomically meaningful, because it is the taxa which are studied in this regard; the categories originate only in man's mind. It is an error to refer to higher taxa, because only categories are higher or lower, although taxa can be placed in higher categories.

It is necessary to make a careful distinction between names, taxa, and categories both in taxonomic work and in nomenclature. The extra word or two required to do this in each case is well rewarded in clarity of thinking and understanding.

Names of higher-category taxa. Because there is no simple way to refer to all these taxa or names at one time, this section will be written about "these names" or "such names." This is to be understood to refer to names of all taxa in any of the categories above the family group, unless otherwise directly limited.

The name of a taxon at any of these levels is a single word. It must denote a group of animals which is directly placed at one of the levels (in one of the categories) of the higher-category group. These groups all together include at least two thousand names, but this number is small enough so that problems of homonymy, synonymy, orthography, and so on, are not so pressing as to force regulation. These names are single words, of classical origin, and treated as if they were Latin. They are assumed to be plural in number and nominative in case. An increasing number are of non-classical origin, partly because virtually all available Greek and Latin words have already been used.

In spite of the lack of rules to govern the use of these names, there is little obvious confusion or lack of uniformity among them. This is partly due to the rather small numbers involved, but it is also in part an illusion. Although most textbooks in one region during a generation or so agree substantially on the names of phyla and classes, there is less agreement on a worldwide basis and still less over a span of several generations.

The number of synonyms is relatively high, but their total number is not great enough to attract attention. Partial synonyms are also very common, where names have been proposed for groups of phyla not now accepted or for sections of phyla or classes not now employed. Many names that have been proposed are now used only informally, as descriptive terms rather than names of taxa; for example, Coelomata, Bilateria, Schizocoela, and Deuterostomia.

Among the differences in usage now found among taxonomists may be cited Coelenterata vs. Cnidaria, Aschelminthes vs. Nemathelminthes & Trochelminthes, Ectoprocta vs. Bryozoa, Pelecypoda vs. Bivalvia, and Chordata vs. Vertebrata. The reasons for the differences in usage among these are not all the same, but there are no rules to lead us to a choice in any case. In general the zoological facts are given most weight in such decisions. Among the possible bases of choice there are the following: (1) priority, (2) extent of prior usage, (3) influence of a preference. All of these have been used as reason for choice between synonyms. None of them has been so consistently used as to be dominant.

Most taxonomists accept priority as a sound basis for choice unless they find the result objectionable for other reasons. Hyman will reject a name because its inclusiveness of reference has been changed. Chitwood will reject a name because of faulty derivation. Shipley will reject a name because it fails to conform to a uniform system. Copeland will reject a name because of priority. Berg will reject a name because it is inappropriate to the better-known included subgroups. There is justification and precedent for all of these.

In the classification adopted in most textbooks, it would be found upon investigation that the accepted names would not all be acceptable under any one system. Some would have priority and some not. Some homonyms would be accepted and also some names that earlier replaced other homonyms. Some would be changed because of removal of part of the contents, and others would be retained in spite of such removal.

Probably the only "rule" that could find general acceptance in regard to names of these taxa is the admonition that any change should be made only after the most careful consideration. The better known a name is, the more reluctance there should be to change it. Even homonymy can occur without serious confusion, as witness the long-time simultaneous use of Decapoda for orders in both the Arthropoda and the Mollusca, and the similar simultaneous use of Cyclostomata for a class or order in both the Vertebrata and the Bryozoa.

Proposal of these names. Any zoologist may propose a new name by distinguishing the taxon and publishing a new name for that taxon. This would usually be done by raising to that level a group (taxon) placed at a lower level, but it is possible to discover a new animal and make it the basis of a new taxon at any level.

Formation of these names. Presumably names at these levels should be governed by the basic rules covering all zoological names:

1. They are to consist of a single word.
2. They must be Latin or latinized and be spelled in the neo-Latin alphabet.
3. They should be euphonious and pronounceable.
4. Some of them will be formed, if they refer to nominate subordinate taxa, from the stem of the name of the basic taxon with a changed ending (distinctive of the category or not). This would usually mean from the name of some other higher-category taxon, such as: Graptolithina–Graptoloidea–Graptozoa; Conularida–Conulata–Conulariida; Hydrozoaria–Hydroida–Hydrida–Hydrariae. Although these names need not be, and normally are not, based on the name of an included genus, some are so based, such as: Arachnida (*Arachne*), Holothurioidea (*Holothuria*), Amoebozoa (*Amoeba*), Hydrozoa (*Hydra*), Graptolithina (*Graptolithus*), and Hyracoidea (*Hyrax*).
5. They should contain none of the proscribed signs (diacritic marks, umlaut, apostrophe, hyphen, and so on).
6. They are to be written with capital initial letters.

Publication of these names. They should be published in the same manner as other zoological names

(1) in acceptable zoological works;
(2) with intent to establish a new zoological name for an actual group of animals;
(3) with a statement of the characters of the taxon;
(4) with indication of the included sub-taxa;
(5) with definite author and date.

There is nothing to prevent their being published anonymously, conditionally, as a junior synonym, or with inappropriate meaning.

Source of these names. It is almost universally the practice to make these names descriptive of the group, such as Coelenterata, Arthropoda, Vertebrata; Bivalvia, Gastropoda, Monoplacophora; Coleoptera, Lepidoptera, Diptera. The name may also be derived from:

1. Latin or latinized Greek; e.g., Myxosporidia, Edentata.
2. Any modern language using the neo-Latin alphabet.
3. Names of people, geologic eras, formations, places, ships, animals, and so on (but these are not recommended); e.g., Burgessiida, from the Burgess Shale; Bassleroceratida, after R. S. Bassler.
4. Arbitrary combinations of letters (not recommended).

Conflicts among these names. There may be conflicts between names

(1) when two are proposed for the same taxon (as Bryozoa and Polyzoa);
(2) when two different spellings are used (as Gastropoda and Gasteropoda, or Endoprocta and Entoprocta);
(3) when the same name is used for two different taxa at the same level (homonymy) (e.g., Proboscidea, an order in Myzostomida and an order in Mammalia);

(4) when the same name is used for a taxon and for one of its subordinate taxa (as Priapuloidea, phylum, and Priapuloidea, class);
(5) when the same name is used for a taxon and for a different taxon at any level (e.g., Gnathostomata, a subclass of vertebrates and an order of echinoderms; Polyzoa, a phylum (Bryozoa) and a class of platyhelminths; Articulata, a class of brachiopods and a subclass of crinoids; Cestoidea, an order of ctenophores and a class of platyhelminths).

None of these is necessarily cause for any change. None causes serious difficulty, partly because of the comparatively small total number of names involved.

Author of these names. The author of a name is the person (or persons) who first publishes it. The author is seldom cited.

1. Changes in spelling or ending should not change the author.
2. Changes of substantial magnitude in the inclusiveness of the taxon have sometimes been used as reason for citing the reviser as author.
3. Changes of rank have sometimes been used as reason for change of author.

Date of these names. There is no official date for these names. The true publication date is the day on which a certain one appeared in a publication. The date is seldom cited, but if it is, the original date should be the one given, not the date of a later change, unless the latter is made clear.

Citation of these names

1. Name. The name should be written in the ordinary text type face, not in italics. It should always be identified as the name of a taxon at a specified level, either as "Phylum Echinodermata" or in a tabular form that makes the rank clear. New names should be indicated as such in the first publication.
2. Author. The author of a name need not be cited, unless there is direct need for a bibliographic reference.
3. Date. The date of a name need not be cited, unless there is need for information on the time of its proposal.

Synonymy. When there are two names for a particular taxon, these are synonyms. Usually there is some difference between the contents of the taxa they represent. For example, the names Coelenterata and Cnidaria may be considered to be synonyms. The Coelenterata of one author may have contained the Ctenophora, whereas the Cnidaria of a certain other author did not. They are nevertheless considered to be synonyms.

Any subsequent writer is free to choose any of the synonyms for his use, regardless of date, priority, spelling, appropriateness, general acceptance, homonymy, or any other factor. He may even propose a new name in place of all the synonyms, although public opinion demands an adequate reason for this.

Priority among these names. Priority among names of higher-category taxa may be used as the basis for a choice, but this is not required. It is not so important in the choice as recognizability, usage, or appropriateness.

Homonymy of these names. When homonymy occurs among these names, it does not force the rejection of the junior homonym. However, at the highest levels, where so few names are involved, zoologists in general would never accept the same name in two different meanings (for example, for two different phyla).

Replacement of these names. Names at these levels do not ever have to be replaced because of any rule. That is, there are no situations in which a zoologist is required by rule to replace them. In the exceptional case indicated above for homonyms, there may be strong feeling in favor of replacement.

Names *may* be replaced for any reason that seems sufficient to the reviser. The following have all been used as reason:

1. Inappropriateness.
2. Synonymy (the junior one being rejected).
3. A change (slight or great) in the contents, such as (a) removal of one or more subgroups, or (b) addition of one or more subgroups, or (c) combining two groups.
4. A different spelling makes the name more distinctive and recognizable.
5. Assignment to a different category—changing the rank.

Subsequent spellings of these names. Changes of spelling are not uncommon at all levels. There is no prohibition and no strong public opinion against changes in spelling if there is justification for them. There have been several attempts at standardizing endings at the upper levels of the hierarchy, but these have not resulted in any widespread uniformity.

Combining taxa. If two such taxa are to be united, the reviser will have to choose a name from among all the synonyms of the two, or propose a new name for the combined taxon. The basis for this choice is left to the reviser.

Dividing a taxon. If a reviser wishes to separate the components of such a taxon into two taxa at the same level, he is free to choose any name for each of the new taxa, rejecting the previous name and its synonyms or retaining one of them for either one of the new groups.

Changing levels. Such a taxon can be elevated to the next higher rank or lowered to the next lower rank. Its name can remain unchanged, or the reviser may select a different name from among the synonyms or propose a new name for the taxon at the new level.

Changes of such names. These may be brought about by

(1) choice of a different spelling;
(2) change of ending in a uniformity system;
(3) choice of a different one of the synonyms;

(4) choice of a different name because of change in the inclusiveness of the taxon;

(5) rejection of all previous names in favor of a distinctive new one.

Change of name at these levels is always unfortunate. It should always be well considered, particularly from the viewpoint of the general zoological user, but it must be remembered that if there are two names in current use for a taxon, the reviser must make a choice between them.

Stability of these names. There are no rules to stabilize these names. Changes at these levels should be avoided whenever possible.

Comments on names of taxa in each category. *Kingdom.* The rules of zoological nomenclature apply only to names of taxa within the animal kingdom. Strictly speaking, then, they do not cover, even by implication, the name of the kingdom itself. No real problems have ever arisen in this regard, although there are two available names for this taxon, Animalia and Zoa. The former is universally used, and the latter has been virtually forgotten.

Subkingdom. The subkingdom category is the rank at which the first subdivisions of the kingdom are placed. Names at this level have almost universally been formed with the ending -*zoa,* as *Eozoa* and *Metazoa.* (This has been more by accident than by stated and planned intent, and names ending in -*zoa* have been used at other levels.)

Subkingdoms can be proposed

(1) by splitting a kingdom into two or more subkingdoms, as occurred when the Protozoa and the Metazoa were distinguished within the kingdom Animalia;

(2) by elevating a former phylum (or superphylum or taxon at lower rank) to subkingdom, as is sometimes done with the Protozoa (to go with the subkingdom Metazoa);

(3) by combining two phyla or superphyla into a subkingdom;

(4) by basing a subkingdom on some animal or group of animals newly discovered or newly segregated from some other subkingdom.

Superphylum. The superphylum category is, like that of the subkingdom, little used and is a relatively unimportant member of the higher categories.

Names at this level are very little used, and in some cases the level between the categories phylum and subkingdom is called the "series" level. If this is used as a formal level, the names of its taxa also would be governed by the suggestions for the names of taxa in the higher categories.

A superphylum can be proposed

(1) by reducing a kingdom or subkingdom to superphylum rank;

(2) by elevating a former phylum (or taxon at lower rank);

(3) by combining two phyla; or

(4) by basing it on some animal or group of animals newly discovered or newly segregated from some other superphylum.

Phylum. The taxa placed in the phylum category are the phyla, subdivisions of the kingdom (or of the subkingdoms, if such are used). They may be assembled into superphyla or subdivided into subphyla. Not more than forty such taxa are recognized in this group, but more than a hundred names are available for them.

The categories dependent upon the phylum category, in a nomenclatural sense, are the subkingdom, the superphylum, and the subphylum. As taxa may be changed at will from any of these levels to any other, the names will be used at whatever rank the taxon is placed. For example, Protozoa has in recent years been used as the name for a phylum and for a subkingdom. Metazoa has been used as the name for a subkingdom, for a superphylum, and for a phylum. The taxon was the same in each case; that is, the contained species were the same, but the taxon was considered to be placed at the different levels indicated.

Phyla can be proposed

(1) by reducing a kingdom, subkingdom, or superphylum to the phylum rank;
(2) by elevating a former subphylum or class to phylum rank, as when Hyman (1951) elevated the former class Acanthocephala to phylum rank; or
(3) by basing it on some animal or group of animals newly discovered, newly segregated from some other phylum, or not previously placed in any phylum, as when Blackwelder (1963) erected Monoblastozoa for the hitherto unattached genus *Salinella.*

Subphylum. The subphylum level has been used in at least four phyla, but it is only rarely used. The need for a category between phylum and class is evidently not widely felt, as the superclass level is also little used.

Subphyla can be proposed

(1) by reducing a phylum to subphylum rank;
(2) by elevating a former class to subphylum rank, as Sporozoa has been recently raised;
(3) by bringing together two classes at a higher level, as Trilobitomorpha was proposed for the Trilobitoidea and Trilobita together; or
(4) by basing it on some animal or group of animals newly discovered or newly segregated from some other subphylum.

Superclass. Superclasses have been used in several phyla, such as the Echinodermata and Arthropoda, but they are not commonly used except by specialists even here. They are formed by grouping classes rather than by dividing a phylum or subphylum, and when they are used the subphylum level is generally not employed. The category is considered to be dependent upon the class level nomenclaturally.

A superclass can be proposed

(1) by reducing a subphylum to superclass rank;
(2) by elevating a class, as was done recently in the Protozoa when the class Sarcodina became the superclass Sarcodina;
(3) by combining two classes, without retaining the two subgroups (a theoretical possibility but an impracticality);
(4) by basing it on some animal or group of animals newly discovered or newly segregated from some other superclass.

Class. The class is the basic category of what has been called the class group, which included also superclass, subclass, and infraclass, as well as any others interpolated among these. In the animal kingdom as a whole, the classes are undoubtedly the best known taxa, even the phyla being subject to more difference of opinion. In virtually every phylum, the class is a level of primary distinction.

A class is generally a subdivision of a phylum. The 35 to 40 phyla are divided into 100 to 130 classes, depending upon the writer involved, but something over 250 names are involved directly.

Classes can be proposed

(1) by reducing a phylum, subphylum, or superclass to class rank, as when Hyman (1940) reduced Acanthocephala from phylum to class;
(2) by elevating a former subclass or order to class rank, as was recently done in the Protozoa when the subclass Phytomastigina became class Phytamastigophorea;
(3) by combining two classes without retaining the subgroups;
(4) by basing it on some animal or group of animals newly discovered or newly segregated from some other class, as the class Cnidospora was separated from the class Sporozoa.

Subclass. Subclasses are commonly used as subdivisions of classes. At least 25 classes are currently subdivided into two or more subclasses, in most classifications. They are formed as divisions of classes, rather than as groups of orders or superorders. The category is subordinate to the class level, and the taxa may be themselves divided into infraclasses.

A subclass can be proposed

(1) by reducing a class to subclass rank, as Polychaeta and Oligochaeta have often been reduced from class rank;
(2) by elevating an order, a superorder, or an infraclass to subclass rank, as order Xiphosurida has been elevated to subclass Xiphosura by some students;
(3) by combining two orders or superorders, without maintaining the subordinate taxa;
(4) by basing it on some animal or group of animals newly discovered or newly segregated from some other subclass.

Infraclass. Infraclasses have apparently been used only in the Mammalia and are currently used at only one point, where superorders are not used (Pantotheria, Metatheria, and Eutheria, in Simpson, 1945). The impossibility of rating these groups as definitely infraclasses rather than superorders or as taxa in the intermediate category of cohort makes it questionable whether there is any need for such a category and its taxa.

Further remarks on the names at this level seem unnecessary, as this subdivision of subclasses is an unnecessary complication and the names of the taxa would be comparable to those of subclasses.

Superorder. The taxa at this level are not often used except in Mammalia, where there are five superorders in one cohort (a taxon in an interpolated category between infraclass and superorder). Superorders are groups of orders. They may be further combined into cohorts, infraclasses, or subclasses, but presumably would not be used unless taxa at one or more of these higher levels were recognized.

A superorder can be proposed

(1) by reducing a subclass, infraclass, or cohort to superclass rank;
(2) by elevating an order, as was done when order Perissodactyla was elevated to superorder Mesaxonia, in the Mammalia;
(3) by combining two classes, without retaining the subgroups (a theoretical possibility but not a reasonable action);
(4) by basing it on some animal or group of animals newly discovered or newly segregated from some other superorder.

Order. The order is the basic category of what has been called the order group, which includes also the superorder, the suborder, the infraorder, and taxa at any other levels interpolated between superfamily and infraclass. In many phyla, orders are very well-known groups, but in some phyla they are less well known than the classes. Whereas classes do have a fairly evident uniformity throughout the animal kingdom, the orders of vertebrates, for example, are scarcely comparable to families in insects, and the levels vary in other groups.

There are something between 600 and 2,000 orders of animals depending upon opinion, with well over 2,000 names, although several thousand groups have at some time been considered to be orders.

Orders can be proposed

(1) by reducing a superorder, subclass, or class to order level;
(2) by elevating a former suborder to order, as when the protozoan suborder Gymnostomata was elevated to be order Gymnostomatida;
(3) by combining two orders without retaining the subgroups, as when orders Coleoptera and Strepsiptera are combined, with the latter being only a family in the former;
(4) by basing it on some animal or group of animals newly discovered or newly segregated.

Suborder. Suborders are commonly used in many phyla. They are formed as subdivisions of orders. Generally they are at the first level above that of superfamily (or family), but in a few cases they are subdivided into infraorders. This category is subordinate to the order level.

A suborder can be proposed

(1) by reducing an order to suborder rank;
(2) by elevating a family, superfamily, or infraorder to suborder rank, as in fishes when family Lycopteridae was made suborder Lycopteroidei;
(3) by combining two families, superfamilies, or infraorders, without maintaining the subordinate taxa;
(4) by basing it on an animal or a group of animals newly discovered or newly segregated from some other suborder.

Infraorder. Like infraclass, taxa at this level have been used principally in Mammalia, where four suborders currently have a total of eleven infraorders. All but two of these also employ superfamilies. They are formed as divisions of suborders and are never used unless the suborder level is also used. They would normally not be subdivided, but the included families may be grouped into superfamilies, achieving the same effect.

An infraorder can be proposed

(1) by reducing a suborder to infraorder rank;
(2) by elevating a family or superfamily, as when the mammal superfamily Amphimerycoidea became also the infraorder Tragulina;
(3) by combining two superfamilies at the infraorder level; or
(4) by basing it on some animal or group of animals newly discovered or newly segregated from some other infraclass.

Interpolated categories. There is no theoretical limit to the number of levels within the higher-category group. *Cohort* has been used between subclass and superorder; *series* has been used between subkingdom and phylum; *branch, division, legion,* and *phalanx* have been used in various levels. *Tribe* has also been used above the family-group levels, but such use is now prohibited by the Code.

Any such category will contain taxa whose names are treated like the names of taxa in the "standard" categories—the nearest ones. Multiplication of categories should be well justified. The use of eleven higher categories to classify 50,000 vertebrate species would seem to be very high, at least if compared to the eight used to classify 600,000 insect species.

CHAPTER 22

Names of the Family-Group Taxa

The names in the family group are the family names themselves, the super-family names, the subfamily names, names of any other taxa placed in any unusual category which may be based on the family (such as infrafamily or sub-superfamily, if such ever occur), and the names of tribes and their derivatives. In many respects these names are alike, used under the same rules.

The family is the basic level or category of the family group. The taxa placed at this level are the families. These are groups of species formed by grouping genera. The families may be brought together into super-families or directly into orders or suborders. They may be subdivided into subfamilies, tribes, and so on. They are identifiable as families by their names, which always end in *-idae*. Superfamilies often end in *-oidea* but may have other endings; subfamilies end in *-inae;* and tribes often end in *-ini* but may have other endings.

History. In the earliest taxonomic works, the comparatively few genera known were grouped directly into orders. As the number increased in the latter part of the eighteenth century, there arose a need to recognize groups of genera in some of the orders. Early in the nineteenth century families were introduced, each name based on a generic name and each one ending in *-idae*, which is a Greek plural meaning "like" (or, more rarely, some comparable ending such as *-ides* or *-aceae*). Thus, the Hominidae are the family of animals like the genus *Homo*.

During the next hundred years various writers used variations of this system, but the desirability of some uniform ending was usually recognized. A brief article (Article 4) in the *Règles* accepted this practice, but many details were not covered by this rule, and considerable difference of opinion has existed as to its meaning. This difference of interpretation has led to real confusion, and several efforts have been made to regulate further details of this problem.

The Copenhagen Decisions recommended a set of rules to alleviate this situation, and the 1961 Code includes many such rules for these names. The basis for this regulation is the use of a type genus for each family,

449

with the name of the family formed from the stem of the name of the type genus. (The 1961 Code does not use the grammatical stem but a nomenclatural stem consisting of the genitive singular without the case ending.)

This apparently simple system is beset with difficulties. Of these the worst is that determination of what is the stem of every chosen generic name is by no means easy. If all generic names were really Latin words and if Latin declensions were simple and recognizable, there would be no such problem. The following list has examples of Latin words (L), latinized Greek words (LG), Greek words transliterated into Latin (G), neo-Latin words, not known to the Romans (NL), and modern non-classical words, arbitrary combinations of letters (NC). These illustrate some of the difficulties of determining the form of a family name.

Genus Name	Origin	Nom.	Gen.	Stem	Family Name
Musca	L	musca	muscae	musc-	Muscidae
Ostrea	L	ostrea	ostreae	ostre-	Ostreidae
Trypanosoma	G	-soma	-somatos	-somat-	Trypanosomatidae
Ostoma	NC	—	—	—	Ostomidae or Ostomatidae
Semele	L	Semele	Semeles	Semel-	Semelidae
Acridium	L	acridium	acridii	acridi-	Acridiidae
Solen	L	solen	solenis	solen-	Solenidae
Procyon	L	procyon	procyonis	procyon-	Procyonidae
Homo	L	homo	hominis	homin-	Hominidae
Teredo	L	teredo	teredinis	teredin-	Teredinidae
Nilio	NC	—	—	—	Nilionidae or Niliidae
Physeter	L	physeter	physeteris	physeter-	Physeteridae
Eugaster	G	eugaster	eugastros	eugastr-	Eugastridae
Tamias	G	tamias	tamiou	tami-	Tamiidae
Anas	L	anas	anatis	anat-	Anatidae
Lampas	G	lampas	lampados	lampad-	Lampadidae
Gigas	G	gigas	gigantos	gigant-	Gigantidae
Termes	L	termes	termitis	termit-	Termitidae
Cupes	NC	—	—	—	Cupesidae or Cupedidae
Vates	L	vates	vatis	vat-	Vatidae
Lepis	G	lepis	lepidos	lepid-	Lepididae
Dinornis	G	dinornis	dinornithos	dinornith-	Dinornithidae
Cis	NC	—	—	—	Cisidae or Cioidae or Ciidae
Aphis	NC	—	—	—	Aphidae or Aphididae
Sus	L	sus	suis	su-	Suidae
Carabus	L	carabus	carabi	carab-	Carabidae
Lanius	L	lanius	lanii	lani-	Laniidae
Basileus	G	basileus	basileos	basil-	Basilidae
Chlamys	L	chlamys	chlamydis	chlamyd-	Chlamydidae

Genus Name	Origin	Nom.	Gen.	Stem	Family Name
Corax	LG	korax	korakos	korak-	Coracidae
Anax	LG	anax	anaktos	anakt-	Anactidae
Cimex	L	cimex	cimicis	cimic-	Cimicidae
Thrix	G	thrix	trichos	trich-	Trichidae
Salpinx	G	salpinx	salpinges	salping-	Salpingidae
Bombyx	L	bombyx	bombycis	bombyc-	Bombycidae
Japyx	NL	Japyx	Japygis	Japyg-	Japygidae

These are actual examples of generic names, and most of the family names have been used. The problem is greatly extended by the possibility that other barbarous (non-classical) names may become the basis for family-group names. The following existing generic names could only be given stems by some arbitrary systems:

Anzac	*Practor*	*Spinello*
Diodon	*Radhica*	*Uanaka*
Fauva	*Ragas*	*Uba*
Jadamga	*Rallicrex*	*Vitracar*
Jemtella	*Rooma*	*Watsa*
Jones	*Sinevra*	*Yunz*

The system chosen for the 1961 Code is to use the Latin genitive, without the case ending, as the stem in all classical words, as shown in an elaborate table and in Latin dictionaries. For example: Latin *ovum*, grammatical stem *ovo-*, genitive singular *ovi*, nomenclatural stem *ov-*. For generic names of non-classical origin, as in the two lists above, the stem arbitrarily selected by the first zoologist to form a family-group name based thereon is to be the nomenclatural stem thereafter.

The superfamily category

The superfamily is the highest level or rank in the family group of categories. It is the next level above the family, upon which it is dependent in most nomenclatural matters. Above it, the next level is the order, or the lowest level used in the order-group categories (suborder or infraorder).

The superfamily taxa

The taxa placed at the superfamily level (in the category superfamily) are the superfamilies. These are groups of species formed by grouping families. They are never directly "combined" into taxa at higher levels, nor are they normally subdivided. They arise as groups of similar families. They are most commonly used in the Insecta, where they are identifiable by the ending *-oidea;* elsewhere the endings are variable or taxa at this level are not employed. The problems of superfamily taxa involve their types and their names.

The types of the superfamilies. The type of a superfamily is a genus. The selection and employment of the type genus of a superfamily are exactly the same as for a family, and the rules and comments cited under the heading The Types of the Families, below, apply equally here, if for "family" one reads "superfamily."

There is one difference. Families are often erected without reference to any groupings (such as subfamilies) between them and the genera, but superfamilies are never used unless there are families. Therefore, every superfamily will be based on the same type genus as some family, and the superfamily and family will have names differing only in termination.

The names of the superfamilies

The rules for the names of superfamilies are included in the 1961 Code in the rules for the family-group names. In this section there will be presented only the ways in which superfamily-name treatment *differs* from that of family names as listed below under the heading The Names of the Families. The word "superfamily" is to be substituted in all cases for "family," and the word "family" for "subfamily."

The name of a superfamily is a single word of Latin form that denotes a group of animals (a taxon) given the rank (category) of superfamily.

Proposal of superfamily names. All applicable.

Formation of superfamily names. Applicable *except that* the name of the superfamily is formed by adding a distinctive ending to the stem of the name of the type genus. In insects the ending *-oidea* is universally used, and the code recommends its use in all groups.

Original spelling of superfamily names. Applicable *except that* the termination is to be *-oidea* or some other one not prescribed for taxa at any other level.

Publication; Conflicts; Author; Date of superfamily names. All applicable.

Citation of superfamily names. Applicable *except that*

(1) in item (8) the ending may be *-oidea* or some other distinctive one;
(2) the notation should be "new superfamily."

Synonyms of superfamily names. Applicable *except that* in the example the ending of the names must be changed: Dipodidae to Dipodoidea, Dipsidae to Dipsoidea, Jaculidae to Jaculoidea, Dipodinae to Dipodidae, and Jaculinae to Jaculidae.

Priority; First reviser; Homonymy; Rejection; Replacement of superfamily names. All applicable.

Subsequent spellings of superfamily names. Applicable *except that* if a change of termination is involved [item (A)(1)(b)], it will be *-idae* to *-oidea* (or other distinctive ending).

Dividing a superfamily. When the families of a superfamily are believed to be separated into two groups so distinct that they should each be ranked as a superfamily, there are two possible actions, in this order:

(1) elevation of a family (taxon) to superfamily rank;

(2) creation of a new superfamily for the segregate group which does not contain the type genus of the original superfamily.

If the new group is already a single family, or if there is any family-group name based on any of its genera, one of such names must be used as basis for the new superfamily name.

If the new group includes several families, its name must be based on the name of one of those families.

For suggestions on how to determine whether family-group names are available for any included genus and whether homonymy involves any of the possible superfamily names, see the remarks on this under the heading Dividing a Family, below.

Combining superfamilies. Applicable *except that* in the example involving Exidae, Exinae, and Zeeinae, these names should read Exoidea, Exidae, and Zeeidae, respectively.

Changing level (rank). Applicable *except that*

(1) the possible changes of rank do not include elevation to a higher rank within the family-group categories, item (1) under Family;

(2) the endings in item (2) should read -*oidea* (or other distinctive ending) to -*idae*.

Changes of superfamily names. All applicable.

Stability of superfamily names. The need for stability at this level is less than for names at the family level. The four items of stabilizing measures are all applicable.

Nominate subordinate taxa. In a sense the family taxa are subordinate to the superfamilies. In each superfamily there will be one family having the same type genus as the superfamily and therefore having a name differing only in ending from that of the superfamily. In reality, the family level is always used, and that of the superfamily may or may not be used. For this reason, the superfamily is actually dependent upon the family. In any case, there is a nominate family within each superfamily, and it has a name differing only in ending.

Each superfamily name must be based on the name of the type genus of some family. If the name of that family is changed, the name of the superfamily will change accordingly.

The family category

The family is the basic level or category of the family group. It is involved with other levels as follows: superfamily; subfamily, supertribe, tribe, subtribe; and any other levels used between superfamily and genus.

The family taxa

The taxa placed at the family level (in the category of family) are the families. These are groups of species formed by grouping genera. They may be brought together into superfamilies or subdivided into subfamilies, tribes, and so on. They are identifiable on reference by their names, which always end in *-idae*. The problems of family taxa involve their types and their names.

The types of the families. The type of a family is a genus. A particular genus is the type of a particular family. The particular family and genus can be identified only by the similarity of their names, each of which is derived from the same Latin word as the other, but they are selected under definite rules. A family cannot exist without a type genus, and its name can only be determined when the type genus has a name.

The 1961 Code says: "The type of each taxon . . . is that nominal genus upon which the . . . name is based." This is a reverse approach, inasmuch as the family name is never formed first and the type recognized afterwards. In reality, the type of each taxon is the genus selected by the taxonomist; the name of the family is then built upon the name of the type genus.

Selection. The genus selected as type

(1) must be one of the genera included in the family (a subgenus could be selected only by first elevating it to genus rank);
(2) need not be the oldest genus; but
(3) should be well known, taxonomically representative of the family, and include widely distributed or common species.

Conflicts among type genera. The following problems may arise in dealing with the type genera of families:

1. Two types selected by different workers at different times. This results in two names for the family—synonyms. (See paragraph on Synonyms of Family Names, below.)
2. Change of type genus can be effected only by action of the Commission. It is not specifically mentioned in the 1961 Code. (See below under Permanence.)
3. Identity of the type genus may be in doubt, although it is assumed to be correctly identified until evidence is found to the contrary. If it so appears, the case is to be submitted to the Commission.
4. If a subsequent fixation of the genotype of a type genus has confused the meaning of the family name, the case is to be submitted to the Commission.

Permanence. In general the family always retains the original type genus, but the Commission has the power to set aside a type genus and designate a different one. (These would be listed in the *Official Index* and in the *Bulletin of Zoological Nomenclature*.)

The names of the families. The name of a family is a single word that denotes a group of animals (a taxon) given the rank (category) of family.

Proposal of family names. A family name can be proposed

(1) by a specified author [Article 50];
(2) by a specified group of authors [Article 50];
(3) by a speaker reported in the minutes of a meeting [Article 50(a)];
(4) by an anonymous writer (if before 1953) [Article 17(7)]; or
(5) by a reviser changing the rank of a taxon from some other level in the family group of categories.

The person proposing the name is its author. (See paragraph on Author of Family Names, below.)

"The name must clearly be used to denote a suprageneric taxon, and not merely be employed as a plural noun or adjective referring to the members of a genus" [Article 11(e)(i)].

Formation of family names. A family name must

(1) be formed from the name of an included genus (its type genus) [Article 11(e)];
(2) be a word that is Latin or can be treated as Latin [Article 11(b)];
(3) after 1899 be in Latin form as prescribed below [Article 11(e)(iii)];
(4) before 1900 need not have been fully latinized if it has later been latinized and has been generally accepted by interested zoologists as dating from the publication in vernacular (non-Latin) form [Article 11(e)(iii)].

The included genus name on which the family name is based must be a valid (correct) one for that genus at the time of proposal of the family name [Article 11(e)]. (Presumably, this means in the opinion of the proposer.)

The name of the family is formed by adding the ending *-idae* to the stem of the name of the type genus [Article 29].

The stem of the name of the type genus is determined as follows [Article 29]:

A. If the name is or ends in a Greek or Latin word, or ends in a Greek or Latin suffix, the stem is found by deleting the case ending of the appropriate genitive singular.
B. If the name is or ends in a Greek word latinized with a change of termination, the stem is that appropriate for the latinized form.
C. If the name is or ends in a word that is not Greek or Latin, or is an arbitrary combination of letters, the stem is determined by the zoologist who first publishes a family group name based on that genus name.

The determination of the stem of all three types of names is illustrated by an extensive table (Table II) in Appendix D, Section VII, of the 1961 Code.

The family name must be based on a supposedly correct name for an included genus, but it may be based on *any* included genus, not necessarily the oldest name nor the name of the oldest genus [Article 64].

A family name proposed at any time without being based on an included genus is unacceptable and has no standing in nomenclature. Ex.: Pilisuctoridae Chatton and Lwoff, 1934, with one included genus (*Conidiophrys*), has no standing and has been properly replaced with Conidiophryidae Kirby, 1941. (It is necessary, however, to cite the unacceptable name in "synonymy" because it is a key to published information.)

Original spelling of family names. The original spelling of a family name

(1) *must have*
   (a) been a correct spelling or been emended [Article 19]; and
   (b) included a capital initial letter [Article 28];
(2) *must not have* (after 1953)
   (a) included letters with diacritic marks, such as accents, umlauts, tilde, cedilla, diaeresis [Article 27 and 32(c)(i)];
   (b) included an apostrophe [Article 27 and 32(c)(i)]; or
   (c) included a hyphen [Article 27 and 32(c)(i)];
(3) *may have* (before 1954) included letters with diacritic marks, apostrophe, hyphen, or ligatures. (The 1961 Code, as well as the Copenhagen Decisions, rule that these signs may not be used *at all*. They *were* legal and *were* used before 1953, so any taxonomist who quotes them without the marks will be making an emendation. The correct form of the name if used for a taxon will be without the marks.);
(4) *should not have* (after 1953) included letter-combinations æ or œ printed as ligatures [Recommendation E3].

The original spelling is to be retained indefinitely unless

(1) it includes diacritic marks, apostrophe, ligatures, or hyphen [Article 32];
(2) it was written with a small initial letter [Article 32];
(3) the termination was other than *-idae* [Article 32(a)(iii)];
(4) the generic name stem is misspelled;
(5) the stem was improperly derived;
(6) there is in the original publication clear evidence of an inadvertent error (lapsus calami, copyist's error, printer's error) [Article 32(a)(ii)].

In case a name is spelled in more than one way in the original publication, only the correct spelling has any status, if it can be recognized. If it cannot, the spelling adopted by the first reviser (see definition) is to be accepted as the correct original spelling [Article 32(b)].

In adding the termination *-idae*, a final *i* on the stem is retained. An *i* not present at the end of the stem is not to be inserted. Example: *Lanius,* lani-, Laniidae *Turdus,* turd-, Turdidae. Similarly, no other letters are to be added to the stem, such as the *-at-* which occasionally occurs normally.

Publication of family names. To be acceptable in nomenclature, family names

(1) *must be published*
- (a) after 1757 [Article 11(a)];
- (b) in an acceptably published work (see Articles 8 and 9);
- (c) in the form prescribed under Formation, above;
- (d) by an author who used binominal nomenclature [Article 11(c)];
- (e) (before 1931) with a description, definition, or indication [Article 12] (automatically provided by type genus) [Article 16(a)(iv)];
- (f) (if after 1930) "either
  - (i) accompanied by a statement that purports to give characters differentiating the taxon; or
  - (ii) accompanied by a definite bibliographic reference to such a statement; or
  - (iii) proposed expressly as a replacement for a pre-existing available name" [Article 13(a)]; or
  - (iv) established expressly by the Commission as a family name displacing any others for that taxon;
- (g) in a form "based on the name then valid for a contained genus" [Article 11(e)];
- (h) As "a noun in the nominative plural" [Article 11(e)]; and
- (i) in a manner clearly "to denote a suprageneric taxon, and not merely be employed as a plural noun or adjective referring to the members of a genus" [Article 11(e)].

(2) *must not be*
- (a) a name first published as a synonym (presumably an unacceptable synonym) [Article 11(d)] (changed in 1964 Code);
- (b) anonymously published if after 1950 [Article 14];
- (c) conditionally proposed if after 1960 [Article 15];
- (d) published in a section of the work that appears before the section that completes the requirements listed in (1) above [Article 10(a)];

(3) *may be*
- (a) published (before 1931) in the index to a work, even if the latter is not binominal, provided it is acceptable in all other respects [Article 11(c)(ii)];
- (b) a junior synonym other than as in (2)(a) above [Article 17];
- (c) incorrectly spelled in proposal, if corrected later [Article 11(e)(ii) and 17(6)];
- (d) anonymous, if published before 1951 [Article 17(7)];
- (e) conditional, if published before 1961 [Article 17(8)];

(f) published (before 1931) without being fully latinized, "provided that it has been latinized by later authors and that it has been generally accepted . ., . as dating from its first publication in vernacular form" [Article 11(e)(iii)];

(4) *should be* accompanied by a statement of characters differentiating the family [Recommendation E1]; and

(5) *should not be* used in a part of a paper or work ahead of the place where it is formally proposed [Recommendation E23].

Sources of family names. Family names are normally derived

(1) from generic names of animals;

(2) by transfer from the Plant Kingdom [Article 2(2)] (presumably very rarely, and not specifically referred to for family names in the 1961 Code);

(3) by elevation of a subfamily to family rank, when the subfamily name becomes a family name (with changed ending);

(4) by lowering of a superfamily to family rank, when the superfamily name becomes a family name (with changed ending).

Conflicts between family names. Several sorts of conflict can arise between names or forms of names:

1. Synonymy. If a family receives more than one name, either because of union of two named taxa or because one author is unaware of or dissatisfied with a previous name, the names are synonyms and one must be selected as the correct name to be used. (See Priority, below.)

2. Homonymy. If the same spelling, except for termination, is used for two taxa in the family group (except for a taxon with its nominate subordinate taxon), the two names are homonyms, and one must be rejected. (See Homonymy, below.)

3. Multiple original spellings (see above).

4. Later spelling alterations, errors or emendations. An emendation, intentional alteration, is justified and acceptable only if it is the correction of an incorrect original spelling [Article 33(a)(i)]. Any other emendation is rejected as a junior synonym [Article 33(a)(ii)]. Any unintentional spelling is an error ("incorrect subsequent spelling") and has no standing in nomenclature [Article 33(b)]. It must be cited in any complete synonymy, however, as a reference to the data published therewith.

5. The same name may be published as new in two or more publications. (This is proscribed by the 1961 Code but when it has occurred in the past the actual dates of each publication must be determined and the subsequent ones treated as later references to the name.)

6. The name of the type genus may be a junior homonym. The name of the family is then unacceptable and must be replaced [Article 39].

7. The name of the type genus may be a junior synonym. After 1960, the family name continues to be acceptable [Article 40]. Before 1960, if such a family name was considered unacceptable and was replaced, and if the replacement name has won general acceptance, it is not to be changed, except by the Commission [Article 40(a)].

8. When a type genus name has been suppressed by the Commission, the status of the family name is not covered by the 1961 Code. Presumably the Commission would decide this status at the same time as that of the type genus name.

9. If it is believed that the type genus has been based on a misidentified type species, and this would affect the stability of a family name, the case is to be submitted to the Commission [Articles 41 and 65].

10. If an overlooked designation of genotype for a type genus threatens the stability of a family name, the case is to be submitted to the Commission [Articles 41 and 65].

Author of family names. The author of a family name is the person or persons who first published it acceptably

(1) if it was first published as a family name;

(2) even if it was first published as the name for a taxon in any of the other family-group categories [Articles 36 and 50(b)];

(3) if it was erroneously spelled and must be emended by a subsequent writer [Article 50(c)];

(4) even if it contained an original spelling error that was corrected by a later writer (that is, if it is a justified emendation) [Article 50(c)];

(5) if, before 1900, it was published in incompletely latinized form but was emended and accepted by interested zoologists as dating from its first publication in vernacular form [Article 11(e)(iii)];

(6) even if it is an unacceptable (unjustified) emendation of an earlier name (the author is the one who published the emendation) [Article 50(d)];

(7) even if the suffix was incorrect [Article 11(e)(ii)].

A different author is credited with the family name

(1) if it is a replacement for one rejected because the name of its type genus is a junior homonym (the author is the author of the original and now rejected name) [Article 39];

(2) if "it is clear from the contents of the publication that only one (or some) of the joint authors, or some other person (or persons) is alone responsible both for the name and [for] the conditions that make it available" (the author is the one actually responsible) [Article 50];

(3) if it "is established by publication in the minutes of a meeting" (the author is "the person responsible for the name, not the secretary or other reporter of the meeting") [Article 50(a)].

Date of family names. The date of a family name is

(1) (in case of replacement of the family name because its type genus is a junior homonym) the date of publication of the name it is replacing [Article 39(a)(i)];

(2) (in case of a name, changed before 1961 because its genotype is a junior synonym, either objective or subjective and which has won general acceptance and is therefore maintained "in the interests of stability" as in Article 40(a)), the date of the rejected name, "of which it is to be considered the senior synonym" [Article 40(b)];

(3) (in all other cases) the day on which it satisfies all the provisions of the Code for availability [Article 10]; even if the form of the suffix were incorrect [Article 11(e)(ii) limited by Article 11(e)(iii)] and whether the date was specified and whether the name was published as a family name or as a name for a taxon in any other family-group category [Article 36].

Citation of family names. Names, authors, and dates are to be referred to or cited in publication as follows:

A. Names

    1. Name of family—to be written in the regular type face, with a capital initial letter.

    2. New names should be indicated by the notation "n. fam." or "new family" [Recommendation E7].

    3. A new form of the name, after change from another categorical level, does not require such notation but should be explained in the text.

B. Authors

    4. The author of the family name may be cited [Article 51(a) and Recommendation E9] but usually is not. If used, his name should follow the family name without any intervening mark of punctuation [Article 51(b)].

    5. A subsequent user of a scientific name, if cited, is to be separated from it in some distinctive manner, other than by a comma [Article 51(b)(i)]. This ruling of the 1961 Code assumes that the author of the taxon is not being cited; if he is, then the subsequent user can be cited after the author, with a comma or other mark between them. Example: *X-idae* (Linnaeus, 1758), Jones, 1964.

    6. Anonymous author: If the name was published anonymously, the author is cited as: "(Anon.)." If the author of such a name is known to subsequent workers, his name, if cited, should be enclosed in brackets to show the original anonymity [Recommendation 51A].

    7. Contributor. If a name is to be credited to someone (Smith) other than the author of the work (Brown or Brown & Smith), he

should be cited by his name alone (Smith). Example: Exidae Smith. (The form Exidae Smith *in* Brown (or *in* Brown and Smith) is a bibliographic citation; it is acceptable and usually appropriate.) [Article 51(c) requires the dual citation.]

8. Where the ending has been changed to -*idae* from some other, the name of the original author may be placed in parenthesis with the name of the reviser (one who transferred it to the family category) immediately following. Example: Fictitiinae Smith, Fictitiidae (Smith) Jones.

C. Dates

9. "The date of publication of a name, if cited, follows the name of the author with a comma interposed" [Article 22].

10. The 1961 Code further specifies: "In citing the date of publication of a name, an author (1) should not enclose the date in either parentheses or brackets if the work containing it specifies the date of publication; (2) should enclose the date, or a part of it, in parentheses if it is determined by evidence derived from the volume concerned other than in (1); or (3) should enclose the date, or a part of it, in square brackets if it is determined only from external evidence" [Recommendation 22A]. This device might be helpful if uniformly followed; the publications of the past have seldom used it, and even for new publications it will not be obvious whether or not the system is being used. This bit of extreme systematization is best forgotten and will doubtless prove to have been stillborn.

Synonyms of family names. Two names for the same family may occur by being

(1) published simultaneously in one work;
(2) published simultaneously in separate works;
(3) published at different times.

Each of these can involve either objective or subjective synonymy (having the same type genus or different type genera, respectively). The name may be published at the same level, or not.

The determination of which name to adopt, in items (1) and (2), is to be made by the first reviser, the first person to deal with both names and choose one of them [Article 24(a)]. In item (3) priority determines the choice, unless the Commission has ruled otherwise.

It must be remembered that if there are synonyms at one rank in the family-group categories, there will be the same synonyms (with changed endings) at lower or higher levels, if the corresponding names are employed there. Example: If the family Dipodidae has synonyms Dipsidae (objective) and Jaculidae (subjective), then a subfamily Dipodinae must have a synonym Dipsinae and may have a synonym Jaculinae. (It will have so long as *Jaculus* and *Dipus* are placed in one subfamily).

Priority among family names. Among family names, priority in original date determines

(1) which of two synonymous names shall be adopted, when they were published at different times in the same rank;
(2) which of two synonymous names shall be adopted, when they were published at different times and in different ranks;
(3) which of two homonyms must be rejected.

The first two items above are both limited by the 50-year rule for nomina oblita (forgotten names). If the senior synonym has been unused in primary literature for more than fifty years, it can be revived only by the Commission. The Commission can similarly conserve names not so long forgotten [Article 23(b)].

Priority does *not* apply to

(1) incorrect original spellings, which are to be corrected automatically, leaving the incorrect spelling with no standing in nomenclature;
(2) multiple original spellings, of which one is selected by the first reviser, and the others have no standing;
(3) incorrect subsequent spellings, which are unintentional and have no standing; and
(4) differences in termination, which do not make different family-group names in the sense of the Code.

First reviser. Where priority is not used as the basis for choice of conflicting names, the choice must be made by the first reviser, the first person to deal with both names and clearly reject one in favor of the other. The first reviser has no power where priority is held to apply; and both priority and the first reviser may be overthrown by action of the Commission.

The first reviser makes the choice

(1) when two or more names are simultaneously proposed for a single taxon [Article 24(a)];
(2) when identical names (homonyms) are proposed simultaneously for different taxa [Article 24(a)];
(3) when a name is spelled in more than one way in the original publication [Article 32(b)]; and
(4) when the generic name is not of classical origin and so has no grammatical stem. ("The stem is determined by the zoologist who first publishes a family-group name based on that nominal genus" [Article 29(b)].

Homonymy among family names. A family name is a homonym

(1) if there is another family name of the same spelling;
(2) if there is another name of the same stem spelling for any taxon at any other level in the family group;

(3) if there have been two identical names based by different authors on the same type genus;

(4) if it is one of two names which differ only in possession by one of an umlaut of non-German origin. (Example: *Holströmidae* would be a homonym of *Holstromidae*.) (Note that umlauts of Swedish origin do not affect homonymy.)

Priority determines which homonym can be retained as an acceptable name. The senior homonym is an available name; the junior homonym must be rejected but not necessarily replaced (see next section).

Item (3) is not mentioned in the 1961 Code and even seems to be excluded by the reiteration that the names must denote "objectively different taxa within . . . the family-group." Nevertheless, these do occur and can otherwise be circumvented only by a pretense that they are the *same* name.

Homonymy does *not* exist or is not relevant

(1) when one of the names is an unintentional misspelling, either original or subsequent [Articles 32(c) and 54(4) and 55(b)];

(2) when one of the names is not acceptable in nomenclature through failure to meet the requirements of availability [Article 54(1)] (except if one of the names is based on a genus exclusively fossil that ends in *-ites, -ytes,* or *-ithes,* which *is* involved in homonymy);

(3) when one of the names has never been applied to a taxon in the animal kingdom [Article 54(2)];

(4) when one of the names represents a hypothetical concept, not an actual taxon [Article 54(3)];

(5) when one of the names is ultimately based only on a teratological specimen as such [Article 54(3)];

(6) when one of the names is ultimately based only on a hybrid individual as such [Article 54(3)];

(7) when one of the names was proposed for, or is based upon a genus name proposed for, some purpose other than taxonomic use [Article 54(3)];

(8) when the names are closely similar but not identical [Article 55(a)] (such a case is to be submitted to the Commission);

(9) when the names are of German origin and differ only in possession of an umlaut (the umlaut is to be transcribed into an *e*) [Article 32(c)(i)]. (Example: *mülleri* and *mulleri* are not homonyms because *mülleri* must be changed to *muelleri*.)

(10) when the names refer to the same family—are both homonyms and synonyms [Article 52 and Glossary]. (See comment on item (3) in previous paragraph.)

If two identical names (homonyms) are published simultaneously, in the same or different works, the first reviser decides which shall be considered senior [Article 24(a)].

Rejection of family names. A family name proposed in a publication is to be rejected

(1) if it does not meet the criteria of availability [Article 11], including
    (a) being Latin or latinized or treatable as Latin,
    (b) being published in a binominal work,
    (c) not being published as a synonym,
    (d) being based on the name then considered correct for a contained genus,
    (e) being a noun in the nominative plural,
    (f) being clearly intended to denote a suprageneric taxon,
    (g) (after 1930) having a description or a reference to one or being a replacement name,
    (h) (after 1950) not being anonymous, or
    (i) (after 1960) not being proposed conditionally;
(2) if it is a nomen oblitum [Article 23(b)(ii)];
(3) if it is a junior homonym [Articles 53 and 60];
(4) if it is an incorrect original spelling (unintentional) [Article 32(c)];
(5) if it is an incorrect subsequent spelling (unintentional) [Article 33(b)];
(6) if it is an unjustified emendation (intentional) [Article 33(a)];
(7) if the name of its type genus is a junior homonym [Article 39];
(8) when it is the junior of two names brought together by the uniting of two families (and subordinate taxa are not used) [Article 23];
(9) if it has been set aside by direct action of the Commission [Article 23(a)(ii)].

A family name is *not* to be rejected

(1) even if (after 1960) its type genus name is rejected as a junior synonym [Article 40];
(2) if before 1961 it is a replacement for a name rejected because the name of its type genus is a junior synonym and has been accepted by interested zoologists [Article 40(a)];
(3) if it has been given acceptance under the Plenary Powers of the Commission;
(4) even if before 1900 it had a non-Latin termination [Article 11(e)-(iii)];
(5) because it is deemed inappropriate (applied by Article 18 only to generic and specific names but also applicable here).

Replacement of family names. A rejected family name needs to be replaced, and the replacement name can be either

(1) a corrected spelling (emendation);
(2) a junior synonym;
(3) a new name based on the replacement name of the type genus;
(4) a new name based on the correct name for any other included genus.

Each of these replacements may be used only if none of the higher ones on the list are available. Thus, a new name is appropriate only if there is no junior synonym or emendation that can be used.

Replacement may be accomplished with an existing name, so proposal of a new name should always wait for assurance that no other name is available, for either nomenclatural or zoological reasons.

Subsequent spellings of family names. Altered spelling of a family name may occur in subsequent works as intentional changes (emendations) or as unintentional variants (errors).

A. An emendation may be
   (1) justified
       (a) if it is a correction of an incorrect original spelling, in which case it is to be used in place of the erroneous spelling and takes the date and authorship of the original spelling [Article 33(a)(i)];
       (b) if it is a change of termination (*-oidea* or *-inae* to *-idae*), which is acceptable and has no effect on the nomenclatural status of the name, other than categorical level [Article 11(e)(ii)].
   (2) unjustified, if it is not a correction of an original error, in which case it is treated as a new name with its own author and date and is available as a replacement and can take part in homonymy [Article 33(a)(ii)].
B. Errors are to be corrected when found and continue in their original status with the corrected spelling; they have no status separate from the correction and take no part in homonymy [Article 33(b)].

Dividing a family. When the genera in a family are believed to be separated into two groups so distinct that they should each be ranked as a family, there are two possible actions, in this order:

(1) elevation of a subfamily (taxon) to family rank;
(2) creation of a new family for the segregate group that does not contain the type genus of the original family.

If the new group is already a subfamily, or if there is any family-group name based on any of its genera, one of such names must be used for the new family.

If there is no subfamily, and if there is no family-group name based on any of its genera, then a new family is erected, a type genus is chosen, and the family name is formed from the name of this type genus.

The determination of whether a family-group name is available for any of a group of genera is a difficult nomenclatural job, as indicated in Chapter 14, The Use of Literature. Not only must one search for such names formed on any of the included generic names but also for any formed on any of the synonyms.

The possibility of homonymy of the family name itself is difficult to eliminate, but it should always be considered. In checking this, one looks in the generic nomenclators for other names of the same spelling, any of which *may* have been the basis for a family-group name, but one must *also* look under all of the other spellings which might yield the same stem. Examples:

Rex	reg-	Regidae	Solex	sol-	Solidae
Regis	reg-	Regidae	Sol	sol-	Solidae
Rega	reg-	Regidae	Sola	sol-	Solidae
Regus	reg-	Regidae	Solis	sol-	Solidae
			Solus	sol-	Solidae

Combining families. When two families are found to be incorrectly separated, not actually distinct at the family level, they are combined into a single family. The name of the combined family must be the oldest available name from all those applied to either of the families or any of their subfamilies [Article 23(d)(i)], *except that*

(1) the ending may be changed [Article 23(d)(i)];
(2) if a generally accepted name would be upset, the case is to be submitted to the Commission, which could choose a name other than the oldest, under its Plenary Powers [Article 23(a)(ii)];
(3) if a family (Exidae) contains two subfamilies (Exinae and Zeeinae) and these are combined, the new united subfamily would take the name based on the oldest subfamily name; if this is not the name previously used for the family name (in this case if it were Zeeinae), presumably the family name also could not stand. (Such a case is not covered by the 1961 Code but presumably should be submitted to the Commission for decision.)

Changing level (rank). When a taxon is raised to a higher rank or lowered to a lower rank, its name must be given the appropriate new ending [Article 34(a)]. The possible changes of family names are

(1) family to superfamily, *-idae* to *-oidea* (or other);
(2) family to subfamily, *-idae* to *-inae;*
(3) family to a family-group rank lower than subfamily, which can be considered to be accomplished by passing to subfamily, see (2) above, and then to the lower rank, see Subfamily Names;
(4) family to a rank above the family group (order, or other), in which case the family name is abandoned and a name formed according to the customs for the higher rank.

It must be recognized in (4) that the new order, for example, will at least theoretically contain a family that includes the type genus of the original family. This family must bear the original family name. Thus, in elevation of family taxa, the names generally do not disappear, just as the taxa don't disappear, at the original level.

Changes of family names. A change in a family name can be caused by any of these circumstances:

1. If it was improperly or incorrectly
    (a) formed (see Formation of Family Names, above);
    (b) published (see Publication of Family Names, above);
    (c) proposed (anonymously after 1950) [Article 17(7)];
    (d) spelled in original (see Original Spelling of Family Names, above);
    (e) emended by a later writer (see Subsequent Spellings of Family Names, above);
    (f) spelled by accident by a later writer (see Subsequent Spellings of Family Names, above).
2. If its type genus
    (a) is suppressed by direct action of the Commission, making its name unavailable;
    (b) is changed by direct action of the Commission.
3. If the name of its type genus
    (a) (before 1961) is a junior synonym, and if the family name has not been accepted by interested zoologists [Article 40];
    (b) is a junior homonym [Article 39] (see Conflicts among Type Genera under The Types of the Families, above).
4. If the family name
    (a) is a junior synonym (unless it has been conserved by the Commission) [Article 23];
    (b) is a junior homonym [Article 55];
    (c) is a nomen oblitum [Article 23(b)].
5. When a proper action by a first reviser has not been followed (see First Reviser, above).
6. When a family is divided into two (see Dividing Families, above).
7. When two families are combined into one (see Combining Families, above).
8. When the level (rank) of the taxon is changed (see Changing Level (Rank), above).

Stability of family names. In some large parts of the animal kingdom families are taxa in daily use by many people. More than any other names in the family group of categories, they must be kept stable if they are to be effective and if taxonomy is to be as useful as possible to non-taxonomists. The present rules, the first to cover the subject broadly, are designed primarily to preserve the stability of these names.

Specifically, the following stipulations are included in the 1961 Code to stabilize family names:

1. Among synonymous family names, if strict application of priority [Articles 23(a) and 36] would upset general usage, the case is to be submitted to the Commission so that the better-known name can be placed on the Official List of arbitrarily accepted names [Article 23(d)(ii)].

2. If the name of a type genus was found, before 1961, to be a junior homonym and thus not acceptable as a type-genus name, and if the family name was replaced under a procedure then acceptable, the replacement name is not to be upset under current rules if it has been widely accepted in the meantime [Article 39(a)(ii)].
3. If, after 1960, a type genus is rejected as a junior synonym, either objective or subjective, a family name based on it is not to be changed [Article 40].
4. If the name of a type genus was (before 1961) found to be a junior synonym and was replaced, the replacement name is not to be upset under current rules if it has been widely accepted in the meantime [Article 40(a)].
5. If a change in the concept of a genus, brought about by discovery that its genotype had been misidentified, or by discovery of an overlooked earlier genotype fixation, would alter a family and threaten its name, the case is to be submitted to the Commission [Article 41].
6. When a family name is changed because of a necessary change in the name of the type genus, to conserve the original taxonomic concept, the new family name (and the new type genus) take the dates of the names replaced [Article 39(a)(i)].

Nominate subordinate taxa. The family is the dominant level in the family group. In some respects all the other levels are dependent upon the family level, even if they are above it (superfamily) or several levels below it (subtribe). Formally the subordinate taxa are those at lower levels which are included in any one family. The nominate subordinate taxa are those whose names are based on the same generic name as the family name but with different endings appropriate to the particular levels (one at each level).

The directly subordinate taxa to a family are as follows, with one nominate taxon in each:

> Subfamilies
> Supertribes
> Tribes
> Subtribes

The name of each family will change when its type genus is changed, and at that time the name of each nominate subordinate taxon will change accordingly.

The subfamily category

The subfamily is the next family-group category below the family. The next category below it is dependent on usage of the author in any given work. It may be the genus level, or it may be the tribe level, or it may be a level between subfamily and tribe, such as supertribe (which has been used) or infrafamily (which does not seem to have been used).

The subfamily taxa

The taxa placed at the subfamily level (in the category of subfamily) are the subfamilies. These are groups of species formed by subdividing families. They are subdivided into genera, tribes, or supertribes. They are identifiable on reference by their names, which always end in *-inae*. The problems of subfamily taxa relevant here are their types and their names.

The types of the subfamilies. The type of a subfamily is a genus. The selection and employment of the type genus of a subfamily are exactly the same as for a family, and the rules and comments cited under the heading The Types of the Families, above, apply equally here, if for "family" one reads "subfamily."

The names of the subfamilies

The rules for the names of subfamilies are included in the 1961 Code in the rules for the family-group names. In this section there will be presented only the ways in which subfamily name treatment *differs* from that of family names as listed above under the heading The Names of the Families. The word "subfamily" is to be substituted in all cases for "family," and the word "tribe" for "subfamily." (Tribe is here understood to be used alternatively for supertribe, which is rare.)

The name of a subfamily is a single word that denotes a group of animals (a taxon) given the rank (category) of subfamily.

Proposal of subfamily names. All applicable.

Formation of subfamily names. Applicable *except that* the name of the subfamily is formed by adding the ending *-inae* to the stem of the name of the type genus [Article 29].

Original spelling of subfamily names. Applicable *except that* the termination is to be *-inae* [Article 29].

Publication; Conflicts; Author; Date of subfamily names. All applicable.

Citation of subfamily names. Applicable *except that* (1) the notation in item (2) should be "new subfamily"; (2) in item (8) the ending would be *-inae.*

Synonyms of subfamily names. Applicable *except that* in the example the endings of the names must be changed: Dipodidae to Dipodinae, Dipsidae to Dipsinae, Jaculidae to Jaculinae, Dipodinae to Dipodini, and Jaculinae to Jaculini.

Priority; First reviser; Homonymy; Rejection; Replacement of subfamily names. All applicable.

Subsequent spellings of subfamily names. Applicable *except that* if a change of termination is involved [item (A)(1)(b)], it will be *-ini* or *-idae* to *-inae.*

Dividing a subfamily. When the subgroups of a subfamily (supertribes, tribes, or genera) are believed to be separable into two groups so distinct that they should each be ranked as a subfamily, there are two possible actions, in this order:

(1) elevation of a sub-taxon (supertribe or tribe) to subfamily rank;
(2) creation of a new subfamily for the segregate group that does not contain the type genus of the original subfamily.

For suggestions on how to determine whether family-group names are available for any included genus and whether homonymy involves any of the possible subfamily names, see the remarks on this under the heading Dividing a Family, above.

Combining subfamilies. Applicable *except that* in the example involving Exidae, Exinae, and Zeeidae, these names should read Exinae, Exini, and Zeeini, respectively.

Changing level (rank). Applicable *except that*

(1) the possible changes of rank are (a) subfamily to family, *-inae* to *-idae;* (b) subfamily to tribe, *-inae* to *-ini;* or (c) subfamily to supertribe, if used, and with no ending prescribed;
(2) the final note is irrelevant.

Changes of subfamily names. All applicable.

Stability of subfamily names. The need for stability at this level is not quite so great as at the family level because fewer people use these names. The six items of stabilizing measures are all applicable.

Nominate subordinate taxa. It is possible to look upon the tribes (or supertribes) as subordinate either to the subfamily or directly to the family. In any case, one of the tribes will have a name differing from that of the subfamily (and family) only in ending. This one is the nominate tribe. There will be one such in every subfamily in which tribes are recognized. The name of the nominate tribe changes whenever the name of the subfamily changes.

The supertribe category, its taxa, and their names

The supertribe is the family-group category next below the subfamily and next above the tribe. This level would never be used unless tribes were also recognized. Taxa are so seldom used at this level that it is scarcely necessary to outline the requirements for their names. They would have type genera and names exactly comparable to those for tribes, taking into account the difference in rank.

Supertribes (the taxa) are groups of tribes, not subdivisions of subfamilies. Each one is a group of species formed by grouping tribes. Inasmuch as the tribes are subdivisions of the subfamilies, the supertribe is more similar to the family than to the genus, both in classification and in nomenclature.

The only specific problems of supertribe names are:

(1) there is no prescribed standard ending, but uniformity in any given work should be maintained;

(2) inasmuch as the rank is not only optional but rarely used, changes of rank from tribe may go direct to subfamily, passing over the possible rank of supertribe.

The tribe category

The tribe is a family-group category between the subfamily and the genus. It is an optional category, and if used may be the only category between subfamily and genus (the lowest of the family-group categories) or it may be used with either supertribe or subtribe levels or both.

The tribe was formerly used in other positions in the hierarchy, above the family level. Beginning with the Copenhagen Decisions (1954), the use of this category name was restricted to the family group.

The next higher category will usually be the subfamily, but supertribe may be used between them. The next lower category will be the subtribe or the genus. When the subtribe category is not used, the tribe will be the lowest of the family-group categories.

The tribe taxa

The taxa placed at the tribe level (in the category of tribe) are the tribes. These are groups of species formed by subdividing subfamilies. They may be united into supertribes or subdivided into subtribes. If the recommendation of the 1961 Code is followed [Recommendation 29A], tribe taxa can be recognized on reference by their names, which should end in *-ini*. The problems of tribe taxa relevant here are their types and their names.

The types of the tribes. The type of a tribe is a genus. The selection and employment of the type genus of a tribe are exactly the same as for a family, and the rules and comments cited under the heading The Types of the Families, above, apply equally here, if for "family" one reads "tribe."

The names of the tribes

The rules for the names of tribes are included in the 1961 Code in the rules for the family-group names. In this section there will be presented only the ways in which tribe-name treatment *differs* from that of family names as listed under the heading The Names of the Families. The word "tribe" is to be substituted in all cases for "family" and the word "subtribe" for "subfamily."

The name of a tribe is a single word that denotes a group of animals (a taxon) given the rank (category) of tribe.

Proposal of tribe names. All applicable.

Formation of tribe names. Applicable *except that* the name of the tribe is formed by adding a special ending, such as *-ini* (recommended), to the stem of the name of the type genus.

Original spelling of tribe names. Applicable *except that* the termination is to be a distinctive one such as *-ini* (recommended).

Publication; Conflicts; Author; Date of tribe names. All applicable.
Citation of tribe names. Applicable *except that*

(1) the notation in item (2) should be "new tribe";
(2) in item (8) the ending should be a distinctive one such as *-ini* (recommended).

Synonymy of tribe names. Applicable *except that* in the example the endings of the names must be changed as required: (as recommended) Dipodinae to Dipodini, Dipsinae to Dipsini, Jaculinae to Jaculini, Dipodini to a subtribe spelling, and Jaculini to a subtribe ending.

Priority; First reviser; Homonymy; Rejection; Replacement of tribe names. All applicable.

Subsequent spellings of tribe names. Applicable *except that* if a change of termination is involved [item (A)(1)(b)], it will be to a subtribe ending or *-inae* to *-ini.*

Dividing a tribe. When the subgroups of a tribe (subtribes or genera) are believed to be separable into two groups so distinct that they should each be ranked as a tribe, there are two possible actions, in this order:

(1) elevation of a subtribe to tribe rank;
(2) creation of a new tribe for the segregate group that does not contain the type genus of the original subfamily.

For suggestions on how to determine whether family-group names are available for any included genus and whether homonymy involves any of the possible tribe names, see the remarks on this under the heading Dividing a Family, above.

Combining tribes. Applicable *except that* in the example involving Exidae, Exinae, and Zeeinae, these names should have the endings appropriate to tribe and subtribe names.

Changing level (rank). Applicable *except that*

(1) the possible changes of rank are
   (a) tribe to subfamily, *-ini* to *-inae;*
   (b) tribe to supertribe, with appropriate endings;
   (c) tribe to subtribe, with appropriate endings;
(2) the final note is irrelevant.

Changes of tribe names. All applicable.

Stability of tribe names. The need for stability at this level is much less than at the family level, but the six items of stabilizing measures are all applicable.

Nominate subordinate taxa. When tribes are used, they may be sub-divided into subtribes, in which case one of the subtribes in each tribe will have a name differing from that of the tribe only in its ending. This is the nominate subtribe. Its name will always change whenever the name of the tribe changes.

The subtribe category, its taxa, and their names

The subtribe is a family-group taxon that has been used as a subdivision of tribes. It is not mentioned directly in the rules but is clearly bracketed into the family-group categories and placed under the rules for the names of the taxa in this group. When used, this is the lowest category above the genus, although it is possible to insert new levels between them. This level would never be used unless tribes were also recognized.

Subtribes (the taxa) are subdivisions of tribes. Each one is a group of species formed by subdividing a tribe. They have type genera and names exactly comparable to those for tribes, taking into account the difference in rank.

The only specific problems of subtribe names are:

(1) there is no prescribed standard ending, but uniformity in any given work should be maintained;

(2) one of the subtribes is always the nominate or type subtribe, and therefore has the same type as the tribe in which it is placed and a name differing from that of the tribe only in ending.

CHAPTER 23

The Names of Genera

The names in the genus group are the names of the genus-group taxa, which are the taxa placed in any of the genus-group categories. The taxa are the genera and the subgenera, and the names are the generic and subgeneric names.

Generic and subgeneric names are of the same general nature. They are alike in most nomenclatural respects, but they represent taxa of quite different sorts and governed by two sets of nomenclature rules that differ in several details.

The genus category

The genus is the basic category of the genus group. It is also the highest in the group. Only one formal category is lower in the group—the subgenus —although sections are also recognized in the code, and collective groups are recognized as equivalent to genera.

The genus taxa

The taxa placed in the genus category are the genera. These are groups of species brought together by the taxonomist as evidenced by the fact that the generic name is part of each name of each of the included species. The problems of genera involve their types, their names, and special problems derived from the inclusion of the generic name in the double name of the species.

The names of the genera

The name of a genus is a single word that denotes a group of animals or species of animals (a taxon) given the rank (category) of genus. Such names are classified in a previous chapter.

Proposal of generic names. A generic name can be proposed

(1) by a specified author;
(2) by a specified group of authors;
(3) by a speaker reported in the minutes of a meeting;
(4) before 1953, by an anonymous writer or writers; or
(5) by a reviser raising a subgeneric taxon to generic rank.

474

Publication requirements of generic names. Generic names, when first published, must have been

(1) published after 1757; and
(2) published in an acceptable publication:
    (a) reproduced in ink on paper by some method that assures numerous identical copies,
    (b) issued for the purpose of scientific, public, permanent record,
    (c) obtainable by purchase or free distribution, and
    (d) not reproduced or distributed solely by a forbidden method (microcards, microfilm, proof sheets, separate photographs with accompanying explanations, reading at a scientific or other meeting, labeling on a museum specimen, mere deposit of a document in a library); and
(3) proposed as a name for a genus or subgenus; and
(4) written in Latin characters, the neo-Latin alphabet; and
(5) a Latin word, or latinized, or so constructed that it can be treated as a Latin word; and
(6) a noun in the nominative singular or be treated as such; and
(7) if before 1931, either in a binominal work or in the index of a work that contains no binomina (therefore is non-binominal) but shows no evidence that the author was not in principle binominal; and
(8) before 1931, accompanied by at least one of the following:
    (A) A description, a statement that purports to give characters differentiating the taxon, or
    (B) a definition, summary of characters, or
    (C) an indication, consisting of
        (a) a bibliographic reference to a previously published description, definition, or figure, or
        (b) the inclusion of a name in an index to a work, provided the provisions of Article 11(c)(ii) of the 1961 Code are satisfied, or
        (c) the substitution of a new name for a previously established name, or
        (d) the citation of one or more available epithets in combination with, or listed under, the new generic name, or
        (e) a single combined description of the genus and one included species, or
        (f) an illustration of an included species, or
        (g) the description of the "work" of an animal (tubes, tracks, feces, nests, and so on); and
(9) after 1930, in a work in which the principles of binominal nomenclature were applied; and
(10) after 1930, in the text of a work rather than only in the index; and
(11) accompanied (after 1930) by a genotype fixation; and

(12) after 1930, shown to contain at least one directly named species; and
(13) after 1930, accompanied by one of the following:
    (a) a statement that purports to give characters differentiating the taxon, or
    (b) a definite bibliographic reference to such a statement, or
    (c) proposal expressly as a replacement for a pre-existing available name; and
(14) after 1950, proposed by an identified author; and
(15) after 1960, published as an acceptable name, not as a synonym; and
(16) after 1960, unconditionally proposed (with an exception cited in Article 48(b) of the 1961 Code).

These provisions apply between certain dates, as follows:

Items	1758–1930	1930–1950	1951–1960	1961–
1–6	X	X	X	X
7–8	X			
9–13		X	X	X
14			X	X
15–16				X

Sources of generic names. At least the following kinds of words can be or could be used as basis for generic names. (Some are not now acceptable; some are recommended against in the 1961 Code; the illustrations are *names* that illustrate past cases made from words of the indicated nature.)

1. Simple ancient Greek nouns transliterated into Latin, either
    (a) retaining the Greek ending; e.g., *lepas* (*Lepas*), *hoplites* (*Hoplites*), *pandrosos* (*Pandrosos*), *pelor* (*Pelor*), *sphinx* (*Sphinx*), *lipothrix* (*Lipothrix*), *neophron* (*Neophron*), or
    (b) with latinized endings; e.g., *Syrphus,* from Greek *syrphos.*
2. Derivative Greek nouns modified by suffixes that change the meaning, transliterated into Latin; e.g., *gaster* + *-odes* to form *Gastrodes.*
3. Compound Greek nouns, either classical or modern, transliterated into Latin; e.g., *Schistosoma,* from Greek *schistos* and Greek *soma; Ptychozoon,* from Greek *ptychos* and Greek *zoon.*
4. Latinized compound Greek words; e.g., *Spanotecnus,* from Greek *spanios* and Greek *teknon.*
5. Latinized derivatives of Greek words; e.g., *Cnodalum,* from Greek *knodalon,* Latin *cnodalum; Empusa,* from Greek *empousa,* Latin *empusa; Chauliodus,* from Greek *chaulios,* Latin *chauliodous; Ceratium,* from Greek *keros,* Latin *ceratus.*
6. Simple Latin nouns; e.g., *discus* (*Discus*), *canis* (*Canis*), *eciton* (*Eciton*).
7. Compounds formed by combining two Latin roots; e.g., *Magnibucca, Calcaritermes.*

8. Derivative Latin names formed by the use of prefixes or suffixes that change the meaning; e.g., *trans-* + *fuga* to form *Transfuga; inter-* + *aulacus* to form *Interaulacus; clamare* + *or* to form *Clamator; auris* + *cula* to form *Auricula.*

9. Latin nouns combined with inseparable particles such as *ambi-, di-, dis-, in-, por-, re-, se-, simi-, ve-;* e.g., *Diloba, Reduvius.*

10. Latin compounds formed by prefixing a preposition or an adverb; e.g., *Bipes, Subursus, Extracrinus.*

11. Latin nouns formed by combining stems; e.g., *Sodaliscala,* from Latin *sodalis* and Latin *scala.*

12. Latin adjectives or participles; e.g., *Rutilus, Dicax, Solivaga.*

13. Mythological names; e.g., *Venus, Danaus, Diana, Hera.*

14. Proper names used by the ancients; e.g., *Diogenes.*

15. Names of modern persons with an appropriate suffix: *-ius, -ia, -ium* if the personal name ends in a consonant; e.g., *Barbouria, Lamarckia, Selysius; -ia,* if it ends in *a;* e.g., *Danaia;* and *-us, -a, -um* if it ends in any other vowel; e.g., *Fatioa, Milneum.* (It is usually assumed that only the surname is used in such generic names, as above, but the following illustrate some less usual forms: *Hughmilleria,* for Hugh Miller; *Ofcookogona,* for O. F. Cook; *Roystonea,* for Roy Stone, in botany; *Embrikia* and *Embrikstrandia,* for Embrik Strand.

16. Latinized names of persons; e.g., *Erichsonius,* from Erichson, Latin root *Erichsoni-.*

17. Names of places; e.g., *Vanikoro, Hollandia, Papuania.*

18. Names of plants, with Latin ending; e.g., *Staphidia,* from Latin *staphidis; Agallia,* from Greek *agallis.*

19. Names of animals, with Latin ending; e.g., *Polyphemus,* from Greek *polyphemos.*

20. Names of ships; e.g., *Challengeria, Vega, Galathea.*

21. Words taken from languages other than classical; e.g., *Fennecus* (moorish), *Okapia* (mbubian, Congo); *Cortaderia* (Spanish).

22. Words formed as arbitrary combinations of letters, with or without original significance; e.g., *Zirfaea, Velletia, Anzac, Neda, Torix, Rimba, Drugia.*

23. Words formed as anagrams of existing names; e.g., *Bathrolium, Lobrathium, Throbalium,* all from *Lathrobium; Dacelo* from *Alcedo; Linospa* from *Spinola.*

24. Words compounded of roots from two languages; e.g., *Pseudogryllus* (Greek and Latin).

25. Compounds formed of a proper name and a Latin stem, prefix, or suffix; e.g., *Möbiusispongia, Heromorpha.*

26. Previously used names to which a prefix is added; e.g., *pseudo-, meta-, Pinnixa, Pseudopinnixa.*

27. Previously used names to which a suffix is added; e.g., *-ops, -ella; Sturnus, Sturnella.*

28. Previously used names to which is added *-ites, -ytes,* or *-ithes,* if the name is not used exclusively for fossils or if there is clear evidence of intent to establish a distinct genus; e.g., *Gomphites,* from *Gomphus + -ites.*

29. Names (from botany) of supposed plants transferred into the animal kingdom.

30. Synonyms (from botany) of plant names so transferred.

Formation of generic names. A generic name must be or must have been formed as a single word (uninominal)

(1) by adopting a Latin or a latinized Greek word in toto;

(2) by adding a suitable Latin ending to any of the other words listed under Sources, above;

(3) by compounding the words, prefixes, and suffixes listed under Sources, above;

(4) by making an arbitrary combination of letters, such as an anagram or a nonsense word. (These may look like Latin words, or they may by accident have the same spelling as a word in some real language. *Funda* Blackwelder, 1952, was intended to be a meaningless combination of letters but does appear to be the Latin word *funda,* a sling; similarly *Manda* Blackwelder happens to be the same as a Spanish word, with which it actually had no connection.)

Compounding Latin and Greek words. A guide to the method of compounding and the use of connective vowels appears in Appendix D of the 1961 Code (pages 114–117). The nature of the stem and a table of grammatical data about Latin and Greek words also appear in Appendix D (pages 118–141).

Latinizing or transliterating Greek words. A guide to these procedures in the formation of names is given in Appendix B of the 1961 Code (pages 95–101).

Recommendations on the formation of generic names are to be found in the 1961 Code in Article 2 (Recommendation 2A) and in Appendix D.

Original spelling of generic names. Generic names are spelled

(1) before 1961, in any letters of languages using the Latin alphabet, even if modified by diacritic marks;

(2) after 1960, in the letters of the neo-Latin alphabet only (*a, b, c, d, e, f, g, h, i, j, k, l, m, n, o, p, q, r, s, t, u, v, w, x, y, z*), with no diacritic marks;

(3) with a capital initial letter [Article 28];

(4) before 1961, with ligature æ or œ if desired;

(5) after 1960, with the ligatures æ and œ spelled out as ae and oe, but with other ligatures (ff, ffi, ffl, etc.) acceptable [Recommendation E3];

(6) before 1961, with compound names divided by a hyphen or as one word, as *Embrik-Strandia* and *Leidynemella;*

(7) after 1960, with compound words written always as one; without a hyphen [Article 32(c)(i)];

(8) before 1961, with an apostrophe in such patronyms as *D'Orbignyia* and *O'Reillia;*

(9) after 1960, with the apostrophe omitted; as *Oreillia* [Article 32(c)(i)];

(10) before 1961, with any appropriate diacritic mark, accent, diaeresis, or umlaut (e.g., *Öesophagostomum, Hǎdziella, Aëthurus, Guérinius*);

(11) after 1960, without diacritic or other marks (the German umlaut is to be spelled as ae, or ue, but all other marks simply deleted) [Articles 27 and 32(c)(i)].

The 1961 Code says, "No diacritic mark, apostrophe, or diaeresis is to be used in a zoological name. . . ." In quoting names published before 1961, however, any zoologist who does *not* copy the name with all its details will be making an emendation. Whether this is considered a justified emendation or not, it can cause confusion. For example, it would be difficult to discuss in print an earlier misspelling of *mülleri* as *mulleri* if one had to spell the first form *muelleri*. Zoologists would do well to apply this ruling only to names proposed after 1960.

Gender of generic names. Because some specific epithets must agree in gender with the generic names to which they are coupled, it may be necessary to know the gender of each generic name.

The gender may be decided by the original author. If he does not do so, it becomes necessary for later students to determine the gender. The original author can indicate the gender by any of the following means:

1. Direct statement of the gender.
2. Direct statement of the origin of the name in Greek, Latin, or any "modern Indo-European language having genders," fixing the gender as that of the original word.
3. Use of a specific name employing a Latin gender ending on the specific epithet, which implies the gender of the generic name.

If the gender is not stated (1) or indicated (2) or (3), the correct gender to be assigned can be determined as the following:

1. By direct action of the Commission under the Plenary Powers, in fixing a gender to avoid confusion [Article 79].
2. If of Greek or Latin origin, or having an apparently Greek or Latin suffix, it takes
   (a) the gender appropriate to its Latin ending [Article 30(a)],
   (b) the gender which is given in the standard Greek or Latin dictionaries,
   (c) the masculine gender if it is a word of variable gender (masculine or feminine) [Article 30(a)(i)(2)].

3. If the name reproduces exactly a word in a modern Indo-European language having genders, it takes the gender of that word [Article 30(b)(i)].

4. If a name not derived from a classical or modern language did not have its gender explicitly fixed originally, it is to be assumed to be masculine, unless its ending is clearly a natural classical feminine or neuter one, in which case it takes that gender [Article 30(b)].

5. Names based on the names of persons, such as *Cummingella* and *Erichsonius*, take the gender of the Latin ending.

Elaborate tables are given in Appendix D of the 1961 Code for determining the genders of classical words.

Subsequent spellings of generic names. The original spelling of a name is to be retained except in certain specified situations, but both errors and deliberate changes do occur. These sources of later spelling variations include

(1) changes required by the 1961 Code, such as
    (a) deletion of a hyphen,
    (b) writing out of a numeral,
    (c) deletion of any diacritic mark, apostrophe, diaeresis, or non-German umlaut,
    (d) conversion of a German umlaut to *e*;
(2) inadvertent errors, such as
    (a) a lapsus calami (slip-of-the-pen) (e.g., a specialist on ants, to whom the genus *Camponotus* was one of the most familiar, in writing about an ant-associated beetle of the genus *Campoporus*, wrote *Camponotus* by mistake),
    (b) a copyist's error (e.g., the inadvertent copying of *Philonthus* as *Philontus*),
    (c) a printer's error (generally not distinguishable from (b) above) (e.g., in *U.S. Nat. Mus. Bull.* 182, it was discovered in page proof that a genus described as new under the name *Eleusinus* had already been named in a different family. This name appeared five times in the manuscript. All places were marked for correction to *Inopeplus*, the older name. The printer changed only four, so that the name *Eleusinus* appeared in print (in a key) and became a stillborn synonym of *Inopeplus*. (The author, not the printer, must carry the blame for this, regardless of responsibility.)
(3) an emendation, or intentional respelling, made either by the same author at a later date or by a later author (e.g., *Ragochila* Motschulsky, 1869, to replace *Rayacheila* Motschulsky, 1845).

Some of these subsequent spellings are the result of errors or supposed errors in the original spelling, such as incorrect transliteration, incorrect latinization, use of an incorrect connective vowel in a compound, or an evident error of any sort (for example, a name spelled *Wooseveltia* but said to refer to a person named Roosevelt), but only the last is to be treated as an error and corrected, under the 1961 Code.

The rules governing the spelling or correction of names in later works are [Article 32(a)]: The original spelling of a name is to be retained except for the following:

A. Correction of original errors:
    1. Incorrect transliteration is not to be corrected. E.g., *Caprocornis* should have been written *Capricornis* but is not to be corrected if the spelling was intentional.
    2. Incorrect latinization is not to be corrected. E.g.: *Aggiosaurus* from the Greek apparently should have been latinized to *Angiosaurus.*
    3. If a compound name is incorrectly compounded, either
        (a) through incorrect use of a connective vowel (not to be corrected; e.g., *Nigrocorax* would not be changed even though it should have been spelled *Nigricorax*), or
        (b) through writing the name as two separate words (to be closed up into a single word), or
        (c) through writing the name with a hyphen (the hyphen is to be deleted and the parts closed up into a single word) [Articles 26(a) and 32(c)(i)].
    4. If a diacritic mark, apostrophe, or diaeresis was used (the mark is to be deleted) [Articles 27 and 32(c)(i)].
    5. If a German umlaut is used in a name (ä, ö, ü), the letter so marked is to be replaced by ae, oe, or ue [Article 32(c)(i)].
    6. Failure to capitalize the initial letter (to be corrected) [Article 28].
    7. Inadvertent error, for which there is clear evidence in the original publication, such as a *lapsus calami,* a copyist's error, a printer's error, is to be corrected in all subsequent use [Article 32(a)(ii)] except where direct reference is being made to this original spelling; the correction replaces the erroneous spelling in all respects, and the error has no standing in nomenclature, does not enter into homonymy, and cannot be used as a replacement name; the error does have to be cited in a complete synonymy, but with notation of its nature.
B. Choice among multiple original spellings:
    8. When several spellings are used in the original publication, the spelling adopted by the first reviser is to be accepted as the correct original spelling [Article 32(b)]; the other original spellings are treated in the same manner as erroneous subsequent spellings, below [Article 32(c)].

C. Acceptance or rejection of emendations, which are demonstrably intentional changes from the original spelling [Article 33(a)] (if not demonstrably intentional, see D below):

9. A "justified emendation" is the correction of an incorrect original spelling and the name thus emended takes the date and authorship of the original spelling and replaces it in all respects [Article 33(a)(i)]. E.g., see *Wooseveltia* in a preceding item.

10. Any other emendation is an "unjustified emendation" [Article 33(a)(ii)] and is treated below under D.

D. Correction of subsequent errors, either inadvertent errors, changes not demonstrably intentional, or unjustified emendations:

11. An incorrect subsequent spelling that is inadvertent has no standing in nomenclature; it does not enter into homonymy and cannot be used as a replacement name [Article 33(b)]; it must be listed in a complete synonymy, however, with notation of its nature. E.g.:

Genus *Homalota*, 1831
Hmalota, 1874
Homalata, 1890
Homalola, 1939
Homalsta, 1865
Homolata, 1875
Homolota, 1858
Ilomalota, 1871
Stonalota, 1846

12. A change of spelling that is not demonstrably intentional is to be treated as an inadvertent error [Article 33(a)].

13. An intentional spelling change (emendation) that is not acceptable (see above) has standing in nomenclature with its own author and date; it is a junior synonym of the correct name; it enters into homonymy and is available as a replacement name [Article 33(a)(ii)]. E.g., *Homalium* Ljungh, 1804, is an intentional emendation of *Omalium* Gravenhorst, 1802; it is unacceptable but has standing and is available as a replacement name.

E. Spelling changes required by new rules in the 1961 Code:

14. (See list at beginning of this section on Subsequent Spellings of Generic Names.)

Author of generic names. The author of a generic name is

(A) the person or persons who was or were responsible for the name in the first publication in which it is acceptably published
(1) if it was first published as the name for a genus;
(2) if it was first published as the name for a subgenus [Articles 43 and 50(b)];

(3) if, before 1961, it was first published as a synonym of a generic or subgeneric name (not accepted in the 1961 Code even for this period but long accepted by many nomenclaturists before 1961);

(4) if it was first published with a spelling error that must be emended by a subsequent writer [Article 32(c)];

(5) if it has been justifiably emended by a subsequent writer (a justified emendation takes the author of the original name) [Article 33(a)(i) and 50(c)];

(6) if it is itself an unjustified emendation of some earlier name (an unjustified emendation takes its own author) [Articles 33(a)(ii) and 50(d)];

(7) even if, after 1960, the spelling has been changed because of new requirements in the 1961 Code [Article 32(c)(i)], such as

(a) deletion of a hyphen,

(b) writing out a numeral,

(c) deletion of a diacritic mark or apostrophe, or

(d) conversion of a German umlaut to e.

(8) even if other persons appear as principal author of the publication (the author of the name is the person actually responsible for the name as published) [Article 50];

(9) even if, before 1951, the author was unknown (anonymous) [Article 9(7)];

(10) even if the only included species was simultaneously published in the same work by another author (e.g., *Tanyrhinus* Mannerheim (1852) published with the one species *T. singularis* Mäklin (1852) in Mannerheim (1852);

(11) even if in the original publication the name itself is credited to some other person (on a museum label, in an unpublished manuscript, and so on) [Article 50];

(B) some other person than the one responsible for its acceptable publication if it is an acceptable emendation of an earlier name. (It takes the author of the original name.) [Articles 33(a)(i) and 50].

Date of a generic name. The date of a generic name is:

A. The date (day, month, or year, as near as is known) on which it was published in complete compliance with all relevant rules:

1. It must conform to the Publication Requirements listed above [Article 10].

2. If publication of the work is interrupted, a new name can date only from the time of the completion of all requirements [Article 10(a)].

3. If the work bears a publication date, that is the date of any included new names unless there is direct evidence to the contrary [Article 21(a)].

4. If the date of a work is not completely specified, it is to be interpreted as the earliest day demonstrated by evidence, but in the absence of such evidence, as the last day of the stated month or year [Article 21(b)].
5. If the date of publication specified in a work is found to be incorrect, the date of included names is to be interpreted as the earliest date demonstrated by any evidence [Article 21(c)].
6. If the date of publication of the work is not indicated, the names are to be given the earliest date demonstrated by external evidence, such as mention in another work [Article 21(f)].
7. When a subgenus is raised in rank to genus, it retains the original date [Article 43].

B. A date determined by the following rule: When the name of the type genus of a family-group taxon is found to be a junior homonym and is replaced, the replacement name is to assume the date of the name being replaced [Article 39(a)(i)].

Citation of generic names. Names, authors, dates, and bibliographic citations of genera are to be cited in publications as follows:

A. Names
1. Name of genus—to be written in italics, or some type face different from the text [Recommendation E2].
2. A new genus should have a notation such as "n. gen." or "new genus" the first time it is published [Recommendation E7], but this notation should never be copied in subsequent publication by the same or other authors.
3. A generic synonym may be cited in parenthesis after a generic name used alone, but, if the generic name is used in a binomen, the generic synonym must not be placed between the generic name and the epithet [Recommendation 44A].

B. Authors
1. "The name of the author does not form part of the name of a [genus] and its citation is optional" [Article 51(a)], but citation of the author at least once whenever the genus itself is being discussed is recommended.
2. "The original author's name, if cited, follows the [generic] name without any intervening mark of punctuation" [Article 51(b)] and is printed in the type face of the text.
3. If, before 1951, a generic name was published anonymously, a notation to this effect is to be used in place of the author's name (e.g., "anonymous").
4. "If the name of a [genus] was published anonymously, but its author is known, his name, if cited, should be enclosed in square brackets to show the original anonymity" [Recommendation 51A].
5. "If a [generic] name and its validating conditions are the responsibility not of the author(s) of the publication containing them,

but only of one (or some) of the authors, or of some other zoologist, the name of the author(s), if cited, is to be stated as 'B in A' or 'B in A & B,' or whatever combination is appropriate" [Article 51(c)]. It is equally appropriate to cite it simply as "B" or "C," omitting the "A" or "A & B," which is merely an abbreviated bibliographic reference.

6. "A zoologist who cites the name of a genus . . . should cite the name of the author . . . at least once in each publication" [Recommendation E10].

7. Citation of the name of a subsequent user (one referring to *X-us* Smith) should always follow the name of the original author and be in parenthesis, or separated by a comma, thus:

X-us Smith, Black (1963)
X-us Smith (Black, 1963)

Article 51(b)(i) appears at first glance to deny this item (7) in part. It reads: "The name of a subsequent user of a scientific name, if cited, is to be separated from it in some distinctive manner, other than by a comma." It is not clear here whether the original author is also being cited. If he is, then his name serves to separate the two effectively, and the subsequent user's name can be set off by parentheses or comma, as shown above.

C. Date

1. "The date of publication of a name, if cited, follows the name of the author with a comma interposed," [Article 22] if that date is directly shown on the publication, as

X-us Smith, 1866 [Recommendation 22A(1)].

2. If the date of the publication is not directly shown on the publication but can be determined from that volume, the date is enclosed in parentheses, as

X-us Smith (1866) [Recommendation 22A(2)].

3. If the date of the publication can be determined only from outside evidence, it is to be enclosed in brackets, as

X-us Smith [1866] [Recommendation 22A(3)].

These three items are the recommendations of the 1961 Code. They can be applied fairly easily to generic names, but they become unduly cumbersome when applied under actual conditions to the names of species, where other uses of parentheses are prescribed. In any case there is no way to tell whether an author using parentheses is using them under this system or simply as a normal punctuation device. It is thus a waste of effort for any writer to use them except in a large paper where their use is stipulated and applied consistently.

4. The date of any generic name separately cited or discussed should be given at least once in each publication [Recommendation E10].

D. Bibliographic Reference: "A zoologist who cites the name of a genus . . . in a taxonomic work should give at least once a full bibliographic reference to the original publication" [Recommendation E14] and to each subsequent use which was separately cited.

Conflicts between generic names. Several sorts of conflicts can arise between generic names or forms of generic names:

1. Multiple original spellings. (The first reviser may select the spelling that is to be adopted.)
2. Subsequent spelling variations. (These may be either errors or emendations; they may be acceptable changes or not.)
3. Multiple publication of a name as new, in two or more separate works. This is proscribed by Recommendation E22 of the 1961 Code, but when it has occurred in the past the actual dates of each publication must be determined and the subsequent ones treated as later references to the name. E.g.:

> *Colpodota* Mulsant & Rey, 1873, *Opusc. Ent.*, vol. 15, p. 156. (New genus)
> *Colpodota* Mulsant & Rey, 1874, *Ann. Soc. Agric. Lyon,* vol. 6, p. 207. (New genus)
> *Colpodota* Mulsant & Rey, 1874, *Hist. Nat. Col. France,* p. 175. (New genus)

4. Synonymy. There may have been two names proposed for the same genus. (See Synonymy, below.)
5. Homonymy. The same generic name (exact spelling) may have been used
   (a) for two different genera by different authors,
   (b) for two different genera by the same author,
   (c) for the same genus by two different authors,
   (d) for one or two subgenera, other than the nominate ones, in any of these situations. (See Homonymy, below.)
6. Misapplication of names,
   (a) originally, through misidentification of the included species,
   (b) subsequently, by inclusion of other species that are not congeneric with the type. (These are taxonomic rather than nomenclatural problems.)
7. Change of rank from genus to subgenus or vice versa. (See Changing the Category, below.)

Synonymy of generic names. When a given genus has two or more applicable names, all the names are synonyms. Only one of the synonyms can be the correct name for the genus at any one time, and the rest are the rejected synonyms (usually, but not always, also junior, or more recent, synonyms). The problems of generic synonymy are greatly complicated by the existence of subgenera with their names, because these names are sometimes categorically separate from generic names and sometimes treated as generic synonyms.

Two or more names for one genus and its parts can occur by being

(A) published simultaneously in one work
  (1) by separate proposal with different genotype (a) with both treated as names of genera (b) with one treated as a genus and one as a subgenus, or (c) with both treated as subgenera;
  (2) (before 1931) by inclusion of an unpublished name as a synonym of a name being proposed as new [1] (a) for a genus, or (b) for a subgenus;
(B) published in separate works
  (3) for the same type species (a) both as generic names, or (b) one as a subgeneric name;
  (4) as a replacement for the original name for (a) a genus, or (b) a subgenus;
  (5) for different species later believed to be congeneric (a) both as generic names or (b) one as a subgeneric name.

Of these five circumstances, (2), (3), and (4) result in objective synonyms (having the same genotype); (1) and (5) result in subjective synonyms (having genotypes that are believed to be congeneric).

Determination of which synonym is to be employed as the correct name depends on the categorical disposition of the taxon (as genus or as subgenus) as well as upon a prescribed basis for choosing:

1. For a genus, the name is to be chosen from all of the following considered together: (a) all the names given to the genus, whether considered "synonyms" or not, and (b) all the names given to all the subgenera, including all of their "synonyms."
2. For a subgenus, the name is to be chosen from the names that have been applied directly to that subgenus.

The prescribed bases are:

(1) the first reviser, in (A)(1), above;
(2) automatic determination by the author (acting as a sort of immediate first reviser) in accepting one name and rejecting the others, in (A)(2) above;

---

[1] Validation of a name in synonymy has never been expressly accepted by any international rules, but it was widely practiced before 1961. From 1960 to 1964 it was explicitly forbidden by Article 11(a): "A name first published as a synonym is not thereby made available." This rule was deleted in its entirety in the 1964 Code.

(3) homonymy, which forces the rejection of any name which has previously been used for another genus or subgenus of animals;

(4) priority among the remaining names.

Subjective synonymy depends on the opinion of the reviser that the genotypes of the names are congeneric (same as zoological synonymy or conditional synonymy) [Article 61(b)].

Objective synonymy [Article 61(b)] (same as absolute or nomenclatural synonymy) exists when

(1) the names have the same species as genotype, independently fixed as such;

(2) one name is proposed as a replacement for another, thereby automatically having the same genotype as the other, whenever the genotype is first fixed for either one of them [Article 67(i)];

(3) one name is an intentional respelling (emendation) of another [Article 67(i)(ii)].

An intermediate position occurs when the genotypes of two names are believed to be one species. This has been called *subjective-objective synonymy*, because the synonymy is objective so long as the identity of the genotypes is (subjectively) maintained. For example: The type of the genus *Anthobium* is the species then called *A. melanocephalus* Fabricius. The type of the genus *Lathrimaeum* is the species then called *L. atrocephalum* Gyllenhal. These two "species" are believed to be one, and so long as this belief is held, the two genera are the same and the two names are synonyms.

The older of two synonyms is the *senior synonym*, and the younger one is a *junior synonym*. These two terms can be misleading because either one can be the correct name for the genus (if the senior one is also a junior homonym). It is more effective to speak of the correct name and its *rejected synonyms.*

Many subjective synonyms were employed for at least partially different groups of species. They are thus in a zoological sense only partially coextensive, and their names are only partially synonymous. In nomenclature, no special account is taken of this partial synonymy—it is simply included under synonymy in general.

Priority among generic names. Priority in date of acceptable publication could be involved

(1) when a genus has two names (see Synonymy, above);

(2) when there are two spellings of the name (see Subsequent Spellings, above);

(3) when two genera are united, bringing together two or more names for one combined genus (see Synonymy, above) [Article 23(e)(i)];

(4) when a genus is reduced to rank of subgenus (there is no effect on priority);

(5) when a subgenus is elevated to rank of genus (there is no effect on priority);
(6) when a genus is divided into two genera (priority may help to determine the names for the segregate genera);
(7) when a genus is divided into subgenera (priority may help to determine the name of the subgenera);
(8) when two genera receive identical names (see Homonymy, below).

The application of priority is limited by the following rules:

1. It applies only to names acceptably published under the provisions of the code then in force. For example, since *Eunonia* of Varnay (1846) was merely a misspelling of *Eunomia* of earlier authors and therefore not an acceptable name, it is not involved in priority with *Eunonia* Casey (1904).
2. It is not applicable to names that have been suppressed by the Commission [Article 23(a)(ii)].
3. It does not apply to a nomen oblitum unless the Commission so directs [Article 23(b)(ii)]. (A nomen oblitum or forgotten name is one that has remained unused as a senior synonym in the primary literature for more than fifty years.)
4. It does not apply to multiple original spellings (see First Reviser, below).
5. It does not apply to erroneous spellings, which do not qualify as separate names [Articles 32(c) and 33(b)].
6. It is not directly affected by change in rank of the name (to or from subgenus) [Article 23(c)], although it may take part in the choice of names for the resulting taxon.

The first reviser in generic nomenclature. The first person to discover certain situations has the right and something of an obligation to make a decision as the "first reviser." The following are the situations:

1. When, before 1931, a name was first published in the synonymy of another name and thereby made available, that author was acting as a sort of first reviser in choosing between the two names.
2. If two or more different names for a single taxon are published simultaneously (simultaneous synonyms), in the same or different works, the first reviser decides which is to be accepted [Article 24(a)]. (Item 1 is a special case of this.)
3. If two or more identical names are published simultaneously for different taxa (simultaneous homonyms), whether in the same or in different works, the first reviser decides which is to be accepted [Article 24(a)].
4. If a name is spelled in more than one way in the original publication, the spelling adopted by the first reviser is to be accepted as the correct original spelling [Article 32(b)].

Homonymy among generic names. When two genus-group names of identical spelling have been published, they are homonyms. Homonyms can arise because of

(1) failure of the second author to know of the prior usage;
(2) belief of the second author
    (a) that the earlier name is not available;
    (b) that the earlier name is to be emended and will thus no longer be a senior homonym;
    (c) that the earlier name is no longer available because the taxon has been removed from the animal kingdom;
    (d) that a rejected synonym does not preoccupy an acceptable name;
    (e) that a subgeneric name does not preoccupy a generic name;
    (f) that a name is a nomen nudum and therefore not available;
    (g) that presence or absence of an umlaut in names of German origin does not effect homonymy;
    (h) that an otherwise unacceptable name has been placed on the Official List by the Commission (with earlier date);
(3) publication of the names so close together in time that neither author had opportunity to know of the intentions of the other.

Only the last of these is an excusable situation. All the rest reflect inadequate preparation by the second author. It is true, however, that the available nomenclators are not adequate to give assurance that a name or a particular spelling have not been used before. Several classes of names have been omitted entirely from recent nomenclators, including unjustified emendations, supposed nomina nuda, and multiple original publications, all of which may later prove to be relevant. The spelling differences that were permissible under the 1901–1953 Rules are also not shown in any of the nomenclators.

Real generic homonyms are to be dealt with as follows:

1. Any junior homonym must be replaced, providing it is the accepted name for a genus or subgenus (So long as it is merely a junior synonym, there is no need to replace it, as it has in effect already been replaced.) [Articles 53 and 60].
2. If homonyms were simultaneously published, either in the same work or in different works,
    (a) the first reviser will choose which is to be accepted and which rejected [Article 24(a)], except
    (b) if one was proposed for a genus and the other for a subgenus, the generic name takes precedence. [Article 56(c)].
3. If the senior homonym has been removed from the animal kingdom, it continues to preoccupy, and later names of the same spelling must be replaced [Article 2(b)].
4. If one of the homonyms was published as a name of a plant genus and later transferred into the animal kingdom, it takes priority on the basis of its date of publication as the name of a plant.

The following situations *do not involve homonymy:*

1. Two names proposed independently *for the same genus,* whether by the same author or by different authors, are excluded from homonymy by Article 52 of the 1961 Code. This may have been an attempt to prevent assumption that every later reference to a name is a homonym of it, but it would result in a ridiculous distinction in such a case as this:

    Ecitobium Wasmann, 1923; new genus for *E. zikáni* n. sp.
    Ecitobium Wasmann, 1925; new genus for *E. zikáni* n. sp.

    These are separate publications of new genera, which are homonyms as well as synonyms.

2. Misspellings, inadvertent errors [Articles 32(c) and 54].

3. Names that are unavailable under the rules [Article 54(1)], *except* that homonymy *does* apply to names rejected because they apply only to fossils but end in *-ites, -ytes,* or *-ithes* [Articles 20 and 56(b)].

4. Names used outside of the animal kingdom (i.e., for plant genera) [Article 54(2)]. (E.g., *Correa* Fauvel, proposed for a species of insect was improperly rejected because of a prior use of *Correa* in plants.) It is recommended that such names not be used in zoology, but, once introduced, they are not to be rejected.

5. Names given
    (a) to hypothetical concepts (imaginary taxa),
    (b) to genera based solely upon species consisting solely of teratological specimens,
    (c) to genera based solely on hybrids between species, or
    (d) for other than taxonomic use [Article 1].

6. Names differing in a single letter or more [Article 56(a)].

7. Names differing only in possession by one of a German umlaut [Article 32(c)(i)].

8. Although Article 35a of the 1905–1953 Rules (requiring that *specific* names of the same origin and meaning and differing only in *ae, oe,* and *e,* or in *ei, i,* and *y,* or in *c* and *k,* or others specified, are to be considered homonyms) was in Opinion 147 (1943) made applicable also to generic names, these differences do prevent homonymy under the 1961 Code. E.g.: *Oncoparia* Bernhauer (1936) is not a homonym of *Oncopareia* Bosquet (1854) under the 1961 Code, but before 1961 it would have been a homonym if the words were shown to be of the same origin and meaning. (The effect of the 1961 Code ruling on names replaced either before 1943 or before 1961 because they differed from older names only in one of the seven specified ways is not covered in the Code.)

Rejection of generic names. A generic name must be rejected for all purposes if

(1) it fails to meet all of the criteria of availability listed in the section on Publication Requirements, above;

(2) it is an incorrect original spelling (unintentional) [Article 32(c)];

(3) it is an incorrect subsequent spelling (unintentional) [Article 33(b)].

Any name *must be rejected* as the correct name but retained in synonymy if

(4) it is a nomen oblitum (see section on Priority, above);

(5) it is a junior homonym;

(6) it is an unjustified emendation (intentional);

(7) it has been set aside by direct action of the Commission, under the Plenary Powers;

(8) it is the junior of two names brought together by the uniting of two genera;

(9) it is the junior of two names brought together in "the synonymy" of a combined genus;

(10) it is the junior of two names brought together by placement of a subgenus within a genus.

A generic name *cannot be rejected* because of

(1) inappropriateness [Article 18(a)]; or

(2) prior use solely in the Plant Kingdom [Article 54(2)]. E.g., *Smilax* Laporte, 1835, has been repeatedly rejected (first by Nordmann, 1837) because of the prior plant name of the same spelling; the name is to be reinstated and the replacement names sunk as "synonyms.")

Replacement of generic names. A rejected generic name needs to be replaced [Article 60], and the replacement name can be either

(1) a corrected spelling (emendation);

(2) a junior synonym;

(3) a junior subgeneric name or its synonym; or

(4) a new name based on the same genus (with either the same genotype or some other).

Replacement can be accomplished with an existing name, if any, so proposal of a new name should always wait for assurance that no other name is available, for either nomenclatural or zoological reasons.

Misapplied generic names. When a writer quotes an earlier genus by name, he is assumed to be referring to the actual group of species properly included by that name. If there is evidence that he has used it for a different group of species, especially if he quotes an erroneous genotype, his misuse is to be corrected. His use of the name, being different from the original, is a sort of homonym. It has no standing in nomenclature but does need to be recorded, usually by being cited in the synonymy of another generic name. The nature of the pseudo-name must be clearly indicated:

X-us Jones (1928)
    Synonym: Y-us of Adams (1900), not Clark (1850).

The Y-us Clark (1850) has no connection with the X-us Jones. The Y-us of 1900 was merely a misapplication of Clark's name by Adams; it was in effect an error for X-us Jones.

For example, *Cotysops* Tottenham (1939) was published as "n.n." for *Hesperophilus* Thomson (1859) (not Curtis, 1829). Thomson proposed no such name, merely referring to *Hesperophilus* Curtis but assuming an erroneous genotype for it. *Cotysops* is not a new name to replace a homonym but a new genus for some species thought to be erroneously assigned by Thomson. The misapplication (Thomson, 1859) has no status in nomenclature.

Dividing a genus. When the species in a genus are believed to be separated into two groups so distinct that they should each be ranked as a distinct genus (rather than subgenus), there are two possible actions:

(1) elevation of a subgenus (taxon) to genus rank;
(2) creation of a new genus for the segregate group that does not contain the genotype species of the original genus.

If the new group is already a subgenus, the name of the subgenus must be used for that new group. If the new group has not previously been named as a subgenus but one or more of its species is genotype of any of the generic synonyms, the new genus must have a name chosen from among those particular synonyms.

If there is no subgenus, and if there is no genus-group name based on any of its species, then a new genus is erected, a type species is chosen, and a generic name is assigned.

The proposal of a new name for the new taxon indicated above involves the same problems of synonymy and homonymy as the proposal of any new name at any level:

1. It must be determined that there is no available genus-group name based on any of the species included in the new taxon.
2. It must be determined that the proposed new name is acceptably formed.
3. It must be determined that this particular name has never been used before in the animal kingdom for any genus-group taxon (homonymy). The only spelling differences that can exist without eliminating homonymy (or in spite of which two generic names are still considered identical) are the German vowels with the umlaut (*ä, ö, ü*) which are considered to be identical with *ae, oe,* and *ue* respectively.

The basic rule for names when a genus is divided is that its correct name must be retained for one of the restricted genera (or subgenera). If the type had been determined for the original genus, the name must be applied to the segregate that contains that species. If the type has never been fixed for the original genus, a division should not be made until the type has been established.

Combining genera. When two genera are found to be incorrectly separated, not actually distinct at the genus level, they are combined into a single genus (with or without subgenera within it). The name of the combined genus must be the oldest available name from all those that have been correctly applied to either of the genera or any of their subgenera [Article 23(e)(i)], *except* if it

(1) has been ruled unacceptable by the Commission, or
(2) is a junior homonym.

Changing the category. The only category change possible for a genus is reduction or lowering to the level of subgenus. The following rules apply:

1. The former generic name becomes the subgeneric name.
2. The genotype remains the same.
3. There is no change in spelling of the name.
4. Citation follows the rules listed under Subgeneric Names.

Changes of generic names. Generic names may be changed (either respelled or replaced) because of a variety of circumstances. In general these changes are of three sorts: (1) changes necessitated by recognition of zoological facts; (2) changes brought about by operation of the arbitrary procedures agreed upon to stabilize names (such as priority or use of the Plenary Powers); and (3) changes caused by inexpert, ignorant, or malicious actions of persons posing as taxonomists but having more or less selfish purposes other than the increase of knowledge.

The first group consists of changes that are not avoidable at all unless zoologists are willing to give up all increase in knowledge of classification, relationships, distribution, and identity of species.

In the second group the changes of name result from (a) changes in procedures between the time of one publication and that of another (such as adoption of new rules), (b) failure of a worker to follow the procedures in force, or (c) actions of the Commission under the Plenary Powers to set aside the rules.

In the third group are the multitudinous errors of inadequately trained persons attempting to use names, as well as the occasional unjustified actions of taxonomists with personal motives and few scruples.

From the point of view of the man using a name for a given genus, a change in the name by which it should be currently known can be produced by any of the following:

1. Discovery that it was improperly or incorrectly
   (a) formed (see Formation of Generic Names, above),
   (b) published (see Publication, above),
   (c) proposed (anonymously after 1950, or as a synonym after 1960, or conditionally after 1960),
   (d) spelled originally (see Original Spelling, above),
   (e) emended by a later writer (see Subsequent Spellings, above).
2. Discovery that it is a junior synonym (there is an earlier generic (or subgeneric) name for the species—priority)
   (a) because another genus with an older name is found to be zoologically the same, or
   (b) because a subgenus of older name has been transferred from some other genus, or
   (c) because a supposedly unavailable older synonym is found to be available, or
   (d) through correction of dates.
3. Discovery that it is a junior homonym (see Homonymy, above).
4. Discovery that it is a nomen oblitum (see Synonymy, above).
5. Discovery that a proper action by a first reviser has not been followed (see First Reviser, above).
6. Action by the Commission to change the name or the applicable rule.
7. Division of the genus into two or more genera (in which case the generic name will change for some of the species) (see Dividing Genera, above).
8. Combining two genera (the generic name will change for one of the groups of species) (see Combining Genera, above).
9. Changing a supposed genus to the rank of subgenus or synonym (in which case the generic name of all included species will change (see Changing the Category, above).
10. Elevation of a subgenus to the status of genus (whereupon the species in all but the nominate subgenus will have a new generic name).
11. Discovery that the wrong species has been accepted as genotype and that the species in hand is not congeneric with the true genotype (transfer of species to a different genus).
12. Discovery that the original publication was invalid and that therefore a later generic name is the correct one.

The action to be taken on these is suggested in previous sections, as indicated here in parenthesis.

Taxa subordinate to the genus. Only the category of subgenus is dependent upon the genus category. Its taxa (the subgenera) are groups of species formed by subdividing the genus.

Collective groups

Certain biological groups which have been proposed distinctly as collective groups rather than as taxa may be treated for convenience as if they were genera [Article 42(c)]. The 1961 Code defines a collective group as "An assemblage of identifiable species of which the generic positions are uncertain; treated as a genus-group for taxonomic convenience" [Glossary].

The rules which apply to names of collective groups in the 1961 Code are these:

1. "Wherever the terms 'taxon' or 'name' are used in this Code at the level of genus, the provision in question is to apply also to a collective group or its name, unless there is a statement to the contrary, or unless such application would be inappropriate" [Article 42(c)(i)].
2. "The provisions of this Section [requiring definite fixation of genotype for every generic name] do not apply to names of collective groups" [Article 13(b)(i)].
3. "The Law of Homonymy applies to all names in the genus-group, including those of collective groups" [Article 56].
4. "The provisions of this Chapter [XV. Types in the Genus-group] apply equally to the categories genus and subgenus, but not to collective groups, which require no type-species" [Article 66].

The following have been cited in the rules as examples of collective groups:

> *Agamodistomum* (Trematoda)
> *Agamofilaria* (Nematoda)
> *Agamomermis* (Nematoda)
> *Amphistomulum* (Trematoda)
> *Cysticercus* (Cestoda)
> *Diplostomulum* (Trematoda)
> *Glaucothoe* (Crustacea or Mollusca)
> *Sparganum* (Cestoda)

These apparently are all used for larvae that cannot be associated with known species. Other such names are numerous in parasitology. Some are listed in nomenclators (as are those cited above) such as *Cercaria* and *Leucochloridium* (Trematoda) and *Ligula* (Cestoda), but many are not so listed, such as *Dubium* and *Plerocercoides* (Cestoda) and *Parapleurolophocerca*, *Stenostoma*, and *Pigmentata* (Trematoda).

It is not clear how collective groups differ from parataxa at the generic level, except that parataxa in paleontology are not known to apply to larval forms but rather to structural fragments that cannot be associated with whole animals. Parataxa are excluded from nomenclature, but collective groups are included in it.

CHAPTER 24

The Names of Subgenera

The subgenus category

The subgenus is a subordinate category to the genus level. It is the lowest rank in the genus-group categories.

The subgenus taxa

The taxa placed in the subgenus category are the subgenera. These are groups of species formed by subdividing the genus. The problems of subgenera are nearly the same as those of genera, involving their types, their names, and the special problems deriving from the inclusion of subgeneric names between the normal two elements of the double name of the species.

The types of subgenera. The problems of genotypes of subgenera are exactly the same as for genotypes of genera, substituting the word subgenus for genus, except as indicated in the following list:

1. Fixation: In original fixation by designation by the original author (a) for the subgenus itself, or (b) if the subgenus is the nominate subgenus of the genus, for the genus.
2. Fixation: In original fixation by automatic rule, there is nothing corresponding to subgeneric monotypy, although normal monotypy does apply.
3. Fixation: In subsequent fixation by automatic rule, there is nothing corresponding to elimination.
4. Conflicts among genotypes of subgenera can occur in the same ways as for genotypes of genera.
5. Permanence of genotypes of subgenera is assured by the same rules as for genotypes of genera.
6. Citation of the genotype of a subgenus is exactly the same as for the genotype of a genus.

The names of the subgenera

The name of a subgenus is a single word that denotes a group of species of animals (a taxon) given the rank (category) of subgenus. The

497

names of subgenera are frequently governed by the same rules as are listed for generic names, with the exceptions noted here.

Proposal of subgeneric names. They are proposed in the same manner as generic names except that (5) should read: ". . . by a reviser reducing a genus taxon to subgeneric rank."

Publication requirements of subgeneric names. They must meet the same requirements as generic names.

Sources of subgeneric names. They are drawn from the same sources as generic names.

Formation of subgeneric names. They are formed in the same manner as generic names.

Original spelling of subgeneric names. Their spelling is regulated by the same rules as generic names; and *in addition* the nominate subgenus must have the identical spelling of the genus.

Gender of subgeneric names. Gender is determined in the same manner as for generic names, but as long as the name applies to a subgenus, the gender is of no interest in nomenclature.

Subsequent spellings of subgeneric names. The same spelling variations can occur as in generic names, and *in addition*

(4) any acceptable change (emendation) in the name of a genus will automatically make the same change in the name of the nominate subgenus.

The same rules control the use and correction of these spelling variations as for those of generic names.

Author of subgeneric names. The author is determined under the same rules as for generic names.

Date of subgeneric names. The date is determined under the same rules as for generic names, except that (7) would read that a genus lowered to the rank of subgenus retains its original date.

Citation of subgeneric names

A. Names
    1. "The name of a subgenus, when used in combination with a generic name and a [specific epithet], is placed in parentheses between these names . . ." [Article 6].
    2. Name of subgenus—to be written in the same type face as the generic name, or, if standing alone, in italics or some type face different from the text [Recommendation E2].
    3. A new subgenus should have a notation such as "n. subg." or "new subgenus" the first time it is published [Recommendation E7], but this notation should never be copied in subsequent publication by the same or any other authors.
    4. A subgeneric synonym must never be cited in connection with the name of a species. (If the subgenus is *Y-us* with a synonym

Z-us: X-us albus Jones, *X-us* (*Y-us*) *albus* Jones, but not any combination of *Z-us*.)

5. The subgeneric name "is not counted as one of the words in the binominal name of a species or trinominal name of a subspecies" [Article 6].

B. Authors: The same rules apply as are listed under generic names.

C. Date: The same rules and recommendations apply as are listed under generic names.

D. Bibliographic Reference: "A Zoologist who cites the name of a [subgenus] in a taxonomic work should give at least once a full bibliographic reference to the original publication" [Recommendation E14].

Conflicts between subgeneric names. Conflicts involving multiple original spellings (1), subsequent spelling variations (2), multiple publication of a name as new (3), misapplication of names (6), and change of rank (7) are covered by the same rules as listed for genera in Chapter 23, under this same heading. Synonymy and homonymy raise problems that are cited in later sections.

Synonymy of subgeneric names. Synonymy of subgeneric names can occur through the same five circumstances as listed for generic names. The synonymy is either objective or subjective depending on whether the genotypes are the same species or are merely believed to be consubgeneric. The problems are simpler because of the absence of any lower category.

When a subgenus has more than one name (synonyms), determination of which one is to be employed as the correct name is made as follows:

1. In case of simultaneous publication of the synonyms, the first reviser will select one to be the correct name.

2. If, before 1931, a subgeneric name was first published in the synonymy of another name and thereby made available, that author was acting as a sort of first reviser in choosing between the two names.

3. Homonymy involving a subgeneric name will cause its rejection if it is the junior homonym.

4. Priority will govern the remaining choice, it being necessary to use the oldest remaining name.

5. The Commission can set aside any of these requirements, removing a particular name from consideration or fixing the name that must be used.

Other remarks under Synonymy of Generic Names apply also to subgeneric names.

Priority among subgeneric names. The eight statements under generic names are all applicable except that

(1) a subgenus cannot be lowered in rank, as there is no subordinate category below subgenus;

(2) a subgenus cannot be further subdivided, at least so far as nomenclature is concerned, but it can be divided into two subgenera.

The six limiting features listed under generic names also apply to subgeneric names.

The first reviser and subgeneric names. The four situations in which a first reviser can make a choice are the same as for generic names. They include names published in synonymy, simultaneous synonyms, simultaneous homonyms, and multiple original spellings.

Homonymy among subgeneric names. In the complex problems of homonymy, subgeneric names are subject to the same rules as generic names.

Rejection of subgeneric names. The problems of synonymy of subgeneric names are dealt with in the same manner as for those of generic names.

Replacement of subgeneric names. A rejected subgeneric name needs to be replaced [Article 60], and the replacement name can be either

(1) a corrected spelling (emendation), or
(2) a junior synonym, or
(3) a new name based on the same subgenus (with either the same genotype or some other), or
(4) in the case of the nominate subgenus, the new name for the genus [Article 44(b)].

Misapplied subgeneric names. These are dealt with in the same manner as cited for generic names.

Dividing a subgenus. Because there is no category below subgenus (except species), it is not usually possible to subdivide a subgenus. Some authors have used an informal category for taxa called Section or Division; these are acceptable only if they are primary subdivisions of a genus; that is, if there are no subgenera between them and the genus level. In this case the sections are treated like subgenera. (It is, of course, possible to split a subgenus into two subgenera, in the same manner as a genus can be split into two genera.)

Combining subgenera. When two subgenera are found to be incorrectly separated, not actually distinct at the subgeneric level, they are combined into a single subgenus. The name of the combined subgenus will be the oldest available name from all those that have been correctly applied to either of the subgenera [Article 23(e)(i)], except if it

(1) has been ruled unacceptable by the Commission, or
(2) is a junior homonym.

Changing the category. The only change of level possible for a subgenus is elevation to the level of genus. The following rules apply:

1. The former subgeneric name becomes the name of the new genus.
2. The genotype remains the same.

3. There is no change of spelling of the name.
4. Citation follows the rules listed under Citation of Subgeneric Names, above.

Changes of subgeneric names. Such names may be changed for any of the twelve reasons cited for generic names in Chapter 23, under this same heading, except that 2(b) is inapplicable. *In addition* if the name of the genus is changed for any reason, this will also change the name of the nominate subgenus.

Taxa subordinate to the subgenus. Inasmuch as the subgenus is itself subordinate to the genus, it has no subordinate categories of its own. The next higher category is always the genus; the next lower category in nomenclature is the species, although informal levels may be used in between.

The names of sections

"A uninominal name proposed for a primary subdivision of a genus, even if the subdivision is designated by a term such as "section" or "division," has the status in nomenclature of a subgeneric name, provided the name satisfies the relevant provisions of Chapter IV" [Article 42(d)].

The relevant provisions of Chapter IV (Criteria of Availability) appear to be:

1. It must have been acceptably published after 1757.
2. It must be Latin, latinized, or treated as if Latin; a noun in the nominative singular.
3. Before 1931, it must have been published with a description, a definition, or an indication.
4. After 1930, it must have been published (a) with differentiating characters stated, or (b) with a bibliographic reference to such, or (c) as a replacement name.
5. It must have a definitely fixed type-species (genotype).
6. After 1950, it must not be anonymously published.
7. After 1960, it must not be conditionally published.

It must be emphasized that to be primary a subdivision of a genus must be a subgenus or must be used in lieu of subgenera. Sections or divisions cannot be used in a genus already divided into subgenera, as they would then be secondary.

No examples of such "sections" are cited in the 1961 Code, and none are known to the writer. The need for such a provision is not evident, as all such "primary" subdivisions can reasonably be called subgenera. The rule does, however, permit the acceptance of such subdivisions in the older works even if the word "subgenus" was not used.

CHAPTER 25

Genotypy and the Types of Genera

The concept of genotypy

When a genus originally including several species is found to be composite according to current standards, it may be divided into two or more genera. The original name must be applied to one of these. It would have been possible to tie the generic name to the first species listed under it or to some other specifically defined species, but the rules instead adopted the method of tying each generic name to a type species, just as each specific name is anchored to a type specimen. This species is called the genotype, type of the genus, or type-species.

The determination of the genotype of a genus is sometimes a very complex problem, but the use of genotypes in nomenclature is very simple. Wherever the type species is placed in the classification, because of its zoological characteristics, the generic name must follow it. For example, if the type species of generic name A is placed in a genus (zoological group or taxon) that has no other generic name, then the name A must be adopted for that genus. If the type species is placed in a genus that already has a name (and possibly synonyms also), the genus must take the oldest available name in the combined list. The genotype in question may be only one of several species being put into the genus at that time, but it is the one that determines the fate of the generic name. For example, a genus A with species 1, 2, 3, 4 has as its genotype species 2. If it is divided for zoological reasons into two groups including species 1, 3, 4 and species 2, respectively, the name A must go with species 2 (its genotype) even if that is placed in a genus with an older name and even if the other group (1, 3, 4) is left entirely without a name.

This is stated in the 1961 Code thus: "The 'type' affords the standard of reference that determines the application of a scientific name. Nucleus of a taxon and foundation of its name, the type is objective and does not change, whereas the limits of the taxon are subjective and liable to change" [Article 61].

The rule of genotypy is therefore this: Every generic name must have a type species (genotype) to determine its zoological application. The

502

disposition of this type species will determine the application of the generic name, but the status of other names applied to the same zoological genus (and its parts) will determine the fate of the name in practice. For example, genus A has as genotype species 1. This species is placed in another genus by a later worker. The name A must now be applied to the second genus, but whether it is the correct name for that second genus depends upon whether there are prior names available. If the second genus already is named B (with genotype species 2), and if B is older than A, then the genus takes the name B (with its genotype species 2), and A becomes a subjective junior synonym (with its genotype species 1). If B were younger than A, the genus would take the name A (with genotype species 1) and B would be the subjective junior synonym (with its genotype species 2).

This rule applies to all names in zoological nomenclature, whether generic or subgeneric, synonyms or homonyms, original spellings or emendations.

The types of genera

What is type of what? It has never been entirely clear how the rules of nomenclature can deal with taxa rather than just their names. It has been found that names can be regulated only under a single system of hierarchic classification. All codes have been involved in this problem, making statements about taxa and how they are handled, instead of solely about their names.

How generic groups are united, divided, subdivided, or changed in rank are taxonomic problems. Any such action may affect the names. The relation between the taxonomy and the nomenclature has never been clear. The *Règles* did not state whether the type of a genus (taxon) is a species (taxon), although it has generally been assumed that this is the correct choice out of the four possibilities:

(a) Type of a genus is a species.
(b) Type of a genus is the name of a species.
(c) Type of a generic name is a species.
(d) Type of a generic name is the name of a species.

Most writers have believed that (a) is correct, but their actions inevitably look more like (b) in certain cases. G. G. Simpson has explicitly stated (1961, p. 30) that (d) is the only possible choice. The 1961 Code had to deal with this question. Its statements are clear-cut in general, but its usage is mixed.

The 1961 Code sidestepped the issue by stating that "The type . . . of a nominal genus is a nominal species. . . ." A "nominal genus" is defined as "a named genus. . . ." In its section on Types in the Genus-Group, the 1961 Code used "nominal genus" and "nominal species" in most of its formal statements but omits the "nominal" in many places.

If the meaning of "nominal genus" is different from "genus" or "name of a genus," the provisions of the code might be intelligible. The following quotations from the Code, some of which refer to the parallel problem in families and in species, seem to the writer to show that the distinction has not been made:

1. Article 42(b) states that each *taxon* of the genus-group (each genus) is defined by its type-species. (This fits with the usual interpretation of the old Rules, and item (a) above.)

2. Article 39 refers to "the name of its nominal type-genus." (This would seem to show that the "nominal type-genus" is not a name but a taxon; item (a) above.)

3. Article 67(e) says: "If a nominal species, type of a genus, is found to be a . . . synonym. . . ." (There cannot be synonymy among species (taxa) but only among their names. From this quotation we can only conclude that a "nominal species" *must* be a name because it *can* be a synonym; item (b) above.)

4. Article 68(a)(i) states: "the 'originally included species' comprise only those actually cited by name . . . , either as valid names (including subspecies, varieties, and forms), as synonyms, or as stated misidentifications. . . ." (The species available as type *must* be cited by name, and any name cited can be the type-species; so it is here necessary to equate "name" with "species"; item (b) above.)

5. Article 40 refers to the possibility that "a nominal type-genus is rejected as a junior synonym." (Only a name can be a synonym, so this would fit (b) or (d) above, depending upon the nature of the taxon typified, a family in this case.

6. Article 67(e) refers to "a nominal species . . . found to be a junior objective synonym." (Item (b) above.)

7. Article 13(b) states that a genus-group *name* must be accompanied by fixation of a type-species. (This seems to tend toward Simpson's interpretation, and fits item (c) above.)

8. Article 67(i) states: "If a zoologist proposed a new generic name expressly as a replacement for a prior name, both nominal genera must have the same type-species. . . ." (There is here only one genus, but with two names. Yet the code says there are two nominal genera, and we can only conclude that nominal genera are the same as names; item (c) above.)

   The 1961 Code expressly states (Article 61) that the type of a genus cannot be a species that is not *named* in the original publication. This seems to put the name in the place of greatest importance, the only feature that must be present, because there need not be any *separate* taxon represented by that name.

9. In Article 67(i): "If a zoologist proposes a new generic name expressly as a replacement for a prior name, both nominal genera must have the same type-species. . . ." (Here, "generic name" and "nominal genera" are directly equated; item (c) or (d) above.)

10. Article 24(a) refers to "more than one name for a single taxon." (A genus can have only one type, but if there are several names, there *will* be a genotype for each. This seems to prove that a taxon (genus) cannot have a type but each of its names can—Simpson's interpretation and item (d) above.)

11. Article 50 refers first to "the author of a scientific name" and second to "the authorship of a nominal taxon." (It is possible but unreasonable to think of the author of a species; this person is the describer, proposer, namer, and so on. Authorship is properly applied only to the name or to the description, not to proposal of the taxon. This seems to equate "nominal taxon" with "name"; item (d) above.)

No consistent picture emerges from these usages and definitions. It is clear in the code that every name represents a nominal taxon. It is also clear that some of the names may be synonyms and still represent different nominal taxa. It seems to follow that nominal taxa may differ only in having different names. It seems also to be clear that nominal genera may be nothing but names and that their types, nominal species, may be nothing but names. This does not seem to be justification for the use of the undefined term "nominal genus."

It is believed here that Simpson's statement, as in (d), represents the only defensible position so far as nomenclature is concerned. Neither the unstated position often ascribed to the *Règles* nor the useless circumvention of the 1961 Code can be logically sustained. No person can honestly use the term "nominal genus" unless he can use it meaningfully and explicitly. The present writer cannot do so. In view of the history of this concept, however, the following discussions of genotypy will continue to refer to the type of a genus as a "species," it being understood that every genus is represented solely by a generic name and every species by a species name.

Genotypes. The 1961 Code does not anywhere use the word "genotype" except in Recommendation 67A: "Only the term 'type-species' or a strictly equivalent form in another language should be used in referring to the type of a genus. The term 'genotype' should never be used for this purpose." Thus does the code perpetuate the self-abnegating attitude of a few persons. The terminology of taxonomy is not within the competence of the International Commission on Zoological Nomenclature, which could properly decide only which terms it would itself employ. The cause of the ruling was the supposed confusion arising through use of the identical spelling for a different concept in genetics. The use of this word in taxonomy preceded its use in genetics by about a hundred years. If confusion is likely to arise, it is the geneticists who should change, especially as their term is both misspelled and inappropriate.

Several persons have suggested that the etymologically proper form of this word is generitype or generotype. In a sense they are right, and in another sense wrong. From the Latin word *genus,* with genitive *generis,* comes *generitype* (or less likely though possible *generotype*). From the Greek word *genos,* with genitive *geneos,* comes *genotype.* Since a large majority of our technical terms come from the Greek, *genotype* is correct and to be preferred. Since some technical terms come from Latin, *generitype* cannot be said to be wrong.

Already discussed above is the curious fact that the types of genera, under whatever name, are called either species, nominal species, or names, without distinction. There should be a difference, but there is none in practice. In any case the type of a genus cannot be a specimen or a structure or feature of a specimen.

Permanence of genotypes. The 1961 Code states [Article 61] that the type of any taxon does not change. This simple declaration is intended to make clear that zoologists are not at liberty to change genotypes to satisfy their own preferences. However, types of taxa in all categories *do change,* because of operation of compulsory rules.

In genotypy, the 1961 Code seems to be trying not to make a distinction between a genus (zoological group or taxon) and a generic name. There is no possible question that the type of a zoological genus can change; for example, if there are a hundred species in the genus *Hypotheticus,* 1850, in North America and one species in a genus *Australis,* 1840, in South America, and if the type of *Hypotheticus* is *H. niger* and the type of *Australis* is *A. albus,* then when these two "genera" are recognized as being a single genus instead of two, the type of the genus containing the hundred species, known previously as *Hypotheticus,* will be changed to *A. albus* when the name *Australis* replaces *Hypotheticus.* It would be ridiculous to say that it is not the same but another genus because of the minor change in its components. So far as taxonomic use of the type is concerned for comparison of species and genera, *it has been changed.*

The type of a generic name (or a nominal genus or a nomenclatural genus) would not change for such a reason as that cited above, and if a genotype has been correctly fixed, by any method, *it will never change.* But the accepted genotype *can be changed at any time* if further information shows it to be in error. Priority of citation has some relevance, but a presumed first fixation can be set aside for a variety of reasons. These include at least the following:

1. If a determination of genotype is made from a publication later found not to be the original, the determination may be erroneous. E.g.: *Colpodota* Mulsant and Rey, 1874, was published as new in 1874 in a well-known multivolume work. Of the twenty-four included species, eight were in the subgenus *Colpodota* s. str., and the first of these, *C. pygmaea* (Gravenhorst), was subsequently cited as genotype in 1916, 1920, and 1949. Unfortunately, in a relatively obscure paper published in 1873, the same authors described a new species as *Colpodota negligens.* Thus *C. pygmaea* was in reality not included and cannot be type. This is a case in which the genotype was understood

for seventy-five years, only to be changed. Nothing but more new discoveries or exercise of the Plenary Powers by the Commission would prevent or reverse this change.

Again, the name *Osorius* was first used in 1821 without either description or acceptable included species. It was therefore a nomen nudum, an unavailable name. Twice in 1825 and once in 1827 the name was similarly cited without being acceptably published. In 1829, the genus was described but with no species mentioned by name. The genus therefore dates from 1829. The first species included was *O. brasiliensis* in 1830. Although the works of this period have been searched for further references, it is not unlikely that others will eventually be found. Until then, the type is *O. brasiliensis;* if any are found it is quite possible that the type will be changed.

2. If the Commission decides that stability of names can best be served thereby, it can set aside all the relevant facts and *change* a genotype, [Article 61].

3. If it can be shown at any time that an accepted genotype was illegally adopted in the first place, it must be changed by later workers. The error may be on the part of the original author (*lapsus*) or on the part of a subsequent designator.

4. If Article 61(a)(i) of the 1961 Code is followed, it will force the changing of genotypes fixed under the *Règles* and under the other provisions of the 1961 Code. (The rule states that the type of a genus is also the type of its nominate subgenus, if there is one, or vice versa; so that the designation of one implies the fixation of the other. But if different types are simultaneously designated for a genus and its nominate subgenus, the designation for the former takes precedence. This *could* result in a subgenus with a genotype that was not originally included. This would be contrary to the spirit of Article 67(h), and presumably one of the designations would be considered illegal and would be rejected.)

Conflicts among genotypes. Conflicts or difficulties arise because of error, because of changes in the rules, or because of the hierarchic system with its subordinate categories. Most of these are discussed elsewhere.

1. Synonymy of generic names involves conflict of genotypes in a sense, although it is the names that are in competition. The genotype is automatically determined by adoption of one name, assuming all the types have been correctly fixed beforehand.

2. There may be two extant subsequent designations of distinct species. Only one can be acceptable, but it is necessary to check the entire history to be certain of which it should be.

3. An original publication may ostensibly fix two types by two different methods simultaneously. This is resolved by a priority sequence of fixation methods.

4. When there are subordinate taxa (subgenera) the nominate subgenus must have the same name and type as the genus. If it is fixed first for either one, it will thereby be fixed for the other also. If simultaneously fixed for both, using different species, the generic fixation has precedence [Article 61(a)(i)].

This last provision, for simultaneous fixation, appears for the first time in the 1961 Code. If taken literally, this rule would take precedence over all other forms of fixation. It would very seldom be called upon, because the situation must be extremely rare and, in fact, can really occur only if there is an error or lapsus. Nevertheless, the rule is inconsistent with the more general rules listed after it. E.g., genus *X-us* has subgenera *X-us*, *Y-us*, and *Z-us;* the subgenus *X-us* contains only species 1; if the original author specifically designates species 6 as type of the genus *X-us*, this cannot stand because he automatically fixed the type of subgenus *X-us* by monotypy and the genus and nominate subgenus must have the same genotype.

The wording and intent of this rule both seem to be faulty.

5. When a zoologist quotes a supposed fixation by an earlier author, where no acceptable one exists, there can be a question about whether the later author may himself have fixed the type unintentionally. [Article 69(a)(iii), discussed below under Methods of Fixation item 14.]

6. In citing the genotype of a genus, it is possible to be objective and unambiguous. The 1961 Code, however, permits certain practices that make it difficult to be sure of the citation, and in fact introduce the possibility of confusion. (See below, under Citation.)

7. In citing a species by name, an author may be thinking of the name or he may be thinking of some specimens which he associates with that name. In the latter case he may have misidentified the specimens. This gives rise to the question of whether the type is the species named or the species in hand. The 1961 Code sensibly rules that it is always the species named, but it offers to consider cases with the possibility of use of its Plenary Powers to prevent confusion that might be caused by following this rule [Article 70(a)].

Citation of the genotype. The *Règles* and the 1961 Code are quite definite about the necessity of the designation being unambiguous. Although the *Règles* made no reference to how the genotype should be cited, it was assumed that this also had to be unambiguous.

Any species cited as the type of a genus is likely to have been represented by more than one binomen (in addition to any synonyms). For example:

Goerius Westwood, 1827.

Genotype:

Staphylinus olens Müller, 1764.
Goerius olens (Müller), Westwood, 1827.
Ocypus olens (Müller), Curtis, 1829.

It is possible to cite this species under any of these binomina. The 1961 Code specifically states that in this example it should be cited as *Staphylinus olens* with *Ocypus olens* also given (or whatever is the current binomen).

Fortunately this instruction is only a Recommendation (69C). It is certainly commendable to cite both the original and the current combinations, that is to show part of the history of the names. It is, however, only while the species was in *Goerius* that it was eligible as genotype, and it has always seemed to the present writer that this is the most important binomen to cite. The history should be given also, in definitive studies.

In the first seventy-five years of taxonomy, it was very common to cite as authority a recent reviser, rather than the original describer. Thus *Staphylinus olens* was more often ascribed to Fabricius than to Müller. The positive identification of this species requires that we know exactly how Westwood wrote it, as well as its correct original form and all important subsequent ones.

The 1961 Code also permits citation of the type species under any other name that is currently believed to be a senior objective synonym [Article 67(e) and Recommendation 69C]. This means that bibliographic research is necessary before *any* designation can be known to represent the binomen actually used, the original binomen of the species, or some other synonymous binomen. Only confusion can result from this laxity of citation.

It is here contended that the stipulation by the *Règles* and the 1961 Code that a designation must be unambiguous applies not only to the form of the statement but to the form of the name of the type species. To be unambiguous, a designation must be of an included species under the exact name by which it was included. Both previous and later combinations or synonymy should also be shown whenever relevant.

At one time there were in use a series of "type" terms to show the nature of the fixation. The *Règles* used autogenotype and orthotype, without definition, and several others have been used (see Frizzell, 1933). These terms are not well enough known to prevent confusion or delay. They are not as explicit as a statement of the actual situation, such as: "By original designation and monotypy," or "through objective synonymy with *X-us*, of which *X. albus* had already been fixed as genotype."

When a genus has several species but no type is fixed, the species may be called genosyntypes. There seems to be little need for the term, which would probably lead to genolectotype, because it is no clearer than the direct expression "originally included species."

Terms. Several terms used in this chapter, and in discussions of this material in other works, require definition or explanation.

Genotype. Recommendation 67A of the 1961 Code exhorts taxonomists to avoid altogether the use of the term "genotype" in referring to the type of a genus. As explained above, the supposed confusion between this and the homonym in use in genetics is a minor problem at most, and the etymology of the two words does not support the conclusion that the taxonomic term should be abandoned.

There are much more compelling reasons for continuing to use genotype for the type of the genus. First of all, the word has already been in use in the literature of most groups of animals for more than a hundred years. No taxonomist can escape dealing with it if his work brings him into contact with the identity of genera or the application of generic names. Not even fifty years from now will these usages be gone; our historical system will keep them with us as long as species and genera of any group are less than completely known.

Second, although there are several alternative terms for genotype—type-species of the genus, type of the genus, genus-type, generitype—none of these lends itself to discussion of general aspects of the study of the basis of genera. Genotypy is a useful word; it is not easily duplicated with any of the above expressions.

Third, the expression adopted in the 1961 Code, "type-species of the nominal genus" involves the user in the problem of whether the type is a species, a nominal species, or the name of a species, as well as in the parallel problem of whether the thing typified is a genus, a nominal genus, or the name of a genus.

The present writer concludes (1) that there is little reason to consider any change from the word long in use, (2) that the Commission is out of order in dealing with the terms to be used by taxonomists outside of nomenclature, (3) that "genotype" is the most suitable term, being indefinite enough to avoid a major difference of opinion over what is typified and by what, and (4) that a long usage cannot be eradicated in any case and argues strongly for continued use of the word.

Genotype is used throughout this book. It is unreservedly recommended to all persons dealing with the names of genera, in the interest of clarity, uniformity, and utility.

Nominal. The use of this word in the 1961 Code is discussed at length above. It is not defined in the code in relation to genus, even in the glossary, except as "a named genus." However, under the word taxon, a nominal taxon is defined as "The taxon . . . to which any given name whether valid or invalid applies." This definition puts a rather different light on the word, because here it becomes clear that a single genus (group of species) may actually be the nominal genus of several names at the same time. If each of these nominal genera must have a genotype, as required by Article 13(b), then the zoological genus has several genotypes simultaneously.

The use of this word appears to solve none of the questions of genotypy and has proven to be a confusing hybrid between "genus" and "name of the genus." It is not used herein, except in direct quotations from the 1961 Code.

Valid and available. These words occur in the rules for genotypes as well as elsewhere in the code. They are discussed in Chapter 19, Rules of Nomenclature.

Fixation of types. As discussed in an earlier chapter, the terminology used in the 1961 Code for the establishment of types of generic names is not entirely satisfactory. The terms used are defined in the Glossary as follows:

"fixation, n. Used in the Code as a general term for the determination of a type-species, whether by designation (original or subsequent), or indication (*q.v.*)"

"designation, n. The act of an author in fixing, by an express statement, the type of a nominal taxon of the genus- or the species-group."

"indication, n. Published information that . . . in the absence of an original designation (*q.v.*) determines the type-species of a nominal genus."

The implications of the word "indication" seem inappropriate to cover both automatic original fixations and subsequent actions that may be inadvertent. There are actually a variety of problems that are not clearly separated by these terms. First, there are recommendations as to how an author should decide on which species to employ as type. This is best called *selection of genotype*. Second, there are prescribed means by which an original author may record which species he is choosing as type. The term *indication of genotype* would be appropriate for this, but such direct action is usually called *designation of genotype*. Third, there are rules to specify circumstances in which a type is fixed originally even when the author does not indicate it as such directly. These are *automatic original fixations*. In the absence of any original designation or fixation, a subsequent writer may make a selection either by direct intentional action or by inadvertence. Fourth, a subsequent writer may take an action which automatically fixes the genotype, perhaps without realizing this. *Automatic subsequent fixation* is the best term for this. Fifth, the subsequent writer may deliberately select a type, an action also usually cited as *designation of genotype*. Sixth, subsequent writers may refer to any of the five above. This would clearly be *citation of the genotype*.

The appropriate terms can be defined thus:

Selection—the choice of type by either original or subsequent writer

Indication—the recording by any means of the selection made by the original author (see also Designation)

Automatic fixation—determination of the type by operation of rules that are automatic in the particular circumstances, either originally or subsequently

Designation—deliberate recording of a selection, either original or subsequent

Citation—reference to a previous fixation or designation

Fixation—when used alone, the general term that refers to all acts that result in a genotype being determined

None of the above terms necessarily carries any implication that the fixation is an acceptable attempt under the rules or that it is the fixation that is to be accepted.

Genotype determination

One of the most detailed and complex sets of rules about any subject in zoological nomenclature governs the determination of genotypes. Even so, the rules fail to answer numerous questions that arise, and in fact leave unstated almost all the underlying basic ideas. These underlying concepts are of the utmost importance and will be discussed below.

There are four terms that are indispensable to a discussion of the fundamentals of genotypy, in addition to the word genotype itself, which is defined above. The general term for the legal establishment of the correct genotype is *fixation*. This fixation of genotype may be accomplished by various means, including designation, automatic fixation, and fixation by special rules. *Designation* is fixation or selection by direct statement, as "I designate the species 1 as type of the genus A" or "Genus A, genotype = species 1." The genotype is automatically fixed by *monotypy* when the genus originally includes a single species. It is automatically fixed by *objective synonymy* when the name is published as nomenclaturally equal to another name, as a new name for it or as a stillborn synonym.[1] (The term "objective" is equivalent to "absolute" and implies that the synonymy is irrevocable and not subject to opinion. The opposite is subjective synonymy, which depends on the judgment of the taxonomist.)

Summary of genotype fixation. The following actions or occurrences may lead to fixation of a genotype:

Original fixation

A. By designation
    (1) if the original author clearly designates the type of the genus, or
    (2) if the original author clearly designates the type of the nominate (typical) subgenus.

B. By indication, if the original author
    (3) before 1931, uses the expression "gen. n., sp. n." (or its exact equivalent) for only one of the included species,

---

[1] The term stillborn signifies that a name was a synonym at the time of its validation. It was first published as a synonym and was in that sense "stillborn." However, such a name *can* be used under certain circumstances, so it is not actually "stillborn." Validation in synonymy is not recognized by the 1961 Code.

(4) uses the name *typicus* or *typus* for one of the included new species,

(5) cites one of the included species or its parts under an epithet identical in spelling to that of the new generic name (whether the trivial name is the accepted name for the species, for a subspecies, or a synonym of either) (tautonymy),

(6) before 1931, included in the synonymy of one of the included species a pre-Linnaean epithet identical with the new generic name (Linnaean tautonymy),

(7) before 1961, clearly sets out a system by which the genotype was to be indicated in each case (special system); e.g., by statement that in that work the first species listed is always the genotype.

C. By automatic rules, if in the original publication

(8) there was included only one species (monotypy),

(9) there were included several species but all but one were shown to be doubtfully included and therefore not available as genotype (virtual monotypy),

(10) before 1961, there were included several species but only one was placed in the nominate (typical) subgenus (subgeneric monotypy), or

(11) the type is fixed for any objective synonym of the genus in question.

Subsequent fixation

D. By automatic rule, if

(12) the first included group consists of only one species (subsequent monotypy),

(13) before 1953, all but one of the originally included species are simultaneously made the types of new genera or subgenera (elimination), or

(14) a genotype is fixed by any method for a subsequent replacement name or other objective synonym (objective synonymy).

E. By subsequent designation, if

(15) a subsequent author clearly designates an included species,

(16) after 1960, a subsequent author chooses a non-included species which he simultaneously synonymizes with one of the included species,

(17) the Commission selects a type.

F. By subsequent indication, if

(18) before 1931, a subsequent author explicitly uses a special system to indicate the types; by

(a) tabulation of genera and genotypes,

(b) use of a special type of description for the genotype species only,

(c) illustration of the genotypes only,

(d) use of a special type face for the names of the genotypes only, or

(e) consistently treating the genotype first; or

(19) accepting a supposed prior designation that is not actually acceptable.

Among these devices, the following are acceptable for any names published during the following periods:

I. During all periods after 1757: 1, 2, 4, 5, 8, 9, 11, 12, 14, 15, 17, 19.

II. Between 1757 and 1931: (A) plus 3, 6, 18.

III. Between 1930 and 1954: (A) plus 13.

IV. Between 1953 and 1961: (A) plus 7, 10.

V. After 1960: (I) plus 16.

Methods of fixation. It is difficult to arrange the methods of fixation in order of importance; yet this is essential because there are cases in which two different species are indicated as genotype by two different methods. One must obviously take precedence over the other. The 1961 Code has two lists in order of precedence (for original and subsequent fixation), but these lists [Articles 68 and 69] together are incomplete, and they are also demonstrably in error. The following appears to be a more satisfactory arrangement:

A. Fixation under the Plenary Powers

    1. Suspension of the Rules [Article 79]

B. Automatic fixation

    2. Objective synonymy [Article 67]

        (a) Isogenotypy

        (b) Objectivity

    3. Monotypy [Article 68]

        (a) Subspecies, varieties, synonyms

        (b) Includes species not named

        (c) Includes hidden species

        (d) Since 1930

        (e) Virtual monotypy

        (f) Synonymy of all original species

        (g) Original name must be available

        (h) Combined description rule [Article 68(a)]

    4. Subgeneric monotypy

    5. Subsequent monotypy (of a genus without originally included species) [Article 69(a)(ii)(2)]

C. Deliberate original fixation

6. By direct statement of designation [Article 68(a)]
7. By use of *typicus* or *typus* as a new specific name [Article 68(b)]
8. By absolute tautonymy of a new specific name [Article 68(d)]
9. By Linnaean tautonymy [Article 68(d)(i)]
10. N.g., n. sp. rule [Article 68(a)(i)]

D. Deliberate subsequent designation

11. Unambiguous designation by direct statement [Article 69(a)]
12. By special system
13. By elimination
14. By designation or acceptance of a designation of a species not originally included but having as synonym the name of one of the originally included species (being designation of the latter) [Article 69(a)(iv)].
15. By acceptance of some supposed prior fixation (but not by mere reference to it) [Article 69(a)(iii)]

These methods are discussed below.

A. *Fixation under the Plenary Powers.* Transcending all the rules of all the codes since 1913, is the power given to the Commission to suspend any rules in a specific case. This power applies to fixation of genotypes.

1. *Suspension of the rules.* At the 9th International Congress of Zoology, at Monaco in 1913, the International Commission on Zoological Nomenclature was granted special Plenary Powers "to suspend the Rules as applied to any given case, where in its judgment the strict application of the Rules will clearly result in greater confusion than uniformity" provided that certain technicalities be complied with. This power has been used many times to legalize generic names that would otherwise have been rejected and to fix as their genotypes species that could not otherwise have been justified.

B. *Automatic fixation.* There are at least five ways in which a genotype can be fixed in certain circumstances by the operation of automatic rules, without any direct designation by the original or subsequent author. These may operate at the time of publication of the genus or at a later time when some other action initiates them.

2. *Objective synonymy.* Two forms of objective synonymy occur, depending only on the sequence of events. The second produces automatic genotype fixation.

(a) Isogenotypy. Two names which have the same species as genotype are objective synonyms. They must always apply to the same genus. They may also be called absolute synonyms or nomenclatural synonyms or they may be said to be isogenotypic.

(b) Objectivity. Conversely, two names which are objective synonyms (such as a junior homonym and the new name proposed to replace it) automatically will have the same genotype, whether it has been fixed or not. This is in every theoretical aspect similar to (a), differing only in the approach. If the genotypes of two names are fixed, and it is then found that they are the same, the two names are isogenotypic synonyms (objective synonyms). If two names are automatically synonymous, they must have the same genotype and are also objective synonyms. A useful distinction can thus be made between isogenotypy and objectivity, even though they are both phases of objective synonymy. Example: *X-us* F., 1792, has as genotype *X-us albus* (L.). *Y-us* Payk., 1800, has as genotype *Y-us albus* (L.). Since the genotypes are the same, *X-us* and *Y-us* are isogenotypic synonyms. Example: *A-us* F., 1792, (not L., 1758) is renamed *B-us*. These two names are objective synonyms, and therefore they must have the same species as genotype. The species will be determined by the first fixation for either name, but it must have been originally included under the older generic name [Article 67(i)].

The only cases of objective synonymy are (1) replacement names for junior homonyms, (2) emendations (intentional), whether justified or unjustified, (3) erroneous (and illegal) replacement of names for reasons not acceptable under the code, and (4) stillborn synonyms (those validated in synonymy of a generic name before 1961). The *Règles* provided for the situation in (1), (2), and (3), but left (4) to the conscience of taxonomists.

The only case of objective synonymy in the one original publication is (4), the publication of names as synonyms of accepted names, stillborn synonyms, validated as being names for the identical genus as the accepted name. Under the *Règles* it was never entirely clear whether such names were thereby published acceptably, but many taxonomists accepted them and used them for replacement names when needed. The 1961 Code clearly makes this impossible after 1960: "A name first published as a synonym is not thereby made available" [Article 11(d)]. In the 1964 Code, Article 11(d) is modified to permit the continued use of such names as were previously accepted. It reads: "A name first published as a synonym is not thereby made available unless prior to 1961 it has been treated as an available name with its original date and authorship, and either adopted as the name of a taxon or used as a senior homonym."

Cases (1), (2), and (3) are all replacement names. The two names are absolute synonyms. If the replaced name has a genotype, the replacement name has the same genotype automatically and originally by objective synonymy. If the replaced name does not have a genotype, it automatically gets one when the replacement name does, or, if neither has one originally, either gets one when the other does, by objective synonymy. These are all three covered by both codes in this manner, for all dates.

This fixation by objective synonymy may be original, as when one of the names already has had its genotype fixed and this same species automatically becomes genotype of the other. It may also be subsequent, as when neither has a genotype until some later author fixes a type for one.

This one automatically becomes the type of the other as well. In this one act, therefore, the later worker has fixed the genotype of one genus by subsequent designation and the genotype of the other genus by the automatic operation of the objective synonymy rule.

3. *Monotypy.* If a new genus is proposed for a single species, that species is automatically the genotype, and the genus is said to be *monobasic.* (The term *monotypic* is sometimes used in this sense, but it is inappropriate and should be avoided.)

This is the simplest of all type fixations and it would seem at first glance that the concept of monotypy—a genus with only one original species, would be easy to apply. Quite the contrary is true, however, for there are two basic points on which nomenclaturists have held widely different views, leading to different conclusions. And there are many individual problems.

The 1903–1951 *Règles* did not directly state that the type of a genus is a species. However, this seems to have been implicit in the rules dealing with the subject. This interpretation was taken by some to mean that only a species (as understood by the original author) included by name can be the genotype. Other concepts that might be included, such as subspecies or synonyms, have no bearing since they were not "species" to the original author. This is thought to be the logical conclusion of the principle of accepting what he said he had rather than requiring detailed subsequent study to determine what he actually did have.

By others it is believed that throughout the *Règles* the word "species" was intended to include subspecies. Support is claimed for this view in the passage in the old Article 11 that "Specific and subspecific names are subject to the same rules and recommendations, and from a nomenclatural standpoint they are co-ordinate, that is, they are of the same value." From this it is held that any name which is included under the genus by the original author is a nomenclatural species and is available as genotype.

In the first of these views, a genus published with one named species which contains two named subspecies is nevertheless monobasic, since the author put only one species into it. That species is therefore the genotype by monotypy. In the second view, the genus would have two "species" available for later genotype selection.

The 1961 Code employs the ambiguous expression "nominal species" ("a named species") in this connection and throughout the code. It includes such statements as, "If a nominal species . . . is found to be a junior objective synonym. . . ." The term synonym can only be applied to a word or name, not to a taxon, so it is by no means clear whether the species (a taxon) or its name is the type of a genus (a taxon) or its name. This ambiguity is not clearly dispelled, but occasional references omitting the word nominal and the exclusion of other names appended to the "nominal species" tend to establish the position that it is not names but taxa which are types and which are typified. (See discussion above.)

The 1961 Code specifically prevents consideration of anything other than "the named species" as genotype: "A genus originally established with a single nominal species takes that species as its type, . . . regardless of cited synonyms, subspecies, unavailable names . . . (etc.)" [Article 68(6)].

The other point involved in this problem which was interpreted in opposite ways is the question of what is "nomenclatural" in the sense of the old Article 11 (quoted above) and what is not. Persons holding the second view described above (all names available) contend that there is nothing but nomenclature involved in the species with two subspecies cited above—that the question of whether there is one "species" or two, for purposes of genotype fixation, is purely nomenclatural. The opposite view is that although it is largely a nomenclatural question, it does contain one zoological factor (the use of two zoological categories) and is therefore no longer entirely nomenclatural. In this view, the old Article 11 is therefore no longer applicable, and only one "species" is present.

The 1961 Code does not contribute directly to the discussion of when a problem is strictly nomenclatural. It does, however, make clear, as quoted above, that the presence of subspecies within the one species does not prevent the genus being monobasic.

These two views, that monotypy requires that there be only one name or merely only one species taxon, are irreconcilable. Some writers have accepted one view and some the other. It is necessary to choose one, and the present writer has believed it proper to follow the original dictum that the type of a genus is a species, which, if taken literally, in the case of monotypy, leaves no room for discussion of the names by which either of these is known. Although the 1961 Code is ambiguous through its use of the meaningless "nominal species," it is clearly on the side of monotypy regardless of synonyms, subspecies, and so on.

The following paragraphs (a–f) are based on the premise that monotypy is not affected by the presence of subgenera, synonyms, unnamed species, unavailable names, or species that are doubtfully included or identified. Before 1961 this was only one of the possible interpretations; under the 1961 Code it is unavoidable.

A genus originally published with more than one species included may nevertheless be monobasic if only one of them was named. For example, *Leucopaederus* Casey, 1905, was founded on *L. ustus* (LeConte), but Casey added, "A few species of *Leucopaederus* occur also in Mexico, one of which has been described by Dr. Sharp." The Mexican species were not mentioned by name, although they can be identified in Sharp's work. They are not available as genotype.

A similar case involved *Onthostygnus* Sharp, 1884, published with two new species, *O. fasciatus* and *O. pallens*. The genus is nevertheless monobasic, because *O. pallens* was on the page following the genus and *O. fas-*

ciatus, which page can be shown to have been actually published at a later date.

a. *Subspecies, varieties, synonyms.* If the single species has named subspecies or varieties, or if it has synonyms that are listed, these have no effect on the monotypy. Only a single species was included from the point of view of the original author, and it is the type under the name by which it was accepted.

Article 11 of the 1901–1953 *Règles* stated: "Specific and subspecific names are subject to the same rules and recommendations, and from a nomenclatural standpoint they are coordinate, that is, they are of the same value." This has been interpreted by some taxonomists as meaning that a named subspecies is of equal rank with a named species and prevents the genus from being monobasic. However, Article 11 restricted its own application to nomenclatural considerations. As long as the specific and subspecific names are being treated merely as names, for validation, orthography, priority, and so on, they are coordinate. When it is stated that one is to apply to a species and the other to a subspecies, a zoological factor has been introduced that removes the problem from the realm of Article 11. Since a species and a subspecies cannot be said to be coordinate, their names cannot either so long as their zoological rank is involved. Article 11 does not say or mean that species and subspecies are coordinate, and it is therefore impossible for the names of species and names of subspecies to be coordinate, except for certain strictly nomenclatural considerations.

Article 46 of the 1961 Code, however, obscures this distinction by stating that, "The categories in the species-group are of co-ordinate status in nomenclature. . . ." Categories cannot possibly be coordinate, as they are defined only by being higher or lower than one another in the hierarchy. They are only indirectly involved in nomenclature, through the taxa. This article, however, clearly applies only to the names of the taxa, in spite of the wording, and says that the names of species and of subspecies are subject to the same rules regardless of the level at which each is placed. It does not say that they are alike in zoological respects.

In the *Règles,* there is ground for argument over whether the subspecies are available as genotypes. The 1961 Code removes all doubt on this point in Article 68(c), where it is categorically stated that a genus is monotypic if it originally contained a single nominal species "regardless of cited synonyms, subspecies," and so on. (However, in subsequent designation, subspecies, varieties, forms, and synonyms are all available for designation as genotypes [Article 69(a)(i)]).

b. *Includes species not named.* In works published before 1931, if a single species was included but not named, it became the genotype if it could later be identified. The genus was monobasic, but the type was made known subsequently. If several species were included but not named, the type must subsequently be designated from among that group of species [Opinion 46]. For example, the new genus *Coproporus* was proposed by Kraatz in 1857 for "the family I of *Tachinus* of Erichson." The eighteen species in this group, readily identified in Erichson's 1840 mono- graph, are the included species from which the genotype must be selected. After 1930, a new genus could not be established without a genotype fixation.

In the 1961 Code a sweeping abolition of this rule was presented with- out reference to the former interpretation or decisions made under it. The new rule, referring to subsequent designation from "the originally in- cluded nominal species," states that these "comprise only those actually cited by name. . . ." This new rule can have no application except before 1931, so it is obviously retroactive. It will force the change of some geno- types and some generic names. There seems to be no purpose served by this new provision.

c. *Includes hidden species.* A new genus published with a single new species is assumed to be monobasic. However, species described or listed in another part of the work (simultaneously published) would also be available and would prevent the genus being monobasic in fact. (This emphasizes the need for careful scrutiny of *all parts* of all relevant works before making decisions on genotypes.

d. Since 1930, when a genus cannot be properly published without "designation" of a genotype, monotypy is accepted as a form of designa- tion. (This use of the word designation in the *Règles* is unfortunate, since fixation would have been more appropriate. Designation is best applied only to selection of a genotype by direct statement.)

Therefore, if an author states that a new genus contains a single species under a particular name, no other names that were then or at any later time applied to the species or any of its parts is of any concern in deter- mining the genotype. If only one named *species* was included, from the point of view of the original author, only that *species* is available as geno- type, and the genus is monobasic. The same arguments apply to originally included synonyms (specific or subspecific) and names of any other rank below species.

e. *Virtual monotypy.* Some genera published with several included spe- cies are nevertheless actually monobasic. Example: The genus *Calophaena* Lynch, 1884, was published with three species. Careful examination of remarks under the genus and the species reveals that two of the species were directly stated to be likely not to belong to the same genus as the other. In effect there was one included species and two doubtfully included species.

Article 30(II)(e) of the *Règles* stated that "species which the author of the genus doubtfully referred to it . . . are excluded from considera- tion in determining the type." Article 67(h) of the 1961 Code states that, "A nominal species . . . that was cited as a species inquirenda (a doubt- fully identified species) or a species incertae sedis (of unknown taxonomic position) . . ." is not available for designation. Therefore only the definitely included one is available as genotype, so the genus is *virtually monobasic*.

f. *Synonymy of all original species*. If all the originally included species are found by the reviser to be synonyms, merely a single species in reality, this subjective synonymy does not make the genus monobasic. All the original species are still available for selection. Neither does the action of the reviser fix the genotype (see method 13, below). The "inclusion of two or more species" means not zoological species in the view of later workers but named species in the original work—species in the belief of the original author as shown by his giving them separate specific names.

g. *Original name must be available*. The species included must be represented by a nomenclaturally available name. Example: *Osorius*, a catalog name, was printed by Dejean in 1821 with one species *O. tardus* Dejean. This is not the genotype because *tardus* was a nomen nudum, even if *tardus* can later be identified with a valid species under the same or another name. The genus, of course, was not valid in 1821 either, but if it had been therein validated by description, it would have been without originally included fixed genotype. If *tardus* had been identified and properly published by the first reviser, it would be credited to the reviser and not to Dejean, 1821, as would *Osorius* itself.

h. *Combined description*. Opinion 43, "on the status of genera the type species of which are cited without additional description," stated that when a description is given for the genus "the characters given for [the genus] cover the genus and type species, and the generic and specific name are published in the sense of the Code." For example, *Telleogmus* Foerster, 1856, with description: genotype *T. orbitalis* Foerster, 1856, merely listed without any descriptive material. The Commission ruled that both the genus and the species were included in the generic description and thereby validated. This is, of course, merely a special case of monotypy.

4. *Subgeneric monotypy*. The genotypes of subgeneric names are fixed and determined in exactly the same manner as those of generic names, from the species originally included in the subgenus or the first group included in it. A question arises here of the status of a genus originally proposed with three species, two of which are originally placed in new subgenera. The genus has three original species, yet the typical subgenus has only one. The typical subgenus must have the same genotype as the genus, and since only one species is available in the subgenus, it must be the type of both. This might be termed subgeneric monotypy. It is not covered by the provisions of any of the codes. For example, *Sableta* Casey, 1910, was published for a group of twelve new species. The species were distributed in four subgenera, of which *Sableta* s. str., the nominate subgenus contained only one species. The genus must have this species as genotype.

5. *Subsequent monotypy*. If a genus is published without included species, there can be no genotype until one or more species has been placed in the genus. If a single species only is placed in the genus, it

thereby automatically becomes the genotype. It is the only species available and has sometimes been called a monotype. However, since this fixation is quite different from the original monotypy described above, it is best to further identify this as subsequent monotypy. This was first stated in the 1961 Code [Article 69(a)(ii)(2)] but has been recognized as automatic under all codes. For example: The genus *Stenus* was published by Latreille in 1796 without mention of species. In 1800 a species was placed in the genus by name by Paykull. This is the only species available as genotype, unless it is found that one or more other species were placed in the genus at an earlier date.

A curious case involves *Borboropora* Kraatz, 1862, which was described by Kraatz. A single species, *B. kraatzi*, was described and named, but the species was directly credited to Fuss (as new). In a sense the genus is monobasic, but it can also be interpreted as a case of subsequent monotypy, with Fuss acting as the first subsequent author to include a species by name.

C. Deliberate original fixation. There were five methods in the *Règles* for "designation" of a genotype by the original author, and there are six such methods in the 1961 Code. All but one of these are special cases which amount to designation only because of specific provisions in the rules. (Not all of them were actually deliberate, but they are treated as if so.)

6. *Originally, by direct statement.* In proposing a name for a supposedly new genus or subgenus, an author has the privilege (and since 1930 the duty) of designating a genotype from among the species he includes in the genus. If none of the previous forms of fixation apply, and if the author has not made an error in his statements, the designation must be accepted. Example: *X-us* Roe, 1880, with species 1 and 2. Roe directly states, "species 2 is the genotype." This is accepted designation. Example: Smith in 1940 finds genus *A-us* is preoccupied and renames it *B-us*. He specifically states that the genotype of *B-us* is *B-us albus*, which was one of three species originally included in *A-us*. However, he failed to note a valid prior fixation of one of the other species as genotype of *A-us* (*A-us niger*). The species *A-us niger* is also type of *B-us*, and Smith's designation is invalid. Example: Jones in 1945 described a new genus *D-us* with three species 1, 2, and 3. He specifically designates a genotype, calling it species 4. It is probable that he changed the name of 4 to 1, 2, or 3, forgetting to change it in the designation. His designation is not valid, and the genotype still is undetermined. The genus is technically invalid also, because of the 1930 genotype rule.

There is no basic difference between the old *Règles* and the 1961 Code [Article 67(b)] in the matter of original designation. The inflexible statement in the 1961 Code, however, makes it highly ambiguous. "If one nominal species is definitely designated as the type-species of a new nominal genus when the latter is established, that species is the type-species, regardless of any other consideration" [Article 68(a)]. For this to be literally true we must (1) assume that the phrase "one nominal species" excludes a named species that is later found to be unacceptable (see example of *D-us,* above), (2) forget the possibility of two types of fixation occurring simultaneously, with an author designating a type for a genus that has one automatically (for example, in explicitly proposing *X-us* as a replacement for the junior homonym *Y-us,* with species 1, 2, and 3, Brown also states that species 2 is the genotype of *X-us,* forgetting the objective synonymy of *X-us* with *Y-us* which forces them to have the same genotype, and that the type of *Y-us* had earlier been fixed as species 1), and (3) forget the possibility that a genus ostensibly published with three species could be monobasic on the only species not doubtfully included although through lapsus one of the others was designated. Article 67(h) of the 1961 Code specifically states the obvious point that "a nominal species that was not included or that was cited as a species inquirendum or a species incertae sedis when a new nominal genus was established, cannot be validly designated or indicated as the type-species of that genus." However, this statement proves that it is a possibility, and Article 68(a) does not clearly deal with it. The expression "definitely designated" should read "acceptably designated," if there are to be no "other considerations."

If this statement in the 1961 Code is taken literally, and if the precedence of fixation methods in Article 68 is followed, it will no longer be possible to say that genotypes are permanent and cannot be changed, at least where homonymy is concerned, as each subsequent proposer of a new name (and there frequently are several) can change the genotype.

It seems to be inescapable to conclude that there *are* other considerations that interfere with this direct statement in the 1961 Code. It seems to the writer that it is also necessary to recognize that all automatic fixations take precedence over all designations (except those by the Commission).

There seems to be no reason for insisting on a categorical statement of designation. All codes have required that the designation be unambiguous. The following have been accepted as such: *Deroderus* Sharp, 1886, with four species: "In the typical species, *D. vestitus. . . ."* *Dinolinus* Casey, 1906, with two species: "This genus is founded upon the large and brilliant blue-green polished species described by Erichson under the name *Xantholinus chalybeus."*

Difficulties can arise through various circumstances. The case of *Colpodota* (cited under Permanence of genotypes, above) shows the necessity of working from the very first publication of the name, not just the work usually accepted as first. Such cases of multiple publication are not infrequent.

Occasionally, through some sort of *lapsus,* an author in citing a genotype will use a name not otherwise mentioned or identifiable. This does not constitute designation. For example, *Oecidiophilus* Silvestri, 1946, with one species (*Oe. mimellus*), has a direct designation of *Oe. oglobinii.* The latter is a nomen nudum, and the genus is monobasic on *Oe. mimellus.*

If, in citing a genotype, the original author credits the name to someone other than the describer of the species, the genotype is the species with its original author. For example, *Laverna* Gistel, 1829, with genotype specified as "*Laverna dilatata* Ill." (Illiger, apparently only in manuscript), which should read *L. dilatata* (Fabricius, 1792). The 1961 Code states this principle for genera whose author is misquoted but does not deal with the problem at the level of genotypes.

Specifically excluded from being accepted as designation of genotype are the following:

1. Mention of a species as an example of a genus [Article 67(c)(i)].
2. Reference to a particular structure as "type" or "typical" of a genus [Article 67(c)(ii)].
3. Any ambiguous or qualified designation [Article 67(c)].
4. Any attempt that involves an error or lapsus that makes the designated species unrecognizable.

Special systems have been employed in citing the genotypes in the original publications. For example:

1.   *Dalotia* **Casey, 1910, p. 106.**
    Genotype: *D. pectorina* (Casey), by implied designation. (On page 90 of the same paper, under the name *Noverota,* Casey writes, "The first species may be regarded as the type, as in all cases where the type is not specifically named." This is definite enough to be accepted in this volume of Casey's work.)
2. By a table of genera and their genotypes (see method 12, below).
3. By a prearranged special type face; e.g., bold face for the name of the type species only (see method 12, below).
4. By illustration of the genotypes only (see method 12, below).
5. By a special type of description for the genotype only (see method 12, below).

Other special systems, accepted in the 1961 Code, are described in 7, 8, 9, and 10, below.

7. *Typicus or typus rule.* Both the *Règles* and the 1961 Code provide for the "designation" of genotype through the device of naming one species *X. typicus* or *X. typus.* The 1961 version is: "If, when a new nominal genus is established, one of the included new species is named *typicus* or *typus,* that species is the type-species" [Article 68(b)].

Note that this applies only if the species so named is a new one and only if there is no other type fixed by direct designation or other means.

8. *Absolute tautonymy.* In both the *Règles* and the 1961 Code identity of generic and trivial names was accepted as "designation" of genotype.

In the 1961 Code it is worded thus: "If a newly established nominal genus contains among its originally included nominal species one possessing the generic name as its specific or subspecific name, either as the valid name or as a cited synonym, that nominal species is ipso facto the type-species (type by absolute tautonymy)" [Article 68(d)].

The species or its names do not have to be new at the time the genus is erected. The species becomes the type, not the tautonymous name. For example:

> *X-us* Smith, 1868 (new genus)
> > Species 1, *X. albus* Smith, 1868 (new species)
> > Species 2, *X. pallidus* (Mann, 1842)
> > > *X. x-us* (Evers, 1850) (synonym)

If no other species is fixed as genotype, the type becomes *X-us pallidus* (Mann), by tautonymy.

This provision opens the door to repeated change of genotype, at least to change of what is cited as genotype, because the *name* of the type species is always subject to change. Furthermore, this provision in the 1961 Code makes it possible for a subspecies to be a type-species of a genus. For example:

> *Y-us* Bell, 1906 (new genus)
> > Species 1, *Y-us novus* Bell, 1906 (new species)
> > Species 2, *Y-us albus* Bell, 1906 (new species)
> > > Subspecies 1, *Y-us albus* s. str.
> > > Subspecies 2, *Y-us albus y-us* Bell, 1906 (new subspecies)

Article 68(d) would hold that the genotype is *Y-us albus* because the name of one of its parts is a tautonym of the generic name. It seems unlikely that taxonomists will be willing to maintain such a fiction, although they might be willing to accept a designation system which consisted of deliberately giving a tautonymous name to one of the subspecies of the intended species. The code probably intended this latter interpretation, but it fails to list it as a "designation" system.

An example involving a synonym rather than a subspecies is:

> *Macropterum* Gistel, 1834 (new genus)
> > Species 1, *Macropterum rufipes* Gistel (new name)
> > > *M. macropterum* (Gravenhorst) (synonym)
> > Species 2, ———

Gistel gave Species 1 a new name because he believed tautonymy to be objectionable. His reason does not really concern us. The 1961 Code would hold that *M. rufipes* is the type, by tautonymy of a synonym. The 1961 Code would also reject *M. rufipes*, however, and replace it with *M. macropterum*. The type would then be *M. macropterum,* an objective synonym of the name under which the type was fixed.

9. *Linnaean tautonymy.* In Opinion 16, the Commission ruled that a species which had among its synonyms a name that was identical with the generic name but was pre-Linnaean in date was the genotype of that genus by tautonymy. This provision was eliminated for genera published after 1930 by the changes in Article 30 made at Budapest in 1927. The 1961 Code accepts this device as fixation for genera published before 1931, as follows: "If, in the synonymy of only one species originally included in a nominal genus established before 1931, there is cited a pre-1758 name of one word identical with the new generic name, that nominal species is construed to be the type-species (type by Linnean tautonymy)" [Article 68(d)(i)].

10. *N.g., n. sp. rule.* Opinion 7 stated, "The expression 'n.g., n. sp.,' used in publication of a new genus for which no other species is otherwise designated as genotype, is to be accepted as designation under Article 30a."

Although this Opinion makes no mention of any of the numerous other forms of this expression which are possible, it is not reasonable to restrict its application to cases appearing *exactly* as stated. For example, if *X-us albus* n.g., n. sp. is acceptable designation, then *X-us albus* n. gen., n. sp. would be equally acceptable. Other forms which seem to be exactly comparable are: *X-us albus* gen. et sp. nov.; *X-us* n.g., *albus* n. sp.; *X-us* (gen. nov.) *albus* sp. nov.; and so forth. A reasonable extension of the principle would cover the following case of a subgeneric name. *X-us* (*Y-us*) *albus* subgen. et sp. nov. or *X-us* (*Y-us* n. subg.) *albus* n. sp. In all these a designation is made that is comparable so that of Opinion 7.

This opinion has been assumed to have been canceled for genera published after 1930 by the new rule requiring a "definite unambiguous designation of the type species." The 1961 Code accepts the broader interpretation of this method of fixation given above by citing it as "The formula 'gen. n., sp. n.,' or its exact equivalent . . . ," but it definitely limits its application to genera published before 1931 [Article 68(a)(i)].

D. *Deliberate subsequent designation.* Several methods are possible for fixation of the genotype in subsequent publications. Three have already been discussed; they are fixation under the Plenary Powers by the International Commission and automatic fixation by either subsequent monotypy or objective synonymy. Others are cited below.

Before any of these subsequent procedures are appropriate, it must be determined (1) whether the genotype has been fixed in the original publication [Article 68] and (2) which species are the originally included ones [Article 69]. "Only the statements or other actions of the original author when establishing the name are relevant in these decisions [Article 67(f)].

11. *Subsequently, by direct designation.* After 1930, no genus could be proposed acceptably without fixation of a genotype. For genera established before 1931 but without a genotype fixation, subsequent writers can and should fix a type according to the rules of Article 69 of the 1961 Code. However, before the 1961 Code, both before and after 1930/31, there

were interpretations of the rules followed faithfully which are not mentioned in the 1961 Code. These are cited below as if they were rules, because they constitute actions which cannot reasonably be reversed because of the 1961 rewording of the rules.

The basic rule was worded in the *Règles* thus: "If an author, in publishing a genus with more than one valid species, fails to designate . . . or to indicate . . . its type, any subsequent author may select the type, and such designation is not subject to change." In the 1961 Code this is worded as: "If an author established a nominal genus but did not designate or indicate its type-species, any zoologist may subsequently designate as the type-species one of the originally included nominal species, or, if there were no original nominal species, one of those first subsequently referred to the genus. . . ." [Article 69(a)].

In spite of the fact that some writers have apparently believed that it is impossible to "select the type" or to "designate" under the *Règles* and the 1961 Code without using the words "type" or "genotype," there are several ways of fixing the genotype in subsequent publication.

First is direct designation as such: Example: Jones in 1910, under the genus *X-us* Smith, 1840, states: "Genotype = *X-us laevis* Smith, 1840." If this was one of the originally included species and there is no prior fixation, *X-us laevis* Smith is the genotype by subsequent designation.

Other methods include

1. special systems: (a) tabulation, (b) special type of description, (c) illustrations, (d) special position, or (e) a special type face;
2. before 1961, one particular form of elimination;
3. acceptance of a supposed prior designation.

These are discussed below, after discussion of some of the problems affecting all forms of subsequent designation.

Problems of subsequent designation. A variety of problems appear in designating or judging the designations of genotypes in subsequent works. Among them are the following:

The included species. The following rules in the 1961 Code determine what species are available for subsequent designation:

1. "Only the statements or other actions of the original author when establishing a new nominal genus are relevant in deciding which species are the originally included species. . . ." [Article 67(f)].
2. "A nominal species that was not included, or that was cited as a species inquirenda or a species incertae sedis when a new nominal genus was established, cannot be validly designated or indicated as the type-species of that genus" [Article 67(h)].
3. ". . . the 'originally included species' comprise only those actually cited by name in the newly established nominal genus, either as valid names (including subspecies, varieties, and forms), as synonyms, or as stated misidentifications of previously established species" [Article 69(a)(i)].

4. "If no nominal species were included at the time the genus was established, the nominal species-group taxa that were first subsequently and expressly referred to it are to be treated as the only originally included species" [Article 69(a)(ii)].

5. "Mere reference to a publication containing the names of species does not by itself constitute the inclusion of species in a nominal genus" [Article 69(a)(ii)(1)].

The eligible species. The following rules determine the eligibility of species for subsequent designation as genotype:

1. "If two or more nominal species were simultaneously referred to a nominal genus, all are equally eligible for subsequent type-designation" [Article 69(a)(ii)(3)].

2. "A nominal species is not rendered ineligible for designation as a type-species by reason of being the type-species of another genus" [Article 69(a)(v)].

3. If all the originally included species are now considered to be but one species, all are still available for designation. (Not in 1961 Code.)

4. Removal of some species from the genus before subsequent designation does not remove them from eligibility to be designated as genotype. (General case of Elimination, but see also Subgeneric Elimination, both below.) (Not in 1961 Code.)

The genus. The following rules apply to the genus and its name:

1. "If, in designating the type-species for a nominal genus, an author refers the generic name to an author or date other than those denoting the first establishment of the genus or the first express reference of nominal species to it, he is nevertheless to be considered, if the species was eligible, to have designated the type-species correctly" [Article 67(g)].

2. If an author quotes an erroneous spelling or unjustified emendation of the generic name when designating a type, he is to be considered to have designated the type for all such spellings. (Not in 1961 Code.)

Fixation. The following rules apply to problems of fixing the genotype:

1. "If an author designates . . . as type-species a nominal species that was not originally included, and if, but only if, at the same time he synonymizes that species with one of the originally included species, his act constitutes designation of the latter as type-species of the genus" [Article 69(a)(iv)].

2. "If a zoologist proposes a new generic name expressly as a replacement for a prior name, . . . the type-species must be a species eligible for fixation as the type of the earlier nominal genus (for which the fixation is subsequent)" [Article 67(i)]. E.g.: *Brundinia* Tottenham, 1949, was proposed to replace *Metaxya* Mulsant and Rey, 1873, which was

preoccupied. Tottenham designated the type of both as *Metaxya me-ridionalis* (M. and R.), but this was not originally included in *Metaxya* and is therefore not available as genotype of the genus under either name.

3. Designation of more than one species as "types" is not acceptable. E.g.: Lucas in 1920 cited six species as "typ." (his usual expression) of *Bolitobius* Leach. The real genotype was not among them. (Not in 1961 Code.)

4. In the case of Westwood's Synopsis . . . (1840), Opinion 71 ruled that the notation "Typical Species" for one species in each genus did constitute acceptable designation. (Not in 1961 Code.)

5. A designation can be unambiguous without use of the word type. E.g.: The type of *Philorinum* Kraatz, 1858, was fixed by Jacquelin du Val in 1859 as follows: "*M. Kraatz . . . a basé . . . son genre Philorinum . . . sur l'A. humile Er.*" (Not in 1961 Code.)

Publication problems. The following rule in the 1961 Code refers to publication of a designation: "A subsequent designation first made in a literature-recording publication is acceptable, if valid in all other respects" [Article 69(a)(vi)]. (This possibility was admitted also in the *Règles*, but Opinion 172 stated that such publication is undesirable.

Misidentified genotypes. In recent years there has been much discussion of the problem of misidentified genotypes—cases in which an author stated that species 1 is the genotype but is afterwards believed to have misidentified species 1 and to have been actually dealing with species 2. The 1961 Code has now clearly ruled that "It is to be assumed that an author correctly identifies the nominal species that he either (1) refers to a new genus when he establishes it, or (2) designates as the type-species of a new or of an established genus" [Article 70]. The Code further rules that when it appears that there has been a misidentification, the case should be submitted to the Commission for ruling. (This is dependent upon the assumption that the "misidentification" was not deliberate; if it was deliberate, see discussion of Article 70 below.)

The writer believes that Article 30 of the *Règles*, as interpreted by Opinion 14, took care of all such cases. The genotype is the species *named*, not some other species that may have been in the author's mind or is now in his collection.

In connection with this last item, it may be pointed out that examination of a man's collection years later has often been used as the basis for a claim that he misidentified the genotype species. This is a most unsatisfactory practice, not justified under the rules, and leading only to confusion. Designations or citations should be based entirely on the literature. No other method can produce sound nomenclatural results in this field. The zoological identity of the various genotype species is another problem entirely.

Under this same heading Article 70 of the 1961 Code disposes of another rare possibility, that of an author who deliberately cites a genotype *in the sense of a subsequent user of that specific name*. The rule is:

"If the type designated for a new nominal genus is a previously established species, but the designator states that he employs its specific name in accordance with the wrong usage of a previous author, the type-species is to be interpreted as the one actually before the designator, not the one that correctly bears the name. (i) In such a case, the author of the new nominal genus is considered to have established also a new nominal species, with the same specific name as the misidentified species, in the new nominal genus. Example.—If Jones, 1900, designates as type-species of *C-us*, gen. n., a species that he cites in some such manner as *A-us b-us* Dupont sensu Schmidt, 1870, the type-species of *C-us* is that which was before Jones, not that named by Dupont, and its name is to be cited as *C-us b-us* Jones, 1900."

12. *Designation by special systems.* The manner of presentation of genotype designation is not prescribed in the 1961 Code except for the statement that it must be unambiguous. Use of a definite system, such as tabulation of the genotypes, use of a special type of description for the genotypes only, illustrations of the genotypes only, or always treating the genotype first, is neither accepted nor rejected by the 1961 Code. Where these systems are explicitly employed for this purpose, and where they are unambiguous, it seems to be proper to accept them, even though it might be thought to be inadvisable to employ them in new studies at this time.

Many writers have designated genotypes for older names without specifically stating their intention *in each case*. This is done by use of a general introductory statement which explains the method employed for indicating the genotypes. For example, in 1810 in the *Considerations Generales* . . . , Latreille included a list of the genera under the following heading: *"Table des genres avec l'indication de l'espèce qui leur sert de type."* Under each name is cited one species (occasionally more than one). In Opinion 11, the Commission declared this list to be acceptable as designation, provided the other requirements are met in each case.

A not uncommon method of indicating (and therefore sometimes designating) genotypes is the use of (1) a prearranged special type face, (2) a special series of illustrations, (3) a special position among species, or (4) a special type of description given to one species only in each genus. To give an example of (1): In 1839 in the *Elements of British Entomology*, Shuckard wrote in a footnote under the first genus, "The type, when British, will be indicated by its being printed in small capitals in the list of species. . . ." By this means he has indicated the types of most of the British genera, without making a specific statement about each.

An example of (2): In 1849 a group of eleven "disciples" of Cuvier issued a new edition of his *Le Règne Animal*. The title page bore the following statement: *"Edition accompagnée de planches gravées, representant les types de tous les genres. . . ."* This has been accepted as designation, although the Commission has never ruled upon it.

An example of (3) is provided in 1910 in volume 1 of the *Memoirs on the Coleoptera*, in which Casey on page 90 under a new genus states, "The first species may be regarded as the type, as in all cases where the type is not specifically named." This would seem to apply to all names in this volume.

The only example of (4) known to me is that of Fabricius in 1792 to 1805. This system (described in detail by Malaise and by Blackwelder) consisted in giving a special description of the mouthparts for one species in each genus. This one species was thereby set apart as the anchor of the genus, the representative of the generic structure—in short, as the genotype. Although this system is not universally accepted as designation, it appears consistent with the principles outlined above.

Other examples of most of these types of designation might be given, along with a few apparently similar ones which do not meet minimum requirements of the rules. An example of the latter is Curtis, 1837, *A Guide to an Arrangement of British Insects* . . . (second edition), in which certain names are proposed for sections of large genera. It is stated that the first species listed after such names is always a "typical species." Since it is always a British species and usually not an originally included one, it is best to consider this as less than unambiguous type selection.

13. *Elimination*. When a genus is founded upon a group of species with no genotype fixed, subsequent workers may remove some of these species from the genus before a fixation is established. Removal of these species does not make them unavailable for subsequent designation. It was once thought that if all but one were removed, this was tantamount to fixation of that remaining one as genotype. This is a logical contention, but it has never been accepted by the rules. This general sort of fixation by elimination is not possible, as all the original species remain available to be selected.

For example, in 1861 Fairmaire and Germain erected a new genus *Oedodactylus* with two species, *Oe. castaneipennis* and *Oe. fusco-brunneus*. In 1869 Fauvel transferred *Oe. castaneipennis* to *Lathrobium*, leaving a single species in *Oedodactylus*. This is not acceptable as genotype fixation. (In this case, as in most similar ones, the species eventually fixed was the one which Fauvel left in the genus, but it did not have to be.)

A second form of elimination is involved in subsequent synonymizing of all the original species. In 1874 the subgenus *Homoeochara* was proposed with three species. Some later writers considered these three to be conspecific, but this did not fix the genotype.

A special case of elimination was outlined in Opinion 6: "When a later author divides the genus A, species Ab and Ac, leaving genus A, only species Ab, and genus C, monotypic with species Cc, the second author is to be construed as having fixed the type of genus A." This special case is not in conformity with the principles of genotype designation employed in most of the rest of the *Règles* and Opinions. It is not to be extended in the logical manner to general cases of elimination (see Article 30.III.k of the *Règles*), although apparently it can reasonably be extended to cases in which more than two species were originally included and *all but one* are simultaneously made types of monobasic new genera by a subsequent author (as suggested in Opinion 154).

This ruling is not directly mentioned in the 1961 Code. However, Article 69(a)(v) states: "A nominal species is not rendered ineligible for designation as a type-species by reason of being the type-species of another genus." If two or more species remain eligible, no genotype has been fixed. Therefore, the 1961 Code removes the possibility of this kind of fixation, but it does not refer to the problem of what to do with fixation made and followed under the explicit earlier rulings of the Commission.

Subgeneric elimination. Article 61(a) of the 1961 Code states: "The type of a taxon is also the type of its nominate subordinate taxon, if there is one, and vice versa. Therefore, the designation of one implies the designation of the other." From this it would appear that only a species available in the nominate subgenus could be designated for the genus (either originally or subsequently).

If in a genus, originally published with several species but no genotype fixation, all but one are subsequently placed in new subgenera, that one would appear to be the only one available for the nominate subgenus and therefore for the genus. It seems that the rejection of elimination as a method of type fixation, in the 1961 Code, is broad enough to deny this appearance. All the original species are still available for selection. (This case seems to be similar to Subgeneric Monotypy in the original publication, but it appears to be clearly rejected by the Code whereas the other is not.)

14. *By designation of a synonym.* "If an author designates (or accepts another's designation) as type-species a nominal species that was not originally included, and if, but only if, at the same time he synonymized that species with one of the originally included species, his act constitutes designation of the latter as type-species of the genus" [Article 69(a)(iv)].

This ruling is in line with the 1961 Code's recommendation that the genotype be cited in its original combination and its present one, but not necessarily in the combination *by virtue of which* it is a type!

The best way to be rigorous in citation of a genotype is to cite it first in the binomen using the generic name of which it is the type, and only second by any other binomina that may be relevant parts of the nomenclatural history. This is just one more rare case to complicate unnecessarily the fixation rules.

15. *Acceptance of a supposed prior designation.* It is, of course, a common occurrence for a writer to quote an earlier worker's attempt at geno-

type fixation. The later writer may accept or reject the earlier citation or he may give no clue to whether he accepts it or rejects it. He may say, "Genotype = X-*us albus* because of designation by Smith 1910," or he may say, "In 1910 Smith stated that the genotype is X-*us albus*." Since it has sometimes happened that the later writer has misquoted the earlier one and no such citation was made, it is necessary to decide whether this quotation by the later author will itself be accepted as type fixation.

It has been claimed that any statement about a prior genotype designation itself constitutes a designation. This leads to several absurdities. If a writer lists all the attempts at fixation by earlier workers and rejects all but one of them, it cannot reasonably be held that he is citing all the various names as genotypes. Again, a legally unacceptable attempt at fixation, such as the use of the word "example" instead of "type," cannot be legalized by the mere quotation of it. And if a writer should quote a previous citation and demonstrate that it is unacceptable, he would nevertheless under this view have himself repeated the designation while at the same time proving that it is unacceptable.

It is therefore concluded that it is necessary to distinguish between acceptance and rejection of the earlier citation by the later writer. If the later writer accepts the citation, he will be credited with fixation if the earlier writer did not in fact make one. But if the later writer rejects the citation or fails to accept it, he does not thereby make a new citation of that same species. For example, if a writer says, "The genotype is X-*us albus* because of designation by Jones in 1842," the fact still remains that the later writer states that "the genotype is X-*us albus*," and this is therefore acceptable as an attempt at designation. On the other hand, if the later writer had said, "The designation of X-*us albus* by Jones in 1842 is not acceptable," he would not thereby be making a designation.

This implies that it is necessary to judge in each case whether the later writer accepted the earlier citation or not. Although this may appear to be a difficult thing to determine, no case has yet come to hand that presented this difficulty. It is usually easy to determine whether the later writer makes a definite statement about the type (perhaps with erroneous reasons) or merely quotes someone else.

The 1961 Code confirms this earlier interpretation in these words: "In the absence of a prior valid type-designation for a nominal genus, an author is considered to have designated one of the originally included nominal species as type-species, if he states that it is the type (or type-species), for whatever reason, right or wrong, and if it is clear that he himself accepts it as the type-species" [Article 69(a)(iii)].

Recommendations of the code

In Recommendation 69A and 69B the 1961 Code lists more than a dozen specific things which it says should be taken into consideration in designating genotypes. This list is said to be in order of precedence. All of these recommendations could lead to fixation of the most useful genotype, but none of them will do so regardless of other circumstances. It is well for zoologists to consider these and other factors, but the repetition of the qualification "other things being equal" makes the order of precedence meaningless, because they never are equal.

The most important of all recommendations is omitted, that no one should attempt to deal with genotypes, and thus with generic synonymy, without a thorough preparation in the relevant rules, in their history and interpretation, and in the use of literature.

CHAPTER 26

The Names of Species

The species category

The species category is the level in the hierarchy at which are placed the taxa known as species. It is the highest level in the species-group categories, in the usage of the 1961 Code. Subordinate to it there is now only one category, that of subspecies, although in the past there have been other levels such as variety, form, and so on.

The species taxa

The species taxa are of course the species. However species are defined, they are taxonomically the groups placed at the level (in the category) called the species category. In general these must be groups of actual individual animals, but various types of fossils are also accepted even though all parts of the animal are now destroyed. These may be replacements, molds, casts, or cores, which yield direct evidence of the nature of the original animal.

Also classed as fossils are trails, tracks, footprints, burrows, borings, tubes, castings, and coprolites. These may be the only evidence of the former existence of an animal and may be distinctive of the extinct species. They have generally been treated as if they were animals, receiving specific names. They are usually classed as "the work of an animal." (Of course, such objects also exist at the present time, but it is customary to leave them undescribed until the animal itself is discovered. Additional "work" of Recent animals include galls, nests, shelters, cases, webs, and honeycombs.)

The 1961 Code (Glossary) defines "work of an animal" as "Result of the activity of an animal, but not a part of the animal, e.g., tracks, galls, worm-tubes, borings," which Article 16(a) accepts in lieu of actual specimens of the animal or parts of it. But the Glossary expressly states that this expression "does not apply to such fossil evidences as internal moulds, external impressions, and replacements." It is unlikely that paleontologists will accept this distinction, and it is certain that many fossils now accepted into taxonomy would require a broader definition. Such "evidence of

535

former existence" was generally believed to be acceptable under earlier rules.

The names of the species taxa

The following provisions apply to the binominal name of the species, herein called the specific name. (The rules applying only to the second word of the name—the specific epithet—are listed in a later chapter.) Most of these apply also to the names of subspecies, as cited in the next chapter.

Proposal of names of species. A name may be proposed for a species through

(1) transfer of the named species into the animal kingdom [Article 2(a)];
(2) direct proposal in a publication
    (a) by a specified author,
    (b) by a specified group of authors,
    (c) by a speaker reported in the minutes of a meeting, or
    (d) before 1951, by an anonymous writer or writers;
(3) a reviser raising a subspecific taxon to species rank;
(4) deliberate designation (as genotype) of a later mis-use of a species name, which action is held by Article 70(b)(i) of the 1961 Code to amount to establishment of a second name identical with the first.

Publication requirements of names of species. There are five general requirements for acceptability:

1. The name must be published. It must be made available to other scientists by being printed or duplicated in approved method and distributed with the intent to make the contents public. At one time this was interpreted by the Commission to mean that the name must be included in a document printed or mechanically reproduced in such a manner that every copy be identical with every other copy *and* that the document be intended for record and general consultation rather than for use by special persons or for a limited time only.

    In 1950 it was added that the document must be reproduced on paper and with ink of a quality and durability such as to offer a reasonable prospect of permanency *and* that some copies must be made available for sale or free of charge to any institution or individual who might desire it.

    All of the rules that have been made on this subject were designed to maintain useful standards of publication and distribution, or to prevent abuses by careless, ignorant, or selfish persons. They are all highly subjective and, in the final analysis, amount to little more than recommendations. There is still no way to determine whether a borderline publication is or is not acceptable except by obtaining a ruling from the Commission.

2. The name must not be based on a fictitious species. Names based on non-existent concepts (hypothetical or mythical animals) are not accepted into zoological nomenclature. For example, a name facetiously proposed for a species said to exist on the planet Mars does not enter into nomenclature. When specimens of animals are obtained from Mars, their names will no longer represent hypothetical species and so will be acceptable.

A name for the unicorn would not be acceptable unless the existence of that mythical animal were demonstrated.

It is possible, however, to establish a name through facetious action and inadvertence. In a famous case, a beetle collector found a specimen of an unknown species and referred to it as "this curious little beetle, this *Ignotus aenigmaticus*. . . ." It was later found that the species was already described, but the names *Ignotus* and *I. aenigmaticus* were published and have been accepted (as synonyms).

3. The name must be given to a taxon based on actual specimens, either whole animals, part of an animal, a fossil of an animal (whether replaced, a cast, a mold, an impression, and so on), or the work of an animal (gall, tube, boring, footprint, and so on).

4. The taxon named must be one of the species-group taxa, species or subspecies, although before 1961 it was often assumed that varieties, races, forms, and so on, could be named also. (Some ant specialists still use variety as an infrasubspecific category.)

5. The names must be binominal, consisting of a generic name and a second word to denote the species.

The following are the detailed requirements of the codes. Any name, to be accepted into nomenclature, must meet the following requirements. It must have been

(1) a binomen [Article 11(g)(ii)] of which the second part (specific epithet) must have been a simple word of more than one letter or a compound word [Article 11(g)(i)], and must have been or been treated as

(a) an adjective in the nominative singular agreeing in gender with the generic part of the binomen,

(b) a noun in the nominative singular in apposition to the generic part of the binomen,

(c) a noun in the genitive case,

(d) an adjective used as a substantive in the genitive case, derived from either part of the name of an organism with which the animal in question is associated,

(e) a participle or a gerund (a verb form used as an adjective or as a noun, respectively) [Article 11(g)];

(2) published after 1757 [Article 11(a)];

(3) published in an acceptable publication [Article 8]:
    (a) reproduced in ink on paper by some method that assures numerous identical copies,
    (b) issued for the purpose of scientific, public, permanent record,
    (c) obtainable by purchase or free distribution, and
    (d) not reproduced or distributed solely by a forbidden method, such as:
       (i) microcards, microfilm, and so on,
       (ii) proof sheets,
       (iii) separate photographs even with accompanying explanations,
       (iv) reading at scientific or other meeting,
       (v) labeling of a museum specimen, or
       (vi) mere deposit of a document in a library;
(4) originally proposed as a name for a species or a subspecies, or, before 1961, for an infrasubspecific taxon [Articles 15 and 17(9)];
(5) written in Latin characters, the neo-Latin alphabet [Article 11(b)(i)];
(6) a Latin word or a latinized word or so constructed that it can be treated as a Latin word [Article 11(b)];
(7) published in a work that has not been ruled unacceptable by the Commission under the Plenary Powers;
(8) before 1931, either in a binominal work or in a work that contains no binomina but shows no evidence that the author was not in principle binominal;
(9) before 1931, in either the text or the index to a non-binominal work, as in (8) above [Article 11(c)(ii)];
(10) before 1931, accompanied by at least one of the following [Article 12]:
    (a) A description, a statement that purports to give characters differentiating the taxon;
    (b) a definition, a summary of characters;
    (c) an indication, consisting of any of the following [Article 16(a)]:
       (i) a bibliographic reference to a previously published description, definition, or figure;
       (ii) the inclusion of a name in an index to a work, provided the stipulations of Article 11(c)(ii) of the 1961 Code are satisfied;
       (iii) the substitution of a new name for a previously established name;
       (iv) a single combined description of the genus and one included species;
       (v) an illustration;
       (vi) the description of the "work" of the animal (tubes, galls, tracks, feces, nests, and so on);

(11) after 1930, in a work in which the principles of binominal nomenclature were applied [Article 11(c)];

(12) after 1930, in the text of a work rather than only in the index
[Article 11(c)];

(13) after 1930, accompanied by one of the following [Article 13(a)]:
    (a) a statement that purports to give characters differentiating the
taxon,
    (b) a definite bibliographic reference to such a statement,
    (c) a statement of proposal expressly as a replacement for a pre-
existing available name;

(14) after 1950, proposed by an identified author [Articles 14 and 17(7)];

(15) after 1960, published as an acceptable name, not as a synonym
[Article 11(d)] (ruling deleted entirely in 1964);

(16) after 1960, unconditionally proposed [Articles 15 and 17(8)].

These provisions apply between certain dates, as shown. The dates are
summarized as follows:

Items	1758–1930	1931–1950	1951–1960	1961–
1–7	X	X	X	X
8–10	X			
11–13		X	X	X
14			X	X
15–16				X

The following are not accepted as means of establishing new species
names:

1. Citation of a type locality [Article 16(b)(i)].
2. Distribution solely by a forbidden method [Article 9], such as:
    (a) reading at a scientific or other meeting,
    (b) mere deposit of a document in a library,
    (c) labeling of a museum specimen [Article 11(b)(ii)].
3. Duplication by a forbidden method [Article 9], such as
    (a) microform of any sort (film, cards, and so on),
    (b) proof sheets,
    (c) separate photographs, even with accompanying printed expla-
nations.

Sources of names of species. The binominal name of a species consists of
a generic name and a second word that denotes the species. The sources
from which this second word may be derived are listed in Chapter 28,
Epithets.

Formation and spelling of names of species. The binomen is formed as
the combination of a generic name with a specific epithet. The formation
and spellings of the epithet are discussed in Chapter 28, Epithets.

Author of the names of species. Authorship, in the 1961 Code, is assumed to be authorship of the second half of the binomen, what is therein

called the specific name (trivial name of most earlier writers and specific epithet in this book). Inasmuch as this word cannot be used by itself, but only in combination with a generic name, such as a binomen, the problems of authorship are more practically applied to the combination, the "name of the species." The rules applying to the binomen are interpreted here; those applicable to the "trivial name" or epithet alone are cited in Chapter 28, Epithets.

The author of a binomen is

(A) the person (or persons) who was responsible for the name in the first publication in which it was acceptably published, regardless of what generic name was employed with the epithet

   (1) if it was first published as the name for a species;

   (2) if it was first published as the name for a subspecies and was later changed to the species rank [Article 50(b)];

   (3) if it was first acceptably published without indication of its infra-subspecific rank and was later changed to the species rank;

   (4) if, before 1961, it was first published as a synonym of the name of a species or subspecies (the restriction after 1960 was deleted in the 1964 Code);

   (5) even if it was first published with a spelling error that must be emended by a subsequent writer;

   (6) even if it has been justifiably emended by a subsequent writer (a justified emendation takes the author of the original name) [Articles 33(a) and 50(c)];

   (7) if it is itself an unjustified emendation of some earlier name (an unjustified emendation takes its own author) [Articles 33(a) and 50(d)];

   (8) if, after 1960, the spelling has been changed because of new requirements of the 1961 Code, such as

     (a) deletion of a hyphen,

     (b) writing out a numeral,

     (c) deletion of a diacritic mark or apostrophe,

     (d) conversion of a German umlaut to *e*, or

     (e) change of *M'*- or *Mc*- or *Mac*- to *mac*- (from 1954 to 1960, but only a recommendation in the 1961 Code);

   (9) even if other persons appear as principal author of the publication (the author of the name is the person actually responsible for the name as published [Article 50];

  (10) even if, before 1951, the author was not recorded (anonymous) [Recommendation 51A];

  (11) even if in the original publication the name itself is credited to some other person (from a museum label, an unpublished manuscript, and so on);

(B) some other person than the one responsible for its acceptable publication:

(1) if it is an acceptable emendation of an earlier name, when it takes the author of the original name [Article 50(d)].

Date of the name of a species. The date of the name of the species is the date of publication of the original combination (binomen) [Article 21], except

(1) when the name is an unjustified emendation (which takes the date of its own publication) [Article 33(a)(ii)];
(2) when the name is a junior homonym (which takes the date of its own publication, not that of the senior homonym, which is an identical binomen).

The date of a species name (the binomen) is the date (day, month, or year, as near as is known) on which

(A) the original combination was published in complete compliance with all relevant rules:

1. It must conform to the Publication Requirements listed above [Articles 1, 2(a), 3, 8, 10, 11, 12, 13, 14, 15, and 21].
2. If publication of the work is interrupted, a new name can date only from the time of the completion of all requirements [Article 10(a)].
3. If the work bears a publication date, that is the date of any included new names unless there is direct evidence to the contrary [Article 21(a)].
4. If the date of a work is not completely specified, it is to be interpreted as the earliest day demonstrated by evidence, but in the absence of such evidence, as the last day of the stated month or year [Article 21(b)].
5. If the date of publication specified in a work is found to be incorrect, the date of included names is to be interpreted as the earliest date demonstrated by any evidence [Article 21(c)].
6. If the date of publication of the work is not indicated, the names are to be given the earliest date demonstrated by external evidence, such as mention in another work [Article 21(f)].
7. When a subspecies is raised to rank of species, it retains the original date [Article 46].

(B) a later action was taken: When an infrasubspecific taxon is raised to rank of subspecies or species, it dates from the time of elevation [Article 10(b)].

Citation of the names of species. Detailed rules are included in the 1961 Code to govern the form of citation of a species name, of its author, and of the date of its publication. These rules apply to all the names of that species, whether they are considered to be the correct name or merely rejected synonyms.

A. Names

1. It is recommended in the 1961 Code, as in previous codes, that the names of species and subspecies, the binomina and trinomina, be printed in italics or some type face different from that used for the surrounding text [Recommendation E7].

2. It is recommended that a new species name be cited in full in the original, as *Hypotheticus affluens,* rather than merely as *H. affluens* somewhere under the heading *Hypotheticus* [Recommendation E8].

3. It is recommended that each new name be accompanied in its first publication (and only in the first) by such a notation as "species novum," "new species," or their customary abbreviations "sp. n." and "n. sp." [Recommendation E7].

4. In citing the name of a species, the subgenus may be inserted in parenthesis between the two words of the name, without affecting any nomenclatural aspects. (A generic synonym may not be so interpolated.)

B. Authors

The author of a name is always the person responsible for its publication in acceptable form. It is either the author of the publication itself or another person therein stated to be responsible for *both* the name and the validation (description or indication) of it. Recent rules tend to show that citation of the author's name is considered to be part of the scientific appellation of the species. However, citation of it is not yet a definite requirement. It is always advisable and often essential to understanding that the author be cited at least once in every publication, and in all cases of formal synonymy or discussion of the names.

The reasons for using the author's name are given very effectively by Mayr, Linsley, Usinger (1953, p. 235): "(1) authority citation makes it possible to distinguish between two or more different species with the same scientific name; (2) it gives an immediate clue to the original description and an indirect clue to the quality of work and to the (possible) location of the type specimen; and (3) it reveals something of the history of the name. In other words, the author's name is a link between nomenclature and classification; it is a tag by which a scientific name may be identified."

It has become almost universal to abbreviate the name of the author, if the name is over five or six letters long. This practice is necessary where space limitations exist. It causes little confusion or inconvenience if it is not overdone. The abbreviations should be widely agreed upon, not different in every work. They should be such as to suggest the complete name and avoid duplication in a field. For example: the abbreviation Bernh. can mean either Bernhardt or Bernhauer. The latter name has been more appropriately abbreviated as Bnhr.

A recent recommendation by the International Commission that all abbreviations be discontinued is doomed to be widely ignored. It is impossible to force all persons to meet such a standard, and it is better to encourage all to avoid any undue inconvenience to others by using abbreviations with care.

There have been, from the earliest days of taxonomy, a few publications without indication of author. A few new names have appeared in such anonymous publications. These were acceptable until 1951, when the Commission ruled that any names published thereafter without a definite author would be unavailable.

It has gradually come to be recognized that the citation of the author of a name is not enough to identify it in all cases. The addition of the date of the original publication helps in this matter, and it also serves as an abbreviated bibliographic citation. It has been widely used and given official acceptance.

It is one of the anomalies of zoological nomenclature that the cited author is not the author of the specific name (the binominal name of the species) but merely the author of the specific epithet (of course in connection with some generic name but not specifying which one). It is possible to cite the author of the new binomen, and this is done occasionally by zoologists (and by rule by botanists) by listing the binomen author (the proposer of the new combination) after the author of the epithet. This is known as double citation of authority.

When this is done, the name of the original author is placed within parentheses (with the date, if used) and the later author follows without parentheses. Thus, *X-us albus* (Smith, 1850) Jones, 1929, indicates that Smith proposed the epithet *albus* in 1850, in connection with some other generic name, and that Jones transferred it to *X-us* in 1929.

The use of parentheses in this way has been widespread but not universal. It was officially sanctioned but not required. Parentheses came to be used even when the second author was not cited, and in this usage gave rise to strong opposition as well as strong support. The situation now is that any subsequent writer is at liberty to use the parentheses around the author's name if he wishes to do so to indicate that the name is now coupled with some other than the original generic name. Any subsequent writer is also at liberty to refrain from using the parentheses. Therefore, it is now the situation that presence of the parentheses gives a definite clue to the history of the name, but absence of the parentheses is ambiguous. In the latter case, one must judge whether the writer concerned is using parentheses or not, to be able to judge what their absence means.

The actual rules now governing the citation of the author's name are these:

1. "The name of the author does not form part of the name of the taxon and its citation is optional" [Article 51(a)], but citation of the author

at least once in each work in which the species is cited is recommended [Recommendation E10].

2. The author's name, if cited, follows the name of the species (binomen) without any intervening mark of punctuation and is printed in the type face of the text [Article 51(b)].

3. "If a scientific name and its validating conditions are the responsibility not of the author(s) of the publication containing them, but only of one (or some) of the authors, or of some other zoologist, the name of the author(s), if cited, is to be stated as 'B in A' or 'B in A & B' [or C in A & B], or whatever combination is appropriate" [Article 51(c)]. It is equally appropriate to cite it simply as "B" or "C," omitting the "A" or "A & B," which is merely an abbreviated bibliographic reference.

4. "The name of a subsequent user of a scientific name, if cited, is to be separated from it in some distinctive manner, other than by a comma. E.g.: Reference to *Cancer pagurus* Linnaeus as used by Latreille may be cited as

Cancer pagurus Linnaeus sensu Latreille,
Cancer pagurus: Latreille,

or in some other distinctive manner, but not as

Cancer pagurus Latreille, nor as
Cancer pagurus, Latreille" [Article 51(b)(i)].

5. "If a species-group taxon was described in a given genus and later transferred to another, the name of the author of the species-group name [the author of the original binomen], if cited, is to be enclosed in parentheses" [Article 51(d)]. "The use of parentheses here applies only to transfers from one nominal genus to another, and is not affected by the presence of a subgeneric name [in either binomen], or by any shifts of rank or position within the same genus" [Article 51(d)(i)].

6. "If it is desired to cite the names both of the original author of a species-group name and of the reviser who transferred it to another genus, the name of the reviser should follow the parentheses that enclose the name of the original author" [Recommendation 51B]. E.g.: *Creophilus maxillosus* (Linnaeus, 1758) Leach, 1819.

7. "When a nominal species has been later divided on taxonomic grounds, the name of the author who restricted the taxonomic species may be cited with a suitable notation, after the name of the original author; e.g., *Taenia solium* Linnaeus, partim Goeze" [Recommendation E12].

8. If, before 1951, a specific name was published anonymously, a notation to this effect is to be used in place of the author's name; e.g., (Anonymous).

9. "If the name of a taxon was published anonymously, but its author is known, his name, if cited, should be enclosed in square brackets to show the original anonymity" [Recommendation 51A].

C. Date

1. "The date of publication of a name, if cited, follows the name of the author with a comma interposed" [Article 22]. (It is customary to consider references to the work in which a name was published as parenthetical remarks. The date is the simplest form of such citation, and as such has frequently been cited in parenthesis. E.g.: *Staphylinus maxillosus* Linnaeus (1758). The following multiple recommendation would result in confusion with this past practice even if uniformly applied by all zoologists from now on. The existence of the past usage and the need for use of parenthetical insertions of many sorts are adequate reasons for abandoning these absurdly detailed recommendations.)

2. "In citing the date of publication of a name, an author (1) should not enclose the date in either parentheses or square brackets if the work containing it specifies the date of publication; (2) should enclose the date, or a part of it, in parentheses if it is determined by evidence derived from the volume concerned other than in (1); or (3) should enclose the date, or a part of it, in square brackets if it is determined only from external evidence" [Recommendation 22A]. (See note under (1) next above.) E.g.: *Hypotheticus ridiculus* Doe, (May)[21], 1958; *Novogeneris ridiculus* (Doe, (May)[21], 1958); but it is cumbersome to cite without confusion an erroneous date shown on the publication: *Hypotheticus confusus* Roe, 1958 (not 1957) would have to read *Hypotheticus confusus* Roe (1958) (not 1957), or *Hypotheticus confusus* Roe ([May] 1958) (not 1957, not (March 1958)).

3. "If the original date of publication is cited for a species-group name in a changed combination, it should be enclosed within the same parentheses as the name of the original author, separated by a comma" [Recommendation 22B].

D. Bibliographic Reference

"A zoologist who cites the name of a genus or taxon of lower rank in a taxonomic work should give at least once a full bibliographic reference to its original publication" [Recommendation E14] and to each subsequent use of that name which is individually cited.

Conflicts between species-group names. Several sorts of conflicts can arise between the names of species-group taxa or their various forms:

1. Synonymy. There may have been two names proposed for the same species or its parts. (See also items (2) to (5), which are special cases of this.) (See discussion under Synonymy, below.)

2. Multiple publication of a name as new, in two or more separate works. This is proscribed by Recommendation E22 of the 1961 Code, but when it has occurred in the past it must be recognized and the subsequent ones treated as later references to the name. E.g.:

A.

Colpodota negligens Mulsant and Rey, 1873, *Opusc. Ent.*, vol. 15,
p. 156 (as new)

Colpodota negligens Mulsant and Rey, 1874, *Hist. Nat. Col.
France*, p. 231 (as new)

B.

Acanthonia gigantea Wasmann, 1916, *Ent. Mitt.*, vol. 5, p. 97
(as new)

Acanthonia gigantea Wasmann, 1917, *Ztschr. wiss. Zool.*, vol.
117, p. 274 (as new)

3. Simultaneous publication. A name may be published as new in several
   publications on the same date, producing names that are both synonyms and homonyms. (See The First Reviser, below.)
4. Multiple original spellings. A name may be spelled in more than one
   way in the original publication. (See Original Spelling in Chapter 28,
   on Epithets.)
5. Subsequent spellings. If subsequent writers use different spellings, either
   intentionally or inadvertently or through interference of a third party,
   a decision must be made as to which one is to be accepted. (See
   Subsequent Spellings, below.)
6. Homonymy. When two names of identical spelling (homonyms) are
   given, at the same or different times, to different species or subspecies
   within a genus, the conflict is resolved by actions prescribed by a
   complex set of rules. (See Homonymy, below.)
7. Tautonymy. When the two terms of the name of a species, or the first
   and last terms in the name of a subspecies, are identical in spelling,
   they are called tautonyms and the situation is tautonymy. (See Rejection of Names of Species, below.)
8. Misidentifications. If a species name is erroneously applied to a certain
   species, or a subspecies name to a certain subspecies, in a misidentification, the error is not to be used for the taxon misidentified, under
   any circumstances. (See Replacement, below.)
9. Change of rank. Species taxa can be lowered to subspecies rank and
   subspecies taxa can be elevated to species rank. (See Change of Rank,
   below.)

Synonymy of names of species. Synonymy of names is a subject curiously
absent from the headings of the 1961 Code. It is discussed therein primarily under the heading Priority.

When a given species or subspecies has two or more applicable names
(binomina or trinomina), all the names are synonyms. Only one of the
synonyms can be the correct name for the taxon at any one time, and the
rest are the rejected synonyms (usually but not always also junior, or more
recent, synonyms).

It is necessary to be consistent about what is involved in synonymy.
Although it is possible for words to be synonyms, zoological nomenclature

is concerned with names, not mere words. Although the 1961 Code illogically calls the second part of the binomen "the specific name," it is not a *name* of anything when it stands alone. These "specific names" cannot be synonyms until they are placed in one genus. At this instant they cease to be single words (whether names or not) and become binomina. It is these that may be synonyms.

Two or more names for one species and its parts can occur by being

(A) published simultaneously in one work
> (1) by separate proposal with different type specimens (a) with both treated as names of species (later found to be zoologically conspecific), (b) with one treated as a species and one as a subspecies (or, before 1961, one as a variety, race, and so on), or (c) with both treated as subspecies or (before 1961) as other infraspecific taxon;
> (2) (before 1931) by inclusion of an unpublished name as a "synonym" of a name being proposed as new (publication in synonymy) (a) for a species, or (b) for a subspecies or other infraspecific taxon;

(B) published in separate works
> (3) for the same type specimen;
> (4) for different specimens later held to be conspecific, and (a) both as names of species, (b) one as a subspecies or other infraspecific taxon, or (c) both as subspecies or other infraspecific taxa;
> (5) the later one as a replacement for the earlier one, (a) as a new name, or (b) as a new spelling.

Determination of which synonym is to be employed as the correct name for a taxon depends on the categorical disposition of the taxon (as species or as subspecies) as well as upon a prescribed basis for choosing:

1. For a species, the name is to be chosen from all of the following considered together:
   (a) All the names given to the species, whether now or then considered "synonyms" or not, and
   (b) All the names given to all the subspecies, including all their "synonyms."
2. For a subspecies
   (a) the name is to be chosen from the names that have been applied directly to that subspecies, or
   (b) for the nominate subspecies, the name is determined for the species and then applies automatically to the nominate subspecies.

The prescribed bases are:

(1) the first reviser, in (A)(1) above;
(2) automatic determination by the author (acting as a sort of immediate first reviser), in (A)(2) above;

(3) homonymy, which forces the rejection of any name which has pre-
    viously been used for another species or subspecies in that genus;
(4) priority among the remaining names.

Synonymy may be subjective or objective. *Subjective synonymy* depends
on the opinion of the reviser that the type specimens of the species are
conspecific (same as zoological synonymy or conditional synonymy). (Items
2, 3, and 5 under A and B above are subjective.) *Objective synonymy*
occurs when the two species have the same type, as when

(1) they were separately proposed for the same type specimen;
(2) the later name was directly proposed as a replacement for the other,
    thereby automatically having the same type; or
(3) the later name is an intentional respelling (emendation) of the other,
    again automatically having the same type.

Priority of names of species. Priority of date of publication could be
involved when a given taxon has received two names (synonymy), because

(1) it has inadvertently been named twice; or
(2) two taxa are united; or
(3) part of the animal is named as well as the entire animal; or
(4) two generations, forms, stages, or sexes are separately named; or
(5) the work of the animal is described (before 1931) as well as the
    animal itself.

All of these result in synonymy, two names for the same species. The
Law of Priority [Article 23] holds that "The valid name of a taxon is the
oldest available name applied to it . . ." except

(1) if the name has been rejected by the Commission under its Plenary
    Powers [Article 79];
(2) if another name has been established in its place by the Commission
    acting under its Plenary Powers [Article 23(a)];
(3) if the name is a nomen oblitum ("a name that has remained unused
    as a senior synonym in the primary zoological literature for more than
    fifty years . . .") [Article 23(b)];
(4) if the two names were simultaneously published (see First Reviser,
    below) [Article 24(a)].

The Law of Priority applies regardless of

(1) rank within the species group [Article 23(c)];
(2) misspelling or (justified) emendation of any of the names [Articles
    31(a) and 32(c) and 33(a)];
(3) transfer to other than the original genus;
(4) what part of the animal is named first [Article 24(b)(i)];
(5) what stage, form, generation, or sex is named first [Article 24(b)(ii)];

(6) whether, before 1931, the work of an animal is named before the animal itself [Article 24(b)(iii)] (after 1930, a name cannot be based solely on the supposed work of an animal) [Article 13(a)];

(7) whether one of the names is an unjustified emendation that has separate status as an objective synonym [Article 33(a)(ii)].

First reviser and names of species. In only two circumstances does a first reviser determine the species name to be adopted. To be a first reviser "an author must have cited two or more such names, must have made it clear that he believes them to represent the same taxonomic unit, and must have chosen one as the name of the taxon" [Article 24(a)(i)].

1. "If more than one name for a single taxon . . . are published simultaneously, whether in the same or different works, their relative priority is determined by the action of the first reviser" [Article 24(a)].

2. "If . . . identical names for different taxa are published simultaneously, whether in the same or different works, their relative priority is determined by the action of the first reviser" [Article 24(a)]. In Article 57(e) there is an exception to this: "Of two homonymous species-group names of identical date, one proposed for a species takes precedence over one proposed for a subspecies." E.g.:

A.
   Oxytelus burgeoni Bernhauer, 1932, p. 80.
   Oxytelus burgeoni Bernhauer, 1932, p. 81.
B.
   Stenus (Nestus) gérardi Bernhauer, 1932, p. 85.
   Stenus (Hypostenus) gérardi Bernhauer, 1932, p. 86.

The first reviser must choose between these in each case. Examples of the exception would be extremely rare in species-group names.

Homonymy of names of species. Homonymy exists when there are two species names of identical (corrected) spelling in one genus. The 1961 Code implies that the homonymy applies only to the second word of the binomen, the "specific name." Reasons are given elsewhere for believing that this word is not a name. It could be involved in homonymy as a word but not as a name. To assume that homonymy involves only a binomen obviates the difficulty over the two species being placed in the same zoological genus but under different generic names, different "nominal genera" in the 1961 Code. For example,

   Pseudomedon ruficolle Casey, 1905
   Lithocharis ruficollis Kraatz, 1858

Is there homonymy when *Pseudomedon* is stated to be a synonym or a subgenus of *Lithocharis?* Of course there is potential homonymy, but this statement implies only that the genotypes belong together; other species may or may not be involved. Until *ruficollis* of Casey is actually put into *Lithocharis,* one cannot say that there are two species with identical names in the genus. This is recognized in the 1961 Code where it is stated that the identical names must be "within the same nominal genus." The following section is based on the assumption that homonymy involves two binomina. There are few ways in which this assumption affects the conclusions on rules of homonymy.

Homonymy of the names of species differs from that of the names of genera because (a) it is limited to one genus, and (b) it involves binomina instead of uninomina. Because of the first, homonymy is much more rare among names of species than among names of genera. Because of the second, homonymy of species names is greatly complicated by interest in the time the binomen was made—in the transfer of species to other genera than the original.

X-us albus Fabricius and *X-us albus* Smith are homonyms. Because neither author's name is in parenthesis, we know that each *albus* was originally described in *X-us.* The names are therefore *primary homonyms* ("originally published in the same genus"). There are three sequences of events that will lead to primary homonymy, as diagrammed in Figures A, B, and C.

X-us albus Smith and *X-us albus* (Latreille) are also homonyms. Because Latreille originally published his species in another genus (*Y-us albus* Latreille), the names are *secondary homonyms* ("later brought together in the same genus"). There are five sequences of events which will lead to secondary homonymy. They may be diagrammed as shown in Figures D–H.

The treatment of primary and secondary homonyms in the 1961 Code is an illogical mixture in which homonymy is recognized not only in A, D, and E but also in B and C. All primary homonyms are recognized and some but not all secondary ones. For discussion of this and other systems, see

Blackwelder, R. E. 1948. An analysis of specific homonyms in zoological nomenclature. *Journ. Washington Acad. Sci.,* vol. 38, pp. 206–213.

The following paragraphs conform exactly to the 1961 Code in distinguishing between primary and secondary homonyms.

The following are not involved in homonymy:

1. "Names that are unavailable in the meaning of the Code" [Article 54(1)].
2. "Names that have never been used for a taxon in the animal kingdom" [Article 54(2)].
3. "Names that are excluded from zoological nomenclature (Art. 1)" [Article 54(3)], including names given
   (a) to hypothetical concepts,

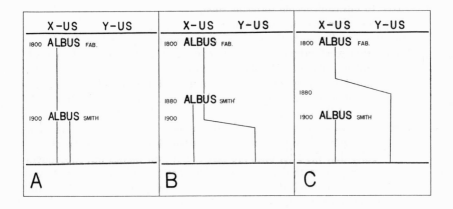

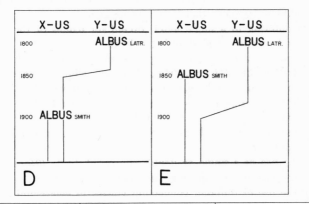

    (b) to teratological specimens as such,

    (c) to hybrids as such,

    (d) to infrasubspecific forms as such,

    (e) for other than taxonomic use.

4. "Incorrect spellings, both original and subsequent" [Article 54(4)].

5. Names that were originally spelled (correctly) with a numeral, symbol, umlaut, diaeresis, hyphen, apostrophe, and so on. (They enter into homonymy only in their altered form. In effect they are emended by the Commission by retroactive rule.) [Articles 26, 27 and 32(c)(i)].

6. Identical names given independently to one species, which are thus synonyms as well as homonyms. (The 1961 Code, in Article 52, rules that the later ones are simply not new names but subsequent references to the original one. This fiction will lead to great confusion in detailed synonymies. It should be ignored, and the homonyms recognized as what they are.)

7. "Identical species-group names originally or subsequently placed in different genera that bear homonymous names" [Article 57(c)].

8. Names differing in a single letter, except for names of the same origin and meaning and differing only in

    (a) the use of *ae, oe,* or *e* (e.g., *caeruleus, coeruleus, ceruleus*);

    (b) the use of *ei, i,* or *y* (e.g., *cheiropus, chiropus, chyropus*);

    (c) the use of *c* or *k* (e.g., *microdon, mikrodon*);

    (d) the aspiration or non-aspiration of a consonant (e.g., *oxyrhynchus, oxyrynchus*);

    (e) the presence or absence of *c* before *t* (e.g., *auctumnalis, autumnalis*);

    (f) the use of a single or double consonant (e.g., *littoralis, litoralis*);

    (g) the use of *f* or *ph* (e.g., *sulfureus, sulphureus*);

    (h) the use of different connecting vowels in compound words (e.g., *nigricinctus, nigrocinctus*);

    (i) the transcription of the semivowel *i* as *y, ei, ej,* or *ij;*

    (j) the termination *-i* or *-ii* in a patronymic genitive (e.g., *smithi, smithii*);

    (k) the suffix *-ensis* or *-iensis* in a geographical name (e.g., *timorensis, timoriensis*);

    (l) being one of three pairs of names treated as special cases: (i) *saghalinensis* and *sakhalinensis,* (ii) *sibericus* and *sibiricus,* and (iii) *tianschanicus* and *tianshanicus* [Article 58];

    (m) having different endings due to gender (which should be corrected in the new genus and will then be identical) [Article 57(b)(i)];

    (n) being respellings caused by elimination of diacritic marks under Article 27 and Article 32(c) (see 5, above).

Names which are homonyms after the above limitations are to be treated as follows:

1. Junior primary homonyms are to be replaced, and the replaced name can never be used again [Articles 53 and 59(a)].
2. Junior secondary homonyms, if the homonymy still exists (diagrams D and E), are to be replaced, but the replaced name is to be reinstated if the homonymous names are later separated into different genera [Articles 53 and 59(b) and (c)].
3. Junior secondary homonyms, if the homonymy no longer exists (diagrams F, G, and H), are not to be replaced [Article 59(b)].

Some additional points with reference to homonymy are the following:

1. Subgeneric name. "The presence of a subgeneric name does not affect homonymy between species-group names within the same genus. Ex.: *A-us* (*B-us*) *intermedius* Pavlov and *A-us* (*C-us*) *intermedius* Dupont are primary homonyms, but *A-us* (*B-us*) *intermedius* Pavlov is not a primary homonym of *B-us intermedius* Black" [Article 57(a)]. This example fails to deal with the question of whether there is secondary homonymy in the last case. In fact the 1961 Code fails entirely to deal with homonymy involving subgenera. There is only one reasonable solution to this, and that is to consider that two binomina are represented by the expression *A-us* (*B-us*) *intermedius*. So long as *B-us* is accepted as a subgenus of *A-us*, all homonymy in *B-us* must be taken into account, as well as all in *A-us*.
2. Differences in generic spelling. "Species-group homonymy within a given nominal genus is not obviated by any emendation or incorrect spelling of the generic name" [Article 57(b)].
3. Gender terminations. "Differences in termination that are due solely to gender are to be disregarded in determining whether adjectival species-group names are homonyms" [Article 57(b)].
4. Precedence of species over subspecies. "Of two homonymous species-group names of identical date, one proposed for a species takes precedence over one proposed for a subspecies" [Article 57(e)].
5. Revival of secondary homonyms. "A name rejected after 1960 as a secondary homonym is to be restored as the valid name whenever a zoologist believes that the two species-group taxa in question are not congeneric, unless it is invalid for other reasons" [Article 59(c)]. This rule is better calculated to cause instability than any rule included in or omitted from any set of rules in this century. Under this rule there can be two entirely correct names for one species in use at one time. E.g.: If Adams believes that *Y-us* is a subgenus of genus *X-us*, then two of the included species must be cited as

1. *X-us* (*Z-us*) *pallidus* Fisher, 1820
2. *X-us* (*Y-us*) *albus* Temple, 1890
   Syn.: *X-us pallidus* Swithen, 1850 (not Fisher, 1820)

At the same time, if Brown believes that *Y-us* is a distinct genus, the same two species must be cited as

1. *X-us* (*Z-us*) *pallidus* Fisher, 1820
2. *Y-us pallidus* (Swithen, 1850)
   Syn.: *X-us* (*Y-us*) *pallidus* Swithen, 1850
   X-us (*Y-us*) *albus* Temple, 1890 (not necessary)

Thus *Y-us pallidus* Swithen and *X-us albus* Temple are both correct names for this species. Of course, *X-us* (*Y-us*) *albus* Temple is also an acceptable name for the species.

6. Rejected replacement names. ". . . the name proposed in replacement of the secondary homonym becomes a junior objective synonym of the latter" when the secondary homonym is restored as in (5) above [Article 59(c)(i)]. (*All* new replacement names are junior objective synonyms of the names they replace, whether they are accepted into use or not.)

7. Replacement of rejected homonyms. "A rejected homonym must be replaced by an existing available name or, for lack of such a name, by a new name" [Article 60]. Such a replacement name may be

   (a) an objective synonym, whether new or not,
   (b) a subjective synonym "only so long as it is regarded as a synonym of the rejected name" [Article 60(a)(ii), which in the 1964 Code becomes Article 60(a)(i)].
   (c) an unjustified emendation of the rejected name.

All of these are to be considered together in light of their priority (dates); "the oldest (available one) of these must be adopted, with its own authorship and date" [Article 60(a)].

Rejection of names of species. A name for a species is to be rejected from nomenclature (in that form and combination)

(1) if the epithet is not an acceptable one, having been given
   (a) to a hypothetical concept, or
   (b) to a teratological specimen as such, or
   (c) to a hybrid as such, or
   (d) to an infrasubspecific form as such, or
   (e) for other than taxonomic use [Article 1];
(2) if the epithet was published before 1758 and never validated since [Article 3];
(3) if the epithet was not acceptably published under Articles 8, 9, and 11;
(4) if it is not binominal [Article 11];

(5) if (after 1960) the epithet was first published in synonymy and has not been subsequently validated [Article 11];

(6) if the epithet is not "a simple word of more than one letter, or a compound word, and . . . treated as

    (a) an adjective in the nominative singular . . . ,

    (b) a noun in the nominative singular . . . ,

    (c) a noun in the genitive case . . . ,

    (d) an adjective used as a substantive in the genitive case . . ." [Article 11(g)];

    (e) a participle used as an adjective;

    (f) a gerund used as a substantive;

(7) if the epithet consists of "words related by a conjunction . . ." [Article 11(g)(iii)];

(8) if the epithet includes "a sign that cannot be spelled out in Latin" [Article 11(g)(iii)];

(9) if, before 1931, the epithet (in combination with a generic name) was published without a description, definition, or indication [Article 12];

(10) if, after 1930, the publication was not

    (a) "accompanied by a statement that purports to give characters differentiating the taxon; or

    (b) accompanied by a definite bibliographic reference to such a statement; or

    (c) proposed expressly as a replacement for a pre-existing available name" [Article 13(a)];

(11) if, after 1950, the epithet was published anonymously [Article 14];

(12) if, after 1960, the epithet was proposed conditionally [Article 15];

(13) if, after 1960, the epithet was proposed explicitly as part of the name of a "variety" or "form" [Articles 15 and 45];

(14) if, after 1930, the epithet was founded on the work of an animal before the animal itself is known [Article 16(a)(viii)];

(15) if it was an erroneous identification (misidentification) rather than a separate name [Article 49];

(16) if there never was a specimen that could be the type [no article explicitly states this, but Article 61 implies that a type is essential, and Article 1 excludes names of hypothetical concepts];

(17) if it has been ruled unacceptable by the Commission;

(18) if the epithet was published in a work that has been ruled unacceptable by the Commission.

A name for a species is to be rejected in that form (and thereupon corrected)

(19) if the epithet is an incorrect spelling, whether original or subsequent [Article 19];

(20) if the epithet
  (a) includes a number that is not written in full as a word and united with the remainder of the epithet without a hyphen [Article 26(b)],
  (b) is spelled with any diacritic mark, apostrophe, diaeresis, or hyphen (except as in Article 26) (presumably only after 1960, when this rule became effective, although it is intended to be retroactive) [Article 27],
  (c) is spelled with a capital initial letter [Article 28],
  (d) is an adjective in the nominative singular but does not agree in gender with the generic name [Article 30],
  (e) is a modern patronymic but does not end in
    (i) *-i* if the personal name is that of a man,
    (ii) *-orum* if of men or of man (men) and woman (women) together,
    (iii) *-ae* if of a woman, or
    (iv) *-arum* if of women [Article 31];
(21) if the original spelling of the epithet is an error as defined by Article 32;
(22) if the epithet is an unjustified emendation, not an acceptable correction of an incorrect original spelling [Article 33];
(23) if the epithet is an incorrect subsequent spelling, either intentional or unintentional [Article 33(b)].

A name for a species is to be rejected as the correct name for the species (but retained as an available synonym)

(24) if it is the junior of two available names (synonymy), unless
  (a) it has been conserved by the Commission [Article 23(a)(i)], or
  (b) the Commission has expressly validated it [Article 23(a)(ii)];
(25) if it is a junior homonym, whether primary or secondary [Articles 53 and 57];
(26) if it is a nomen oblitum [Article 23(b)].

The name of a species is *not* to be rejected (in any sense) solely because

(1) the epithet was originally given
  (a) to part of the animal rather than an entire animal [Article 25(b)(i)];
  (b) to a fossil [Article 1];
  (c) to an immature stage or unusual form rather than to the adult [Article 17(4)];
  (d) to "an organism now but not then considered an animal" [Article 17(5)];
  (e) to an organism then but not now considered an animal [Article 2(b)];
  (f) to an original series containing more than one species [Article 17(2)];

(g) to an animal or animals later found to be hybrid [Article 17(2)];

(2) it included a generic name that is invalid or unavailable [Article 17(3)] (a most peculiar interpretation!);

(3) it becomes a junior synonym [Article 17(1)];

(4) the epithet was originally misspelled [Article 17(6)];

(5) the epithet was, before 1951, published anonymously [Article 17(7)];

(6) the epithet was, before 1961, proposed conditionally [Article 17(8)];

(7) the epithet was, before 1961, proposed as a "variety" or "form" [Article 17(9)];

(8) it is deemed to be inappropriate [Article 18(a)];

(9) it is tautonymous; that is, because the two parts of the name are identical in spelling [Article 18(b)];

(10) it is an emendation, whether justified or unjustified [Article 19]:

(11) the taxon has been changed in rank (from species to subspecies or from subspecies to species, or from variety or form to either) [Article 23(c)];

(12) the epithet was, before 1931, founded on the work of an animal rather than on the animal itself [Article 24(b)];

(13) the epithet is the same as the epithet of one of its subspecies [Article 47];

(14) it is cited with erroneous author or date.

Replacement of names of species. The names that must be rejected, as listed in the first part of the previous section, are to be treated or replaced as follows:

(1–18) are not acceptable in zoological nomenclature; they are to be replaced by the oldest available synonym that is not a senior homonym or otherwise eliminated.

(19–23) are erroneous spellings (at least under the 1961 Code) and are to be rejected in their original form and replaced with a corrected spelling.

(24) are junior synonyms to be replaced with an older synonym.

(25) are junior homonyms and are to be replaced with their next oldest synonyms, or, if there are none, with a new name.

(26) are to be left unused, until the Commission rules on their status.

Of these, (1) to (18) and (25) may require a new name as replacement. A name can be in (24) only if it has an available senior synonym. Those in (19) to (23) are replaced with correct spellings. (Those in (26) are replaced by the oldest available synonym.)

It must be remembered that the "synonyms" from which the name of a species is to be selected include the names of all subspecies (written as binomina) and their synonyms.

Misapplied names of species. When a writer quotes an earlier species by name, he is assumed to be referring to the actual zoological species properly indicated by that name. If there is evidence that he has used it by mistake for some other species, such as by erroneous synonymizing of it with some extraneous species, his misuse of it is to be corrected. His use of the name, being different from the original, is a sort of homonym. It has no standing in nomenclature but does need to be recorded as the correction of an error, usually by being cited in the synonymy of the name of another species. The nature of this pseudo-name must be clearly indicated: For example,

1. *X-us albus* Linnaeus, 1758
2. *Z-us pallidus* Fabricius, 1801
   Syn.: *X-us albus* of Stephens, 1829 (not Linnaeus, 1758)

Misidentification of a species that is actually new with some already named species has been dealt with in both *Règles* and the 1961 Code in this manner (the latter quoted):

1. "The specific name used in an erroneous specific identification cannot be retained for the species to which the name was wrongly applied, even if the two species in question are in, or are later referred to, different genera" [Article 49].
2. An exception is cited in Article 70(b)(i) in the case of a species which is genotype of a genus, in which case there has been "established also a new nominal species, with the same specific name as the misidentified species, in the new nominal genus."

Combining species. "A species-group taxon formed by the union of two or more species-group taxa takes the oldest valid name among those of its components" [Article 23(e)(ii)]. Such a species can be formed by

(1) the uniting of two or more former species;
(2) the transfer of two or more subspecies out of one species and their establishment as a separate species (with or without subspecies);
(3) the transfer of one or more subspecies into a different species where they join with the original subspecies or are merged with them.

Dividing a species. A species can be divided by

(1) recognizing subspecies within it; or
(2) splitting it into two species; or
(3) removing one of its subspecies to another species or to stand as a separate species.

In each case the names follow their type specimens, as indicated in a later section.

Changing the category. Supposed species can be reduced to subspecies or supposed subspecies can be elevated to status of species. In the first class of cases, there is no change of name except for its citation as a trinomen: *X-us albus* becomes *X-us fuscus albus*. In the second class of cases, there is no change of name except for its citation as a binomen: *Y-us pulcher sinensis* becomes *Y-us sinensis.*

Taxa subordinate to the species. Only the category of subspecies is dependent upon the species category. "Superspecies" is not used. Variety, form, aberration, and so on have been used as infraspecific categories and were not specifically proscribed by the *Règles.* All such infraspecific categories, except the subspecies, are directly ruled out of zoological nomenclature by the 1961 Code.

The codes have always stated that the names of species and the names of subspecies are governed by the same rules. The 1961 Code has also stated that the categories of species and subspecies are of co-ordinate status in nomenclature. As is the case with genera and subgenera, this ruling is inconsistent with the clear position of subspecies taxa as *subordinate* to species taxa (in a lower category); they cannot then also be co-ordinate.

Categories cannot be co-ordinate in any sense because they exist solely by virtue of being different in the only feature they possess—position in the hierarchy. Taxa also cannot be co-ordinate at two levels. A species cannot be co-ordinate with one of its own subspecies, which is at the lower rank for the very reason that it is not *co-ordinate* but *subordinate.*

The names of the taxa at the species and subspecies levels are alike in most respects but different in some. They can sometimes be described as co-ordinate, but at other times they are not co-ordinate and cannot be dealt with in the same manner. Furthermore, there are several rules applying to names of taxa at one of these levels but not at the other.

The names of species and subspecies are *similar* in the following ways:

1. They are the names of taxa.
2. The taxa represented by the names have types, governed by a single set of rules.
3. The names can be proposed in the same manner.
4. They are published under the same requirements.
5. They are formed in the same way from the same source words and are spelled under the same rules.
6. They must both agree with the generic name in gender (if adjectival) and use the corresponding gender endings (if appropriate).
7. Original and subsequent spelling variations are subject to the same rules.
8. The author of the name is determined in the same manner.
9. The date of publication is determined in the same manner.
10. The author and date are cited in the same manner.
11. Both may have synonyms (but there are differences in the treatment of them).
12. Both may be homonyms and are then treated under the same rules.

Names of species and subspecies are *different* in the following ways:

1. There are no categories subordinate to the subspecies level, so no names are transferred from below.
2. The species is the basic category in the group and the one which cannot be dispensed with; the subspecies can, at any time, be reduced to the status of a rejected taxon with its name a synonym of the name of the species, or mention of it can be eliminated altogether.
3. Subspecific names are trinominal instead of binominal. They cannot be properly cited without indication of the species.
4. In case of synonymy, the name of the species is selected from a combined list that includes the name of the species, its synonyms, the names of all the subspecies, and all their synonyms. The name of the subspecies is selected from the names applied only to that one subspecies. Thus, priority applies under partly different rules to the two sorts of names.
5. If the name of a species and name of a subspecies are homonyms, and they were published simultaneously (either in the same work or in different works), the one proposed for the species takes precedence [Article 57(e)].
6. A species name which must be rejected can be replaced with any of its subspecies names, but a subspecies name (other than that of the nominate subspecies) cannot be replaced with a species name.
7. A species can be divided into two species or into two subspecies; a subspecies can only be divided into two subspecies.
8. Species can be internally subdivided into subspecies; subspecies are not normally subdivided, and never within the framework of the standard hierarchy.

It can only be concluded here, as for genera, that Article 43 of the 1961 Code, referred to above, is a statement only of a similarity in general treatment but not of identity of status. The specific statement that the categories are subject to the same rules is completely irrelevant, and the implication that the names of the taxa at these categorical levels are governed by the same rules is a considerable exaggeration.

Causes of changes of names of species. Basically there are four causes of changes of names. They are

(1) increase in knowledge about the animal or its nomenclatural history, making the name in use no longer acceptable;
(2) actions of the Commission under the Plenary Powers to suppress a name or establish a name (in contravention of the rules);
(3) change of the rules themselves, as making previous decisions no longer acceptable; and
(4) changes caused by inexpert, ignorant, or malicious actions of persons posing as taxonomists but having more or less selfish purposes other than the increase of knowledge.

The first group consists of changes that are unavoidable unless we are willing to give up all increase in knowledge of classification, relationships, distribution, and identity of species.

Actions under (2) are supposedly taken only for the purpose of preventing change of names or contributing to stability of names. It is seldom, however, that such an action is taken before there are two names involved, and at that time any action will result in a change of one of the names.

The changes in (3) result from changes in procedures between the time of one publication and another (that is, adoption of new rules). Any set of rules will force certain name changes; changing the rules always results in additional changes of names, as well as the saving of others. If there were enough taxonomists to find out all the facts of taxonomy before the rules are changed, many difficulties would be eliminated, but with more than a million species to deal with, taxonomists have scarcely begun the task of recording the data.

During most of the life of regulated nomenclature, changes in the rules have been few. Even at the 1930/1931 deadline, few changes of names from the past were produced, when making more rigid requirements for the future. The Copenhagen Decisions made substantial changes in the rules, with many of them retroactive. Although it is doubtful if many taxonomists followed these rules closely, because they were extremely verbose in presentation and excessively complicated in detail, the rules did introduce many changes.

The 1961 Code rejected some of the Copenhagen Decisions but ended up with a large number of rule changes of which many were retroactive in effect and themselves produced name changes.

In the fourth group are the multitudinous errors made by inadequately trained persons attempting to use or apply names, as well as the occasional unjustified action of taxonomists with personal motives and inadequate scruples. Taxonomists are held responsible for these changes, whether justifiably or not.

The changes under (1) above are the normal ones due to growth of the science. Some changes are inevitable, but some are the result of inadequate work by taxonomists [see item (4)]. It is the changes in (1) and (4) together that have given taxonomy a bad name because of the changing of familiar names, although the changes caused by the new rules (3) will now have to take a large part of the blame for quite a few years to come.

From the point of view of the man using a name for a given animal, a change in the name by which it may be currently known can be produced (under the 1961 Code) by any of the following:

1. Discovery that the same specific name had earlier been given to a different species (for simplicity, assume a case of primary homonymy still existing).

2. Discovery that the species has an older name:
   (a) through uniting with it another species with an older name, or
   (b) through recognition of a previously unaccepted synonym, or
   (c) through placing in it a subspecies or variety with a name of older date, or
   (d) through correction of publication dates.
3. Subdivision of the species.
4. Discovery that the original species has been misunderstood by later authors, a different species being currently designated by that name.
5. Discovery that the supposed original publication is technically invalid and therefore a later name is the correct one.
6. Discovery that the name was incorrectly formed originally and must be emended.
7. Discovery that the original name has been misspelled by later authors.
8. Through removal of a secondary homonym from the genus in which it was a homonym, followed by automatic restoration of the former name.
9. Discovery that the generic name had previously been used for some other genus (homonymy).
10. Discovery of an earlier generic name for the same species (priority):
    (a) because another genus with an older name is found to be zoologically the same, or
    (b) because a subgenus of older name has been transferred from some other genus, or
    (c) because a supposedly invalid older synonym is found to be valid, or
    (d) through correction of dates.
11. Subdividing the genus into two or more genera, with this species falling in a new genus.
12. Elevation of a subgenus to the status of genus, so that its name becomes the generic name.
13. Changing a supposed genus to the rank of subgenus or synonym, so that its name disappears, in effect.
14. Discovery that the wrong species has been accepted as genotype, and that the species under consideration is not congeneric with the true genotype.
15. Discovery that the original publication was invalid and that therefore a later name is the correct one.
16. Discovery that the name was incorrectly formed originally and must be emended.

This is not an exhaustive list of the causes of name changes but includes the common ones. If stability of names is considered the most important aspect of taxonomy, then most of these causes can be made inoperative by changing the rules. Numbers 1, 3, 8, 10, 11, and 12 cannot be so eliminated; they persist as causes of changes under any system we are likely to consider—any system that does not stifle increase of knowledge.

However, the number of *causes* is no gauge of the number of *cases*. It is possible that most changes result from one or two of the causes, making the others relatively unimportant. There have been very few tabulations from which we may judge what causes predominate. Non-taxonomists seem to believe that most changes are caused by the rules, because taxonomists have claimed to them that the actions are not arbitrary or personal. Many taxonomists think that a majority of the changes are unavoidable.

In one family of animals there has recently been a study of all the generic names and also a study of the species of one part of the world. About 2,500 generic names were studied in the one; about 500 species in the other. In the case of the generic names there were about 150 changes necessitated by the facts brought out—almost entirely nomenclatural facts, inasmuch as the zoological status of the names was not directly studied. In the case of the species over 200 changes were necessitated by the revisionary study.

According to the above definition of changes that are unavoidable, 65 percent of the generic changes would have been necessary under any system of rules. Only 25 percent of the listed changes could be prevented by any changes in the rules. However, even in some of the latter cases, a change of one name from current usage (in which both names are employed) would result from any possible rule.

In the case of the species, exactly 50 percent of the changes were due to synonymy, the sinking of names as synonyms. This cannot be avoided by rule, since it is merely recognition of fact. In addition 25 percent more are in the category of unavoidable changes, bringing the total to 75 percent.

If the generic name study had included examination of the zoological status of the names, instead of just their nomenclatural status, bringing out new subjective synonymy, the percentage of unavoidable changes would doubtless have been much greater. The specific name study covered a region in which half the species were still undescribed. Had it been in a well-known area, the percentage of subjective synonyms might well have been even larger.

These estimates indicate that more than 75 percent of name changes are unavoidable no matter what we do with the rules. That the estimates are representative and can be duplicated in many other fields is substantiated by available estimates of other zoologists. This large majority should give pause to anyone who is tempted to blame the taxonomists because they change so many names. There can be no possible justification for so sweeping a criticism. But non-taxonomists are seldom concerned over the *quantity* of changes but rather over *certain* changes. And the cases that annoy them sometimes involve avoidable changes.

It is here that taxonomists fail to realize their responsibility to biology in general. Changes in names of species used by non-taxonomists in other sciences, in commerce, or in everyday life should be made only after exhaustive investigation has substantiated two facts: first, that the change is necessary; and second, that no other change will be required by foreseeable studies on other aspects of the species or its genus. For example, in the work on generic names alluded to above, a subgeneric name found to be pre-occupied was renamed. No investigation was made to determine whether the group of species involved really constituted a subgenus or not. If it should later be discovered that it does not, then the subgenus falls and its name becomes a synonym of something else. The new name would then have been unnecessary. If this had been an animal dealt with by other biologists or laymen, such halfway treatment would have been very undesirable. In the present case it can at most inconvenience a mere handful of specialists. Too many taxonomists forget to distinguish between the cases that will cause non-taxonomic difficulties and those that will not.

Taxonomists thus should be very careful in making changes in the names of species affecting non-taxonomists. All phases of the applicability of the name should be thoroughly investigated. It may be well to recall that these include at least the following:

1. Specific synonymy, whether the species is the same as any other known species.
2. Specific homonymy, whether the name has been given to any other species (according to whatever rules are then in effect).
3. Generic assignment, whether the species is correctly assigned to the genus.
4. Generic synonymy, whether the oldest available name is being applied to the genus and there are no undetected synonyms.
5. Generic synonymy, whether the name used for the genus is available, because of possible prior use for some other genus.
6. Publication, that the supposed original publication of the specific name was valid and was actually the first such publication.
7. Spelling, that it is correctly spelled according to the current rules.
8. Identification, that the specimens currently at hand are actually conspecific with the original ones, and not merely the same as those of some intermediate worker who may have been misidentifying his.
9. Homogeneity, whether the species as now understood is actually one species or should be divided into more than one.
10. Status of genus, whether the genus is a distinct genus or merely a synonym or subgenus of some other one.

If all these factors are checked, so far as is possible, before any much-used name is changed, there will be fewer opportunities for other scientists to feel that taxonomists are heedless of the general scientific welfare, or that they are merely playing games among themselves. It may then be possible to see in true perspective the tremendous size and complexity of the field of taxonomy and its true position as an indispensable part of biology.

The "work" of an animal

There are times, especially in paleontology, when the "specimens" available are not the animals themselves but some objects produced by the animals. Often these are entirely distinctive and may even be identifiable with known species. There are a wide variety of these objects, as listed below. Some of them are covered by the general regulations in the 1961 Code, some are excluded from nomenclature, and some are covered by the following special rules.

1. Definition: "Work of an animal. Result of the activity of an animal, but not a part of the animal, e.g., tracks, galls, worm-tubes, borings; does not apply to such fossil evidences as internal molds, external impressions, and replacements" [Glossary].
2. (Article 1 merely refers to Article 16.)
3. In satisfying the general availability requirements of Article 11, a name published before 1931 might have been accompanied by an "indication," of which one of the possibilities is "the description of the work of an animal, even if not accompanied by a description of the animal itself" [Article 16(a)(viii)].
4. "The Law of Priority applies . . . when, before 1931, a name is founded on the work of an animal, before one is founded on the animal itself" [Article 24(b)(iii)].

Neither Article 1 nor the Glossary makes it clear what sorts of objects are acceptable as the basis of names. The animals may be "living or extinct." The "work of an animal" is acceptable in lieu of the animal itself up to 1931. This is all that is definite. (The only other references to fossils, or objects that never were part of a living animal, are concerned only with generic names [Article 20 and 56(b)] and with citation of the geologic age of a new species [Recommendation 73C].)

The Glossary seems to imply either that internal molds, external impressions, and replacements cannot be named, either as animals or as the work of animals, or that they may be treated as if they were animals. Inasmuch as a large percentage of fossils are replacements, and many others are molds or impressions, it is certain that paleontologists will reject any such implication as the first. This leaves no clear solution in the 1961 Code to the question of what can be named, at least after 1930, and even after 1960.

Among the few references listed above, there is one error. Galls are not the work of animals but the work of plants. The animal does not mold the gall; its presence stimulates the plant tissues to grow in that particular form. The inclusion of this item among the work of animals further confuses the problems of naming the traces left by animals.

A major group of items that are literally the work of animals are omitted almost entirely. These are the lifeless items made by animals, including nests, dams, honeycomb, cases, tubes, webs, and so on. There is no indication of how most of these are to be dealt with nomenclaturally if they cannot be associated with actual animals.

The variety of these objects may be illustrated by the following list, in which the treatment under the Code is suggested for each.

Traces of animals

"Work" of animals

    Tracks, footprints; which show the original shape of a body part [Glossary: up to 1931]

    Trails; in which the path of the activity of the animal is shown [No mention]

    Coprolites, faecal pellets, castings [No mention]

    Borings; holes whose walls are principally the material of the matrix, sometimes lined with silk or extraneous matter; including worm tunnels, insect tunnels and mines, holes in trees, burrows in earth, borings in rocks, wood, and so on [Glossary; up to 1931]

    Constructions; separate objects constructed by the animal

        (a) of secreted or internally worked material alone; webs, honeycomb, cocoons, pupal cases [No mention]

        (b) of outside materials held together by secretions or internally worked materials; tunnels of termites, organic tubes of many animals, cases (of sand, plant fragments, and so on), mud nests, paper nests, trash heaps of ants [No mention]

        (c) of outside materials alone; nests of birds, dams, sand traps [No mention]

"Work" of host organism

    Galls; produced by the host plant (or rarely host animal) under the stimulus of the presence of the parasite [Glossary, and apparently mis-classified; up to 1931]

Substitutions

    Impressions, molds [Article 1, not clear]

    Casts, steinkern [No mention]

    Replacements; by carbonization, pyritization, dolomitization, calcification, silicification, leaching, distillation, and so on [Article 1, not clear]

Many of these have been considered distinctive of their species. Some have been the basis of described and named species. The only clear guide at present is the agreement of workers in the field as to what is necessary and appropriate.

Names of hybrids

An animal that is known to be a hybrid between individuals of two species cannot be named under the 1961 Code, but under the *Règles* it could be so named under the old Article 18. An animal not recognized as a hybrid could be given a specific name under either of the codes.

The relevant provisions of the 1961 Code are these:

1. "Names given . . . to hybrids as such . . . are excluded" from zoological nomenclature [Article 1].
2. "A name is or remains available even though . . . it is found that the original description relates . . . to an animal or animals later found to be hybrid" [Article 17(2)].

CHAPTER 27

The Names of Subspecies

The subspecies category

The subspecies level is the first level below the species level. It is, under the 1961 Code, the only level recognized below the species level, as all other possible levels (infrasubspecific) are explicitly ruled out of zoological nomenclature.

This level is quite definitely comparable to the other levels in the hierarchy. In it are placed groups of individuals, but there is some possibility that these groups are of different nature than the groups placed at other levels.

The subspecies taxa

The taxa placed in the subspecies category are the subspecies. These are groups of specimens based on geographical or stratigraphical distribution in addition to features held in common. Subspecies are populations, and in this regard they seem to be substantially different in nature from species or other taxa, but this does not affect their names, which are regulated in much the same way as those of species.

The relationship between species and subspecies, as members of what the 1961 Code calls the "species-group taxa," is discussed in the previous chapter.

Among every group of subspecies there is one that contains the type of the species. This is the nominate subspecies [Article 47(a)], previously known as the typical or nominotypical subspecies. In fact, every species theoretically contains at least one subspecies, the nominate one, whether subspecies are recognized in it or not.

The aspects of the subspecies taxa that are concerned with nomenclature are their types and their names.

The types of subspecies

The rules governing the types of subspecies are cited in Chapter 29, Types of Species and Subspecies. There is only one aspect of typification in which the taxa at these two levels are *not* identical, and that is in the nature of the nominate subspecies.

568

The names of subspecies

Much of the discussion of species-group names in the preceding chapter is relevant to the names of subspecies. Some additional problems are discussed here.

Proposal of names of subspecies. A name may be proposed for a subspecies through

(1) transfer of the named subspecies into the animal kingdom; or
(2) direct proposal for a subspecies in a publication
    (a) by a specified author,
    (b) by a specified group of authors,
    (c) by a speaker reported in the minutes of a meeting, or
    (d) before 1951, by an anonymous writer or writers; or
(3) a reviser reducing a named species to subspecies rank.

Before 1961, a name could also be proposed by elevating to subspecies rank a named variety or form.

Publication requirements of names of subspecies. These are the same as cited for species names in the preceding chapter, except that the names may originally have been either trinominal or binominal but must now be cited as trinomina.

Sources of names of subspecies. The trinominal name of a subspecies consists of a specific name (binomen) and a third word (subspecific epithet) that denotes the subspecies. The sources from which specific and subspecific epithets together may be derived are listed in Chapter 28, Epithets.

The subspecific epithet of one subspecies in every species (the nominate subspecies) will be identical with the specific epithet of that species.

Formation and spelling of names of subspecies. These aspects of the subspecific epithet are discussed in Chapter 28, Epithets.

Author of the name of a species. The author of a subspecific name is the person who first publishes the subspecific epithet (the third part of the name) in any acceptable trinomen or binomen. Nevertheless, the name exists only as a trinomen (or a binomen) and is cited only when the epithet is part of such a name.

The rules applying to the trinomen are interpreted here: those applicable only to the epithets (the second and third parts of the trinomen) are cited in Chapter 28, Epithets.

The author of a subspecific name (trinomen) is

(A) the person who was responsible for the name in the first publication
    in which it is acceptably published:
    (1) if it was first published as the name for a subspecies;
    (2) if it was first published as the name for a species (binomen)
        and was later changed to the subspecies level;
    (3) if it was first acceptably published without indication of its
        infraspecific rank and was later specified to be at the sub-
        species level;

(4) if it was first published as a variety, form, aberration, or other infraspecific taxon not now acceptable under the 1961 Code but was later elevated to the rank of subspecies (the first *acceptable* publication is the one at time of elevation);

(5) if, before 1961, it was first published as a synonym of the name of a species or subspecies;

(6) even if it was first published with a spelling error that must be emended by a subsequent writer;

(7) even if it has been justifiably emended by a subsequent writer (a justified emendation takes the author of the original—misspelled—name);

(8) even if it is itself an unjustified emendation of some earlier name (an unjustified emendation takes its own author);

(9) even if, after 1960, the spelling has been changed because of new requirements of the 1961 Code, such as
    (a) deletion of a hyphen,
    (b) writing out of a numeral,
    (c) deletion of a diacritic mark or apostrophe,
    (d) conversion of a German umlaut to *e*, or
    (e) change of *M'*- or *Mc*- or *Mac*- to *mac*- (from 1954 to 1960, but only a recommendation in the 1961 Code);

(10) even if other persons appear as principal author of the publication (the author of the name is the person actually responsible for the name as published);

(11) even if, before 1951, the author was unknown (anonymous);

(12) even if in the original publication the name itself is credited to some other person (from a museum label, an unpublished manuscript, correspondence, and so on);

(13) even after the epithet is combined with a different generic name or a different specific name (binomen);

(B) some other person than the one responsible for its acceptable publication *if* it is an acceptable emendation of an earlier name (it then takes the author of the original—erroneous—spelling).

Date of the name of a subspecies. The date of the name of the subspecies is the date of publication of the original trinomen (or binomen), except

(1) when the name is an unjustified emendation (which takes the date of its own publication);

(2) when the name is a junior homonym (which takes the date of its own publication, not that of the senior homonym, which is an identical trinomen).

The actual date of the subspecies name (the trinomen) is the date (day, month, or year, as near as is known) on which

(A) the original combination (trinomen or binomen) was published in complete compliance with all relevant rules:

1. It must conform to the Publication Requirements listed above [Articles 1, 2(a), 3, 8, 10, 11, 12, 13, 14, 15, and 21].
2. If publication of the work is interrupted, a new name can date only from the time of the completion of all requirements [Article 10(a)].
3. If the work bears a publication date, that is the date of any included new names unless there is direct evidence to the contrary [Article 21(a)].
4. If the date of the work is not completely specified, it is to be interpreted as the earliest day demonstrated by evidence, but in the absence of such evidence, as the last day of the stated month or year [Article 21(b)].
5. If the date of publication specified in a work is found to be incorrect, the date of included names is to be interpreted as the earliest date demonstrated by any evidence [Article 21(c)].
6. If the date of publication of the work is not indicated, the names are to be given the earliest date demonstrated by external evidence, such as mention in another work [Article 21(f)].
7. When a species is lowered to the rank of subspecies, it retains the original date.

(B) a later action was taken: when an infrasubspecific taxon is raised to rank of subspecies, it then dates from the time of elevation [Article 10(b)].

Citation of the names of subspecies. General aspects of the citation of names, applicable to both subspecies and species, are discussed under this heading in Chapter 26, The Names of Species. These refer to citation of the trinomen itself, its author, its date, and its bibliographic reference.

The actual rules covering these four aspects are intended to apply to both binomina and trinomina. They are cited in Chapter 26, The Names of Species, where one should read subspecies for species and trinomen for binomen.

Conflicts between species-group names. Several sorts of conflicts can arise between the names of species-group taxa or their various forms. The conflicts may involve only subspecific names (trinomina) or both binomina and trinomina at once. They are listed and exemplified under this heading in Chapter 26, The Names of Species.

Synonymy of names of subspecies. When two or more names are applied to a single subspecies, the names are all synonyms. One of these synonyms will be the accepted name and the others will be the rejected synonyms. Some of these synonyms may originally have been in the form of trinomina, some of binomina. Thus, the problems of synonymy of subspecies names are independent of the specific epithet in some respects. The synonymy of the subspecies names is governed by the same general rules as the synonymy of the names of species. A real difference appears, however, because the subspecific names may enter into the synonymy of the specific names but not vice versa. The synonyms of a subspecific name (trinomen) consist of those names actually applied to that one subspecies.

Two or more names for one subspecies can occur by being

(A) published simultaneously in one work
    (1) by separate proposal with different type specimens
        (a) with both treated as subspecies or (before 1961) as other infraspecific taxon;
        (b) with one treated as a subspecies and one as a species (later reduced to subspecies),
        (c) with both treated as species (both later reduced to subspecies), or
        (d) before 1961, with one or both treated as a variety, race, or other (later elevated to subspecies);
    (2) (before 1931) by inclusion of an unpublished name as a "synonym" of a name being proposed as new (publication in synonymy)
        (a) for a subspecies, or
        (b) for some other infraspecific taxon;
(B) published in separate works
    (3) for the same type specimen;
    (4) for different specimens later held to be consubspecific, and
        (a) both as subspecies or other acceptable infraspecific taxon, or
        (b) both as names of species (later reduced to subspecies), or
        (c) one as name of a subspecies and one as name of a species (later reduced to subspecies), or
        (d) before 1961, with one or both treated as a variety, race, and so on (later elevated to subspecies);
    (5) the later one as a replacement for the earlier one,
        (a) as a new name, or
        (b) as a new spelling.

Determination of which synonym is to be employed as the correct name for the subspecies depends on the following basis for choosing:

(a) The name is to be chosen from the names that have been applied directly to that subspecies, or
(b) for the nominate subspecies, the name is determined for the species and then applies automatically to the nominate subspecies [Article 47(a)].
(c) In (A)(1) above, the first reviser chooses the name [Article 24(a)].
(d) In (A)(2) above, there is automatic determination by the author (acting as a sort of immediate first reviser).
(e) If either name is itself a junior homonym, it will be automatically eliminated [Article 53].
(f) In all other bases priority will determine the name to be accepted.

Remarks on subjective and objective synonymy, under Species above (Chapter 26), also apply here.

It must also be noted that the subgeneric names and their synonyms are all part of the synonymy of the species in which the subgenera are placed. Any one of them may, under certain circumstances, become the name of the species (except for junior homonyms). They are thus simultaneously part of two synonymies, that of the species and that of one of its subspecies. And all the synonyms of the species are also synonyms of some of the subspecies.

Priority among names of subspecies. When a subspecies has received two acceptable names (synonyms) and neither one is a junior homonym or has been rejected by the Commission, and if they were not published simultaneously, choice between them is made on the basis of priority (prior or earlier date). This is exactly the same as the priority among the synonyms of a species, described in more detail in the previous chapter.

Priority also affects subspecific names in another way. As mentioned earlier in this chapter, subspecific names are part of the synonymy of some specific name, for which purpose they are cited in binominal form. For example:

A. Subspecies
    X-us albus Petrof
        X-us albus albus Petrof
        X-us albus pallidus Sigfried
        X-us albus brunneus Pacheco
        (*X. a. niger* John =)

B. Synonymy
    X-us albus Petrof
        X-us albus Petrof, 1849
        X-us pallidus Sigfried, 1921 (as subspecies)
        X-us brunneus Pacheco, 1944 (as subspecies)
        X-us niger John, 1945 (as subspecies; same as *X. brunneus* Pacheco)

The name of the species will be determined by priority among the four names. If *X-us albus* is discovered to be a junior homonym, *X-us pallidus* would become the name of the species and the nominate subgenus, with subspecies and synonymy now appearing like this:

C.
    X-us pallidus Sigfried, 1921
        X-us pallidus pallidus Sigfried (s. str.)
        X-us pallidus novus Author, 1966
        (*X-us albus* Petrof, 1849; not Curtis, 1835)
        X-us pallidus brunneus Pacheco, 1944
        (*X-us niger* John, 1945)

Such things as misspellings of the subspecies name, or the stage or sex of the specimens, do not affect priority; they are listed under species names in the preceding chapter.

First reviser and names of subspecies. The two circumstances in which a "first reviser" may select the subspecies name to be used are cited under the counterpart Species section (Chapter 26). For this purpose, the names of subspecies are written as binominals and treated exactly like names of species.

Homonymy of names of subspecies. For purposes of homonymy all subspecific names are treated like binomina and are governed by the rules for specific homonyms. For example: *X-us albus pallidus* L. is treated as if written *X-us pallidus* L.; it is a homonym of *X-us pallidus* F. The rules for species apply equally to subspecies.

Rejection of names of subspecies. In Chapter 26 there are twenty-six reasons listed for rejecting the name of a species. All of these except (4) are fully applicable to subspecies names, although these are for this purpose *written* as binomina. Item (4) states that the names of species must be binominal; it would read trinominal for subspecies, with the exception just noted.

In the same section there is also a list of fourteen circumstances which do *not* require rejection. All of these are fully applicable to subspecies.

Replacement of names of subspecies. Under the counterpart Replacement section in Chapter 26, the twenty-six reasons for rejecting names are again listed to show the action to be taken in replacing the rejected name. All are applicable also to subspecies.

It must be noted that the "synonyms" from which the name of a subspecies is to be selected include only the names directly applied to that one subspecies. The name of the nominate subspecies is never selected as a subspecies name but is derived automatically from the name of the species.

Misapplied names of subspecies. When a writer quotes an earlier subspecies by name, he is assumed to be referring to the actual zoological subspecies properly indicated by that name. If there is evidence that he has used it by mistake for some other subspecies, such as by erroneous synonymizing of it with some extraneous subspecies, his misuse of it is to be corrected. His use of the name, being different from the original, is a sort of junior homonym. It has no standing in nomenclature but does need to be recorded as the correction of an error, usually by being cited in the synonymy of the name of another subspecies. The nature of this pseudo-name must be clearly indicated:

> *X-us albus* Linnaeus, 1758
>> *X-us albus albus* Linnaeus s. str.
>> *X-us albus pallidus* Fabricius, 1801
>>> (*X-us albus albus* of Doe, 1925; not Linnaeus, 1758)

Misidentification of an actually new species with some already named species is covered by the codes, but subspecies are not directly mentioned. A paraphrase of Article 49 would certainly be applicable:

1. The subspecific name used in an erroneous identification cannot be retained for the subspecies to which the name was wrongly applied, even if the two subspecies in question are in, or are later referred to, different species.
2. The exception cited for genotype species is not applicable to subspecies.

Combining subspecies. Two subspecies within one species can be combined if taxonomic facts so indicate. The combined subspecies "takes the oldest valid name" among those applied to it [Article 23(e)(ii)].

Dividing a subspecies. A subspecies can be divided only by separating its components into two separate subspecies. The resulting position of the type specimens of all available names will determine the synonyms of each new subspecies, and priority will determine which is to be adopted for each subspecies. If there are no names for one of the subspecies, it must receive a new name.

Changing the category. A supposed subspecies can be elevated to the rank of species. There is no change of name except for citation as a binominal. *X-us albus fuscus* becomes *X-us fuscus*. A supposed species can be reduced to the rank of subspecies. There is no change of name except for citation as a trinominal. *Y-us sinensis* becomes *Y-us pulcher sinensis.*

Taxa subordinate to the subspecies. The 1961 Code accepts no category below subspecies after 1960. Before 1961, variety and form could also be used as infraspecific categories, and in a few cases one of them was used as an infrasubspecific category. It is best to assume that there is no level below subspecies, as far as nomenclature is concerned.

The levels of species and subspecies are sometimes said to be co-ordinate —governed by the same rules. This is an exaggeration. The similarities and differences in status and treatment are listed under the counterpart Species section (Chapter 26).

Causes of changes of names of subspecies. In discussing the causes of the changes of names of subspecies, the names are generally written as binomina and then are subject to the causes discussed under this heading in Chapter 26, The Names of Species. The only inapplicable items are 2(c) and 3 (p. 562). One additional item may cause a change in a subspecies name. This would be item 17, thus:

17. any change in the specific epithet.

The extensive remarks, under the counterpart Species section (Chapter 26), on these changes and their avoidance, apply equally to subspecies.

Categories below the subspecies. All subdivisions of subspecies or groupings of animals at levels below the subspecies level are specifically ruled out of zoological nomenclature since 1960. In the past they were occasionally used like taxa and given trinominal or quadrinominal names. Presumably, under the 1961 Code, such previous uses are not to be recognized now, unless they are now considered to represent subspecies. It is possible, however, to use the subdivisions and their names outside of nomenclature, in the same manner as parataxa, although use of any names that look like zoological names (and are not) would be likely to cause confusion.

The relevant provisions of the 1961 Code are these:

1. Definition: "Infrasubspecific, a. Of a category or name, of lower rank than the subspecies, and, as such, not subject to regulation by the Code" [Glossary].
2. "Names given . . . to infrasubspecific forms as such . . . are excluded" from zoological nomenclature [Article 1].
3. "A name first established with infrasubspecific rank becomes available if the taxon in question is elevated to a rank of the species-group" [Article 10(b)].
4. "After 1960, a new name . . . proposed explicitly as the name of a 'variety' or 'form' . . . is not available" [Article 15].
5. "Infrasubspecific forms are excluded from the species-group and the provisions of this Code do not apply to them" [Article 45(c)].
6. Determination of whether a given name is subspecific or infrasubspecific is made under Article 45(d) and (e), as cited in Chapter 20 under The Infraspecific Categories.
7. A named variety or form is, in genera published before 1931, to be considered one of "the originally included species" for purposes of genotype fixation by subsequent designation [Article 69(a)(i)].

The "work" of an animal. Subspecies are named on the same basis as species, except that the distinguishing features may be of a different sort. They can thus be proposed for objects classed as "the work of an animal," rather than specimens of the animal itself. This is discussed in Chapter 26, The Names of Species.

Names of hybrids between subspecies. This subject is not directly covered in the codes. Hybrids were at one time named by a formula, but all such "names" are ruled out by the 1961 Code. If specimens were considered to be hybrids between two subspecies, they could not be given a name in zoological nomenclature [Article 1]. However, if a subspecies is founded on certain specimens, and these are later discovered to be hybrids, the name retains its nomenclatural status [Article 17(2)] but falls into the synonymy of both the subspecies.

CHAPTER 28

Epithets[1]

During the preparation of these discussions of nomenclature, it has been found to be impractical to do without a word to denote the second part of the name of a species, to enable us to talk about the word *sapiens* in the name *Homo sapiens*. In the 1961 Code this word is called the specific name, to contrast with the generic name *Homo*. This usage forces us to make a distinction between the adjective expressions "name of the species" and "specific name." The expressions have identical inherent grammatical meanings, and it would seem to be very difficult to maintain the distinction.

Although this usage may have been common in the past, as suggested by the fact of its adoption at the Copenhagen and London Congresses and by its use during the nineteenth century, it had been abandoned by many nomenclaturists both in the United States and abroad (for example, Richter, 1948). This abandonment took the form of return to the Linnaean term "trivial name," which would cover both the second part of the name of the species and the third part of the name of the subspecies. It could be modified to refer to either one of these, as "specific trivial name" or "subspecific trivial name." Trivial name was the term adopted by the Paris Congress in 1948, but its action was reversed at Copenhagen in 1953 by a narrow vote.

The two sets of terms, in relation to the species and the subspecies *Mus musculus domesticus*, are as follows:

	1948	1961 Code
Name of genus	*Mus*	*Mus*
Generic name	*Mus*	*Mus*
Name of species	*Mus musculus*	*Mus musculus*
Specific name	*Mus musculus*	*musculus*
Name of subspecies	*Mus musculus domesticus*	*Mus musculus domesticus*
Subspecific name	*Mus musculus domesticus*	*domesticus*
Trivial names	*musculus, domesticus*	——

[1] This chapter covers epithets at both the species and the subspecies levels.

577

	1948	1961 Code
Specific trivial name	*musculus*	——
Subspecific trivial name	*domesticus*	——
Species-group names	——	*musculus, domesticus*

These can be arranged differently to emphasize the names rather than the terms:

	Up to 1953	1961 Code
Mus	Name of genus	Name of genus
	Generic name	Generic name
Mus musculus	Name of species	Name of species
	Specific name	
Mus musculus domesticus	Name of subspecies	Name of subspecies
	Subspecific name	
musculus	Specific trivial name	Specific name
		Species-group name
domesticus	Subspecific trivial name	Subspecific name
		Species-group name
musculus, domesticus	Trivial names	Species-group names

The two systems differ in two ways. First, the 1948 system is grammatically defensible whereas the 1961 system requires us to distinguish in usage between expressions of identical inherent meaning. Second, the older system provides a term for use in referring to the second and third parts of all names together when discussing the features they have in common (for example, spelling, establishment, synonymy, and homonymy).

As shown above, the 1961 Code does offer an alternative to the expression "trivial names." This alternative is "species-group names." The species group includes the taxa in the categories species and subspecies. By rule, the names of species are binominal and those of subspecies trinominal, so it is impossible for the words *musculus* and *domesticus* to be names of taxa in the species group. The 1961 Code is thus inconsistent, and it does not seem to be possible to deal with nomenclatural problems in the detail used in this book by using "species-group names" for both *Mus musculus* and *musculus* as well as both *Mus musculus domesticus* and *domesticus*.

It appears that the drafters of the 1961 Code were not entirely satisfied with their system for referring to the second half of the binomen ("specific name" or "species-group name"), because on page 45 of both editions they refer to "the generic and specific elements of a binomen."

Neither of these sets of terms is entirely satisfactory. They both tend to maintain the fiction that the second word in the name of the species is itself a name. It is only part of a name. This difficulty has been solved in the botanical code by the use of the term "epithet" to refer to this word. Specific and subspecific epithets can thus be dealt with individually or both can be referred to at once by the unmodified term—epithets. This last solution is adopted here as the most appropriate and effective one.

In this book the following scheme of terms is used:

1. Name of genus	*Mus*
2. Generic name	*Mus*
3. Name of species	*Mus musculus*
4. Specific name	*Mus musculus*
5. Name of subspecies	*Mus musculus domesticus*
6. Subspecific name	*Mus musculus domesticus*
7. Specific epithet	*musculus*
8. Subspecific epithet	*domesticus*
9. Epithets	*musculus, domesticus*

The sources and grammar of epithets

Inasmuch as the problems of generic names are dealt with elsewhere, this section applies only to specific epithets—the second part of the binomen or specific name, and to subspecific epithets—the third part of the trinomen or subspecific name.

Grammatical relations. The words used in names are Latin or considered as latinized, and there is a definite grammatical relationship between the two parts of the binomen. The generic name is assumed to be a noun in the nominative singular, and the specific epithets are modifiers of that noun. Such a modifier can be:

(1) an adjective in the nominative singular modifying the generic name [Article 11(g)(i)(1)],
(2) an adjective used as a substantive in the genitive case, derived from the epithet of an organism with which it is associated [Article 11(g)(i)(4)],
(3) a participle (adjectival verb form),
(4) a noun in the nominative singular in apposition to the generic name [Article 11(g)(i)(2)],
(5) a noun in the genitive case, singular or plural [Article 11(g)(i)(3)],
(6) a gerund (substantive verb form).

1. Adjectives. Latin adjectives as a class cannot be distinguished by their endings, but there are a limited number of possible endings. E.g., *aureus, bona, perditum, asper, gibbera, glabrum, palustre, laevis, anceps, atrox,* and *cantans.* Thus, Latin adjectives can end only in the letters *a, e, m, r, s,* or *x.*

Very rarely a Greek adjective (presumably carried through Latin) is used as an epithet. The genders of these are rather diverse and must be determined in a dictionary of classical Greek. A list of examples is given in R. W. Brown, *Composition of Scientific Words,* pp. 29–30. Such Greek adjectives can end in *a, e, n, r, s,* or *y.*

An epithet which is a Latin adjective (or treated as such if of barbaric origin—non-classical) must agree grammatically with the generic name. It must agree in gender and in number [Article 30]. Inasmuch as all generic names must be singular, most epithets will also be singular in number. For agreement in gender, it is necessary to determine the declension of the adjective, which can be done in a Latin dictionary. If the declension has different endings for the genders, one must be used to agree with the gender of the generic name (discussed in Chapter 23, The Names of Genera). The endings (masculine, feminine, and neuter) of the possibilities in the various Latin declensions are as follows (plurals not being used with adjectives):

-us, -a, -um; bonus, bona, bonum
-r, -ra, -rum; asper, aspera, asperum; glaber, glabra, glabrum
-r, -ris, -re; paluster, palustris, palustre
-is, is, -e; cornis, cornis, corne; tristis, tristis, triste
-ps, -ps, -ps; anceps, anceps, anceps
-x, -x, -x; atrox, atrox, atrox; felix, felix, felix; tenax, tenax, tenax
-or, -or, -or; bicolor, bicolor, bicolor
-ns, -ns, -ns; cantans, cantans, cantans; repens, repens, repens.
-pes, -pes, -pes; longipes, longipes, longipes
-us, -us, -us; vetus, vetus, vetus

The actual rules on gender of epithets are these:

1. "A species-group name, if an adjective in the nominative singular, must agree in gender with the generic name with which it is at any time combined, and its termination must be changed, if necessary, when the species is transferred to another genus" [Article 30].
2. "In names of the species-group, the ending must be changed, if necessary, to conform with the gender of the generic name with which the species-group name is at any time combined" [Article 34(b)].

The recognition of gender in epithets (as in generic names) is not easy. The ending -*a* will normally be the feminine of the Latin first declension, a very common ending, but it may also be a Greek neuter retained in Latin or improperly treated as if Latin. Examples are compound epithets ending in *-derma, -soma, -stoma, -chora, -gramma*. It may also be an arbitrary combination of letters, to which any gender may be assigned.

An Appendix to the 1961 Code (pp. 118–139) contains a long table in which the gender and endings are shown for both generic names and epithets.

2. Adjectives derived from the names of associated organisms. E.g., *Lernaea lusci,* a copepod parasite on *Gadus luscus,* a fish; *Chernes cimicoides,* louse-like.

3. Participles may be present participles, ending in *-ans* or *-ens* in all genders, or they may be perfect participles ending in *-atus, -etus, -itus, -sus, -tus,* or *-xus,* with gender endings *-us, -a, -um.* E.g.,

(*salto*, to leap)   *saltans*, leaping, *saltatus, -a, -um*, leapt
(*plecto*, to braid)   *plectens*, braiding, *plexus, -a, -um*, braided
(*salio*, to leap)   *saliens*, leaping, *saltus, -a, -um*, leapt
(*vincio*, to bind)   *vinciens*, binding, *vinctus, -a, -um*, bound
(*volo*, to fly)   *volitans*, flying
(*obsolesco*, to wear out) *obsolescens*, becoming obsolete

4. Nouns are frequently used as epithets in apposition with the generic name, as *Orca gladiator* (Orca, the swordsman). These do not change ending, although they do have gender. They can be identified as nouns in the Latin dictionary. They are always in the nominative singular, because they must agree in number with the generic name which is always singular.

5. Nouns may also be in the genitive to show possession, origin, commemoration, and so on. E.g., *Sorex granti* (masculine), *Falco eleonorae* (feminine). These include the only exceptions to the rule of agreement in number, because some possessives are in plural form. E.g., *Bombus muscorum* (of the mosses), *Pieris brassicae* (of the cabbages), *Helix desertorum* (of the deserts), *Spirodinium equi* (of horses). A few place names, which are personifications, are genitive nouns. E.g., *galliae, britannicae, sanctipauli*. The suffixes based on *-ensis* (*-iensis, -ense, -iense*) convert proper names into nouns in the genitive. E.g., *canadense* (of Canada), *arvensis* (of the field). Many other suffixes convert words into nouns, but these may not be in the genitive and must be determined in the Latin dictionary.

6. Gerunds are noun forms of verbs and are treated like nouns. They are indistinguishable from adjectives except in being nouns rather than modifiers. In Latin they always end in *-andus, -endus*, or *-iendus*. They correspond to English verb forms in *-ing*, when used as a noun rather than as a verb or an adjective modifier, as "the dancing of a dancing girl," in which the first "dancing" is a gerund and the second is a participle. E.g., *saltandus*, a leaping; *abdendus*, a hiding; *operiendus*, a covering. They are declined like adjectives and so have gender endings, *-us, -a, -um*, as required for agreement.

The grammatical sources of epithets. Epithets may be taken from either Latin or Greek or from a non-classical language. They must all be treated as Latin words. They may be:

1. Greek words. These of course are ancient Greek words. The epithets are transliterated into Latin, but they may retain the Greek endings. E.g., *gigas, phalanx, daimon, onyx, sialis*.
2. Latin words. These may be ancient, medieval, or "modern" Latin (neo-Latin). E.g., *dorsalis, comes, virgo, aureus, acer, vivax*.
3. Latinized Greek words. E.g., *taurus*, from Greek *tauros*, Latin *taurus; organum*, from Greek *organon*, Latin *organum; cyclops*, from Greek *kyklops*, Latin *cyclops*.

4. Latinized derivatives of Greek words. E.g., *giganticus*, from Greek *gigantos*, plus Latin suffix *-icus*.
5. Compound Greek words. E.g., *ichthopoles*, *dermopteros*. In addition to such compounds of two Greek roots, there may be compounded one root with a prefix or suffix. Greek prefixes include: *a-*, *amphi-*, *an-*, *anti-*, *di-*, *ecto-*, *endo-*, *epi-*, *hemi-*, *hyper-*, *hypo-*, *meta-*, *para-*, *peri-*, *pro-*, *syn-*. Greek suffixes include: *-ia*, *-icos*, *-idae*, *-ides*, *-ion*, *-ist*, *-ize*, *-oma*, *-sis*, *-tes*.
6. Compound Latin words. E.g., *avicularia*, *grandidentatus*. In addition to such compounds of two Latin roots, there may be compounded one root with a prefix or suffix. Latin prefixes include: *ab-*, *ambi-*, *antero-*, *bi-*, *circum-*, *co-*, *demi-*, *e-*, *in-*, *juxta-*, *pro-*, *semi-*, *sub-*, *super-*, *trans-*. Latin suffixes include: *-aceus*, *-ad*, *-alis*, *-anus*, *-arium*, *-ate*, *-bilis*, *-men*, *-or*, *-osus*, *-tia*, *-trum*, *-ulus*.
7. Words from other languages. E.g., *goliath* (Hebrew), *zigzag* (French), *iguana* (aboriginal American), *bohor* (african), *pongo* (Bornean).
8. Names of places. E.g., *magellanicus*, *cubensis*, *arctica*.
9. Names of ships. E.g., *galatheae*.
10. Names of persons. E.g., *smithi*, *elisabethae*, *bohartorum*, *mohammed*.
11. Names of geological periods or formations. E.g., *liasicus*, *miocenica*, *etchegoinensis* (for Etchegoin formation).
12. Anagrams. The letters of one epithet may be rearranged to form a new one. E.g., *rolum*, *mulor*; *diana*, *idana*.
13. Mythological names. E.g., *diana*.
14. Latinized non-classical words. E.g., *ntlakapamuxanus*.
15. Latinized names of persons. E.g., *erichsonii*, from Erichson, Latin Erichsonius.
16. Names of associated organisms. E.g., *lusci*, for a parasite of *G. luscus*; *querci*, for a species living on oak.
17. Arbitrary combinations of letters, with or without original meaning. E.g., *rulomus*, *nosa*, *loonae*, *fandana*, *gandana*, *handana*, *kandana*, and so on to *zandana*.

The meaning conveyed by epithets. Epithets are commonly words conveying meanings such as these:

1. Descriptive words. Historically these were the first ones used (by Linnaeus), and they are still the most common. They refer to size (*minuta*), to shape (*rotundus*), to similarity to something else (*simulatrix*, *lithopis*), to position of a part (*dextratus*), to number of parts (*trifida*), to abundance (*profusus*), to color (*viridis*), to toxicity (*toxicodendron*), to hardness (*durum*), to strength (*validus*), to nearness (*contiguus*), to arrangement of parts (*spiralis*), to transparency (*pellucidus*), to activeness (*impiger*), to persistence (*indefatigabilis*), and to many other attributes.

2. Ecological words. The habitat or manner of living can be shown by descriptive adjectives or nouns which are not clearly distinguishable from

1 above. They include such epithets as *subterraneus, xerophila, arboricola, marinus, montanus, indigenus, arcticus, parasitus, galligena, troglodytes, aggregatus,* and so on.

3. Geographical proper names. In order to emphasize the geographical range of a species, epithets may be based on geographic proper names of any sort, resulting in such epithets as *arizonicus, sanctaehelenae, panamensis, eriensis, atlanticus,* and so on. Geographical directions may also be the basis, as in such epithets as *orientalis, meridionalis, thula, borealis, hesperius,* and *occidentalis.* The name may be derived from a geographical part of the earth, as *insularis, peninsularis, isthmicus, littoralis, interioris,* and *mundanus.*

4. Patronyms, names of persons. Epithets may be based on the names of persons—usually the surname, as *smithi.* A variety of unusual patronyms have been proposed, of doubtful appropriateness, such as *austinclarki, embrikstrandi,* and *embriki.*

5. Classical words without descriptive meaning, such as *affinis, perplexus,* and *validus.*

6. Barbaric or non-classical words. Originally all taxonomists used Latin and all names were of Latin or Greek origin. Gradually there have come to be an increasing number of non-classical names proposed, and these have been acceptable under recent rules. Some of these are of Latin form and look like Latin names, such as *mexicanus* (modern geography) or *rulomus* (arbitrary combination of letters). Some are letter combinations that would have been entirely foreign to the Latin language, such as *pongo* or *ziczac.*

Some taxonomists still believe that such non-classical names are objectionable. Others have found them helpful, especially in large genera where the classical names have been virtually all pre-occupied or where transfers may make them so.

Orthography of epithets

Under this heading will be discussed the rules governing formation of epithets, those relating to the problems of original spellings, and those governing the emendation of epithets.

Formation of epithets. The following rules govern the formation and spelling of epithets (called "specific names" or "species-group names" in the 1961 Code):

1. "The name must be either Latin or latinized, or, if an arbitrary combination of letters, must be so constructed that it can be treated as a Latin word" [Article 11(b)]. (No examples are cited of a word which cannot be treated as a Latin word; perhaps this is intended merely to prevent the use of signs or symbols as part of names. As long as anagrams and arbitrary combinations of letters are permitted, any word can be used as an epithet, even *anzac* (the World War I acronym of Australian New Zealand Army Corps), *aguti,* or *zigzag.*)

2. "A species-group name must be a simple word of more than one letter, or a compound word . . ." [Article 11(g)(i)]. For example, the epithet in *Macrodonia van de polli* must be written without spaces, as *vandepolli.*

3. "A species-group name must not consist of words related by a conjunction . . ." [Article 11(g)(iii)].

4. "A species-group name must not . . . include a sign that cannot be spelled out in Latin" [Article 11(c)(iii)]. This apparently means that the sign must be a Latin letter, as in *c-album, x-maculatum,* because all "signs" which *can* be spelled out in Latin letters *must* be so spelled out [Article 26(b) in item (7) below]. This presumably is intended principally to prevent use of such an epithet as *?-album.*

5. "Zoological names must be formed in accordance with the provisions of Articles 26 to 31" [Article 25]. The purpose of this article, quoted in full, is in question.

6. "If a name based on a compound name is published as two separate words in a work in which the author duly applied the principles of binominal nomenclature, the component words are to be united without a hyphen, and the name is to be treated as though it had been originally published in that form" [Article 26(a)]. There is no time limit in this rule, which therefore makes it possible to publish a name as two words. This is in direct contradiction of Article 11(g)(i) in item (2) above. It is also not clear what is meant by "a name based on a compound name."

7. "A number or numerical adjective or adverb forming part of a compound name is to be written in full as a word and united with the remainder of the name (e.g., *decemlineata,* not *10-lineata)*" [Article 26(b)].

8. "If the first element of a compound species-group name is a Latin letter used to denote a character of the taxon, it is connected to the remainder of the name by a hyphen (e.g., *c-album)*" [Article 26(c)]. This is the only acceptable use of a hyphen under the 1961 Code.

9. "No diacritic mark, apostrophe, or diaeresis is to be used in a zoological name" [Article 27]. Being retroactive, this rule will require changes in the spelling of hundreds of names. Detailed rules are given in a later article for the respelling of names acceptably published under the *Règles* and using such marks. "Examples.—The name . . . *d'urvillei* [is to be corrected] to *durvillei,* and *nuñezi* to *nunezi;* but *mülleri* becomes *muelleri* and is not a homonym of *mulleri*" [Article 32(c)(i)].

10. ". . . the hyphen is to be used only as specified in Article 26c" [Article 27]. This single permitted use of the hyphen is described in item (8) above. "Examples.—The name *terrae-novae* is [to be] corrected to *terraenovae* . . ." [Article 32(c)(i)].

11. ". . . names of the species-group [must be printed] with a lower-case initial letter" [Article 28]. Apparently this rule is also retroactive.

Previous codes were not clear on this point, and some names based on proper names have been spelled with a capital initial letter.

12. "A species-group name, if an adjective in the nominative singular, must agree in gender with the generic name with which it is at any time combined, and its termination must be changed, if necessary, when the species is transferred to another genus" [Article 30]. Presumably this rule applies also to declinable participles and gerunds.

13. "A species-group name, if a noun formed from a modern personal name, must end in -i if the personal name is that of a man, -orum if of man (men) and woman (women) together, -ae if of a woman, and -arum if of women" [Article 31]. Such a name may also end in -ii, if the stem of the name already ends in -i. This Article 31 was entirely deleted in the 1964 Code, where it is replaced by the single Recommendation 31A, whose wording is similar to the rules quoted except that instead of "must end in i" it reads "should usually end in i."

14. "In names of the species-group, the ending must be changed, if necessary, to conform with the gender of the generic name with which the species-group name is at any time combined" [Article 34(b)]. This rule duplicates Article 30 exactly.

15. There are twenty-six recommendations in Appendix D on the formation of names. They include instructions to spell as "*maccoyi*" all names based on the personal names MacCoy, McCoy, and M'Coy. Also to use the mediaeval form *burdigalensis* in preference to *bordeausiacus* in a geographical name based on the city name Bordeaux.

16. In Appendix E, Recommendation 3, it is recommended that vowels should not be linked together in printing diphthongs. F, r example, *amoeba* should not be written as *amœba*, regardless of its origin.

Original spellings of epithets. The original spelling of an epithet is the one used in the publication in which it was first acceptably published in conformance with the rules and, of course, in combination with a generic name. Problems arise here only when (1) the original publication included more than one spelling, thereby demonstrating that all but one were errors, and (2) there is other evidence that the original spelling was erroneous.

1. Multiple original spellings. "If a name is spelled in more than one way in the original publication, the spelling adopted by the first reviser is to be accepted as the correct original spelling, unless the adopted spelling is subject to emendation under the provisions of Articles 26 to 31" [Article 32(b)]. (For definition of "first reviser," see the section entitled First Reviser and Names of Species in Chapter 26, The Names of Species. The rules governing emendation of epithets are cited below under Subsequent Spellings.)

2. Erroneous original spelling. After establishment of a name, the original spelling (orthography) of the epithet is to be preserved unless it is later possible to demonstrate that an error was made in the original

spelling. If the original author said that he was naming a species after the famous evolutionist Charles Darwin, but the epithet actually appeared as *darbini*, it would be evident that the "b" was an error for "w." This could be a typographical or printer's error, or it could be a *lapsus calami*, a slip-of-the-pen.

The rules have never been clear on the details of whether a certain spelling was intentional or an error. It is necessary for the sake of stability of names to require that the evidence for the error be found in the original publication. It is not enough to show that the original author *should* have followed some rule of Latin spelling which he did not follow. (For further discussion of these concepts, see Blackwelder, Knight, and Sabrosky, 1948; and for examples see Mayr, Linsley, and Usinger, 1953, p. 231.)

The 1961 Code defines "incorrect original spelling" thus: "An original spelling that does not satisfy the provisions of Articles 26 to 31 [on formation of names], or that is an inadvertent error (Art. 32a(ii)), or that is one of the multiple spellings not adopted by a first reviser (Art. 32b), is an 'incorrect original spelling.' . . ." Such an incorrect spelling exists if "there is in the original publication clear evidence of an inadvertent error, such as a lapsus calami, or a copyist's error (incorrect transliteration, improper latinization, and use of an inappropriate connecting vowel are not to be considered inadvertent error) . . ." [Article 32].

Subsequent spellings of epithets. In subsequent works spellings different from the original arise in several ways. They may be

(1) inadvertent misspellings, such as
  (a) a lapsus calami or slip-of-the-pen, in which, e.g., the name *X-us albifrons* is inadvertently written as *X-us albicollis;*
  (b) a typist's or copyist's error, uncorrected before publication;
  (c) a printer's error, sometimes made after final proof;
(2) intentional respellings or emendations, made either by the same author at a later date or by a later author, for
  (a) correction of an incorrect transliteration, or
  (b) correction of an incorrect latinization, or
  (c) correction of an incorrect connective vowel, or
  (d) correction of an evident error of whatever sort, such as an epithet spelled *caligorniensis* but given to a species said to be from California, or
  (e) correction of an initial capital letter, or
  (f) personal preference, whim, or false idea of origin or correct spelling;
(3) changes required by a change in the rules, as by the 1961 Code, including
  (a) deletion of any hyphen,
  (b) writing out of a numeral,
  (c) correction of a patronym used as a noun in apposition by adding the appropriate genitive ending,

(d) deletion of any diacritic mark, apostrophe, diaeresis, or non-German umlaut,

(e) conversion of a German umlaut to *e*,

(f) change of *m'-* and *mc-* to *mac-*, as in *m'coyi* and *mccoyi* to *maccoyi* [in Copenhagen Decisions but merely recommended in 1961 Code], and

(g) closing up over all spaces, as *van de polli* to *vandepolli*.

Emendation of epithets. Subsequent changes in the spelling of epithets (or names) may be either inadvertent errors or intentional emendations. The *errors* are to be corrected whenever found, according to the 1961 Code, but it is advisable to record them in the synonymy. They have no separate standing in nomenclature and do not pre-occupy a later usage of the same name [Article 19]. *Emendations* are intentional changes in spelling. If they are corrections of original errors, they are "justified" and replace the error in all respects, taking the date and author of the erroneous form. For example: *X-us walacei* Pliny (1884), published to honor the famous zoogeographer, would be corrected to *X-us wallacei* Pliny (1884) whenever used. (It is advisable, in major synonymies, to refer also to the error, its location, and its correction. If the later author was mistaken and there actually was no error, his "corrected" form would still be classed as an emendation but an unjustified one.)

Any other emendation is "unjustified," unacceptable under the rules. These name forms have separate status, with their own author and date. They are junior objective synonyms of the older name. They enter into homonymy and are available as replacement names [Article 19].

The rules governing the spelling or correction of epithets in later works are:

1. "The original spelling of a name is to be retained as the 'correct original spelling,' unless (i) it contravenes a mandatory provision of Article 26 to 31 [compounding, diacritic marks, initial capitals, agreement in gender, and modern patronyms [2]]; or (ii) there is in the original publication clear evidence of an inadvertent error, such as a lapsus calami, or a copyist's or printer's error . . ." [Article 32(a)].

2. If an epithet is spelled in more than one way in the original publication, the first reviser may choose which one is to be the correct original spelling [Article 32(b)].

3. "An original spelling that does not satisfy the provisions of Articles 26 to 31, or that is an inadvertent error [Art. 32a(ii)], or that is one of the multiple spellings not adopted by a first reviser [Art. 32b], is an 'incorrect original spelling' and is to be corrected wherever it is found; the incorrect spelling has no separate status in nomenclature, and

---

[2] The requirement in Article 31 for the emendation of certain modern patronyms was removed in the 1964 Code when that article was replaced by Recommendation 31A.

therefore does not enter into homonymy and cannot be used as a replacement name" [Article 32(c)].

4. "A name published with a diacritic mark, apostrophe, diaeresis, or hyphen is to be corrected by the deletion of the mark concerned and any resulting parts are to be united, except for one specified use of the hyphen [Art. 26c], and except that when, in a German word, the umlaut sign is deleted from a vowel, the letter 'e' is to be inserted after that vowel" [Article 32(c)(i)].

5. "Any demonstrably intentional change in the original spelling of a name is an 'emendation.' (i) A 'justified emendation' is the correction of an incorrect original spelling and the name thus emended takes the date and authorship of the original spelling. (ii) Any other emendation is an 'unjustified emendation'; the name thus emended has status in nomenclature with its own date and author, and is a junior objective synonym of the name in its original form" [Article 33(a)(ii)].

6. "Any change in the spelling of a name, other than an emendation, is an 'incorrect subsequent spelling'; it has no status in nomenclature and therefore does not enter into homonymy and cannot be used as a replacement name" [Article 33(b)].

7. "In names of the species-group, the ending must be changed, if necessary, to conform with the gender of the generic name with which the species-group name [epithet] is at any time combined" [Article 34(b)].

8. "A species-group name [epithet], if an adjective in the nominative singular, must agree in gender with the generic name with which it is at any time combined, and its termination must be changed, if necessary, when the species is transferred to another genus" [Article 30].

9. "If a species-group name based on a modern personal name is treated as a noun in apposition to the generic name, it is to be corrected by adding the appropriate genitive termination [Article 31(a)]. (This automatic correction was eliminated in the 1964 Code when this article became merely a recommendation [Recommendation 31A].)

CHAPTER 29

Types of Species and Subspecies

Certain specimens have special value in the taxonomy of species and sub-species. These specimens are called "types." Their value may be nomenclatural or taxonomic or both, but the establishment and control of the important types is governed by the rules of nomenclature.

The nomenclatural problems of types have no helpful connection with the taxonomic subject of typology. The nomenclatural types do serve important taxonomic functions, especially in identification, but even these have no firm contact with typology—the idea of an idealized morphological concept to which the species must conform.

In taxonomy, types have served several purposes:

1. As base for description. The type is often the principal source of the features given in the description. The type will also serve for later reference to show whether the published description was accurate or not. It may serve for later examination to add some further data to the original description.
2. As a standard for comparison. The specimens originally included in the species were usually so included because they agreed with the type in what the taxonomist deemed to be the pertinent features. Subsequently the same sort of comparison will serve to permit other specimens to be associated with the species.
3. As a source of data not shown by the type. In most cases the type is capable of showing only part of the basic attributes of the species. Other examples may show additional features. The allotype represents the opposite sex from the holotype. A paratype may show variation, but frequently does not.

In nomenclature, the use of species types is very similar to the use of genotypes. When a species is divided into two, because it is believed to have been composite, the specific name is applied to that one of the resulting species which contains the type. The purpose of the type is to show where the name must be applied. There is almost no difficulty about this, and rules about types as such are few.

589

The *Règles* did not, in fact, make any direct reference to type specimens. Several proposals intended to clarify their use were made at recent Congresses, and some rules and recommendations are included in the 1961 Code.

Nomenclaturally there can be only one type. If there were more, they could be in conflict as to the disposition of the name. *The type* is the only concept that can be used in nomenclature. The one nomenclatural purpose is to show to what the name shall be applied. (This one type can be a holotype, a lectotype, or a neotype, as outlined below.)

What is the type? The type of a species is a specimen specifically chosen as such, whether by the original author or a later writer. The relevant rules are:

1. "The type of a nominal species is a specimen" [Article 61].
2. "The type of each taxon of the species-group is a single specimen, either the only original specimen or one designated from the type-series (holotype, lectotype) or a neotype" [Article 72(a)].
3. "The type-series of a species consists of all the specimens on which its author bases the species, except any that he refers to as variants, or doubtfully associates with the nominal species, or expressly excludes from it" [Article 72(b)].
4. "type-specimen, n. The single specimen (holotype, lectotype, or neotype) that is the type of a taxon in the species-group" [Glossary].

The type of what? There is some question of what it is that is represented by a type. Under the *Règles*, it was taken for granted that the chosen specimen was the type of the *species*. In the 1961 Code it is the type of a *taxon,* or of a *nominal species,* or of a *replacement nominal species* (name?). Some writers have stated that it is the type of the *name* of the species.

The usage of the 1961 Code with "nominal species" is so similar to "name" that no distinction has been possible. A single zoological species, a group of specimens or populations, may have several names. For each name there is a nominal species. If each nominal species then has a type, the zoological species can and often does have several types. In this sense the type would be the type of a name. In taxonomy, however, a species is held to have only one correct name, and *the type associated with that name* is the type of the species. When the name of the species is changed, the type may also change. In this sense the species has a type, but that type is determined by its association with a name.

There is no question that the type is coupled to the name rather than to the zoological species. There is also no question that the type of the zoological species exists and is useful in taxonomy as well as in nomenclature.

The type of a "nominal species," which is the type associated with a particular name, may not be *the type* of any zoological species. It is at least a potential type of the species, in case its name ever becomes the accepted name of the species. (For discussion of "nominal taxa" see Chapter 25, Genotypy and the Types of Genera).

Kinds of species types

A variety of names have been applied to types, such as holotype, metatype, and plastotype. These specimens are of very different sorts and serve very different purposes. The general taxonomic utility of certain types has lent to these other terms an aura of importance which is sometimes undeserved.

These "types" have been classified into three groups: (1) primary types, the original specimens used by an author in describing or illustrating a new species, (2) supplementary types, the specimens serving as a basis for descriptions or illustrations which supplement or correct knowledge of a previously described species, and (3) typical specimens, ones of varying degrees of authenticity which have not been used in connection with published descriptions or illustrations.

This classification is clear and logical, but it does not seem to result in the most useful grouping. It is of little interest whether *the one basic specimen* was chosen originally (holotype) or subsequently (lectotype) from original material, or from later material (neotype). What counts is that these all represent the specimen of primary importance in taxonomy and nomenclature.

This viewpoint leads to a different classification of these specimens (as represented by the terms):

1. Primary types—the single nomenclatural types
    Holotype, lectotype, neotype
2. Secondary types—the specimens from which the primary type must be selected
    Syntypes (cotypes), paralectotypes
3. Tertiary types—other specimens originally set aside as of special taxonomic interest to supplement the primary type
    Paratype, allotype
4. Specimens identified as of special origin
    Topotypes
5. Specimens identified as to time or person of identification
    Metatypes, homotypes (homoeotypes), and so on
6. Specimens identified as to special treatment or use
    Plesiotypes, hypotypes, and so on
7. Replicas of type specimens
    Plastotypes

Relationship of terms. If a species was based on a single specimen, that specimen is the holotype [Article 73(a)]. If the original author had several specimens but stated that one particular one is the type, it is also called the holotype [Article 73(b)]. If the species was represented by more than one specimen and none was specified as type (that is, if the species has no holotype), all the specimens of the original type-series are syntypes [Article 73(c)]. Any zoologist may designate one of the syntypes as a lectotype (a subsequent single type) [Article 74(a)]. The remaining members of the type-series are paralectotypes [Recommendation 74E]. These are the terms and usages of the 1961 Code.

To be acceptable at all, a species must have either a holotype or syntypes. (The latter should subsequently be reduced to a lectotype and paralectotypes, but this may not have been done.) But there are two kinds of holotypes, unique original specimens and originally specified single examples from a type-series. These will sometimes be distinguished below as "unique holotypes" and "specified holotypes" respectively.

Primary types. The real type, for nomenclatural use, is the holotype, lectotype, or neotype. These are the primary types, and there can be only one per name. It will be the type of a species only if species are considered to have only one name.

Eligibility of specimens. The type of a species can only be a specimen. The type specimen can be either

(1) "the only original specimen" [Article 72(a)]; or
(2) "one designated from the type-series . . . all the specimens on which its author bases the species, except any that he refers to as variants, or doubtfully associates with the nominal species, or expressly excludes from it" [Article 72(a) and (b)]; or
(3) in the case of a replacement name, the type of the replaced name [Article 72(d)]; or
(4) in the case of the nominate subspecies, the type of the species [Article 61(a)]; or
(5) any non-original specimen, in the case of a neotype, providing it is shown to be
 (a) "consistent with what is known of the original type-material . . . ," except for sex or developmental stage [Article 75(c)(4)],
 (b) from as nearly as practicable the original type locality and the same geological horizon or host species [Article 75(c)(5)], and
 (c) "the property of a recognized scientific or educational institution, cited by name, that maintains a research collection, with proper facilities for preserving types, and that makes them accessible for study" [Article 75(c)(6)].

Selection of type specimen. Although there are almost no specific rules governing the selection of holotypes, there are recommendations applicable to original selection of the holotype, as well as rules and recommendations applying to lectotype and neotype selection.

In selecting a type specimen from the series before him, an author should make certain that his choice serves the purposes of types in the most satisfactory manner possible under the circumstances that exist in that group of animals. He should consider the effect of his choice on future workers. This generally means that the specimen should be in good condition, as complete as possible, and show the maximum number of the features that are thought to distinguish that species.

It is impossible for one specimen to be typical of the species in a biological sense, because the species is not something with a definite set of characters. It is a range of variation, a series of populations in which most characters vary. It is, however, usually desirable for the type to be midway in the range of most characters—in other words, to be a normal rather than an extreme member of the species.

Inasmuch as the place at which the type was collected is given special significance as the type locality, it is well always to choose as type a specimen for which the locality data are as complete and definite as possible.

Of holotype. The following provisions and recommendations in the 1961 Code apply to selection of holotypes:

1. Definition: "holotype, n. The single specimen designated or indicated as 'the type-specimen' of a nominal species-group taxon at the time of the original publication", [Glossary].
2. "If a new nominal species is based on a single specimen, that specimen is the 'holotype' " [Article 73(a)].
3. "If an author states in the description of a new nominal species that one specimen and only one is 'the type' or uses some equivalent expression, that specimen is the holotype" [Article 73(b)].
4. "A zoologist when describing a new species should clearly designate a single specimen as its holotype" [Recommendation 73A].
5. "A zoologist in establishing a new species should publish at least the following data concerning its holotype, in so far as they are relevant and known to him: (1) the size; (2) the full locality, date and other data on the labels accompanying the holotype; (3) the sex, if the sexes are separated; (4) the developmental stage, and the caste, if the species included more than one caste; (5) the name of the host species; (6) the name of the collector; (7) the collection in which it is situated and any collection- or register-number assigned to it; (8) in the case of a living terrestrial species, the elevation in metres above sea-level at which it was taken; (9) in the case of a living marine species, the depth in metres below sea-level at which it was taken; (10) in the case of a fossil species, its geological age and stratigraphical position, stated, if possible, in metres above or below a well-established plane" [Recommendation 73C]. (Other facts, such as host, may be equally desirable.)

Of lectotype. The following rules and recommendations from the 1961 Code apply to lectotype selection:

1. Definition: "lectotype, n. One of several syntypes designated after the original publication of a species-group name, as 'the type-specimen' of the taxon bearing that name" [Glossary].
2. "If a nominal species has no holotype, any zoologist may designate one of the syntypes as the 'lectotype' " [Article 74(a)].
3. "The first published designation of a lectotype fixes the status of the specimen, but if it is (later) proved that the designated specimen is not a syntype, the designation is invalid" [Article 74(a)(i)].
4. "Designation of a figure as lectotype is to be treated as designation of the specimen represented by the figure; if that specimen is one of the syntypes, the designation as lectotype is valid from the nomenclatural standpoint" [Article 74(b)].
5. "Lectotypes must not be designated collectively by a general statement; each designation must be made specifically for an individual nominal species, and must have as its object the definition of that species" [Article 74(c)].
6. "In designating a lectotype, a zoologist should in general act consistently with, and in any event should give great weight to, previous valid restrictions of the taxonomic species, in order to preserve stability of nomenclature" [Recommendation 74A].
7. "A zoologist should choose as lectotype a syntype of which a figure has been published, if such exists" [Recommendation 74B].
8. "A zoologist who designates a lectotype should publish the data listed in Recommendation 73C, besides describing any individual characteristics by which it can be recognized" [Recommendation 74C].
9. "When possible, a lectotype should be chosen from syntypes in the collection of a public institution, preferably of the institution containing the largest number of syntypes of the species, or containing the collection upon which the author of the nominal species worked, or containing the majority of his types" [Recommendation 74D].

Of neotype. The rules and recommendations applying to selection of a neotype in the 1961 Code are these:

1. Definition: "Neotype, n. A single specimen designated as the type-specimen of a nominal species-group taxon of which the holotype (or lectotype), and all paratypes, or all syntypes are lost or destroyed" [Glossary].
2. "Subject to the following limitations and conditions, a zoologist may designate another specimen to serve as the 'neotype' of a species if, through loss or destruction no holotype, lectotype, or syntype exists" [Article 75].
3. "A neotype is to be designated only in connection with revisory work, and then only in exceptional circumstances when a neotype is necessary in the interests of stability of nomenclature. (i) The words 'exceptional circumstances' refer to those cases in which a neotype is

essential for solving a complex zoological problem, such as the confused or doubtful identities of closely similar species for one or more of which no holotype, lectotype, or syntype exists" [Article 75(a)].

4. "A neotype is not to be designated for its own sake, or as a matter of curatorial routine, or for a species of which the name is not in general use either as a valid name or as a synonym" [Article 75(b)].

5. "A neotype is validly designated only when it is published with the following particulars: (1) a statement of the characters that the author regards as differentiating the taxon . . . , or a bibliographic reference to such a statement; (2) data and description sufficient to ensure recognition of the specimen designated; (3) the author's reasons for believing all of the original type-material to be lost or destroyed, and the steps that have been taken to trace it; (4) evidence that the neotype is consistent with (the relevant part of) what is known of the original type-material . . . ; (5) evidence that the neotype came as nearly as practicable from the original type-locality, and where relevant, from the same geological horizon or host-species . . . ; (6) a statement that the neotype is, or immediately upon publication has become, the property of a recognized scientific or educational institution . . ." [Article 75(c)].

6. "The first neotype-designation published for a given nominal species in accordance with the provisions of this Article is valid, and any subsequent designation has no validity unless the first neotype is lost or destroyed" [Article 75(d)].

7. "Before designating a neotype, a zoologist should satisfy himself that his proposed designation does not arouse objections from other specialists in the group in question" [Recommendation 75A].

8. "A zoologist who published an invalid neotype-designation before 1961 should be given opportunity to validate it before another zoologist designates a neotype for the same nominal taxon" [Recommendation 75B].

9. "If an invalid neotype-designation was published before 1961, the specimen then designated should be given preference when a neotype for the same nominal taxon is validly designated" [Recommendation 75C].

10. "A neotype-designation published before 1961 takes effect from the time when it fulfils all the provisions of this Article" [Article 75(e)].

11. "If after the designation of a neotype, original type-material is found to exist, the case is to be referred to the Commission" [Article 75(f)].

Conflicts among types. Conflicts may arise between or over type specimens because of any of the following situations. (The provisions of the 1961 Code governing these are quoted or cited in each case.)

1. No single type has yet been selected from the original series. "The type of each taxon of the species-group is a single specimen, either the

only original specimen or one designated from the type-series (holotype, lectotype), or a neotype" [Article 72(a)]. The "single specimen" would be the holotype [Article 73(a)]. The "one designated" may be chosen either by the original author in the original publication (also a holotype) [Article 72(2)] or by a later writer (lectotype chosen from syntypes) [Article 74].

2. Two or more names are applied to the same species, each bringing to it a potential type. (No rule details the effect of synonyms, each with a type, because the 1961 Code uses the fiction that "nominal species" are real species rather than names of species. Determination of which name is to be applied to the taxon will automatically determine which "type" is to be accepted.)

3. Different specimens may be selected as type from the original series by different people. (No rule tells what to do if different lectotypes are selected by different writers for one species under a single name, but priority would surely apply. If these are published simultaneously, the case should be submitted to the Commission.)

4. The species and the nominate subspecies must have the same type, but determination can be at either level. Simultaneous designation of *different* specimens for such a species and its subordinate taxon is possible. "The type of a taxon is also the type of its nominate subordinate taxon, if there is one, and vice versa. Therefore, the designation of one implies the designation of the other. (i) If different types are designated simultaneously for a nominal taxon and for its nominate subordinate taxon, the designation for the former takes precedence" [Article 61(a)].

5. A single specimen may be selected as the type of two or more "species." "The fact that a specimen is already the type of one nominal species does not prevent its designation as the type of another" [Article 72(c)]. The two "species" thereupon become objective or absolute synonyms. A selection is made between them in the usual manner, generally on priority.

6. When a name is replaced by another (as in case of homonymy), the species under the new name must have a type. (No rules relate directly to this, but the species takes the type brought to it by the name that is accepted for it.)

7. When the holotype, syntypes, or lectotype are lost or destroyed. "Subject to the following limitations and conditions, a zoologist may designate another specimen to serve as the 'neotype' of a species if, through loss or destruction no holotype, lectotype, or syntype exists" [Article 75]. The conditions refer to

(a) what cases are to be admitted [75(a)],
(b) what cases are to be excluded [75(b)],
(c) qualifying conditions of publication [75(c)],
(d) priority of neotype selections [75(d)],

(e) status of previously designated neotypes (before 1961) [75(e)], and (f) status of rediscovered type material [75(f)].

(Several possible complications in the loss of types are not referred to in the 1961 Code. Some of these are discussed in this chapter, under Selection of Type Specimens.)

8. If a person wishes to change the type, for any reason. "The type . . . once fixed . . . is not subject to change except by exercise of the plenary powers of the Commission" (as outlined in Article 79) [Article 61].

9. If a replacement name is given a different type from that of the name it replaces. "If an author proposes a new specific name expressly as a replacement for a prior name, but at the same time applies it to particular specimens, the type of the replacement nominal species must be that of the prior nominal species, despite any contrary designation of type-specimen or different taxonomic usage of the replacement name" [Article 72(d)].

Objective synonymy. Although the 1961 Code says that the type of a species is a single specimen, a species which consists of more than one specimen can have several types—even several holotypes. This happens because every name applied to the species will bring a type to it. The 1961 Code presumably intends that the type referred to as "the type" is the one associated with the one correct name for the species. Nevertheless, the types are so closely tied to the names that we must think of them as types of the names for most purposes, rather than as types of the species.

Under rare and unusual circumstances it may happen that a specimen, already the type of a species, is not recognized as such and is made the type of a new "species." There is then one zoological species, described under two names, with one specimen serving as type of both. The names are synonyms, and, because they are represented by the same type specimen, they are objective synonyms. (Objective synonyms are ones with exactly the same basis, which cannot ever be applied to different things.)

Much more commonly a name, which for some reason cannot be used, must be replaced with a "new name." The type of this new name (called a "replacement nominal species" in the 1961 Code) is the same specimen as that of the replaced name ("prior nominal species" of the 1961 Code). They are the same by rule [Article 72(d)]. They also are objective synonyms.

Thus, objective synonymy can occur in two ways: (1) through two unconnected names (or nominal species) being applied to one type specimen, or (2) through a second name being proposed to replace an existing name and thus automatically having the same type.

Suspension of the rules. The powers of the Commission to suspend the rules in individual cases are practically unlimited. Anything that can affect nomenclature, if covered by the code, can be set aside under the Plenary

Powers [Article 79]. It is hardly conceivable that cases should arise in which original selection of a holotype could be challenged, but designation of lectotypes and neotypes offer opportunities for actions which the Commission might wish to reverse. Especially the application of modern rules to types employed before such rules were adopted gives opportunity for questions that require action by the Commission.

Every taxonomist must keep track of any suspensions of the rules that affect his groups. This can be done only by use of the eight *Official Lists* and constant search of the issues of the *Bulletin of Zoological Nomenclature*.

Replacement of lost types. It inevitably happens that some types become permanently unavailable. There have been entire collections destroyed by violent means; specimens may be destroyed inadvertently during use. Unmarked types in older collections may be unknowingly lost. Because the use of a type in connection with each name is an inescapable necessity in our system, these lost specimens must be replaced. This can be done from specified sources, as follows:

1. If a unique holotype is lost, it can be replaced only with some non-original specimen (a neotype; see Article 75).
2. If a specified holotype is lost, it could be replaced from the other members of the original type-series (paratypes), although the code does not refer directly to this possibility.
3. If all the syntypes are lost, a single type can be chosen only from non-original material (a neotype; see Article 75).
4. If a lectotype is lost, presumably a second lectotype can be selected from the paralectotypes, although the code does not refer directly to this possibility.
5. If the last possible lectotype is lost, it can be replaced only with a neotype (see Article 75).
6. If a neotype is lost, presumably another neotype can be selected under the same conditions.

When a lost type is "replaced" with a paratype or a paralectotype, it is possible to look upon this as mere *substitution* of one original specimen for another. When the unique holotype or the last paratype or the last paralectotype is lost, the type series must be *replaced* with a new specimen or series.

Rules for replacement of lost types were not included in the *Règles,* but other specimens from the original series were commonly used in place of lost types. No serious problems arose as long as any original material was left, but when all the original type series was lost, if serious identification problems arose, the need for a replacement specimen was so great that a few taxonomists selected neotypes from subsequently collected material.

Neotypes. There was public discussion of the circumstances and precautions involved in neotype use, and the 1953 Copenhagen Colloquium studied the problems. A set of rules was presented in the Copenhagen Decisions. Other selections of neotypes were made under these rulings, so that by 1960 there were already many "neotypes" in existence, chosen under various procedures. In the 1961 Code, for the first time, a definite set of rules for neotypes was promulgated as Article 75.

A *neotype* is a single specimen selected as the type of a species of which all original specimens (holotype, paratypes, syntypes, lectotype, or paralectotypes) are lost or destroyed.

Article 75 specifies

(1) the conditions under which a neotype can be designated;
(2) circumstances which do not justify such action;
(3) requirements of publication of relevant data;
(4) application of priority if two designations are made;
(5) the status of neotype designations made before 1960; and
(6) the status of supposedly lost but rediscovered original type material.

Among these rules and the accompanying definitions some problems arise:

1. The relationship between a neotype and any remaining "syntypes" is not clear. The glossary says that the syntypes must all be lost, but if a lectotype was once chosen from the syntypes, there cease to be any syntypes—the specimens are now either lectotype or paralectotypes. Nowhere does the 1961 Code specify that successive lectotypes may be chosen from among the paralectotypes.
2. The 1961 Code says [Article 75] that a neotype may be designated "if, through loss or destruction no holotype, lectotype, or syntype, exists." It should include also "no neotype." It should make clear that this means also no paratypes and no paralectotypes.
3. The definition, in the Glossary of the 1961 Code, for "neotype" specifies that "the neotype (or lectotype), and all paratypes, or all syntypes are lost. . . ." This should include paralectotypes.

Permanence of types. "The type is objective and does not change" [Article 61]. This simple statement rules out the possibility of change of type because of choice by some later writer. In this direct form, however, the rule is much too sweeping. The type of a zoological species *can* change, and it *may* change every time the name of the species is changed.

This article itself goes on to cite two further exceptions to this general statement, but they are not clearly associated with it. They are: "except by exercise of the plenary powers of the Commission [Art. 79], or, exceptionally . . . , under the provisions of Article 75" (regarding neotypes).

Once properly established, the type remains the same unless and until

(1) the name of the species is changed to a name attached to some different type specimen, or
(2) the type is lost and must be replaced, or

(3) the Commission uses its Plenary Powers to displace the original type specimen and fix another in its stead.

Secondary types. As defined above, these are the specimens (always more than one) from which the lectotype must be selected. They are the syntypes (or cotypes) and the paralectotypes. The following provisions of the 1961 Code relate to the function of secondary types:

1. Definitions: "Syntype, n. Every specimen in a type-series in which no holotype has been designated" [Glossary]. "Cotype, n. A term formerly used for either syntype or paratype" [Glossary] but "To avoid misunderstandings, a zoologist should not use the term 'cotype'" [Recommendation 73E]. "Paralectotype, n. Any one of the original syntypes remaining after the selection of a lectotype" [Glossary].
2. "If a new nominal species has no holotype under the provisions of (a) and (b), all the specimens of the type-series are 'syntypes,' of equal value in nomenclature" [Article 73(c)].
3. When a syntype is chosen, the remaining lectotypes become paralectotypes. "A zoologist who designates a lectotype should clearly label any remaining syntypes with the designation 'paralectotype'" [Recommendation 74E].

Although it is not directly stated in the 1961 Code, the syntypes are available not only for one selection of a lectotype but for repeated selections if the selected lectotype is itself lost or destroyed. In each case a new lectotype would be selected from the remaining paralectotypes. Furthermore, it is also not directly stated that all paralectotypes must be used in succession as lectotypes (and lost or destroyed) before a neotype can be selected.

Tertiary types. The specimens originally set aside in addition to the holotype (primary) or syntypes (secondary) as of special taxonomic interest may be called tertiary types. They have no nomenclatural functions. They are all paratypes, but some of them are also called allotypes.

The functions of tertiary types include two very different things. The first is to supplement the holotype in showing the features of the species. They thus may show sexual characters of the opposite sex from that of the holotype (as the allotype). The second function is to serve as duplicates of the holotype, to be distributed to other collections to supplement the published statements in a concrete way where the holotype cannot be seen.

These two functions are mixed in the customary usage, because the type series usually contains all the original specimens of either sex definitely believed to belong to the species even if not identical in all respects. The only provisions in the 1961 Code are these:

1. Definition: "Paratype, n. Every specimen in a type-series, other than the holotype" [Glossary].

2. "After the holotype has been labelled, each remaining specimen (if any) of the type-series should be conspicuously labelled 'paratype,' in order clearly to identify the components of the original type-series" [Recommendation 73D].

3. "The type-series of a species consists of all the specimens on which its author bases the species, except any that he refers to as variants, or doubtfully associates with the nominal species, or expressly excludes from it" [Article 72(b)].

Other types or specimens. One other object which can reasonably be given a type term is a cast of a type, especially in fossils where such a cast may show all the visible features of the original specimen. Such "plasto-types" are not recognized in the 1961 Code, and it is better to use the expression "cast of the type" or other appropriate expression which can be more explicit.

All of the scores of other terms proposed to identify certain specimens are best abandoned. They serve only to delay recognition of facts, being in general less identifiable and explicit than a descriptive statement.

Type localities. The place where the type was collected is the type locality. This is such a simple and straightforward idea that it is a surprise to find that there are strongly contrasting views held as to how it is to be applied or interpreted. Some persons believe that the type locality is the locality cited in the original description. If this citation is to North America, then North America is the type locality.

Others believe that this is a view that makes the type locality sometimes useless, leads to the need for additional rules on the restriction of the type locality, and represents in reality not the type locality but the recorded locality. They believe that the type locality is the spot from which the type came, whether or not this is known accurately or at all. Later information can then clarify the type locality without the need for rules on methods of restriction. The present writer obviously adopts this latter view.

The 1961 Code implies support to both these views, without recognizing the problem explicitly. The relevant provisions are:

1. Definition: "Type-locality, n. The geographical place of origin of the type-specimen of a species-group taxon" [Glossary].

2. "An author who either designates or restricts a type-locality should base his action on one or more of the following criteria: (1) the original description of the taxon; (2) data accompanying the original material; (3) collector's notes, itineraries, or personal communications; and (4) as a last resort, localities within the known range of the species or from which specimens identified with the species have been taken" [Recommendation 72E].

3. "If a type-locality was erroneously designated or restricted, it should be corrected" [Recommendation 72E].

4. In proposing a neotype, Article 75 requires that there be evidence that the specimen "came as nearly as practicable from the original type-locality. . . ." Many of the species requiring neotypes are the older ones with the type locality stated only in general terms. It would be meaningless to require that the neotype be from "North America," if the species was actually restricted to a few localities in one mountain range.

Item (1) above seems to refer to the actual "place of origin" rather than the recorded region. Items (2) and (3) use the term "restricted" and therefore seem to equate the type locality with the published locality. Item (4) would be meaningless unless the actual place was intended.

This subject is obviously not effectively covered by the code, which would have done better to omit all mention of this technicality unless it intended to deal with it fully.

Subspecific types

Although not repeatedly stated above, all the rules and discussions on types of species apply also to types of subspecies [Article 71]. One additional problem arises for subspecies that cannot occur among species.

Each species that is divided into subspecies must have a nominate or nominotypical subspecies—one that has the subspecific epithet identical with the specific epithet (for example, *Peromyscus leucopus leucopus*). The type of this particular subspecies must be the same as the type of the species [Articles 61(a) and 72(e)]. Furthermore, any objective synonyms of the specific name are also synonyms of the nominate subspecific name, although subjective synonyms may not apply to that particular subspecies. Any synonyms applied directly to this subspecies will also be synonyms of the species in the same manner.

Ownership and preservation

The importance of types in both taxonomy and nomenclature may diminish with time in well-known groups. In general, however, the historical and practical importance of types is so great that their preservation and availability are matters of concern to taxonomists in general.

The specimens are generally the property of some individual or institution. There is a strong opinion that they should always be deposited in a public collection where they can be consulted by any qualified person. The 1961 Code recommends this for all holotypes and lectotypes [Recommendation 72A] but makes it mandatory in the case of a neotype [Article 75(c)(6)].

Preservation of such specimens is, of course, a special task for the persons or institutions possessing them. The permanent nature of their importance to science makes it essential to preserve them in the best possible condition. But preservation will not by itself suffice if the identity of the specimen is not assured, as many valuable types have been temporarily or even permanently lost through later lack of ability to identify that particular specimen as a type.

Thus, both the storage and the marking of these specimens becomes of considerable importance. Some of the requirements are cited in the 1961 Code, under Institutional Custody [Recommendation 72A], Labelling [Recommendation 72B], Information on Labels [Recommendation 72C], and Institutional Responsibility [Recommendation 72D].

Bibliography

Bibliography

This bibliography is designed to bring together under appropriate headings both works referred to in the text and other works relevant to the subjects treated. The headings parallel those in the text and are arranged by chapters or groups of chapters. Lists of references in the text are cross-referenced to the chapter in which they appear.

General Works

Blackwelder, R. E.
   1940. Some aspects of modern taxonomy. Journ. New York Ent. Soc., vol. 48, pp. 245–257.
   1951. Systematics in zoology. Science, vol. 113, p. 3.
   1960. The present status of systematic zoology. Syst. Zool., vol. 8, pp. 69–75.
Blackwelder, R. E. and Boyden, A. A.
   1952. The nature of systematics. Syst. Zool., vol. 1, pp. 26–33.
Davis, P. H. and Heywood, V. H.
   1963. Principles of angiosperm taxonomy. Edinburgh; Oliver and Boyd.
Emerson, A. E. and Schmidt, K. P.
   1960. Taxonomy. *In* Encyclopaedia Britannica, 1960 ed., 4 pages.
Ferris, G. F.
   1928. The principles of systematic entomology. Stanford, California; Stanford University Press.
Gier, L. J.
   1965. Principles of taxonomy. Liberty, Missouri; by the author. Mimeographed.
Gunter, G.
   1946. The ills of taxonomy. Proc. Trans. Texas Acad. Sci., vol. 30, pp. 69–73.

608   BIBLIOGRAPHY

Mayr, E.

1953. Concepts of classification and nomenclature in higher organisms and microorganisms. Ann. New York Acad. Sci., vol. 56, pp. 391–397.

1954. Notes on nomenclature and classification. Syst. Zool., vol. 3, pp. 86–89.

Mayr, E., Linsley, E. G., and Usinger, R. L.

1953. Methods and principles of systematic zoology. New York; McGraw-Hill Book Co.

Myers, G. S.

1952. The nature of systematic biology and of a species description. Syst. Zool., vol. 1, pp. 106–111.

Pearl, R.

1922. Trends of modern biology. Science, vol. 56, pp. 581–592.

Rollins, R. C.

1954. Plant taxonomy today. Syst. Zool., vol. 2, pp. 180–190. (Reprinted from Rhodora, January 1952.)

Schmitt, W. L. (ed.)

1953. Conference on importance and needs of systematics. Washington; Nat. Acad. Sci./Nat. Res. Counc.

Simpson, G. G.

1945. The principles of classification and a classification of mammals. Bull. American Mus. Nat. Hist., vol. 85, pp. 1–350.

1961. Principles of animal taxonomy. New York; Columbia University Press.

Thompson, W. R.

1937. Science and common sense. London; Longmans Green.

1958. The interpretation of taxonomy. Proc. 10th Internat. Congr. Ent. Montreal 1956, vol. 1, pp. 61–73.

Young, F. N.

1953. The approach to taxonomic problems. Proc. Indiana Acad. Sci. for 1952, vol. 62, pp. 172–175.

PART I

Introduction

Chapters 1–3

General

Blackwelder, R. E. and Boyden, A. A.

1952. The nature of systematics. Syst. Zool., vol. 1, pp. 26–33.

Ferris, G. F.

1928. The principles of systematic entomology. Stanford, California; Stanford University Press.

1942. The needs of systematic entomology. Journ. Econ. Ent., vol. 35, pp. 732–738.

Lamanna, C. and Mallette, M. F.
1953. Basic bacteriology. Its biological and chemical background. Baltimore; Williams and Wilkins.

Leone, C. A. (ed.)
1964. Taxonomic biochemistry and serology. New York; Ronald Press.

Mainx, F.
1955. Foundations of biology. In International Encyclopedia of Unified Science, vol. 1, pt. 2, pp. 567–654.

Mason, H. L.
1950. Taxonomy, systematic botany and biosystematics. Madroño, vol. 10, pp. 193–208.

Munroe, E. G.
1964. Problems and trends in systematics. Canadian Ent., vol. 96, pp. 368–377.

Pearl, R.
1922. Trends of modern biology. Science, vol. 56, pp. 581–592.

Simpson, G. G.
1945. The principles of classification and a classification of mammals. Bull. American Mus. Nat. Hist., vol. 85, pp. 1–350.
1961. Principles of animal taxonomy. New York; Columbia University Press.

Thorpe, W. H.
1940. Ecology and the future of systematics. In The new systematics, ed. by J. S. Huxley. London; Oxford Univ. Press. Pp. 341–364.

Vavilov, N. I.
1940. The new systematics of cultivated plants. In The new systematics, ed. by J. S. Huxley. London; Oxford Univ. Press. Pp. 549–566.

History and prophecy

Brues, C. T.
1929. Present trends in systematic entomology. Psyche, vol. 36, pp. 13–20.

Davis, P. H. and Heywood, V. H.
1963. Principles of angiosperm taxonomy. Edinburgh; Oliver and Boyd.

Ehrlich, P. E.
1961. Systematics in 1970: some unpopular predictions. Syst. Zool., vol. 10, pp. 157–158.

Ferris, G. F.
1955. The contribution of natural history to human progress. In A century of progress in the natural sciences, ed. by E. L.

Kessel. San Francisco; California Academy of Sciences. Pp. 75–87.

Gardner, E. J.
1960. History of life science. Outlines and references. Minneapolis; Burgess Publishing Co. Pp. 65–73.

Gurney, A. B.
1962. What of taxonomy in 1970? Syst. Zool., vol. 11, pp. 92–93.

Horn, W.
1928. The future of insect taxonomy. Trans. 4th Internat. Congr. Ent. Ithaca 1928, vol. 2, pp. 34–51.

Huxley, J. S. (ed.)
1940. The new systematics. London; Oxford Univ. Press. ("Introductory: Towards the new systematics," pp. 1–46)

Jahn, T. L.
1961. Man versus machine: a future problem in protozoan taxonomy. Syst. Zool., vol. 10, pp. 179–192.

Mayr, E.
1942. Systematics and the origin of species. New York; Columbia Univ. Press. (Reprinted, 1965.)

Mayr, E., Linsley, E. G., and Usinger, R. L.
1953. Methods and principles of systematic zoology. New York; McGraw-Hill Book Co. (Pp. 5–12)

Ross, E. S.
1955. Systematic entomology. In A century of progress in the natural sciences, ed. by E. L. Kessel. San Francisco; California Academy of Sciences. Pp. 485–495.

Weaver, C. E.
1955. Invertebrate paleontology and historical geology from 1850 to 1950. In A century of progress in the natural sciences, ed. by E. L. Kessel. San Francisco; California Academy of Sciences. Pp. 689–745.

Applied taxonomy

Chadwick, C. E.
1955. The practical entomologist and taxonomy. Journ. Australian Inst. Agric. Sci., vol. 21, pp. 230–238.

Clausen, C. P.
1942. The relation of taxonomy to biological control. Journ. Econ. Ent., vol. 35, pp. 744–748.

Essig, E. O.
1942. The significance of taxonomy in the general field of economic entomology. Journ. Econ. Ent., vol. 35, pp. 739–743.

Frison, T. H
1942. The significance of economic entomology in the field of insect taxonomy. Journ. Econ. Ent., vol. 35, pp. 749–752.

Keifer, H. H.
    1944. Applied entomological taxonomy. Pan-Pacific Ent., vol. 20, pp. 1–6.
Mayr, E., Linsley, E. G., and Usinger, R. L.
    1953. Methods and principles of systematic zoology. New York; McGraw-Hill Book Co. ("The relation of taxonomy to other branches of biology," pp. 19–22)
Merrill, E. D.
    1943. Some economic aspects of taxonomy. Torreya, vol. 43, pp. 50–64.
Muesebeck, C. F. W.
    1942. Fundamental taxonomic problems in quarantine and nursery inspection. Journ. Econ. Ent., vol. 35, pp. 753–758.
Schmitt, W. L.
    1953. Applied systematics. In Conference on importance and needs of systematics. Washington; Nat. Acad. Sci./Nat. Res. Counc. Pp. 4–12.
    1954. Applied systematics: the usefulness of scientific names of animals and plants. Smithsonian Ann. Rep. for 1953, pp. 323–337.
Silvestri, F.
    1929. The relation of taxonomy to other branches of entomology. 4th Internat. Congr. Ent., vol. 2, pp. 52–54.

Professional taxonomy
Ball, C. R.
    1946. Why is taxonomy ill supported? Science, vol. 103, p. 713.
Blackwelder, R. E.
    1955. The open season on taxonomists. Syst. Zool., vol. 3, pp. 177–181.
Dobzhansky, T.
    1961. Taxonomy, molecular biology, and the peck order. Evolution, vol. 15, pp. 263–264.
Parr, A. E.
    1958. Systematics and museums. Curator, vol. 1, No. 2, pp. 13–16.

Organizations
Gressitt, J. L.
    1958. Our knowledge of the insects of the Pacific Islands. Proc. 10th Internat. Congr. Ent. Montreal 1956, vol. 1, pp. 431–433. (List of museums)
Museums directory of the United States and Canada. Washington; American Assoc. Mus.
Sabrosky, C. W.
    1956. Entomological societies. Bull. Ent. Soc. America, vol. 2, No. 4, pp. 1–22.
Wallace, G. J.
    1955. An Introduction to ornithology. New York; Macmillan. (Ornithological organizations)

PART II

Practical Taxonomy

Chapters 4–7

General

Davis, P. H. and Heywood, V. H.
>1963. Principles of angiosperm taxonomy. Edinburgh; Oliver and Boyd.

Grinnell, J.
>1921. The museum conscience. Museum Work, IV, pp. 62–63.

Mayr, E.
>1942. Systematics and the origin of species. New York; Columbia Univ. Press.

Munroe, E. G.
>1964. Problems and trends in systematics. Canadian Ent., vol. 96, pp. 368–377.

Parr, A. E.
>1958. Systematics and museums. Curator, vol. 1, No. 2, pp. 13–16.

Simpson, G. G.
>1961. Principles of animal taxonomy. New York; Columbia University Press.

Glossaries of terms
>(See list in text of Chapter 5.)

Identification, general

Barr, A. R.
>1954. Punch-card taxonomy. Syst. Zool., vol. 3, p. 143.

Bowers, D. E.
>1957. A study of methods of color determination. Syst. Zool., vol. 5, pp. 147–160, 182.

Camp, C. L. and Hanna, G. D.
>1937. Methods in paleontology. Berkeley, California; Univ. of California Press.

de Laubenfels, M. W.
>1955. Unprecedented for zoologists. Syst. Zool., vol. 3, p. 181. (Fees for identifying)

Mayr, E., Linsley, E. G., and Usinger, R. L.
>1953. Methods and principles of systematic zoology. New York; McGraw-Hill Book Co. (Identification, pp. 72–78)

Rand, A. L.
>1948. Probability in subspecific identification of single specimens. Auk, vol. 65, pp. 416–432.

Identification, keys
(See list in text of Chapter 5.)

Identifiers, directory
(See reference in text of Chapter 5.)

Curating

Anderson, R. M.
    1948. Methods of collecting and preserving vertebrate animals. 2nd ed. Bull. Nat. Mus. Canada, Dept. Mines, No. 69, Biol. Ser., pp. 1–162.

Anthony, H. E.
    1945. The capture and preservation of small mammals for study. American Mus. Nat. Hist., Sci. Guide, No. 61, pp. 1–54.

Beer, J. R. and Cook, E. F.
    1958. A method for collecting ectoparasites from birds. Journ. Parasit., vol. 43, p. 445.

Bigelow, R. S.
    1957. Museum taxonomy and taxonomic research. Ann. Soc. Ent. Quebec, vol. 3, pp. 68–74.

British Museum (Natural History)
    1936, etc. Instructions for collectors. (Several animal groups and several editions)

Burt, W. H.
    1957. Mammals of the Great Lakes Region. Ann Arbor, Michigan; Univ. Michigan Press. (Collecting and preserving, pp. 160–178)

Camp, C. L. and Hanna, G. D.
    1937. Methods in paleontology. Berkeley; Univ. of California Press.

Casanova, R.
    1957. An illustrated guide to fossil collecting. San Martin, California; Naturegraph Co.
    1960. Fossil collecting: an illustrated guide. London; Faber and Faber.

Cockrum, E. L.
    1962. Laboratory and field manual for introduction to mammalogy, 2nd ed. New York; Ronald Press.

Davis, D. H. S.
    1960. Making the most of mammal specimens. Bull. Transvaal Mus., No. 4, pp. 4–5.

Duellman, W. E.
    1962. Directions for preserving amphibians and reptiles. Univ. Kansas Mus. Nat. Hist. Misc. Publ. 30, pp. 37–40.

Hall, E. R.
    1955. Handbook of mammals of Kansas. Lawrence, Kansas; Museum of Natural History. (Univ. of Kansas Mus. Nat. Hist. Misc. Pub. No. 7)

Macfayden, A.

1955. A comparison of methods for extracting soil arthropods. Soil Zool., 1955, pp. 315–332.

Mayr, E. and Goodwin, R.

1956. Biological materials. Part I. Preserved materials and museum collections. Wash. D.C.; Publ. 399, NAS/NRC, Bio. Council, Div. Biol. Agric.

Meryman, H. T.

1960. The preparation of biological museum specimens by freeze-drying. Curator, vol. 3, pp. 5–19.

Miller, R. R.

1956. Plastic bags for carrying and shipping live fish. Copeia, 1956, pp. 118–119.

Myers, G. S.

1956. Brief directions for preserving and shipping specimens of fishes, amphibians and reptiles. Circ. No. 5 (mimeo), Nat. Hist. Mus. Stanford Univ., 2nd ed.

1956. Manual of tropical herpetological collecting. Circ. No. 4 (mimeo), Nat. Hist. Mus., Stanford Univ., 2nd ed.

Newell, I. M.

1955. An autosegregator for use in collecting soil-inhabiting arthropods. Trans. American Microscop. Soc., vol. 74, pp. 389–392.

Oman, P. W. and Cushman, A. D.

1946. Collection and preservation of insects. U.S. Dept. Agric. Misc. Pub. 601, pp. 1–42.

Peterson, A.

1934. A manual of entomological equipment and methods. Part I. Ann Arbor, Michigan; by the author.

1937. Part 2. St. Louis, Missouri; by the author.

Raw, F.

1955. A flotation extraction process for soil micro-arthropods. Soil. Zool., 1955, pp. 341–346.

Stohler, R.

1959. How to build a private collection which is scientifically valuable. Veliger, vol. 2, No. 2, pp. 39–40.

Storey, M. and Wilimovsky, N. J.

1955. Curatorial practices in zoological research collections. 1. Preliminary report on containers and closures for storing specimens preserved in liquid. Circ. Nat. Hist. Mus. Stanford Univ. No. 3, pp. 1–22.

Wagstaffe, R. and Fidler, J. H.

1955. The preservation of natural history specimens. Vol. 1. Invertebrates. London; H. F. and G. Witherby Ltd.

Recording the data

Mayr, E., Linsley, E. G., and Usinger, R. L.
   1953. Methods and principles of systematic zoology. New York; Mc-
   Graw-Hill Book Co. (Cataloging of specimens, pp. 66–67; publications, pp. 155–198.)

Steyskal, G. C.
   1949. An indexing system for taxonomists. Coleopt. Bull., vol. 3, pp.
   65–71. (Examples of the actual cards used in a literature catalog)

PART III

The Diversity to Be Classified

Chapters 8-9

General

Calman, W. T.
   1940. A museum zoologist's view of taxonomy. *In* The new systematics,
   ed. by J. S. Huxley. London; Oxford Univ. Press. Pp. 455–459.

Hyman, L. H.
   1940. The invertebrates: Protozoa through Ctenophora. New York;
   McGraw-Hill Book Co.
   1951. The invertebrates: Platyhelminthes and Rhynchocoela. The
   acoelomate Bilateria, volume II. New York; McGraw-Hill Book
   Co.
   1951. The invertebrates: Acanthocephala, Aschelminthes, and Entoprocta. The pseudocoelomate Bilateria, volume III. New York;
   McGraw-Hill Book Co.
   1955. The invertebrates: Echinodermata. The coelomate Bilateria,
   volume IV. New York; McGraw-Hill Book Co.
   1959. The invertebrates: Smaller coelomate groups. Chaetognatha,
   Hemichordata, Pogonophora, Phoronida, Ectoprocta, Brachiopoda,
   Sipunculida. The coelomate Bilateria, volume V. New York; Mc-
   Graw-Hill Book Co.

Linsley, E. G. and Usinger, R. L.
   1961. Taxonomy. *In* The encyclopedia of the biological sciences, ed.
   by P. Gray. New York; Reinhold Publ. Corp. (Variation, p. 995)

Mayr, E., Linsley, E. G., and Usinger, R. L.
   1953. Methods and principles of systematic zoology. New York; Mc-
   Graw-Hill Book Co.

616 BIBLIOGRAPHY

Prosser, C. L. et al.
1950. Comparative animal physiology. Philadelphia; W. B. Saunders Co.

Prosser, C. L. and Brown, F. A., Jr.
1961. Comparative animal physiology, 2nd ed. Philadelphia; W. B. Saunders Co.

Sonneborn, T. M.
1957. Breeding systems, reproductive methods, and species problems in Protozoa. *In* The species problem, ed. by E. Mayr. Washington, D.C.; American Assoc. Adv. Sci. Publ. 50.

The number of kinds

Hyman, L. H.
1955. How many species? Syst. Zool., vol. 4, pp. 142–143.

Muller, S. W. and Campbell, A.
1955. The relative number of living and fossil species of animals. Syst. Zool., vol. 3, pp. 168–170.

Sabrosky, C. W.
1952. How many insects are there? Insects. Yearb. of Agric. 1952. Pp. 1–7.

1953. How many insects are there? Syst. Zool., vol. 2, pp. 31–36.

1953. An animal census for 1958. Syst. Zool., vol. 2, pp. 142–143.

Variability

Bateson, W.
1894. Materials for the study of variation, treated with special regard to discontinuity in the origin of species. London; Macmillan.

Bond, E. B.
1957. Polymorphism in plants, animals and man. Nature, vol. 180, pp. 1315–1319.

Cappe de Baillon, P.
1927. Recherches sur la teratologie des insectes. Encyclopèdie Entomologie, (A), vol. 8, pp. 5–291.

Coker, R. E.
1939. The problem of cyclomorphosis in Daphnia. Quart. Rev. Biol., vol. 14, pp. 137–148.

Ebeling, W.
1938. Host-determined morphological variation in Lecanium corni. Hilgardia, vol. 11, pp. 613–631.

Ford, E. B.
1940. Polymorphism and taxonomy. *In* The new systematics, pp. 493–513.

1945. Polymorphism. Biol. Rev., vol. 20, pp. 73–88.

Fulton, B. B.
1925. Physiological variation in the snowy tree-cricket, Oecanthus niveus DeGeer. Ann. Ent. Soc. America, vol. 18, pp. 363–383.

Goldschmidt, R.
  1945. Mimetic polymorphism, a controversial chapter of Darwinism. Quart. Rev. Biol., vol. 20, pp. 147–164, 205–230.
Kerr, W. E.
  1950. Genetic determination of castes in the genus Melipona. Evolution, vol. 4, pp. 7–13.
Kinsey, A. C.
  1937. Supraspecific variation in nature and in classification from the viewpoint of zoology. American Nat., vol. 71, pp. 206–222.
Prosser, C. L.
  1955. Physiological variation in animals. Biol. Rev., vol. 30, pp. 229–262.
Robson, G. C. and Richards, O. W.
  1936. The variation of animals in nature. London, Longmans Green.
Salt, G.
  1927. The effects of stylopization on aculeate Hymenoptera. Journ. Exper. Zool., vol. 48, pp. 223–231.
  1941. The effects of hosts upon their insect parasites. Biol. Rev., vol. 16, pp. 239–264.

Group diversity
Boyden, A. A.
  1954. The significance of asexual reproduction. Syst. Zool., vol. 3, pp. 26–37, 47.
Cutress, C. E.
  1955. An interpretation of the structure and distribution of cnidae in Anthozoa. Syst. Zool., vol. 4, pp. 120–137.
Matthiesen, F. A.
  1962. Parthenogenesis in scorpions. Evolution, vol. 16, pp. 255–256.
Novikoff, M. M.
  1952. Regularity of form in organisms. Syst. Zool., vol. 2, pp. 57–62.
Rothschild, (Lord)
  1956. Unorthodox methods of sperm transfer. Sci. American, vol. 195, No. 5, pp. 121–132.

PART IV

Classification, Naming, Description

General

Burma, B. H.
  1954. Reality, existence, and classification: A review of the species problem. Madroño, vol. 12, pp. 193–224.

Ferris, G. F.
    1928. Principles of systematic entomology. Stanford, California; Stanford University Press.
van Steenis, C. G. G. J.
    1957. Specific and infraspecific delimitation. Flora Malesiana, ser. 1, vol. 5, No. 3, pp. clxvii–ccxxix.

Chapter 10. Comparative Data

Blackwelder, R. E.
    1962. Introduction. Symposium: The Data of Classification. Syst. Zool., vol. 11, pp. 49–52.
Cain, A. J. and Harrison, G. A.
    1958. An analysis of the taxonomist's judgment of affinity. Proc. Zool. Soc. London, vol. 131, pp. 85–98.
Cox, L. R.
    1960. General characteristics of Gastropoda. *In* Treatise on Invertebrate Paleontology, ed. by R. C. Moore. Part *I*, pp. *I*84–*I*169. New York and Lawrence, Kansas; Geological Society of America and University of Kansas Press.
Davis, P. H. and Heywood, V. H.
    1963. Principles of angiosperm taxonomy. Edinburgh; Oliver and Boyd.
Lankester, E. R.
    1878. Preface. *In* Elements of comparative anatomy, by C. Gegenbaur (translated). London; Macmillan.
Leone, C. A. (ed.)
    1964. Taxonomic biochemistry and serology. New York; Ronald Press.
Mayr, E., Linsley, E. G., and Usinger, R. L.
    1953. Methods and principles of systematic zoology. New York; McGraw-Hill Book Co.
Newell, I. M.
    1953. The natural classification of the Rhombognathinae (Acari, Halacaridae). Syst. Zool., vol. 2, pp. 119–135.
Robson, G. C. and Richards, O. W.
    1936. The variation of animals in nature. London; Longmans Green.
Simpson, G. G.
    1961. Principles of animal taxonomy. New York; Columbia University Press.
Spector, W. S. (ed.)
    1956. Handbook of biological data. Philadelphia; W. B. Saunders. (And collateral volumes: Standard values in blood; Standard values in nutrition and metabolism; and Handbook of toxicology.)

Stoll, N. R.
    1961. Introduction. *In* International code of zoological nomenclature.
    London; Internat. Trust Zool. Nomencl. Pp. vii–xvii.

Correspondence, homology
Boyden, A. A.
    1935. Genetics and homology. Quart. Rev. Biol., vol. 10, pp. 448–451.
    1943. Homology and analogy: A century after the definitions of
        "homologue" and "analogue" of Richard Owen. Quart. Rev. Biol.
        vol. 18, pp. 228–241.
    1947. Homology and analogy. A critical review of the meanings and
        implications of these concepts in biology. American Midl. Nat.,
        vol. 37, pp. 648–669.
Davis, D. D.
    1949. Comparative anatomy and the evolution of vertebrates. *In*
        Genetics, paleontology and evolution, ed. by G. L. Jepsen et al.
        Princeton, New Jersey; Princeton Univ. Press. Pp. 64–89.
Hubbs, C. L.
    1944. Concepts of homology and analogy. American Nat., vol. 78,
        pp. 289–307.
Huxley, J.
    1943. Evolution. The modern synthesis. New York; Harper and Bros.
        Publ. (London; George Allen and Unwin Ltd., 1942)
Ferris, G. F.
    1928. The principles of systematic entomology. Stanford, California;
        Stanford University Press.
Needham, J.
    1942. Biochemistry and morphogenesis. London; Cambridge Univer-
        sity Press.
Romer, A. S.
    1955. The vertebrate body, 2nd ed. Philadelphia; W. B. Saunders Co.
Schoute, J. C.
    1949. Biomorphology in general. Amsterdam; North-Holland Publ. Co.
        (Verh. Kon. Nederland Akad. Wetensch., Afd. Natuurkunde,
        Tweede Sectie, Deel XLVI, No. 1, pp. 1–93)
Simpson, G. G.
    1961. Principles of animal taxonomy. New York; Columbia University
        Press.
Spencer, W. P.
    1949. Gene homologies and the mutants of *Drosophila hydei*. *In*
        Genetics, paleontology and evolution, ed. by G. L. Jepsen et al.
        Princeton, New Jersey; Princeton Univ. Press. Pp. 23–44.

Smith, H. M.
    1962. Classification of structural and functional similarities in biology.
        Syst. Zool., vol. 11, pp. 45–47.
Tait, J.
    1928. Homology, analogy and plasis. Quart. Rev. Biol., vol. 3, pp.
        151–173.
Woodger, J. H.
    1937. The axiomatic method in biology. Cambridge; Cambridge Uni-
        versity Press.

Comparative data

Blair, W. F.
    1962. Non-morphological data in Anuran classification. Syst. Zool.,
        vol. 11, pp. 72–84.
Burch, J. Q.
    1957. Taxonomic characters in Mollusca. Syst. Zool., vol. 5, p. 144.
Burch, J. B. and Thompson, F. G.
    1958. Taxonomic characters and infraspecific variation in mollusks.
        Syst. Zool., vol. 7, p. 48.
Evans, J. W.
    1958. Character selection in systematics with special reference to the
        classification of leafhoppers (Insecta, Homoptera, Cicadelloidea).
        Syst. Zool., vol. 7, pp. 126–131.
Johnson, C. G.
    1939. Taxonomic characters, variability, and relative growth in Cimex
        lectularius L. and C. columbarius Jenyns (Heteropt. Cimicidae).
        Trans. Roy. Ent. Soc. London, vol. 89, pp. 543–577.
Kinsey, A. C.
    1930. The gallwasp genus Cynips. Indiana Univ. Studies, vol. 16,
        pp. 1–577.
    1937. Supraspecific variation in nature and in classification from the
        viewpoint of zoology. American Nat., vol. 71, pp. 206–222.
Michener, C. D.
    1949. Parallelisms in the evolution of the saturniid moths. Evolution,
        vol. 3, pp. 129–141.
Newell, I. M.
    1953. The natural classification of the Rhombognathinae (Acari,
        Halacaridae). Syst. Zool., vol. 2, pp. 119–135. (List of criteria
        in evaluating the systematic significance of character variants)
Sabrosky, C. W. and Bennett, G. F.
    1958. The utilization of morphological, ecological, and life history
        evidence in the classification of Protocalliphora (Diptera: Cal-
        liphoridae). Proc. 10th Internat. Congr. Ent. Montreal 1956,
        vol. 1, pp. 163–164 (abstract only).

Simpson, G. G.
    1943. Criteria for genera, species, and subspecies in zoology and paleontology. Ann. New York Acad. Sci., vol. 44, pp. 145–178.
Smith, R. I.
    1952. Cooperation between systematists and experimental biologists. Science, vol. 116, pp. 152–153.
Vladykov, V. D.
    1934. Environmental and taxonomic characters of fishes. Trans. Roy. Canadian Inst., vol. 20, pp. 99–140.

Morphological data

Amadon, D.
    1943. Bird weights as an aid in taxonomy. Wilson Bull., vol. 55, pp. 164–177.
Bailey, R.
    1960. Effect of age on the gaster shape of females of the Eurytomidae, Hymenoptera. Entomologist, vol. 93, pp. 44–48.
Hubbs, C. L. and Hubbs, L. C.
    1945. Bilateral asymmetry and bilateral variation in fishes. Pap. Michigan Acad. Sci., vol. 30, pp. 239–310.
Sibley, C. G.
    1957. The evolutionary and taxonomic significance of sexual dimorphism and hybridization in birds. Condor, vol. 59, pp. 166–191.

Structural or anatomical data

Adams, C. T.
    1955. Comparative osteology of the night herons. Condor, vol. 57, pp. 55–60.
Barlow, George W.
    1961. Causes and significance of morphological variation in fishes. Syst. Zool., vol. 10, pp. 105–117.
Berger, A. J.
    1952. The comparative functional morphology of the pelvic appendage in three genera of Cuculidae. American Midl. Nat., vol. 47, pp. 513–605.
    1959. Leg-muscle formulae and systematics. Wilson Bull., vol. 71, pp. 93–94.
Brower, L. P.
    1959. Speciation in butterflies of the Papilio glaucus group. I. Morphological relationships and hybridization. Evolution, vol. 13, pp. 40–63.
Chandler, A. C.
    1916. A study of the structure of feathers, with reference to their taxonomic significance. Univ. California Publ. Zool., vol. 13, pp. 243–446.

622   BIBLIOGRAPHY

Fisher, H. I.
    1946. Adaptations and comparative anatomy of the locomotor appa-
        ratus of New World vultures. American Midl. Nat., vol. 35, pp.
        545–727.
    1955. Avian anatomy, 1925–1950, and some suggested problems. In
        Recent Studies in Avian Biology. Urbana, Illinois; Univ. of Illi-
        nois Press.
Frizzell, D. L. and Exline, H.
    1955. Micropaleontology of holothurian sclerites. Micropaleontology,
        vol. 1, pp. 335–342.
Kelton, L. A.
    1959. Male genitalia as taxonomic characters in the Miridae (Hemip-
        tera). Canadian Ent., vol. 91, Suppl. 11, pp. 3–72.
Maslin, T. P.
    1952. Morphological criteria of phyletic relationships. Syst. Zool., vol.
        1, pp. 49–70.
Matsuda, R.
    1960. Morphology, evolution and a classification of the Gerridae
        (Hemiptera-Heteroptera). Kansas Univ. Sci. Bull., vol. 41, pp.
        25–632.
Raup, D. M.
    1962. Crystallographic data in echinoderm classification. Syst. Zool.,
        vol. 11, pp. 99–108.
Robertson, J. G.
    1960. Ovarioles as a meristic character in Coleoptera. Nature, vol.
        187, pp. 526–527.
Stains, H. J.
    1959. Use of the calcaneum in studies of taxonomy and food habits.
        Journ. Mamm., vol. 40, pp. 392–401.
    1962. Osteological data used in mammal classification. Syst. Zool.,
        vol. 11, pp. 127–130.
Svetovidov, A. N.
    1959. The structure of the brain of fishes in relation to classification
        and habits. Proc. 15th Intern. Congr. Zool., pp. 406–409.
Tonapi, G. T.
    1958. A comparative study of the respiratory system of some Hymen-
        optera. Part III. Apocrita-Aculeata. Indiana Journ. Ent., vol. 20,
        pp. 245–269.
Tordoff, H. B.
    1954. A systematic study of the avian family Fringillidae based on
        the structure of the skull. Univ. Michigan Mus. Zool. Misc. Publ.
        No. 81, pp. 1–41.
Whitten, J. M.
    1959. The tracheal system as a systematic character in larval Diptera.
        Syst. Zool., vol. 8, pp. 130–139.

Wille, A.
    1958. A comparative study of the dorsal vessels of bees. Ann. Ent.
        Soc. America, vol. 51, pp. 538–546.

Cytological data

de Azevedo, J. F. and Goncalves, M. M.
    1956. Ensaios sobre o cotulo da numeraçao cromosomica de algunas
        especies de moluscos de agua doce. Anal. Inst. Med. Trop., vol.
        13, pp. 569–577.
Cain, A. J.
    1958. Chromosomes and taxonomic importance. Proc. Linn. Soc. Lon-
        don, vol. 169, pp. 125–128.
Hubbs, C. L. and Hubbs, L. C.
    1932. Experimental verification of natural hybridization between dis-
        tinct genera of sunfishes. Pap. Michigan Acad. Sci. Arts Lett.,
        vol. 15, pp. 427–437.
Johnsgard, P. A.
    1960. Hybridization in the Anatidae and its taxonomic implications.
        Condor, vol. 62, pp. 25–33.
Maeki, K.
    1958. A use of chromosome numbers in the study of taxonomy of the
        Lepidoptera. . . . Lep. News, vol. 11, pp. 8–9 (1957)
Moore, J. A.
    1959. Hybridization as an adjunct to the systematics of amphibians.
        Proc. 15th Internat. Congr. Zool., pp. 117–119.
Smith, S. G.
    1958. Animal cytology and cytotaxonomy. Proc. Genet. Soc. Canada,
        vol. 3, pp. 57–64.
White, M. J. D.
    1957. Cytogenetics and systematic entomology. Ann. Rev. Ent., vol.
        2, pp. 71–90.
Yamashina, Y.
    1952. Classification of the Anatidae based on cytogenetics. Coord.
        Comm. Res. Genet., Yamashina Ornith. Inst., vol. 3, pp. 1–34.

Developmental data

de Beer, G. R.
    1940. Embryology and taxonomy. In The New Systematics, ed. by
        J. Huxley. London; Oxford University Press. Pp. 365–393.
Eaton, T. H., Jr.
    1953. Pedomorphosis: an approach to the chordate-echinoderm prob-
        lem. Syst. Zool. vol. 2, pp. 1–6.
van Emden, F. I.
    1957. The taxonomic significance of the characters of immature in-
        sects. Ann. Rev. Ent., vol. 2, pp. 91–106.

Gordon, I.

 1955. Importance of larval characters in classification. Nature, vol. 176, pp. 911–912.

Hartman, C. G.

 1925. On some characters of taxonomic value appertaining to the egg and ovary of rabbits. Journ. Mamm., vol. 6, pp. 114–121.

Levi, C.

 1958. Ontogeny and systematics in sponges. Syst. Zool., vol. 6, pp. 174–183. (1957)

Michener, C. D.

 1953. Life-history studies in insect systematics. Syst. Zool., vol. 2, pp. 112–118.

Mosher, E.

 1916. A classification of the Lepidoptera based on characters of the pupa. Bull. Illinois State Lab. Nat. Hist., vol. 12, pp. 17–159.

Noble, G. K.

 1926. The importance of larval characters in the classification of South African Salientia. American Mus. Nov., No. 237, pp. 1–10.

 1927. The value of life-history data in the study of the evolution of the Amphibia. Ann. New York Acad. Sci., vol. 30, pp. 31–128.

Orton, G. L.

 1953. The systematics of vertebrate larvae. Syst. Zool., vol. 2, pp. 63–75.

 1955. The role of ontogeny in systematics and evolution. Evolution, vol. 9, pp. 75–83.

Parr, A. E.

 1949. An approximate formula for stating taxonomically significant proportions of fishes with reference to growth changes. Copeia, 1949, pp. 47–55.

Stunkard, H. W.

 1940. Life-history studies and the development of parasitology. Journ. Parasit., vol. 26, pp. 1–15.

 1953. Life histories and systematics of parasitic worms. Syst. Zool., vol. 2, pp. 7–18.

Thompson, W. R.

 1922. On the taxonomic value of larval characters in tachinid parasites (Dipt.) Proc. Washington Ent. Soc., vol. 24, pp. 85–93.

Biochemical data

Ball, G. H. and Clark, E. W.

 1953. Species differences in amino acids of Culex mosquitoes. Syst. Zool., vol. 2, pp. 138–141.

Bergman, W. and Domsky, I. I.

 1960. Sterols of some invertebrates. Ann. New York Acad. Sci., vol. 90, pp. 906–909.

Boyden, A. A.
1943. Serology and animal systematics. American Nat., vol. 77, pp. 234–255.
1951. A half-century of systematic serology. Serological Mus. Bull., No. 6, pp. 1–3.
1959. Serology as an aid to systematics. Proc. 15th Internat. Congr. Zool., pp. 120–122.

Boyden, A. A. and Paulsen, E.
1959. Serology as an aid to systematics. Bull. Serological Mus., No. 22, pp. 4–8.

Brezner, J. and Enns, W. R.
1958. Preliminary studies utilizing paper electrophoresis as a tool in insect systematics. Journ. Kansas Ent. Soc., vol. 31, pp. 241–246.

Deutsch, H. F. and Goodloe, M. B.
1945. An electrophoretic survey of various animal plasmas. Journ. Biol. Chem., vol. 161, pp. 1–20.

Florkin, M. and Mason, H. S., eds.
1960–1965. Comparative biochemistry. A comprehensive treatise. 7 vols. New York; Academic Press.

Haslewood, G. A. D.
1960. Steroids in marine organisms. Ann. New York Acad. Sci., vol. 90, pp. 877–883.

Hughes-Schrader, S.
1958. The DNA content of the nucleus as a tool in the cytotaxonomic study of insects. Proc. 10th Internat. Congr. Ent. Montreal 1956, vol. 2, pp. 935–944.

Leone, C. A. (ed.)
1964. Taxonomic biochemistry and serology. New York; Ronald Press.

Micks, P. A. and Doniger, D. E.
1956. Serological light on porcupine relationships. Evolution, vol. 10, pp. 47–55.

Moore, D. H.
1945. Species differences in serum protein patterns. Journ. Biol. Chem., vol. 161, pp. 21–32.

van Sande, J. and Karcher, D.
1960. Species differentiation of insects by hemolymph electrophoresis. Science, vol. 131, pp. 1103–1104.

Sibley, C. G.
1960. The electrophoretic patterns of avian egg-white proteins as taxonomic characters. Ibis, vol. 102, pp. 215–284.
1962. The comparative morphology of protein molecules as data for classification. Syst. Zool., vol. 11, pp. 108–118.

Sibley, C. G. and Johnsgard, P. A.
1959. Variability in the electrophoretic patterns of avian serum proteins. Condor, vol. 61, pp. 85–95.

Stephen, W. P.
  1958. Hemolymph proteins and their use in taxonomic studies. Proc.
    10th Internat. Congr. Ent. Montreal 1956, vol. 1, pp. 395–400.
  1961. Phylogenetic significance of blood proteins among some or-
    thopteroid insects. Syst. Zool., vol. 10, pp. 1–9.
Stephen, W. P. and Steinhauer, A. L.
  1957. Sexual and developmental differences in insect blood proteins.
    Physiol. Zool., vol. 30, pp. 114–120.
Stormont, C., Miller, W. J., and Suzuki, Y.
  1961. Blood groups and the taxonomic status of American buffalo
    and domestic cattle. Evolution, vol. 15, pp. 196–208.
van Thoai, N. and Roche, J.
  1964. Diversity of phosphagens. *In* Taxonomic biochemistry and se-
    rology, ed. by C. A. Leone. New York; Ronald Press. Pp. 347–362.
Wolfe, H. R.
  1939. Serologic relationships among Bovidae and Cervidae. Zoologica,
    vol. 24, pp. 309–321.
Zweig, G. and Crenshaw, J. W., Jr.
  1957. Differentiation of species by paper electrophoresis of serum
    proteins of Pseudemys turtles. Science, vol. 126, pp. 1065–1066.

Genetical data
(See also Cytological data, above)

Blair, W. F.
  1943. Criteria for species and their subdivisions from the point of
    view of genetics. Ann. New York Acad. Sci., vol. 44, pp. 179–188.

Ethological data

Alexander, R. D.
  1962. The role of behavioral study in cricket classification. Syst.
    Zool., vol. 11, pp. 53–72.
Amadon, D.
  1959. Behaviour and classification. Some reflections. Vjschr. Naturf.
    Ges. Zürich, vol. 104, Festchr. Steiner, pp. 73–78.
Boudreaux, H. B.
  1963. Reproductive data in classification. Syst. Zool., vol. 11, pp.
    145–150.
Cooper, D. M.
  1960. Food preferences of larval and adult Drosophila. Evolution,
    vol. 14, pp. 41–55.
Downey, J. C.
  1963. Host-plant relations as data for butterfly classification. Syst.
    Zool., vol. 11, pp. 150–159.
Evans, H. E.
  1954. Comparative ethology and the systematics of spider wasps.
    Syst. Zool., vol. 2, pp. 155–172.

Petrunkevitch, A.
  1926. The value of instinct as a taxonomic character in spiders. Biol.
    Bull., vol. 50, pp. 427–432.
Schneirla, T. C.
  1952. A consideration of some conceptual trends in comparative psy-
    chology. Psych. Bull., vol. 49, pp. 559–597.
Smith, R. I.
  1958. On the reproductive pattern as a specific characteristic among
    nereid polychaetes. Syst. Zool., vol. 7, pp. 60–73.

Parasitological data
Amsel, —.
  1950. (History of the human race in relation to the systematics of
    lice.) Kosmos, vol. 46, pp. 507–509.
Emerson, A. E.
  1935. Termitophile distribution and quantitative characters of physi-
    ological speciation in British Guiana termites (Isoptera). Ann.
    Ent. Soc. America, vol. 28, pp. 369–395.
Hopkins, G. H. E.
  1949. The host-associations of the lice of mammals. Proc. Zool. Soc.
    London, vol. 191, pp. 387–604.
Linsley, E. G.
  1937. The effect of stylopization on *Andrena porterae*. Pan-Pacific
    Ent., vol. 13, p. 157.
Metcalf, M. M.
  1929. Parasites and the aid they give in problems of taxonomy, geo-
    graphical distribution, and paleo-geography. Smithsonian Misc.
    Coll., vol. 81, No. 8, pp. 1–36.
Salt, G.
  1927. The effects of stylopization on aculeate Hymenoptera. Journ.
    Exp. Zool., vol. 48, pp. 223–331.
  1941. The effects of hosts upon their insect parasites. Biol. Rev., vol.
    16, pp. 239–264.
Vanzolini, P. E. and Guimaraes, L. R.
  1955. South American land mammals and their lice. Evolution, vol. 9,
    pp. 345–347.

Ecological data
Emerson, A. E.
  1941. Taxonomy and ecology. Ecology, vol. 22, p. 213.
Kohn, A. J. and Orians, G. H.
  1962. Ecological data in the classification of closely related species.
    Syst. Zool., vol. 11, pp. 119–127.
Sabrosky, C. W.
  1950. Taxonomy and ecology. Ecology, vol. 31, pp. 151–152.

Paleontological data

Newell, N. D.
   1959. The nature of the fossil record. Proc. American Philos. Soc., vol. 103, pp. 264–285.
Nicol, D.
   1958. Taxonomy versus stratigraphy. Journ. Washington Acad. Sci., vol. 48, pp. 113–114.

Miscellaneous data

Amadon, D.
   1943. Bird weights as an aid in taxonomy. Wilson Bull., vol. 55, pp. 164–177.
Blackwelder, R. E.
   1962. Symposium: The data of classification. Introduction. Syst. Zool., vol. 11, pp. 49–52.
Blair, W. F.
   1962. Non-morphological data in Anuran classification. Syst. Zool., vol. 11, pp. 72–84.
Brower, L. P.
   1958. Bird predation and food-plant specificity in closely related procryptic insects. American Nat., vol. 92, pp. 183–187.
Johnson, C. G.
   1939. Taxonomic characters, variability, and relative growth in *Cimex lectularius* L. and *C. columbarius* Jenyns (Heteropt. Cimicidae). Trans. Roy. Ent. Soc. London, vol. 89, pp. 543–577.
Manning, R. B. and Kumpf, H. E.
   1959. Preliminary investigation of the fecal pellets of certain invertebrates of the South Florida area. Mar. Sci. Gulf. Caribbean, vol. 9, pp. 291–309.
Munroe, E. G.
   1960. An assessment of the contribution of experimental taxonomy to the classification of insects. Rev. Canadienne Biol., vol. 19, pp. 293–319.
Raup, D. M.
   1962. Crystallographic data in echinoderm classification. Syst. Zool., vol. 11, pp. 99–108.
Sabrosky, C. W. and Bennett, G. F.
   1958. The utilization of morphological, ecological, and life history evidence in the classification of Protocalliphora (Diptera: Calliphoridae). Proc. 10th Internat. Congr. Ent. Montreal 1956, vol. 1, pp. 163–164 (abstract only).
Simpson, G. G.
   1941. Range as a zoological character. American Journ. Sci., vol. 239, pp. 785–804.

Smith, R. I.
  1952. Cooperation between systematics and experimental biologists. Science, vol. 116, pp. 152–153.
Troelsen, J. C.
  1955. On the value of aragonite tests in the classification of the Rotaliidae. Contr. Cushman Found., vol. 6, pp. 50–51.

Comparative zoology
(See also comparative biochemistry list in Chapter 10.)
Amadon, D.
  1959. Behaviour and classification. Some reflections. Vjschr. Naturf. Ges. Zurich, vol. 104, Festschr. Steiner, pp. 73–78.
Bier, M.
  1959. Electrophoresis: theory, methods, and applications. New York; Academic Press.
Blackman, P. D. and Todd, R.
  1959. Mineralogy of some foraminifera as related to their classification and ecology. Journ. Paleont., vol. 33, pp. 1–15.
Boyden, A. A.
  1942. Systematic serology: a critical appreciation. Physiol. Zool., vol. 15, pp. 109–145.
  1943. Serology and animal systematics. American Nat., vol. 77, pp. 234–255.
  1951. A half-century of systematic serology. Serolog. Mus. Bull., vol. 6, pp. 1–3.
  1953. Fifty years of systematic serology. Syst. Zool., vol. 2, pp. 19–30.
  1959. Serology as an aid to systematics. Proc. 15th Internat. Congr. Zool., pp. 120–122.
Boyden, A. A. and Paulsen, E.
  1959. Serology as an aid to systematics. Bull. Serol. Mus., No. 22, pp. 4–8.
Buzzati-Traverso, A. A. and Rechnitzer, A. B.
  1953. Paper partition chromatography in taxonomic studies. Science, vol. 117, pp. 58–59.
Evans, H. E.
  1954. Comparative ethology and the systematics of spider wasps. Syst. Zool., vol. 2 (1953), pp. 155–172.
Fox, A. S.
  1956. Application of paper chromatography to taxonomic studies. Science, vol. 123, p. 143.
Harvey, W. E.
  1955. Paper chromatography. Tuatara, vol. 5, pp. 100–110.
Johnson, M. L. and Wicks, M. J.
  1959. Serum protein electrophoresis in mammals—taxonomic implications. Syst. Zool., vol. 8, pp. 88–95.

Lorenz, K. Z.
1950. The comparative method in studying innate behavior patterns. *In* Symp. Soc. Exp. Biol. No. 4, Physiological mechanisms in animal behavior. New York; Academic Press. Pp. 221–268.

Metcalf, M. M.
1929. Parasites and the aid they give in problems of taxonomy, geographical distribution, and paleo-geography. Smithsonian Misc. Coll., vol. 81, no. 8, pp. 1–36.

Munroe, E. G.
1957. Comparison of closely related faunas. Science, vol. 126, pp. 437–439.

Schneirla, T. C.
1952. A consideration of some conceptual trends in comparative psychology. Psych. Bull., vol. 49, pp. 559–597.

Sibley, C. G.
1962. The comparative morphology of protein molecules as data for classification. Syst. Zool., vol. 11, pp. 108–118.

Smith, S. G.
1958. Animal cytology and cytotaxonomy. Proc. Genet. Soc. Canada, vol. 3, pp. 57–64.

Thompson, R. R.
1960. Species identification by starch gel zone electrophoresis of protein extracts. I. Fish. Journ. Assoc. Agric. Chem., vol. 43, p. 763.

White, M. J. D.
1957. Cytogenetics and systematic entomology. Ann. Rev. Ento., vol. 2, pp. 71–90.

Methods of study
Ansell, W. F. H.
1958. On the study of African mammalia. Journ. Mamm., vol. 39, pp. 577–581.

Blackwelder, R. E. and Hoyme, L. E.
1955. Statistics, experiment, and the future of biology. Sci. Monthly, vol. 80, pp. 225–229.

Cain, A. J.
1959. Deductive and inductive methods in post-Linnaean taxonomy. Proc. Linn. Soc. London, vol. 170, pp. 185–217.

Cassie, R. M.
1957. The sampling problem, with particular reference to marine organisms. Proc. New Zealand Ecol. Soc., No. 4, pp. 37–39.

Cochran, W. G.
1959. Sampling techniques. New York; John Wiley and Sons.

Ginsburg, I.
1940. Divergence and probability in taxonomy. Zoologica, vol. 25, pp. 15–31.

Lorenz, K. Z.
   1950. The comparative method in studying innate behavior patterns, pp. 221–268. *In* Symp. Soc. Exp. Biol., No. 4, Physiological mechanisms in animal behavior, New York; Academic Press.
Miller, R. R.
   1957. Utilization of X-rays as a tool in systematic zoology. Syst. Zool., vol. 6, pp. 29–40.
Munroe, E. G.
   1957. Comparison of closely related faunas. Science, vol. 126, pp. 437–439.
Sokal, R. R.
   1961. Distance as a measure of taxonomic similarity. Syst. Zool., vol. 10, pp. 70–79.
Womble, W. H.
   1951. Differential systematics. Science, vol. 114, pp. 315–322.
Zangerl, R.
   1948. The methods of comparative anatomy and its contribution to the study of evolution. Evolution, vol. 2, pp. 351–374.

Graphic methods
(See also list in text of Chapter 10.)
Dice, L. R. and Leraas, H. J.
   1936. A graphic method for comparing several sets of measurements. Contr. Lab. Vert. Gen. Univ. Michigan, No. 3.
Hubbs, C. L. and Hubbs, C.
   1953. An improved graphical analysis and comparison of series of samples. Syst. Zool., vol. 2, pp. 49–59, 92.
Hubbs, C. L. and Perlmutter, A.
   1942. Biometric comparison of several samples, with particular reference to racial investigations. American Nat., vol. 76, pp. 582–592.
Parr, A. E.
   1949. An approximate formula for stating taxonomically significant proportions of fishes with reference to growth changes. Copeia, 1949, pp. 47–55.

Numerical methods
(Anonymous)
   1961. The mathematical assessment of taxonomic similarity, including the use of computers. Taxon, vol. 10, pp. 97–101.
Barnett, J. A.
   1961. *In* The mathematical assessment of taxonomic similarity, including the use of computers. Taxon, vol. 10, pp. 99–100.
Blackwelder, R. E. and Hoyme, L. E.
   1955. Statistics, experiment, and the future of biology. Sci. Monthly, vol. 80, pp. 225–229.

Burma, B. H.
  1948. Studies in quantitative paleontology. 1. Some aspects of the
  theory and practice of quantitative invertebrate paleontology.
  Journ. Paleo., vol. 22, pp. 725–761.

Cain, A. J. and Harrison, G. A.
  1960. Phyletic weighting. Proc. Zool. Soc. London, vol. 135, pp. 1–31.

Cain, A. J.
  1961. *In* The mathematical assessment of taxonomic similarity, including the use of computers. Taxon, vol. 10, p. 98.

Cazier, M. A. and Bacon, A.
  1949. Introduction to quantitative systematics. Bull. American Mus.
  Nat. Hist., vol. 93, pp. 347–388.

Crawshay-Williams, R.
  1961. *In* The mathematical assessment of taxonomic similarity, including the use of computers. Taxon, vol. 10, pp. 98–99.

Davis, P. H. and Heywood, V. H.
  1963. Principles of angiosperm taxonomy. Edinburgh; Oliver and
  Boyd.

Denmark, H. A., Weems, H. V., and Taylor, C.
  1958. Taxonomic codification of biological entities. Science, vol. 128,
  pp. 990–992.

Dice, L. R.
  1952. Quantitative and experimental methods in systematic zoology.
  Syst. Zool., vol. 1, pp. 97–104.

Dixon, W. J. and Masser, F. J., Jr.
  1957. Introduction to statistical analysis, 2nd ed. New York; McGraw-
  Hill Book Co.

Edwards, J. G.
  1962. Morphological statistics versus biology. Syst. Zool., vol. 11, pp.
  142–143.

Ehrlich, P. R.
  1961. Systematics in 1970: some unpopular predictions. Syst. Zool.,
  vol. 10, pp. 157–158.

Evans, J. W.
  1958. Character selection in systematics with special reference to the
  classification of leafhoppers (Insecta, Homoptera, Cicadelloidea).
  Syst. Zool., vol. 7, pp. 126–131.

Fisher, R. A.
  1936. The use of multiple measurements in taxonomic problems. Ann.
  Eugenics, vol. 7, pp. 179–188.
  1950. Statistical methods for research workers. 11th ed. New York;
  Hafner Publ. Co.

Ginsburg, I.
  1954. Certain measures of intergradation and divergence. Zoologica,
  vol. 39, pp. 31–35.

Haas, O.
   1962. Comment on numerical taxonomy. Syst. Zool., vol. 11, p. 186.
Hampton, J. S.
   1959. Statistical analysis of holothurian sclerites. Micropaleontology, vol. 5, pp. 335–349.
Heslop-Harrison, J.
   1961. *In* The mathematical assessment of taxonomic similarity, including the use of computers. Taxon, vol. 10, p. 97.
Hubbs, C. L. and Perlmutter, A.
   1942. Biometric comparison of several samples, with particular reference to racial investigations. American Nat., vol. 76, pp. 582–592.
Huxley, J. S.
   1939. Clines: an auxiliary method in taxonomy. Bijdr. Dierk., vol. 27, pp. 491–520, 624, 626, 627.
Imbrie, J.
   1956. Biometrical methods in the study of invertebrate fossils. Bull. American Mus. Nat. Hist., vol. 108, pp. 211–252.
Inger, R. F.
   1959. Reply to Sokal. Evolution, vol. 13, p. 423.
Jahn, T. L.
   1961. Man versus machine: a future problem in protozoan taxonomy. Syst. Zool., vol. 10, pp. 179–192.
Kempthorne, O. et al. (eds.)
   1954. Statistics and mathematics in biology. Ames, Iowa; Iowa State College Press.
Li, J. C. R.
   1957. Introduction to statistical inference. Ann Arbor, Michigan, Edwards Bros.
Maccacaro, G. A.
   1961. *In* The mathematical assessment of taxonomic similarity, including the use of computers. Taxon, vol. 10, p. 99.
Mather, K.
   1947. Statistical analysis in biology. New York; Interscience Publ.
Michener, C. D. and Sokal, R. R.
   1957. A quantitative approach to a problem in classification. Evolution, vol. 11, pp. 130–162.
Parker-Rhodes, A. F.
   1961. *In* The mathematical assessment of taxonomic similarity, including the use of computers. Taxon, vol. 10, p. 100.
Rao, C. R.
   1948. The utilization of multiple measurements in problems of biological classification. Journ. Roy. Statist. Soc., B, vol. 10, pp. 159–203.

Richens, R. H.
    1961. *In* The mathematical assessment of taxonomic similarity, including the use of computers. Taxon, vol. 10, p. 100.
Rogers, D. J. and Tanimoto, T. T.
    1960. A computer program for classifying plants. Science, vol. 132, pp. 1115–1118.
Rohlf, F. J. and Sokal, R. R.
    1962. The description of taxonomic relationships by factor analysis. Syst. Zool., vol. 11, pp. 1–16.
Russell, N. H.
    1961. The development of an operational approach in plant taxonomy. Syst. Zool., vol. 10, pp. 159–167.
Seitner, P. G.
    1960. Biology code of the Chemical-Biological Coordination Center. Publ. 790, Nat. Acad. Sci./Nat. Res. Counc.
Seitner, P. G., Livingston, G. A., and Williams, A. S.
    1960. Key to the biological code of the Chemical-Biological Coordination Center. Publ. 790K, Nat. Acad. Sci./Nat. Res. Counc.
Simpson, G. G.
    1959. Note on biometry and systematics. Bull. American Mus. Nat. Hist., vol. 117, pp. 51–54.
Simpson, G. G. and Roe, A.
    1939. Quantitative zoology. Numerical concepts and methods in the study of Recent and fossil animals. New York; McGraw-Hill Book Co.
Simpson, G. G., Roe, A., and Lewontin, R. C.
    1960. Quantitative zoology, rev. ed. New York; Harcourt, Brace.
Sneath, P. H. A.
    1957. Some thoughts on bacterial classification. Journ. Gen. Microbiol., vol. 17, pp. 184–200.
    1957. The application of computers to taxonomy. Journ. Gen. Microbiol., vol. 17, pp. 201–226.
    1961. *In* The mathematical assessment of taxonomic similarity, including the use of computers. Taxon, vol. 10, p. 98.
Sneath, P. H. A. and Cowan, S. T.
    1958. An electro-taxonomic survey of Bacteria. Journ. Gen. Microbiol., vol. 19, pp. 551–565.
Sneath, P. H. A. and Sokal, R. R.
    1962. Numerical taxonomy. Nature, vol. 193, pp. 855–860.
Snedecor, G. W.
    1946. Statistical methods. 4th ed. Ames, Iowa; Iowa State College Press.
Sokal, R. R.
    1958. Quantification of systematic relationships and of phylogenetic trends. Proc. 10th Internat. Congr. Ent. Montreal 1956. vol. 1, pp. 409–415.

1959. Comments on quantitative systematics. Evolution, vol. 13, pp. 420–423.

1961. Distance as a measure of taxonomic similarity. Syst. Zool., vol. 10, pp. 70–79.

1961. *In* The mathematical assessment of taxonomic similarity, including the use of computers. Taxon, vol. 10, pp. 97–98.

Sokal, R. R. and Michener, C. D.

1958. A statistical method for evaluating systematic relationships. Univ. Kansas Sci. Bull., vol. 38, pp. 1409–1437.

Sokal, R. R. and Rohlf, F. J.

1962. The comparison of dendrograms by objective methods. Taxon, vol. 11, pp. 33–40.

Sokal, R. R. and Sneath, P. H. A.

1963. Principles of numerical taxonomy. San Francisco; W. H. Freeman and Co.

Stroud, C. P.

1953. An application of factor analysis to the systematics of Kalotermes. Syst. Zool., vol. 2, pp. 76–79.

Wardle, R. A.

1954. The impact of mathematics upon zoology. Trans. Roy. Soc. Canada, ser. 3, vol. 48, No. 5, pp. 73–76.

Chapter 11. Species and subspecies

The nature of species

Amadon, D.

1950. The species—then and now. Auk, vol. 67, pp. 492–498.

Beaudry, J. R.

1960. The species concept: its evolution and present status. Rev. Canadienne Biol., vol. 19, pp. 219–240.

Burma, B. H.

1949. The species concept: a semantic review. Evolution, vol. 3, pp. 369–370.

1954. Reality, existence, and classification; a discussion of the species problem. Madroño, vol. 12, pp. 193–209.

Clark, R. B.

1956. Species and systematics. Syst. Zool., vol. 5, pp. 1–10.

Dunbar, C. O.

1950. The species concept: further discussion. Evolution, vol. 4, pp. 175–176.

Durrant, S. D.

1959. The nature of mammalian species. Journ. Arizona Acad. Sci., vol. 1, pp. 18–21.

Ehrlich, P. R.

1961. Has the biological species concept outlived its usefulness. Syst. Zool., vol. 10, pp. 167–176.

Grant, W. F.
  1960. The categories of classical and experimental taxonomy and the species concept. Rev. Canadienne Biol., vol. 19, pp. 241–262.
Hairston, N. G.
  1958. Observations on the ecology of Paramecium, with comments on the species problem. Evolution, vol. 12, pp. 440–450.
Hatch, M. H.
  1941. The logical basis of the species concept. American Nat., vol. 75, pp. 193–212.
Huxley, J. S. (ed.)
  1940. The new systematics. London; Oxford University Press.
Löve, Á.
  1960. Species concept and taxonomy—a prelude. Rev. Canadienne Biol., vol. 19, pp. 216–218.
Maslin, T. P.
  1959. The nature of amphibian and reptilian species. Journ. Arizona Acad. Sci., vol. 1, pp. 8–17.
Mayr, E.
  1949. The species concept: semantics versus semantics. Evolution, vol. 3, pp. 371–373.
  1955. The species as systematic and as a biological problem. 16th Ann. Biol. Coll., pp. 3–12.
Meglitsch, P. A.
  1954. On the nature of the species. Syst. Zool., vol. 3, pp. 49–65.
Murray, D.
  1955. Species revalued. London: Blackfriars Publications.
  Nitecki, M. H.
  1957. What is a paleontological species? Evolution, vol. 11, pp. 378–380.
Phillips, A.
  1959. The nature of avian species. Journ. Arizona Acad. Sci., vol. 1, pp. 22–30.
Rising, G. R.
  1959. The species concepts in ornithology. Part 1. Kingbird, vol. 9, pp. 111–114.
Robson, G. C.
  1928. The species problem. London; Oliver and Boyd.
Senn, H. A.
  1960. The species concept and taxonomy—a summary. Rev. Canadienne Biol., vol. 19, pp. 320–325.
Simpson, G. G.
  1952. The species concept. Evolution, vol. 5, pp. 285–298.
  1961. Principles of animal taxonomy. New York; Columbia University Press.

Smith, H. M.

1952. Definition of species. Turtox News, vol. 30, pp. 110–112, 180–182.

1955. The perspective of species. Turtox News, vol. 33, pp. 74–77.

1958. The synthetic natural populational species in biology. Syst. Zool., vol. 7, pp. 116–119.

Warren, B. C. S.

1958. On the recognition of the species. Proc. 10th Internat. Congr. Ent. Montreal 1956, vol. 1, pp. 111–123.

The nature of subspecies

Amadon, D.

1950. The species—then and now. Auk, vol. 67, pp. 492–498.

Bogert, C. M.

1954. The indication of infraspecific variation. Syst. Zool., vol. 3, pp. 111–112.

Bright, P. M. and Leeds, H. A.

1938. A monograph of the British aberrations of the Chalk-Hill blue butterfly *Lysandra coridon* (Poda, 1761). Bournemouth, England; Richmond Hill Printing Works.

Burt, W. H.

1954. The subspecies category in mammals. Syst. Zool., vol. 3, pp. 99–104.

Doutt, J. K.

1955. Terminology of microgeographic races in mammals. Syst. Zool., vol. 4, pp. 179–185.

Durrant, S. D.

1955. In defense of the subspecies. Syst. Zool., vol. 4, pp. 186–190.

Edwards, J. G.

1954. A new approach to infraspecific categories. Syst. Zool., vol. 3, pp. 1–20.

1956. Clarification of certain aspects of infraspecific systematics. Syst. Zool., vol. 5, pp. 92–94.

1956. What should we mean by subspecies? Turtox News, vol. 34, pp. 200–202, 230–232.

Fox, R. M.

1955. On subspecies. Syst. Zool., vol. 4, pp. 93–95.

Goodnight, C. J. and Goodnight, M. L.

1954. Taxonomic recognition of variation in Opiliones. Syst. Zool., vol. 2, pp. 173–179, 192. 1953.

Gosline, W. A.

1954. Further thoughts on subspecies and trinomials. Syst. Zool., vol. 3, pp. 92–94.

Hagmeier, E. M.

1958. Inapplicability of the subspecies concept to North American marten. Syst. Zool., vol. 7, pp. 1–7.

Hubbell, T. H.
　　1954. The naming of geographically variant populations. Syst. Zool., vol. 3, pp. 113–121.
Kiriakoff, S. G.
　　1956. On the subspecies concept in taxonomy. Lep. News, vol. 10, pp. 207–208.
Lidicker, W. Z., Jr.
　　1963. The nature of subspecies boundaries in a desert rodent and its implications for subspecies taxonomy. Syst. Zool., vol. 11, pp. 160–171.
Linsley, E. G.
　　1944. The naming of infra-specific categories. Ent. News, vol. 55, pp. 225–232.
Mayr, E.
　　1963. Animal species and evolution. Cambridge; Harvard University Press.
Meikle, R. D.
　　1957. "What is the subspecies?" Taxon, vol. 6, pp. 102–105.
Moore, I. M.
　　1954. Nomenclatural treatment of specific and infraspecific categories. Syst. Zool., vol. 3, pp. 90–91.
Newell, N. D.
　　1947. Infraspecific categories in invertebrate paleontology. Evolution, vol. 1, pp. 163–171.
Peters, J. A. et al.
　　1954. Symposium: subspecies and clines. Syst. Zool., vol. 3, pp. 97–126, 133.
Pimentel, R. A.
　　1958. Taxonomic methods, their bearing on subspeciation. Syst. Zool., vol. 7, pp. 139–156.
　　1959. Mendelian infraspecific divergence levels and their analysis. Syst. Zool., vol. 8, pp. 139–159.
Smith, H. M.
　　1956. A case for the trinomen. Syst. Zool., vol. 5, pp. 183–190.
van Son, G.
　　1955. A proposal for the restriction of the use of the term subspecies. Lep. News, vol. 9, pp. 1–3.
Starrett, A.
　　1958. What *is* the subspecies problem? Syst. Zool., vol. 7, pp. 111–115.
Stirton, R. A.
　　1955. Specific and infraspecific categories in fossil mammals. 16th Ann. Biol. Colloq., pp. 26–37.
Sylvester-Bradley, P. C.
　　1951. The subspecies in palaeontology. Geol. Mag., vol. 88, pp. 88–102.

Tilden, H. J. W.
    1961. Certain comments on the subspecies problem. Syst. Zool., vol. 10, pp. 17–23.
Wilson, E. O. and Brown, W. L., Jr.
    1953. The subspecies concept and its taxonomic application. Syst. Zool., vol. 2, pp. 97–111.

Chapter 12. The Practice of Classification

General

Ball, G. H.
    1960. Some considerations regarding the Sporozoa. Journ. Protozool., vol. 7, pp. 1–6.
Blackwelder, R. E.
    1959. The functions and limitations of classification. Syst. Zool., vol. 8, pp. 202–211.
    1964. Phyletic and phenetic vs. omnispective classification. *In* Phenetic and phylogenetic classification. Systematics Assoc. Publ. No. 6, pp. 17–28.
Borgmeier, T.
    1957. Basic questions of systematics. Syst. Zool., vol. 6, pp. 53–69.
Cain, A. J.
    1959. Taxonomic concepts. Ibis, vol. 101, pp. 302–318.
Cain, A. J. and Harrison, G. A.
    1960. Phyletic weighting. Proc. Zool. Soc. London, vol. 135, pp. 1–31.
Camp, W. H.
    1951. Biosystematy. Brittonia, vol. 7, pp. 113–127.
Daly, H. V.
    1962. Phenetic classification and typology. Syst. Zool., vol. 10, 1961, pp. 176–179.
Davis, P. H. and Heywood, V. H.
    1963. Principles of angiosperm taxonomy. Edinburgh; Oliver and Boyd.
Downey, J. C. and Strawn, M. A.
    1963. Sound-producing mechanisms in pupae of Lycaenidae. (Abstract) Bull. Ent. Soc. America, vol. 9, p. 161.
Edmunds, G. F., Jr.
    1962. The principles applied in determining the hierarchic level of the higher categories of Ephemeroptera. Syst. Zool., vol. 11, pp. 22–31.
Ehrlich, P. R.
    1959. Problems of higher classification. Syst. Zool., vol. 7 (1958), pp. 180–184.
Felt, E. P.
    1934. Classifying symbols for insects. Journ. New York Ent. Soc., vol. 42, pp. 373–392.

Gates, R. R.

> 1951. The taxonomic units in relation to cytogenetics and gene ecology. American Nat., vol. 85, pp. 31–50.

Gilmour, J. S. L.

> 1940. Taxonomy and philosophy. *In* The new systematics, ed. by J. S. Huxley. London; Oxford University Press. (Pp. 461–474)

Gregg, J. R.

> 1954. The language of taxonomy: an application of symbolic logic to the study of classificatory systems. New York; Columbia University Press.

Hargis, W. J., Jr.

> 1956. A suggestion for the standardization of the higher systematic categories. Syst. Zool., vol. 5, pp. 42–46.

Hoedemann, J. J.

> 1960. Remarks on the classification and phylogeny of teleostean fishes. Bull. Aquat. Biol., vol. 2, pp. 50–51.

Holloway, B. A.

> 1960. Taxonomy and phylogeny in the Lucanidae (Insecta: Coleoptera). Rec. Dom. Mus. Wellington, vol. 3, pp. 321–365.

Horn, W.

> 1929. On the splitting influence of the increase of entomological knowledge and on the enigma of species. Trans. 4th Internat. Congr. Ent. Ithaca 1928, vol. 2, pp. 500–507.

Kevan, D. K. M.

> 1961. Current tendencies to increase the number of higher taxonomic units among insects. Syst. Zool., vol. 10, pp. 92–103.

Kinsey, A. C.

> 1936. Origin of the higher categories in Cynips. Indiana Univ. Publ., Sci. Ser., no. 4, pp. 1–334.

Mayr, E.

> 1942. Systematics and the origin of species from the viewpoint of a zoologist. New York; Columbia University Press.
>
> 1953. Concepts of classification and nomenclature in higher organisms and microorganisms. Ann. New York Acad. Sci., vol. 56, pp. 391–397.

McAtee, W. L.

> 1926. Insect taxonomy: preserving a sense of proportion. Proc. Ent. Soc. Washington, vol. 28, pp. 68–70.

Michener, C. D.

> 1944. Comparative external morphology, phylogeny, and a classification of the bees (Hymenoptera). Bull. American Mus. Nat. Hist., vol. 82, pp. 157–326.
>
> 1957. Some bases for higher categories in classification. Syst. Zool., vol. 6, pp. 160–173.

Munroe, E. G. and Ehrlich, P. R.

  1960. Harmonization of concepts of higher classification of the Papilionidae. Journ. Lep. Soc., vol. 14, pp. 169–175.

Nicol, D.

  1956. The taxonomic significance of gaps in pelecypod morphology. Syst. Zool., vol. 5, p. 143.

Petrunkevitch, A.

  1952. Principles of classification as illustrated by studies of Arachnida. Syst. Zool., vol. 1, pp. 1–19.

du Rietz, G. E.

  1930. The fundamental units of biological taxonomy. Svensk. Bot. Tidskr., vol. 24, pp. 333–428.

Robson, G. C. and Richards, O. W.

  1936. The variation of animals in nature. London; Longmans Green.

Rohlf, F. J. and Sokal, R. R.

  1962. The description of taxonomic relationships by factor analysis. Syst. Zool., vol. 11, pp. 1–16.

Schindewolf, O. H.

  1950. Grundfragen der Palaeontologie. Stuttgart; Schweizerbart.

Simpson, G. G.

  1945. The principles of classification and a classification of mammals. Bull. American Mus. Nat. Hist., vol. 85, pp. 1–350.

  1961. The principles of animal taxonomy. New York; Columbia University Press.

Thompson, W. R.

  1937. Science and common sense. An Aristotelian excursion. London; Longmans Green.

Whittaker, R. H.

  1959. On the broad classification of organisms. Quart. Rev. Biol., vol. 34, pp. 210–226.

Genera

Bartlett, H. H. et al.

  1940. Concept of the genus. Bull. Torrey Bot. Club, vol. 67, pp. 349–389.

Cain, A. J.

  1956. The genus in evolutionary taxonomy. Syst. Zool., vol. 5, pp. 97–109.

Edwards, F. W.

  1953. Genera and subgenera. Syst. Zool., vol. 2, p. 135.

Hubbs, C. L.

  1943. Criteria for subspecies, species and genera, as determined by researches on fishes. Ann. New York Acad. Sci., vol. 44, Art. 2, pp. 109–121.

Inger, R. F.
1958. Comments on the definition of genera. Evolution, vol. 12, pp. 370–384.

James, M. T.
1953. An objective aid in determining generic limits. Syst. Zool., vol. 2, pp. 136–137.

Nicol, D.
1958. Trends and problems in pelecypod classification (the genus and subgenus). Journ. Washington Acad. Sci., vol. 48, pp. 285–293.

Paclt, J.
1957. A protest against the atomizing of genera, with special reference to the Hepialidae (Lepidoptera). Syst. Zool., vol. 6, pp. 51–52.

Sumner, F. B.
1915. Some reasons for saving the genus. Science, vol. 41, pp. 899–902.

Groups

Brown, W. J.
1959. Taxonomic problems with closely related species. Ann. Rev. Ent., vol. 4, pp. 77–98.

Cain, A. J.
1955. Not the superspecies. Syst. Zool., vol. 4, pp. 143–144, 139.

Dobzhansky, T.
1955. The genetic basis of systematic categories. 16th Ann. Biol. Colloqu., pp. 37–44.

Dodson, E. O.
1956. A note on the systematic position of the Mesozoa. Syst. Zool., vol. 5, pp. 37–40. (Definition of phylum)

Muller, S. W.
1958. Korobkov's proposal of a new morphologic entity, *forma accommodata,* in Mollusca. Syst. Zool., vol. 7, pp. 89–92.

Reed, C. A.
1960. Polyphyletic or monophyletic ancestry of mammals, or: what is a class? Evolution, vol. 14, pp. 314–322.

Simpson, G. G.
1943. Criteria for genera, species, and subspecies in zoology and paleozoology. Ann. New York Acad. Sci., vol. 44, pp. 145–178.
1959. The nature and origin of supraspecific taxa. Cold Spring Harbor Symp. Quant. Biol., vol. 24, pp. 255–271.

Sylvester-Bradley, P. C.
1954. Form-genera in paleontology. Journ. Paleont., vol. 28, pp. 333–336.
1954. The superspecies. Syst. Zool., vol. 3, pp. 145–146, 173.

Wainstein, B. A.
  1960. On the criteria of taxonomic categories. Zool. Zhurnal, vol. 39, pp. 1774–1778.

History of classification
Bather, F. A.
  1927. Biological classification: past and future. Quart. Journ. Geol. Soc. London, vol. 83, pp. lxii–civ.
Cain, A. J.
  1958. Logic and memory in Linnaeus's system of taxonomy. Proc. Linn. Soc. London, vol. 169, pp. 144–163.
  1959. Deductive and inductive methods in post-Linnaean taxonomy. Proc. Linn. Soc. London, vol. 170, pp. 185–217.
  1959. The post-Linnaean development of taxonomy. Proc. Linn. Soc. London, vol. 170, pp. 234–244.

Classifications
Berg, L. S.
  1947. Classification of fishes, both Recent and fossil. Ann Arbor, Michigan; Edwards Bros.
Blackwelder, R. E.
  1963. Classification of the Animal Kingdom. Carbondale, Illinois; Southern Illinois University Press.
Copeland, H. F.
  1956. The classification of lower organisms. Palo Alto, California; Pacific Books.
Honigberg, B. M. et al.
  1964. A revised classification of the phylum Protozoa. Journ. Protozool., vol. 11, pp. 7–20.
Jordan, D. S.
  1917–1920. The genera of fishes (1758–1920, in 4 pts.) Leland Stanford Jr. University Publ., Univ. Series. (Reprinted 1963, with additions.)
Jordan, D. S., Evermann, B. W., and Clark, H. W.
  1930. Check list of the fishes and fishlike vertebrates of North and Middle America. Rep. U.S. Comm. Fish. for 1928, pt. II.
Simpson, G. G.
  1945. The principles of classification and a classification of mammals. Bull. American Mus. Nat. Hist., vol. 85, pp. 1–350.

Types
Hansen, E. D.
  1961. Animal diversity. New York; Prentice-Hall.
Simpson, G. G.
  1940. Types in modern taxonomy. American Journ. Sci., vol. 238, pp. 413–431.

Chapter 13. The Use of Names

General

Blanchard, R.
>1905. Règles internationales de la nomenclature zoologique adoptées par les congrés internationaux de zoologie. Paris; F. R. de Rudeval.

Hemming, F. (ed.)
>1953. Copenhagen decisions on zoological nomenclature. . . . London; International Trust for Zoological Nomenclature.

Lamanna, C. and Mallette, M. F.
>1953. Basic bacteriology. Baltimore; Williams and Wilkins.

Simpson, G. G.
>1961. The principles of animal taxonomy. New York; Columbia University Press.

Stoll, N. R. et al. (eds.)
>1961. International Code for Zoological Nomenclature. . . . London; International Trust for Zoological Nomenclature.

Names of forms, ages, sexes, assemblages

Davies, R. G.
>1958. The terminology of the juvenile phases of insects. Trans. Soc. Brit. Ent., vol. 13, pp. 25–36.

Gray, P. (ed.)
>1961. The Encyclopedia of the biological sciences. New York; Reinhold Publ. Corp.

Common names

(See also list in text of Chapter 13.)

Dirst, J. et al.
>1957. Liste des noms français et espagnols des mammifères d'Europe. Mammalia, vol. 21, pp. 258–266.

Frommhold, E.
>1954. Heimische Lurche und Kriechtiere. Wittenberg/Lutherstadt; Die neue Brehm-Bucherei.

Gay, F. J.
>1955. Common names of insects and allied forms occurring in Australia. Bull. Commonw. Sci. Industr. Res. Org., No. 275, pp. 1–32.

Manville, R. H.
>1959. Nomenclature of mammals. Journ. Mamm., vol. 40, p. 477.

McAtee, W. L.
>1957. Folk-names of Canadian birds. Bull. Nat. Mus. Canada, No. 149, pp. 1–74.

Muesebeck, C. F. W.
>1942. Common names of insects approved by the American Association of Economic Entomologists. Journ. Econ. Ent., vol. 35, pp. 83–101.

Roscoe, E. J.
   1955. [Letter to editor on stability of common names.] Syst. Zool.,
      vol. 3, 1954, p. 182.
Thomas, I. and Janson, H. W.
   1957. Common names of British insects and other pests. Tech. Bull.
      Min. Agric. Fish. Food, No. 6, pp. 1–49.
Zimmermann, K. et al.
   1958. Zur Liste der deutschen Namen für deutsche Säugetiere. Säuge-
      tierk. Mitt., vol. 6, pp. 124–126.

The use of names
Adler, K. K.
   1957. Reptiles: Use of scientific names. Position of reptiles in the
      animal kingdom. Ohio Herp. Soc. Spec. Publ. No. 1, pp. 1–7.
Blackwelder, R. E.
   1949. Synonyms and genotypes. Coleopt. Bull., vol. 3, pp. 73–75.
   1954. The open season on taxonomists. Syst. Zool., vol. 3, pp. 177–
      181.
Dall, W. H.
   1877. Nomenclature in zoology and botany. Proc. American Assoc.
      Adv. Sci., 1877, pp. 7–56.
Felt, E. P. and Bishop, S. C.
   1926. Science and scientific names. American Nat., vol. 55, pp. 275–
      281.
Gould, S.
   1954. Permanent numbers to supplant the binominal system. Ameri-
      can Sci., vol. 12, pp. 269–274.
Griffiths, R. J.
   1960. Names and what they stand for. The Cowry, vol. 1, pp. 2–7.
Landin, B. O.
   1957. Critical comments upon some nomenclatorial and synonymical
      questions. Ent. Tidskr., vol. 78, pp. 101–114.
de Laubenfels, M. W.
   1958. Zoological science and the problem of names. Turtox News, vol.
      35, pp. 258–261.
   1959. (Continuation). Vol. 36, pp. 28–30.
Macan, T. T.
   1955. A plea for restraint in the adoption of new generic names. Ent.
      Monthly Mag., vol. 91, pp. 278–282.
Mayr, E.
   1954. Notes on nomenclature and classification. Syst. Zool., vol. 3,
      pp. 86–89.
Mayr, E., Linsley, E. G., and Usinger, R. L.
   1953. Methods and principles of systematic zoology. New York; Mc-
      Graw-Hill Book Co.

Moore, R. C.
1955. Treatise on invertebrate paleontology. Editorial preface to Part E. Geological Society of America and University of Kansas Press.

Needham, J. G.
1910. Practical nomenclature. Science, vol. 32, pp. 295–300.

Sailer, R. I.
1961. Utilitarian aspects of supergeneric names. Syst. Zool., vol. 10, pp. 154–156.

Pronunciation of names
(See also list in text of Chapter 13.)

Baker, J. P.
1956. The pronunciation of scientific terms. School Sci. Rev., vol. 37, pp. 201–205.

Flood, W. E.
1957. The pronunciation of scientific terms. New Scientist, vol. 3, pp. 19–20.

Henderson, I. F.
1929. A dictionary of scientific terms. Pronunciation, derivation, and definition of terms in biology, botany, zoology, anatomy, cytology, embryology, physiology, 2nd ed. New York; D. Van Nostrand Co.

Hopper, H. P.
1959. The pronunciation and derivation of the names of genera and subgenera of the family Ichneumonidae found in North America north of Mexico (Hymenoptera). Proc. Ent. Soc. Washington, vol. 61, pp. 155–171.

Jaeger, E. C.
1960. The biologist's handbook of pronunciations. Springfield, Illinois; Chas. C. Thomas.

Melander, A. L.
1916. The pronunciation of insect names. Bull. Brooklyn Ent. Soc., vol. 11, pp. 93–101.

Chapter 14. The Use of Literature

Nomenclators; higher groups
(See lists in text of Chapter 14.)

Nomenclators; generic names
(See lists in text of Chapter 14.)

Nomenclators; specific names
(See lists in text of Chapter 14.)

Directories
(See list in text of Chapter 14.)

Bibliographies—General

(See also items cited in the text of Chapter 14, under the heading Interpreting the literature.)

Besterman, T.

1947–1949. A world bibliography of bibliographies . . . , 2nd ed., 3 vols. London; T. Besterman.

Blackwelder, R. E.

1949. An adventure in biblio-chronology. Journ. Washington Acad. Sci., vol. 39, pp. 301–305.

Esdaile, A.

1931. A student's manual of bibliography. London; Geo. Allen and Unwin.

Journal of the Society for the Bibliography of Natural History, vol. 1, 1936– .

Bibliographies by author

(See lists in text of Chapter 14.)

Bibliographies by animal groups

(See lists in text of Chapter 14.)

Subject bibliographies

(See lists in text of Chapter 14.)

Periodical lists

(See list in text of Chapter 14.)

Chapter 15. Descriptive taxonomy

General

Bowers, D. E.

1956. A study of methods of color determination. Syst. Zool., vol. 5, pp. 147–160, 182.

Cain, A. J. and Harrison, G. A.

1958. An analysis of the taxonomist's judgment of affinity. Proc. Zool. Soc. London, vol. 131, pp. 85–98.

Chamberlin, J. C.

1931. The arachnid order Chelonethida. Stanford Univ. Publ., Univ. Series, Biol. Sci., vol. 7, No. 1, pp. 1–284.

Ferris, G. F.

1928. Principles of systematic entomology. Stanford, California; Stanford University Press.

Goto, H. E.

1955. On the need for detailed descriptions of species of Collembola. Ent. Monthly Mag., vol. 91, pp. 238–239.

Kohn, A. J.

1960. Graduating the novices, or the description of new species. Hawaiian Shell News, vol. 9, pp. 3–4.

Mayr, E., Linsley, E. G., and Usinger, R. L.
  1953. Methods and principles of systematic zoology. New York; Mc-Graw-Hill Book Co.
Michener, C. D.
  1958. Morphologically meaningful vs. descriptive terminologies for use by taxonomists. Proc. 10th Internat. Congr. Ent. Montreal 1956, vol. 1, pp. 583–586.
Rensch, B.
  1958. Die ideale Artbeschreibung. Uppsala Univ. Årsskr., vol. 6, pp. 91–103.
Rhodes, F. H. T.
  1954. The value of a diagnosis in systematic paleontology. Journ. Paleo., vol. 28, pp. 487–488.
Ridgway, R.
  1912. Color standards and color nomenclature. Washington, D.C.; A. Hoen Co.
Riemer, W. J.
  1954. Formulation of locality data. Syst. Zool., vol. 3, pp. 138–140.
Ross, E. S.
  1956. What is species describing? Syst. Zool., vol. 5, pp. 191–192.
Simpson, G. G.
  1961. Principles of animal taxonomy. New York; Columbia University Press.
Simpson, G. G., Roe, A., and Lewontin, R. C.
  1960. Quantitative zoology, 2nd ed. New York; Harcourt, Brace and World.
Sylvester-Bradley, P. C.
  1958. The description of fossil populations. Journ. Paleo., vol. 32, pp. 214–235.

Descriptive characters
Burch, J. B. and Thompson, F. G.
  1958. Taxonomic characters and infraspecific variation in mollusks. Syst. Zool., vol. 7, p. 48.
Burch, J. Q.
  1957. Taxonomic characters in Mollusca. Syst. Zool., vol. 5, p. 144.
Simpson, G. G.
  1943. Criteria for genera, species, and subspecies in zoology and paleontology. Ann. New York Acad. Sci., vol. 44, pp. 145–178.
Vladykov, V. D.
  1934. Environmental and taxonomic characters of fishes. Trans. Roy. Canadian Inst., vol. 20, pp. 99–140.

Key construction
(See list in text of Chapter 15.)

Chapter 16. The Publication of Data

Descriptions
(See list in text of Chapter 16.)

Style manuals
(See list in text of Chapter 16.)

Terms; word books
(See list in text of Chapter 16.)

Technical writing manuals
(See list in text of Chapter 16.)

Preparation of illustrations
(See list in text of Chapter 16.)

PART V

Theoretical Taxonomy

Chapters 17–18

Bader, R. S.
1958. Similarity and recency of common ancestry. Syst. Zool., vol. 7, pp. 184–187.
Bigelow, R. S.
1957. Monophyletic classification and evolution. Syst. Zool., vol. 5 (1956), pp. 145–146.
1959. Classification and phylogeny. Syst. Zool., vol. 7, pp. 49–59.
1959. Similarity, ancestry, and scientific principles. Syst. Zool., vol. 8, pp. 165–168.
1961. Higher categories and phylogeny. Syst. Zool., vol. 10, pp. 86–91.
Blackwelder, R. E.
1962. Animal taxonomy and the new systematics. *In* Survey of Biological Progress, ed. by B. Glass, vol. 4, pp. 1–57.
1964. Phyletic and phenetic versus omnispective classification. *In* Phenetic and phylogenetic classification, ed. by V. H. Heywood and J. McNeill. Systematics Association Publ. No. 6, pp. 17–28.
Blackwelder, R. E. and Boyden, A. A.
1952. The nature of systematics. Syst. Zool., vol. 1, pp. 26–33.
Blair, W. F. and Hubbs, C.
1961. Biological species and phylogenetic taxonomy. Syst. Zool., vol. 10, pp. 42–43.
Borgmeier, T.
1955. Grundfragen der Systematik. Studia Ent., No. 3, pp. 22–51.
1957. Basic questions of systematics. Syst. Zool., vol. 6, pp. 53–69.

Boyden, A. A.

    1952. (See Blackwelder, R. E. and Boyden, A. A.)

    1960. A brief commentary on Simpson's Anatomy and morphology: Classification and evolution. 1859 and 1959. Syst. Zool., vol. 9, pp. 44–46.

Bruner, J. S. et al.

    1956. A study of thinking. New York; John Wiley and Sons.

Calman, W. T.

    1940. A museum zoologist's view of taxonomy. In The new systematics, ed. by J. S. Huxley. London; Oxford University Press. Pp. 455–459.

    1949. The classification of animals. London; Methuen and Co.

Danser, B. H.

    1950. A theory of systematics. Bibl. Biotheor., vol. 4, pp. 117–180.

Davis, P. H. and Heywood, V. H.

    1963. Principles of angiosperm taxonomy. Edinburgh; Oliver and Boyd.

Dobzkansky, T.

    1941. Genetics and the origin of species, 2nd ed. New York; Columbia University Press.

Ferris, G. F.

    1928. The principles of systematic entomology. Stanford, California; Stanford University Press.

Gerard, R. W.

    1957. Units and concepts in biology. Science, vol. 125, pp. 429–433.

Gilmour, J. S. L.

    1940. Taxonomy and philosophy. In The new systematics, ed. by J. S. Huxley. London; Oxford University Press.

Grant, V.

    1957. The plant species in theory and practice. In The species problem, ed. by E. Mayr. Washington, D.C.; American Association for the Advancement of Science.

Gregg, J. R.

    1950. Taxonomy, language, and reality. American Nat., vol. 84, pp. 419–435.

    1954. The language of taxonomy. An application of symbolic logic to the study of classificatory systems. New York; Columbia University Press.

Harper, J. L. et al.

    1961. The evolution and ecology of closely related species living in the same area. Evolution, vol. 15, pp. 209–227.

Heslop-Harrison, J.

    1956. New concepts in flowering-plant taxonomy. Cambridge, Massachusetts; Harvard University Press. (Also published in London, 1953, 1955, 1960, 1963.)

Hogben, L.

    1940. Problems of the origins of species. In The new systematics, ed. by J. S. Huxley. London; Oxford University Press.

Hubbs, C. L.
    1934. Racial and individual variation in animals, especially fishes. American Nat., vol. 68, pp. 115–128.
Huxley, J. S.
    1940. Introductory: Towards the new systematics. In The new systematics, ed. by J. S. Huxley. London; Oxford University Press.
    1942. Evolution. The modern synthesis. New York; Harper.
Keck, D. D.
    1957. Trends in systematic botany. In Survey of biological progress, ed. by B. Glass, vol. 3, pp. 47–107.
Kiriakoff, S. G.
    1959. Phylogenetic systematics versus typology. Syst. Zool., vol. 8, pp. 117–118.
    1963. On the neo-adansonian school. Syst. Zool., vol. 11, pp. 180–185.
Lamanna, C. and Mallette, M. F.
    1953. Basic bacteriology. Baltimore; Williams and Wilkins.
Lenzen, V. F.
    1955. Procedures of empirical science. In International Encyclopedia of Unified Science, vol. 1, pt. 1, pp. 279–339.
Mayr, E.
    1942. Systematics and the origin of species from the viewpoint of a zoologist. New York; Columbia University Press.
    1955. Systematics and modes of speciation. 16th Ann. Biol. Colloqu., 1955, pp. 45–51.
Mayr, E., Linsley, E. G., and Usinger, R. L.
    1953. Methods and principles of systematic zoology. New York; McGraw-Hill Book Co.
Michener, C. D. and Sokal, R. R.
    1957. A quantitative approach to a problem in classification. Evolution, vol. 11, pp. 130–162.
Myers, G. S.
    1930. [Critical comments on Hall and Clements, The phylogenetic method in taxonomy.] Micropaleo. Bull., vol. 2, pp. 55–58.
    1952. The nature of systematic biology and of a species description. Syst. Zool., vol. 1, pp. 106–111.
    1960. Some reflections on phylogenetic and typological taxonomy. Syst. Zool., vol. 9, pp. 37–41.
Pearl, R.
    1922. Trends of modern biology. Science, vol. 56, pp. 581–592.
Petrunkevitch, A.
    1952. Principles of classification as illustrated by studies of Arachnida. Syst. Zool., vol. 1, pp. 1–19.
Popper, K. R.
    1959. The logic of scientific discovery. New York; Basic Books.
Richards, O. W.
    1940. Phylogeny and taxonomy. Proc. Linn. Soc. London, vol. 152, p. 241.

Robson, G. C.

1928. The species problem. An introduction to the study of evolutionary divergence in natural populations. London; Oliver and Boyd.

Roget, P. M.

1933. Thesaurus of English words and phrases. . . . New York; Grosset and Dunlap. (First American edition)

Simpson, G. G.

1945. The principles of classification. . . . Bull. American Mus. Nat. Hist., vol. 85, pp. 1–350.

1961. Principles of animal taxonomy. New York; Columbia University Press.

Simpson, G. G., Roe, A., and Lewontin, R. C.

1960. Quantitative zoology. New York; Harcourt, Brace and World.

Sneath, P. H. A.

1961. Recent developments in theoretical and quantitative taxonomy. Syst. Zool., vol. 10, pp. 118–139.

Sokal, R. R.

1958. Quantification of systematic relationships and of phylogenetic trends. Proc. 10th Internat. Congr. Ent. Montreal 1956, vol. 1, pp. 404–415.

1963. Typology and empiricism in taxonomy. Journ. Theor. Biol., vol. 3, pp. 230–267

Sonneborn, T. M.

1957. Breeding systems, reproductive methods, and species problems in Protozoa. In The species problem, ed. by E. Mayr. Washington, D.C.; American Association for the Advancement of Science.

Thompson, W. R.

1937. Science and common sense. An Aristotelian excursion. London; Longmans Green.

1952. The philosophical foundations of systematics. Canadian Ent., vol. 84, pp. 1–16.

1955. Systematics: The ideal and the reality. Bull. Lab. Zool. Gen. Agrar. Portici, vol. 33, pp. 320–329.

1958. The interpretation of taxonomy. Proc. 10th Internat. Congr. Ent. Montreal 1956, vol. 1, pp. 61–73.

1962. Evolution and taxonomy. Studia Ent., vol. 5, pp. 549–570.

Turrill, W. B.

1940. Experimental and synthetic plant taxonomy. In The new systematics, ed. by J. S. Huxley. London; Oxford University Press.

Weller, J. M.

1955. Fatuous species and hybrid populations. Journ. Paleontol., vol. 29, pp. 1066–1069.

Woodger, J. H.

1937. The axiomatic method in biology. Cambridge; Cambridge University Press.

1952. Biology and language. An introduction to the methodology of the biological sciences including medicine. Cambridge; Cambridge University Press

PART VI

Zoological Nomenclature

Chapters 19–29

General
(Anonymous)
1952. The law of priority. Syst. Zool., vol. 1, pp. 34–35. (Nomenclature Committee, Society of Systematic Zoology)
Allen, J. F.
1954. Stability in taxonomy. Syst. Zool., vol. 3, pp. 140–142.
Amadon, D.
1955. In defence of the principle of the "First Reviser." Bull. Brit. Ornith. Club, vol. 75, pp. 21–23.
Balfour-Browne, F.
1958. The failure of the International Commission of Zoological Nomenclature. Ent. Monthly Mag., vol. 93, pp. 274–278.
Blackwelder, R. E.
1946. Misuse of the Linnaean system of nomenclature. Science, vol. 104, pp. 277–278.
1948. The principle of priority in biological nomenclature. Journ. Washington Acad. Sci., vol. 38, pp. 306–309.
Blackwelder, R. E., Knight, J. B., and Smith, H. M.
1950. Categories of availability or validity of scientific names. Science, vol. 111, pp. 289–290.
1950. Categories of availability or validity of scientific names. Bull. Zool. Nom., vol. 8, pp. 27–28.
Dall, W. H.
1877. Nomenclature in zoology and botany. Proc. American Assoc. Adv. Sci., 1877. Pp. 7–56.
Emiliani, C.
1956. Nomenclature and grammar. Journ. Washington Acad. Sci., vol. 42, pp. 137–141.
Felt, E. P. and Bishop, S. C.
1926. Science and scientific names. American Nat., vol. 60, pp. 275–281.
Ferris, G. F.
1928. The principles of systematic entomology. Stanford, California; Stanford University Press.

Follett, W. I.

1955. An unofficial interpretation of the International Rules of Zoological Nomenclature as amended by the XIII International Congress of Zoology, Paris, 1948 and by the XIV International Congress of Zoology, Copenhagen, 1953. San Francisco; privately published.

Gould, S.

1954. Permanent numbers to supplement the binominal system. American Sci., vol. 42, pp. 269–274.

Hiltermann, H.

1956. Ten rules concerning the nomenclature and classification of the Foraminifera. Micropaleontology, vol. 2, pp. 296–298.

Hubbs, C. L.

1956. Ways of stabilizing zoological nomenclature. XIV Internat. Congr. Zoology, pp. 548–553.

Jordan, D. S.

1926. Scientific names and their convenience. Science, vol. 64, pp. 575–576.

Keen, A. M. and Muller, S. W.

1948. Revised edition of Schenk and McMasters, Procedure in Taxonomy. . . . Stanford, California; Stanford University Press.

1956. Third edition enlarged and in part rewritten of Schenk and McMasters, Procedure in Taxonomy. . . . Stanford, California; Stanford University Press.

Landin, B.-O.

1957. Critical comments upon some nomenclatorial and synonymical questions. . . . Ent. Tidskr., vol. 78, pp. 101–114.

Mayr, E.

1954. Notes on nomenclature and classification. Syst. Zool., vol. 3, pp. 86–89.

Mayr, E., Linsley, E. G., and Usinger, R. L.

1953. Methods and principles of systematic zoology. New York; Mc-Graw-Hill Book Co.

Moore, R. C.

1955. Treatise on invertebrate paleontology. Part E. Editorial preface. New York; Geological Society of America. (Pp. viii–xviii.)

Schenk, E. T. and McMasters, J. H.

1936. Procedure in taxonomy. Stanford University, California; Stanford University Press.

Smith, H. M.

1962. The hierarchy of nomenclatural status of generic and specific names in zoological taxonomy. Syst. Zool., vol. 11, pp. 139–142.

Stiles, C. W.

1927. Underlying factors in the confusion in zoological nomenclature with a definite practical suggestion for the future. Science, vol. 65, pp. 194–199.

Stohler, R.
   1959. Why rules of priority? Veliger, vol. 1, pp. 17–18.
de la Torre, L. and Starrett, A.
   1959. Name changes and nomenclatural stability. Nat. Hist. Misc.,
      No. 167, pp. 1–4.

Nomenclature rules
(Anonymous)
   1926. International rules of zoological nomenclature. Proc. Biol. Soc.
      Washington, vol. 39, pp. 75–104.
   1961. International code. . . . (See Stoll, N. R., 1961.)
   1964. International code of zoological nomenclature adopted by the
      XV International Congress of Zoology. London; International
      Trust for Zoological Nomenclature.
Blanchard, R. (ed.)
   1905. Règles internationales de la nomenclature zoologique adoptées
      par la congrés internationaux de zoologie. Paris; F. R. de Rudeval.
Collin, J. E.
   1958. A critical note on two new "rules" proposed to be added to the
      revised code of rules for zoological nomenclature. Ent. Monthly
      Mag., vol. 94, pp. 61–62.
Ferris, G. F.
   1928. The principles of systematic entomology. Stanford, California;
      Stanford University Press. [Includes reprint of International Rules
      and Opinions.]
Follett, W. I.
   1955. An unofficial interpretation of the International Rules of Zoolog-
      ical Nomenclature as amended by the XIII International Congress
      of Zoology, Paris, 1948 and by the XIV International Congress
      of Zoology, Copenhagen, 1953. San Francisco; privately published.
   1963. New precepts of zoological nomenclature. A.I.B.S. Bull., vol. 13,
      pp. 14–18. (American Institute of Biological Sciences.)
Hemming, F. (ed.)
   1943. Bulletin of Zoological Nomenclature. Vol. 1, 1943– London;
      International Trust for Zoological Nomenclature.
   1953. Copenhagen decisions on zoological nomenclature, additions to,
      and modifications of, the Règles internationales de la nomencla-
      ture zoologique. Approved and adopted by the Fourteenth Inter-
      national Congress of Zoology, Copenhagen, August, 1953. Lon-
      don; International Trust for Zoological Nomenclature.
   1958. Official index of rejected and invalid family-group names in
      zoology. First instalment: Names 1–273. London; International
      Trust for Zoological Nomenclature.
   1958. Official index of rejected and invalid generic names in zoology.
      First instalment: Names 1–1169. London; International Trust for
      Zoological Nomenclature.

1958. Official index of rejected and invalid specific names in zoology. First instalment: Names 1–527. London; International Trust for Zoological Nomenclature.

1958. Official index of rejected and invalid works in zoological nomenclature. First instalment: Names 1–58. London; International Trust for Zoological Nomenclature.

1958. Official list of generic names in zoology. First instalment: Names 1–1274. London; International Trust for Zoological Nomenclature.

1958. Official list of family-group names in zoology. First instalment: Names 1–236. London; International Trust for Zoological Nomenclature.

1958. Official list of specific names in zoology. First instalment: Names 1–1525. London; International Trust for Zoological Nomenclature.

1958. Official list of works approved as available for zoological nomenclature. First instalment: Names 1–38. London; International Trust for Zoological Nomenclature.

Keen, A. M. and Muller, S. W.

1948. Revised edition of Schenk and McMasters, Procedure in Taxonomy. Stanford, California; Stanford University Press.

1956. Third edition enlarged and in part rewritten of Schenk and McMasters, Procedure in Taxonomy. Stanford, California; Stanford University Press.

Schenk, E. T. and McMasters, J. H.

1936. Procedure in taxonomy. Stanford University California; Stanford University Press. (See Keen and Muller above.)

Smith, H. M.

1962. Commentary on the 1961 Code of Zoological Nomenclature. Syst. Zool., vol. 11, pp. 85–91.

1962. The nomen oblitum rule of the 1961 International Code of Zoological Nomenclature. Herpetologica, vol. 18, pp. 11–13.

Stoll, N. R. et al. (eds.)

1961. International code of zoological nomenclature adopted by the XV International Congress of Zoology. London; International Trust for Zoological Nomenclature.

Van Cleave, H. J.

1933. An index to the International Rules of Zoological Nomenclature. Trans. American Microscop. Soc., vol. 52, pp. 322–325.

Formation of names

Blackwelder, R. E.

1941. The gender of scientific names in zoology. Journ. Washington Acad. Sci., vol. 31, pp. 135–140.

Blackwelder, R. E., Knight, J. B., and Sabrosky, C. W.

1948. A revised proposal for errors and emendations in the rules of zoological nomenclature. Science, vol. 108, pp. 37–38.

1953. A revised proposal for errors and emendations in the rules of zoological nomenclature. Bull. Zool. Nom., vol. 10, pp. 129–134.

Brown, R. W.
   1954. Composition of scientific words. A manual of methods and a lexicon of materials for the practice of logotechnics. Washington; by the author.
Buchanan, R. E.
   1956. Transliteration of Greek to Latin in the formation of names of zoological taxa. Syst. Zool., vol. 5, pp. 65–67.
Follett, W. I.
   1952. Emendation of zoological names. Syst. Zool., vol. 1, pp. 178–181.
Grensted, L. W.
   1944. The formation and gender of generic names. Ent. Monthly Mag., vol. 80, pp. 229–233.
Schenkling, S.
   1922. Nomenclator coleopterologicus. Eine etymologische Erklärung sämtlicher Gattungs- und Artnamen der Käfer der deutschen Fauna so wie der angrenzenden Gebiete. Zweite Auflage. Jena; Verlag von Gustav Fischer.
Sinclair, G. W.
   1952. Diacritical marks in zoological names. Syst. Zool., vol. 1, pp. 84–86.

Name publication problems
Blackwelder, R. E.
   1947. The dates and editions of Curtis' British Entomology. Smithsonian Misc. Coll., vol. 107, no. 5, pp. 1–27.
   1949. An adventure in biblio-chronology. Journ. Washington Acad. Sci., vol. 39, pp. 301–305.
   1952. Preprints of Proc. U.S. National Museum, 1890–1897. Syst. Zool., vol. 1, pp. 86–89.
Grant, C. H. B.
   1958. New names published privately. Bull. Brit. Ornith. Club, vol. 78, p. 98.
Laffoon, J. L.
   1957. On the inadvisability of including new nomenclatural information in abstracts. Syst. Zool., vol. 6, pp. 109–111.
Parrott, A. W.
   1956. A note on whether A. A. Girault's privately printed leaflets are scientific publication in a technical sense. Proc. Roy. Zool. Soc. New South Wales, 1954–55, pp. 66–67.

Names of groups above genus
Arkell, W. J.
   1954. Nomenclature of families and superfamilies. Journ. Paleo., vol. 28, p. 218.
Baker, H. B.
   1956–1960. Family names in Pulmonata, 1. Nautilus, vol. 69, pp. 128–139 (1956); 2, vol. 70, pp. 34–35 (1956); 3, vol. 70, pp. 141–142 (1957); 4, vol. 73, pp. 114–117 (1960).

Bradley, J. C.
  1963. Difficulty arising from compulsory application of the Law of Priority to Family-group names, with proposed amendments. Syst. Zool., vol. 11, pp. 178–179.
Brown, W. L., Jr.
  1957. The ending for subtribal names in zoology. Syst. Zool., vol. 6, pp. 193–194.
Corliss, J. O.
  1958. Problems in selections of types for higher categories within the subphylum Ciliophora. Journ. Protozool., vol. 5 (suppl.), p. 17.
  1958. Proposed type genera for higher taxa within the sub-phylum Ciliophora. Bull. Zool. Nom., vol. 15, pp. 520–522.
  1958. Order/class names: problems in the selection of type genera. Bull. Zool. Nom., vol. 15, pp. 1073–1079.
Grensted, L. W.
  1947. On the formation of family names. A note on the implications of Opinion 143 of the International Commission on Zoological Nomenclature. Ent. Monthly Mag., vol. 83, pp. 137–141.
Hopkins, G. H. E.
  1958. Order-group and family-group names for the fleas. Ann. Mag. Nat. Hist., ser. 13, vol. 1, pp. 478–487.
Reed, C. A.
  1960. Polyphyletic or monophyletic ancestry of mammals, or: what is a class? Evolution, vol. 14, pp. 314–322.

Uniform endings
Alvarez, J.
  1951. Sobre la terminacion uniforme de los nombres de ordenes zoologicas. Ciencia, vol. 11, pp. 279–280.
Hemming, F.
  1953. Copenhagen decisions on zoological nomenclature. London; International Trust for Zoological Nomenclature. (Pp. 33–36)
Hubbs, C. L.
  1952. On uniform endings. Syst. Zool., vol. 1, pp. 146–147.
Levine, N. D.
  1958. Uniform endings for the names of higher taxa. Journ. Protozool., vol. 5 (suppl.), pp. 16–17.
  1958. Uniform endings for the names of higher taxa. Syst. Zool., vol. 7, pp. 134–135.
Moment, G. B.
  1953. Uniform endings and the revival of interest in taxonomy. Syst. Zool., vol. 2, pp. 191–192.
Pearse, A. S.
  1949. Zoological names. A list of phyla, classes, and orders. Durham, N.C.; American Association for the Advancement of Science.
Rogers, D. J.
  1952. On uniform group endings. Syst. Zool., vol. 1, pp. 147–148.

Shipley, A. E.
　1904. The orders of insects. Zool. Anz., vol. 27, pp. 259–262.
Simpson, G. G.
　1952. For and against uniform endings in zoological nomenclature. Syst. Zool., vol. 1, pp. 20–23.
Stenzel, H. B.
　1950. Proposed uniform endings for names of higher categories in zoological systematics. Science, vol. 112, p. 94.

Family names
Bradley, J. C.
　1963. Difficulty arising from compulsory application of the Law of Priority to family-group names, with proposed amendments. Syst. Zool., vol. 11, pp. 178–179.
Brown, W. L., Jr.
　1958. The ending for subtribal names in zoology. Syst. Zool., vol. 6, pp. 193–194.
Grensted, L. W.
　1947. On the formation of family names. A note on the implications of Opinion 143 of the International Commission on Zoological Nomenclature. Ent. Monthly Mag., vol. 83, pp. 137–141.
McMichael, D. F.
　1956. Problems of family nomenclature. Syst. Zool., vol. 5, pp. 141–142.
Oberholzer, H. C.
　1920. The nomenclature of families and subfamilies in zoology. Science, vol. 53, pp. 142–147.
Sabrosky, C. W.
　1939. A summary of family nomenclature in the order Diptera. Verh. 7 Internat. Kongr. Ent., vol. 1, pp. 599–612.
　1947. Stability of family names, some principles and problems. American Nat., vol. 81, pp. 153–160.
　1954. Nomenclature of families and superfamilies. Journ. Paleo., vol. 28, pp. 489–490.
Van Duzee, E. P.
　1916. Priority in family names and related matters. Ann. Ent. Soc. America, vol. 9, pp. 89–93

Generic names and genotypes
Blackwelder, R. E.
　1946. Fabrician genotype designation. Bull. Brooklyn Ent. Soc., vol. 41, pp. 72–78.
　1949. Synonyms and genotypes. Coleopt. Bull., vol. 3, pp. 73–75.
　1952. The generic names of the beetle family Staphylinidae. With an essay on genotypy. Bull. United States Nat. Mus., 200, pp. 1–483.
Conisbee, L. R.
　1960. Newly proposed genera, 1952–56. Journ. Mamm., vol. 41, pp. 112–113.

Grensted, L. W.

1944. The formation and gender of generic names. Ent. Monthly Mag., vol. 80, pp. 229–233.

Malaise, R.

1937. Fabricius as the first designator and original inventor of genotypes. Ent. News, vol. 48, pp. 130–134.

Richter, R. and Schmidt, H.

1956. Kann eine Unterart Genotypus sein? Senckenbergiana, vol. 37, pp. 543–546.

Sabrosky, C. W.

1957. Two overlooked sources of type designations for genera. Proc. Ent. Soc. Washington, vol. 59, pp. 171–172.

Specific names

Blackwelder, R. E.

1948. An analysis of specific homonyms in zoological nomenclature. Journ. Washington Acad. Sci., vol. 38, pp. 206–213.

Blackwelder, R. E., Knight, J. B., and Smith, H. M.

1950. Categories of availability or validity of scientific names. Science, vol. 111, pp. 289–290.

1953. Categories of availability or validity of scientific names. Bull. Zool. Nom., vol. 8, pp. 27–28.

Brown, W. L., Jr. and Wilson, E. O.

1954. The case against the trinomen. Syst. Zool., vol. 3, pp. 174–176.

Fennah, R. G.

1955. Subspecific nomenclature: the proposed method of Wilson and Brown. Syst. Zool., vol. 4, p. 140.

Grant, C. H. B.

1955. Page and line priority in ornithological nomenclature. Bull. Brit. Ornith. Club, vol. 75, p. 12.

Harrison, J. M.

1945. Races, intermediates and nomenclature—a suggested modification of the trinomial system. Ibis, vol. 87, pp. 49–51.

Hershkovitz, P.

1958. Type localities and nomenclature of some American primates, with remarks on secondary homonyms. Proc. Biol. Soc. Washington, vol. 71, pp. 53–56.

Hopkins, G. H. E.

1955. Homonyms for subspecies and aberrations. Entomologist, vol. 88, pp. 64–65.

de Laubenfels, M. W.

1953. Trivial names. Syst. Zool., vol. 2, pp. 42–45.

Sabrosky, C. W.

1940. Entomological usage of subspecific names. Ent. News, vol. 51, pp. 159–164.

1955. Postscript to a survey of infraspecific categories. Syst. Zool., vol. 4, pp. 141–142.

Smith, H. M.
   1945. Categories of species names in zoology. Science, vol. 102, pp.
      185–189.
   1947. (Comments on the terms "availability" and "validity.") Science,
      vol. 106, p. 11.
   1949. Some principles of taxonomy: The meaning of "occupancy" and
      "validity." Herpetologica, vol. 5, pp. 11–18.
Smith, H. M. and White, F. N.
   1956. A case for the trinomen. Syst. Zool., vol. 5, pp. 183–190.
Thalmann, H. E.
   1959. Foraminiferal homonyms. Contr. Cushman Found., vol. 10, pp.
      127–129.
de la Torre, L. and Starrett, A.
   1959. Name changes and nomenclatural stability. Nat. Hist. Misc.,
      no. 167, pp. 1–4.

Types in specific nomenclature
Fennah, R. G.
   1957. A guiding principle for lectotype selection. Syst. Zool., vol. 6,
      pp. 47–48.
Fernald, H. T.
   1939. On type nomenclature. Ann. Ent. Soc. America, vol. 32, pp.
      689–702.
Frizzell, D. L.
   1933. Terminology of types. American Midl. Nat., vol. 14, pp. 637–
      668.
Simpson, G. G.
   1940. Types in modern taxonomy. American Journ. Sci., vol. 238, pp.
      413–431.
   1961. Principles of animal taxonomy. New York; Columbia University
      Press.
Usinger, R. L.
   1952. Neotypes. Syst. Zool., vol. 1, pp. 171–173.
Williams, C. B.
   1940. On "type" specimens. Ann. Ent. Soc. America, vol. 33, pp. 621–
      624.
Young, D. A., Jr.
   1958. On lectotype proposals. Syst. Zool., vol. 7, pp. 120–122.

Type localities
Dunn, E. R. and Stuart, L. C.
   1951. On the legality of restriction of type locality. Science, vol. 113,
      pp. 677–678.
Smith, H. M.
   1953. Revision of type localities. Syst. Zool., vol. 22, pp. 37–41.

Author Index

Because the bibliography is arranged by subject, all entries therein are indexed here, as well as all in the text. Cited publications are indicated by the date, followed by the page references to that publication. Citations without dates are to general references to the man or his effect on taxonomy. Names are alphabetized under the first capital letter.

Adams, C. T., *1955*, 621
Adanson, M., 151, 188
Adler, K. K., *1957*, 645
Agassiz, L., 15
  1846, 243, 258
  1848, 243, 258
  1848–1854, 250, 253
Alexander, G., *1956*, 339
Alexander, R. D., *1962*, 626
Allen, J. F., *1954*, 653
Alvarez, J., *1951*, 658
Amadon, D., *1943*, 621, 628
  1950, 635, 637
  1955, 653
  1959, 626, 629
Amsel, —, *1950*, 627
Anderson, R. M., *1948*, 613
Ansell, W. F. H., *1958*, 630
Anthony, H. E., *1945*, 613
Arkell, W. J., *1954*, 657
Arkell, W. J. et al., *1957*, 63
Arnett, R. H., Jr., *1963*, 298, 315
de Azevedo, J. F. & Goncalves, M. M., *1956*, 623

Bader, R. S., *1958*, 649
Bailey, R. *1960*, 621
Baker, E. W. & Wharton, G. W., *1952*, 241
Baker, H. B., *1956–1960*, 657

Baker, J. P., *1956*, 646
Balfour-Browne, F., *1958*, 653
Ball, C. R., *1946*, 611
Ball, G. H., *1960*, 639
Ball, G. H. & Clark, E. W., *1953*, 624
Barlow, G. W., *1961*, 621
Barnett, J. A., *1961*, 631
Barr, A. R., *1954*, 612
Bartlett, H. H. et al., *1940*, 641
Bassler, R. S., 441
  1955, 63
Bateson, W., *1894*, 616
Bather, F. A., 15
  1927, 643
Beaudry, J. P., *1960*, 635
de Beer, G. R., *1940*, 623
Beer, J. R. & Cook, E. F., *1958*, 613
Bennett, G. F., *1958*, 620, 628
Berg, L. S., 440
  1947, 241, 643
Berger, A. J., *1952*, 621
  1959, 65, 621
Bergman, W. & Domsky, I. I., *1960*, 624
Besterman, T., *1947–1949*, 248, 647
Bier, M., *1959*, 155, 629
Bierce, A., *1909*, 319
Bigelow, R. S., *1957*, 613, 649
  1959, 649
  1961, 649

Binney, W. G., *1864*, 250
Blackman, P. D. & Todd, R., *1959*, 629
Blackwelder, R. E., 531
   1940, 607
   1941, 656
   1942, 249
   1946, 653, 659
   1947, 657
   1948, 550, 653, 660
   1949, 257, 322, 645, 647, 657, 659
   1950, 249
   1951, 607
   1952, 98, 259, 657, 659
   1954, 645
   1955, 611
   1957, 249, 254, 318, 322
   1959, 260, 639
   1960, 607
   1962, 618, 628, 649
   1963, 240, 445, 643
   1964, 190, 191, 639, 649
Blackwelder, R. E. & Blackwelder, R. M., *1961*, 74, 246
Blackwelder, R. E. & Boyden, A. A., *1952*, 4, 607, 608, 649
Blackwelder, R. E. & Hoyme, L. E., *1955*, 630, 631
Blackwelder, R. E., Knight, J. B., & Sabrosky, C. W., *1948*, 586, 656
   1953, 656
Blackwelder, R. E., Knight, J. B., & Smith, H. M., *1950*, 396, 653, 660
   1953, 660
Blair, W. F., *1943*, 626
   1962, 620, 628
Blair, W. F. et al., *1957*, 64
Blair, W. F. & Hubbs, C., *1961*, 649
Blanchard, R., *1905*, 644, 655
Bogert, C. M., *1954*, 637
Bolton, H. C., *1897*, 253
Bond, E. B., *1957*, 616
Borgmeier, T., *1957*, 173, 184, 639, 649
Borror, D. J., *1960*, 62, 238
Boudreaux, H. B., *1963*, 626
Boulenger, G. A., 15
Bøving, A. G., 283
Bowdler Sharpe, R., 15
Bowerman, E. G., *1947*, 254

Bowers, D. E., *1956*, 647
   1957, 612
Boyden, A. A., 4
   1935, 619
   1942, 629
   1943, 155, 619, 625, 629
   1947, 619
   1951, 625, 629
   1952, 4, 607, 608, 649
   1953, 629
   1954, 617
   1959, 625, 629
   1960, 650
Boyden, A. A. & Paulsen, E., *1959*, 629
Bradley, J. C., *1930*, 100, 180
   1963, 658, 659
Branson, C. C., *1948*, 251
Brennan, J. M., *1959*, 247
Brennecke, E., Jr. & Clark, D. L., *1942*, 319
Brezner, J. & Enns, W. R., *1958*, 625
Bright, P. M. & Leeds, H. A., *1938*, 175, 637
Brower, L. P., *1958*, 628
   1959, 621
Brown, R. W., *1954*, 62, 237, 318, 324, 579, 657
Brown, W. J., *1959*, 642
Brown, W. L., Jr., *1953*, 639
   1957, 658
   1958, 659
Brown, W. L., Jr. & Wilson, E. O., *1954*, 660
Brues, C. T., *1929*, 609
Brues, C. T., Melander, A. L., & Carpenter, F. M., *1954*, 241
Bruner, J. S. et al., *1956*, 354, 650
Buchanan, L. L., *1935*, 249
Buchanan, R. E., *1956*, 657
Bulman, O. M. B., *1955*, 64
Burch, B. L., *1950*, 63
Burch, J. B. & Thompson, F. G., *1958*, 620, 648
Burch, J. Q., *1957*, 620, 648
   1958, 247
Burma, B. H., *1948*, 632
   1949, 635
   1954, 137, 617, 635
Burt, W. H., *1954*, 637
   1957, 65, 252, 613

Buzzati-Traverso, A. A. & Rechnitzer, A. B., *1953*, 629

Cain, A. J., 161
1955, 642
1956, 641
1958, 623, 643
1959, 630, 639, 643
1961, 632
Cain, A. J. & Harrison, G. A., *1958*, 618, 647
1960, 632, 639
Calman, W. T., 15
1940, 341, 615, 650
1949, 650
Camp, C. L. et al., *1928–1961*, 248
Camp, C. L. & Hanna, G. D., *1937*, 612, 613
Campbell, A., *1955*, 616
Campbell, A. S., *1954*, 63
Cannon, H. G., *1936*, 319
Cappe de Baillon, P., *1927*, 616
Carpenter, M. M., *1945*, 248
1953, 248
Casanova, R., *1957*, 613
1960, 613
Casey, T. L., 268
1910, 531
Cassie, R. M., *1957*, 630
Cassino, M. E., 247
Cazier, M. A. & Bacon, A., *1949*, 632
Chadwick, C. E., *1955*, 610
Chamberlin, J. C., *1931*, 283, 647
Chamberlin, R. V. & Hoffman, R. L., *1958*, 241
Chamberlin, W. J., *1952*, 254, 298
Chandler, A. C., *1916*, 621
Chitwood, B. G., 440
Claassen, E. S., *1945*, 254
Claassen, P. W., *1940*, 245
Clapp, W. F. & Kenk, R., *1963*, 248
Clark, R. B., *1956*, 635
Clausen, C. P., *1942*, 610
Cochran, W. G., *1959*, 630
Cockrum, E. L., *1962*, 252, 613
Coker, R. E., *1939*, 616
Collin, J. E., *1958*, 655
Conisbee, L. R., *1960*, 659
Cooper, D. M., *1960*, 626
Cooper, K. W., 77

Copeland, H. F., 440
1956, 241, 300, 301, 643
Corliss, J. O., *1958*, 658
1959, 63
Cox, E. G., *1935–1949*, 252
Cox, L. R., *1955*, 63
1960, 63, 618
Crawshay-Williams, R., *1961*, 632
Crouch, W. G. & Zetler, R. L., *1954*, 319
Curtis, J., *1837*, 531
Cushman, A. D., *1946*, 614
Cushman, J. A., 27, 28
Cutress, C. E., *1955*, 617
Cuvier, G. C. L. D., 256, 530

Dall, W. H., *1877*, 645, 653
Dallas, W. S., 12, 15
von Dalla Torre, K. W. (C.G.), *1892–1902*, 245
Daly, H. V., *1962*, 639
Danser, B. H., *1950*, 650
Darwin, C., 9, 15, 22, 333, 339
1859, 7, 333
Davies, R. G., *1958*, 644
Davis, D. D., *1949*, 619
Davis, D. H. S., *1960*, 613
Davis, P. H. & Heywood, V. H., *1963*, 49, 148, 149, 150, 170, 182, 190, 194, 203, 329, 607, 609, 612, 618, 632, 639, 650
Dawes, B., *1956*, 241
Dean, B., *1916–1923*, 251
Denmark, H. A. et al., *1958*, 632
Deutsch, H. F. & Goodloe, M. B., *1945*, 625
Dice, L. R., *1952*, 632
Dice, L. R. & Leraas, H. J., *1936*, 631
Dillon, E. S. & Dillon, L. S., *1961*, 64
Dirst, J. et al., *1957*, 644
Dixon, W. J. & Masser, F. J., Jr., *1957*, 632
Dobzhansky, T., *1941*, 650
1955, 642
1961, 611
Dodson, E. O., *1956*, 642
Doutt, J. K., *1955*, 637
Downey, J. C., xi, 86
1963, 200, 626
1964, 247

Downey, J. C. & Strawn, M. A., *1963*, 639
Dreisbach, R. R., *1952*, 319
Driver, E. C., *1942*, 70
Duellman, W. E., *1962*, 613
Dunbar, C. O., *1950*, 635
Dunn, E. R. & Stuart, L. C., *1951*, 661
Durrant, S. D., *1955*, 637
1959, 635

Eaton, T. H., Jr., *1953*, 623
Ebeling, W., *1938*, 616
Eddy, S. & Hodson, A. C., *1955*, 71
Edmondson, C. H., *1959*, 71
Edmunds, G. F., Jr., *1962*, 639
Edwards, F. W., 15
1953, 641
Edwards, J. G., *1954*, 637
1956, 637
1962, 632
Ehrlich, P. R., *1959*, 639
1960, 641
1961, 609, 632, 635
Einstein, A., 332
Elton, C., [*1927*] *1949*, 26
Emberger, M. R. & Hall, M. R., *1955*, 319
van Emden, F. I., *1957*, 623
Emerson, A. E., *1935*, 627
1941, 627
Emerson, A. E. & Schmidt, K. P., *1960*, 607
Emiliani, C., *1956*, 653
Endo, R., *1951*, 251
Engelmann, W. et al., *1846–1923*, 249
Enns, W. R., *1958*, 625
Erichson, W. F., 12
1840, 257
Esdaile, A., *1931*, 647
Essig, E. O., *1942*, 22, 610
Evans, C. L., *1956*, 260
Evans, H. E., *1954*, 626, 629
Evans, J. W., *1958*, 620, 632
Exline, H., *1955*, 622

Fabricius, J. C., 261
1792–1805, 531
Falleroni, —, 31
Felt, E. P., *1934*, 639

Felt, E. P. & Bishop, S. C., *1926*, 645, 653
Fennah, R. G., *1955*, 660
1957, 661
Fernald, H. T., *1939*, 661
Ferris, G. F., 337
1916, 245
1928, 4, 137, 255, 291, 292, 298, 315, 320, 336, 364, 373, 607, 608, 618, 619, 643, 650, 653, 655
1955, 609
Fisher, H. I., *1946*, 622
1955, 622
Fisher, R. A., *1936*, 632
1950, 632
Flood, W. E., *1957*, 646
Florkin, M. & Mason, H. S., *1960–1965*, 155, 625
Follett, W. I., xi, 377
1952, 657
1955, 7, 255, 654, 655
1963, 399, 655
Ford, E. B., *1940*, 616
1945, 616
Fowler, H. W., *1965*, 319
Fox, A. S., *1956*, 629
Fox, R. M., *1955*, 637
Frison, T. H., *1942*, 610
Frizzell, D. L., *1933*, 661
Frizzell, D. L. & Exline, H., *1955*, 622
Frommhold, E., *1954*, 644
Fulton, B. B., *1925*, 616

Gallagher, J. J., *1959*, 247
Gardner, E. J., *1960*, 610
Gates, R. R., *1951*, 640
Gay, F. J., *1955*, 644
Gemminger, M. & Harold, E., *1868–1876*, 12, 245
Gerard, R. W., *1957*, 650
Gerstaecker, C. E. A., 12
Gier, L. J., *1965*, 607
Gilmour, J. S. L., *1940*, 190, 205, 346, 347, 363, 640, 650
Ginsburg, I., *1940*, 630
1954, 632
Goldschmidt, R., *1945*, 617
Gooding, R. U., *1964*, 247
Goodnight, C. J. & Goodnight, M. L., *1954*, 637

Gordon, I., *1955*, 624
Gosline, W. A., *1954*, 637
Goto, H. E., *1955*, 647
Gould, S., *1954*, 645, 654
Graham, D. C., 77
Grant, C. H. B., *1955*, 660
 1958, 657
Grant, V., *1957*, 363, 650
Grant, W. F., *1960*, 636
Gray, P., *1961*, 212, 644
Gregg, J. R., *1950*, 347, 357, 650
 1954, 347, 640, 650
Gregory, W., *1943*, 254
Grensted, L. W., *1944*, 657, 660
 1947, 658, 659
Gressitt, J. L., *1958*, 611
Griffin, F. J., 257
Griffiths, R. J., *1960*, 645
Grinnell, J., *1921*, 612
Gunter, G., *1946*, 607
Günther, A. C. L., 15
 1859–1870, 245
Gurney, A. B., *1962*, 610

Haas, O., *1962*, 633
Hagen, H. A., *1862–1863*, 250
Hagmeier, E. M., *1958*, 637
Hairston, N. G., *1958*, 636
Hall, E. R., *1946*, 65
 1955, 65, 252, 613
 1957, 213
Hampton, J. S., *1959*, 633
Hanna, G. D., *1931*, 320
Hansen, E. D., *1961*, 169, 643
Hargis, W. J., Jr., *1956*, 640
Harper, J. L. et al., *1961*, 650
Harrington, H. J. et al., *1959*, 63
Harrison, G. A., *1958*, 618, 647
 1960, 632, 639
Harrison, J. M., *1945*, 660
Harrison, R. A., *1959*, 64
Hartman, C. G., *1925*, 624
Harvey, W. E., *1955*, 629
Haslewood, G. A. D., *1960*, 625
Hass, W. H., *1962*, 64
Hatch, M. H., *1941*, 636
Hay, O. P., *1902, 1929, 1930*, 248
Hedicke, H., *1935–*, 245
Hemming, F., 382, 383, 384, 386, 387, 388

1943–, 655
1953, 255, 644, 655, 658
1958, 655, 656
Henderson, I. F., *1929*, 646
Herrmannsen, A. N., *1847–1852*, 251
Hershkovitz, P., *1958*, 660
Heslop-Harrison, J., *1956*, 336, 650
 1961, 633
Heyne & Taschenberg, *1893–1908*, 257
Heywood, V. H., *1963*, 49, 148, 149, 150, 170, 182, 190, 194, 203, 329, 607, 609, 612, 618, 637, 639, 650
Hiatt, R. W., *1954*, 247
Hiltermann, H., *1956*, 654
Hirst, L. F., 32
Hoedemann, J. J., *1960*, 640
Hoffman, R. L., *1958*, 241
Hoffmann, A., *1930*, 247
Hogben, L., *1940*, 365, 650
Holloway, B. A., *1960*, 640
Honigberg, B. M. et al., *1964*, 300, 301, 643
Hopkins, G. H. C. & Clay, T., *1952*, 245
Hopkins, G. H. E., *1949*, 627
 1955, 660
 1958, 658
Hopper, H. P., *1959*, 646
Horn, W., *1928*, 610
 1929, 640
Horn, W. & Kahle, I., *1935–1937*, 252
Horn, W. & Schenkling, S., *1928–1929*, 250
Hornaday, W. T., *1891*, 252
Horvath, G. et al., *1927–*, 245
Hubbs, C., *1953*, 631
 1961, 649
Hubbs, C. L., *1934*, 342, 651
 1943, 641
 1944, 619
 1952, 658
 1956, 654
Hubbs, C. L. & Hubbs, C., *1953*, 631
Hubbs, C. L. & Hubbs, L. C., *1932*, 623
 1945, 621
Hubbs, C. L. & Perlmutter, A., *1942*, 631, 633
Hubbs, L. C., *1932*, 631
 1945, 621

Hughes-Schrader, S., *1958*, 625
Huxley, J. S., *1939*, 633
  1940, 9, 170, 334, 341, 342, 347, 610, 636, 651
  1942, 329, 342, 651
  1943, 619
Huxley, T. H., 347
Hyman, L. H., 440
  1940, 97, 98, 300, 301, 446, 615
  1951, 445, 615
  1955, 241, 615, 616
  1959, 615

Imbrie, J., *1956*, 633
Inger, R. F., *1958*, 642
  1959, 633
Ivanov, A. V., *1963*, 241

Jaeger, E. C., *1950*, 62, 318
  1960, 238, 646
Jahn, T. L., *1961*, 610, 633
James, M. T., *1953*, 642
Jaques, H. E. ed., *1947–1960*, 71
Johnsgard, P. A., *1960*, 623
Johnson, C. G., *1939*, 620, 628
Johnson, M. L. & Wicks, M. J., *1959*, 629
Jordan, D. S., *1917–1920*, 643
  1926, 654
Jordan, D. S. et al., *1930*, 643

Keck, D. D., *1957*, 336, 372, 651
Keen, A. M. & Muller, S. W., *1948*, 255, 654, 656
  1956, 255, 259, 265, 315, 654, 656
Keifer, H. H., *1944*, 611
Kelton, L. A., *1959*, 622
Kempthorne, O. et al., *1954*, 633
Kenk, R., *1963*, 248
Kerr, W. E., *1950*, 617
Kertesz, K., *1902–1910*, 245
Kevan, D. K. M., *1961*, 640
Kierzek, J. M., *1954*, 319
King, R. R. et al., *1940–1959*, 250
Kinsey, A. C., *1930*, 620
  1936, 640
  1937, 617, 620
Kirby, W. F., 12, 15
  1904–1910, 245

Kirby, W. & Spence, W., *1815–1826*, 258
Kiriakoff, S. G., *1956*, 638
  1959, 651
  1963, 651
Kirkaldy, G. W., *1909–*, 245
Knight, J. B., *1941*, 98
  1948, 586, 656
  1950, 396, 653, 660
  1953, 656, 660
Kohn, A. J., *1960*, 647
Kohn, A. J. & Orians, G. H., *1962*, 627
Korschefsky, R., *1939*, 249

Laffoon, J. L., *1957*, 657
Lamanna, C. & Mallette, M. F., *1953*, 210, 338, 609, 644, 651
Landin, B.-O., *1957*, 645, 654
Lankester, E. R., *1878*, 153, 618
Latreille, P. A., *1810*, 530
de Laubenfels, M. W., *1953*, 660
  1955, 63, 612
  1958, 645
  1959, 645
Lenzen, V. F., *1955*, 332, 651
Leone, C. A., *1964*, 23, 609, 618, 625
Lethierry, L. F. & Severin, G., *1893–1896*, 245
Levi, C., *1958*, 624
Levi, H. W., *1958*, 247
Levine, N. D., *1958*, 658
Li, J. C. R., *1957*, 633
Lidicker, W. Z., Jr., *1963*, 638
Light, S. F. et al., *1954*, 71
Linnaeus, C., 8, 9, 10, 167, 207, 215
  1758, 6, 11, 13
Linsley, E. G., *1937*, 627
  1944, 638
  1953, see Mayr, Linsley, Usinger
Linsley, E. G. & Usinger, R. L., *1961*, 106, 615
Lorenz, K. Z., *1950*, 630, 631
Löve, A., *1960*, 636
Lydekker, R., 15

Macan, T. T., *1955*, 645
McAtee, W. L., *1926*, 640
  1957, 644
Maccacaro, G. A., *1961*, 633
Macfayden, A., *1955*, 614

McMasters, J. H., *1936*, 256, 259, 265, 654, 656
McMichael, D. F., *1956*, 659
Maeki, K., *1958*, 623
Mainx, F., *1955*, 5, 609
Malaise, R., *1937*, 531, 660
Manning, R. B. & Kumpf, H. E., *1959*, 628
Manville, R. H., *1959*, 644
Marschall, A., *1873*, 243
de Marschall, A., *1873*, 15
Martinez Fontes, E. & Parodiz, J. J., *1949*, 246
Maslin, T. P., *1952*, 622
  1959, 636
Mason, H. L., *1950*, 609
Mather, K., *1947*, 633
Mathiesen, F. A., *1962*, 617
Matsudo, R., *1960*, 622
Matthew, W., *1907*, 319
Mayr, E., *1942*, 7, 8, 342, 610, 612, 640, 651
  1949, 636
  1953, 608, 640
  1954, 608, 645, 654
  1955, 636, 651
  1957, 129, 366
  1963, 172, 638
Mayr, E. & Goodwin, R., *1956*, 614
Mayr, E., Linsley, E. G., & Usinger, R. L., *1953*, 73, 106, 148, 158, 256, 286, 292, 298, 315, 360, 542, 586, 608, 610, 611, 612, 615, 618, 645, 648, 651, 654
Meglitsch, P. A., *1954*, 636
Meikle, R. D., *1957*, 638
Meisel, M., *1924–1929*, 250, 253, 255
Melander, A. L., *1916*, 646
  1940, 62, 318
Mendel, G., 7
Merrill, E. D., *1943*, 611
Meryman, H. T., *1960*, 614
Metcalf, M. M., *1929*, 627, 630
Metcalf, Z. P., *1954*, 299
Michener, C. D., *1944*, 640
  1949, 620
  1953, 624
  1957, 640
  1958, 635, 648

Michener, C. D. & Sokal, R. R., *1957*, 363, 633, 651
Micks, D. W., *1956*, 155
Micks, P. A. & Doniger, D. E., *1956*, 625
Miller, R. R., *1956*, 614
  1957, 631
Mills, H. G. & Walter, J. H., *1954*, 319
Moment, G. B., *1953*, 658
Moore, D. H., *1945*, 625
Moore, I. M., *1954*, 638
Moore, J. A., *1959*, 623
Moore, R. C., *1955*, 646, 654
  1961, 63
Moore, R. C., Lalicker, C. G., & Fischer, A. G., *1952*, 178
Moore, R. C. & Sylvester-Bradley, P. C., 176
Mosher, E., *1916*, 624
Muesebeck, C. F. W., *1942*, 169, 611, 644
Muir-Wood, H. M., 15
Muller, S. W., *1948*, 255, 654, 656
  1956, 255, 259, 265, 315, 654, 656
  1958, 642
Muller, S. W. & Campbell, A., *1955*, 616
Mulsant, E. & Rey, C., *1839–1886*, 260
  1872–1873, 259
Munroe, E. G., *1957*, 630, 631
  1960, 628
  1964, 49, 609, 612
Munroe, E. G. & Ehrlich, P. R., *1960*, 641
Murray, D., *1955*, 636
  1957, 636
Musgrave, A., *1932*, 250, 251
Myers, G. S., *1930*, 651
  1952, 608, 651
  1956, 614
  1960, 651

Neave, S. A., 15, 16
  1939–1950, 241, 243
Needham, J., *1942*, 619
Needham, J. G., *1910*, 646
Newell, I. M., 148
  1953, 618, 620
  1955, 614

Newell, N. D., *1947,* 638
  1959, 628
Newton, A., 15
Nicol, D., *1956,* 641
  1958, 628, 642
Noble, G. K., *1926,* 624
  1927, 624
Novikoff, M. M., *1952,* 617
Nybakken, O. E., *1959,* 62, 238, 318

Oberholzer, H. C., *1920,* 659
Okulitch, V. J., *1955,* 63
Oldroyd, H., *1958,* 252
Olsen, T. H. & Morrow, J. E., *1959,*
  320
Oman, P. W. & Cushman, A. D., *1946,*
  614
Orton, G. L., *1953,* 624
  1955, 624

Paclt, J., *1957,* 642
Palmer, E. L., *1949,* 71
Papp, C. S., *1963,* 320
Parker-Rhodes, A. F., *1961,* 633
Parodiz, J. J., *1949,* 246
Parr, A. E., *1949,* 624, 631
  1958, 611, 612
Parrott, A. W., *1956,* 657
Patrick, R., 25
Paulsen, E., *1959,* 629
Pearl, R., 22, 27
  1922, 363, 608, 609, 652
Pearse, A. S., *1949,* 658
Pennak, R. W., *1953,* 71
  1964, 240
Perlmutter, A., *1942,* 631, 633
Perrin, P. G., *1950,* 319
Peters, J. A., *1964,* 64
Peters, J. A. et al., *1954,* 638
Peters, J. L., *1931–,* 245
Peterson, A., *1934,* 252, 614
  1937, 252, 614
Peterson, R. T., 71
Petrunkevitch, A., *1926,* 627
  1952, 329, 641, 652
  1955, 64
Phillips, A., *1959,* 636
Pimentel, R. A., *1958,* 638
  1959, 638
Poche, F., *1938,* 243, 244

Popper, K. R., *1959,* 333, 652
Pratt, H. S., *1935,* 63, 71
Procter, W., 77
Prosser, C. L., *1955,* 617
Prosser, C. L. & Brown, F. A., Jr.,
  1961, 616
Prosser, C. L. et al., *1950,* 616

Quenstedt, W., *1913–1935,* 245

Rand, A. L., *1948,* 612
Rao, C. R., *1948,* 633
Raup, D. M., *1962,* 622, 628
Raw, F., *1955,* 614
Reed, C. A., *1960,* 642, 658
Reeside, J. C., Jr., *1930,* 320
Regan, C. T., 15
Rensch, B., *1934,* 315
  1958, 648
Rhodes, F. H. T., *1954,* 648
Richards, O. W., *1936,* 138, 194, 617,
  618, 641
  1940, 651
Richens, R. H., *1961,* 634
Richter, R., 383
  1948, 256, 577
Richter, R. & Schmidt, H., *1956,* 660
Ridgway, J. L., *1938,* 320
Ridgway, R., *1912,* 648
Riemer, W. J., *1954,* 648
du Rietz, G. E., *1930,* 641
Riley, N. D., 15
Rising, G. R., *1959,* 636
Robertson, J. G., *1960,* 622
Robson, G. C., 337
  1928, 338, 636, 652
Robson, G. C. & Richards, O. W., *1936,*
  138, 194, 617, 618, 641
Rogers, D. J., *1952,* 658
Rogers, D. J. & Tanimota, T. T., *1960,*
  634
Roget, P. M., *1931,* 319
  1933, 319, 347, 652
Rohlf, F. J., *1962,* 635
Rohlf, F. J. & Sokal, R. R., *1962,* 634,
  641
Rollins, R. C., *1954,* 608
Romer, A. S., 241
  1955, 619
Romer, A. S. et al., *1962,* 248

Roscoe, E. J., *1955*, 645
Ross, E. S., *1955*, 610
  1956, 648
Rothschild, Lord, *1956*, 617
Rothschild, N. C., 32
Russell, N. H., *1961*, 634
Rye, E. C., 15

Sabrosky, C. W., 383
  1939, 659
  1940, 660
  1947, 659
  1948, 586, 656
  1950, 627
  1952, 616
  1953, 616, 656
  1954, 659
  1955, 660
  1956, 611
  1957, 660
Sabrosky, C. W. & Bennett, G. F.,
  1958, 620, 628
Sailer, R. I., *1961*, 646
Salt, G., *1927*, 617, 627
  1941, 617, 627
van Sande, J. & Karcher, D., *1960*,
  625
Savory, T., *1962*, 256
Schenk, E. T. & McMasters, J. H.,
  1936, 256, 259, 265, 654, 656
Schenkling, S., *1910–1941*, 245
  1922, 657
Schindewolf, O. H., *1950*, 194, 641
Schmidt, H., *1956*, 660
Schmidt, K. P., *1960*, 607
Schmitt, W. L., v, xi, 22, 23, 74
  1953, 608, 611
  1954, 611
Schneirla, T. C., *1952*, 627, 630
Schönherr, C. J., *1817*, 260
Schoute, J. C., *1949*, 619
Schrödinger, E., 332
Schulze, F. E. et al., *1926–1954*, 15,
  243
Sclater, W. L., 15
Scudder, S. H., 12, 15
  1879, 253
  1882, 243
Sears, P. B., 24
Seitner, P. G. et al., *1960*, 634

Selander, R. B. & Vaurie, P., *1962*,
  253
Senn, H. A., *1960*, 636
Shannon, R., 32
Sharp, D., 15
Sherborn, C. D., 15, 257
  1902, 243, 244
  1922–1933, 243, 244
  1940, 252
Shipley, A. E., 440
  1904, 659
Shuckard, W. E., *1839*, 530
Sibley, C. G., *1957*, 621
  1960, 625
  1962, 155, 625, 630
Sibley, C. G. & Johnsgard, P. A., *1959*,
  625
Silvestri, F., 15
  1929, 611
Simpson, G. G., 36, 149
  1940, 643, 661
  1941, 628
  1943, 621, 642, 648
  1945, 1, 22, 98, 182, 241, 339, 357,
    362, 363, 608, 609, 641, 643, 652
  1952, 636, 659
  1959, 634, 642
  1961, 4, 138, 141, 142, 166, 167,
    168, 169, 170, 171, 188, 193, 195,
    295, 503, 505, 608, 609, 612, 618,
    619, 636, 641, 644, 648, 652, 661
Simpson, G. G. & Roe, A., *1939*, 634
Simpson, G. G., Roe, A., & Lewontin,
  R. C., *1960*, 158, 287, 634, 648,
  652
Sinclair, G. W., *1952*, 657
Slastenenko, E. P., *1958*, 64
Smith, H. M., *1945*, 661
  1947, 661
  1949, 661
  1950, 396, 653, 660
  1952, 637
  1953, 660, 661
  1955, 637
  1956, 64, 638
  1958, 637
  1962, 620, 654, 656
Smith, H. M. & White, F. N., *1956*,
  661
Smith, R. C., *1962*, 240, 265

Smith, R. I., *1952*, 621, 629
 1958, 627
Smith, S. G., *1958*, 623, 630
Sneath, P. H. A., *1957*, 634
 1961, 634, 652
 1963, 635
Sneath, P. H. A. & Cowan, S. T., *1958*,
 634
Sneath, P. H. A. & Sokal, R. R., *1962*,
 634
Snedecor, G. W., *1946*, 634
Snyder, T. E., *1949*, 245
Sokal, R. R., *1957*, 363, 633, 651
 1958, 634, 652
 1959, 635
 1961, 631, 635
 1962, 634, 641
 1963, 652
Sokal, R. R. & Michener, C. D., *1958*,
 635
Sokal, R. R. & Rohlf, F. J., *1962*, 635
Sokal, R. R. & Sneath, P. H. A., *1963*,
 635
van Son, G., *1955*, 638
Sonneborn, T. M., *1957*, 115, 129, 359,
 369, 370, 616, 652
Soule, R., *1938*, 319
Soulsby, B. H., *1933*, 251
Spector, W. S., *1951–1959*, 154, 618
Spencer, W. P., *1949*, 619
Sprague, T. A., *1940*, 339
Stains, H. J., *1959*, 622
 1962, 622
Starrett, A., *1958*, 638
Stebbins, R. C., *1951*, 64
 1954, 64
Stenzel, H. B., *1950*, 659
Stephen, W. P., *1958*, 626
 1961, 626
Stephen, W. P. & Steinhauer, A. L.,
 1957, 626
Steyskal, G. C., *1949*, 615
Stiles, C. W., 382
 1927, 654
Stirton, R. A., *1955*, 638
Stohler, R., *1959*, 614, 655
Stoll, N. R., ed., *1961*, 255, 380, 619,
 644, 656
 1964, 255, 380

Storey, M. & Wilimovsky, N. J., *1955*,
 614
Størmer, L., *1955*, 64
Stormont, C. et al., *1961*, 626
Strand, E., *1911–1935*, 245
Stroud, C. P., *1953*, 635
Stunkard, H. W., *1940*, 624
 1953, 624
Sumner, F. B., *1915*, 642
Sumney, G., Jr., *1949*, 318
Svetovidov, A. N., *1959*, 622
Sylvester-Bradley, P. C., 176
 1951, 638
 1954, 642
 1958, 648

Tait, J., *1928*, 620
Tanner, Z. L., *1897*, 252
Terek, E., *1952*, 252
Thalmann, H. E., *1959*, 661
van Thoai, N. & Roche, J., *1964*, 626
Thomas, I. & Janson, H. W., *1957*, 645
Thompson, R. R., *1960*, 630
Thompson, W. R., 348
 1922, 624
 1937, 204, 332, 608, 641, 652
 1952, 652
 1955, 329, 652
 1958, 608, 652
 1962, 652
Thomson, J. A., 15
Thorpe, W. H., *1940*, 22, 609
Tilden, H. J. W., *1961*, 639
Tonapi, G. T., *1958*, 622
Tordoff, H. B., *1954*, 622
de la Torre, L. & Starrett, A., *1959*,
 655, 661
de la Torre Bueno, J. R., 324
 1937, 64
Totton, A. K., 15
Townsend, C. H., *1901*, 252
Trelease, S. F., *1958*, 319
Troelsen, J. C., 629
Tryon, G. W., Jr., *1861*, 247
Turrill, W. B., 337
 1940, 341, 652
Tuxen, S. L., ed., *1956*, 64

Uchida, T., *1954*, 247

Usinger, R. L., *1952*, 661
1953, *see* Mayr, Linsley, Usinger
1956, 64
1961, 106, 615

Van Bentham Jutting, W. SS. & Van
Regteren Altena, C. O., *1958*, 254
Van Cleave, H. J., *1933*, 656
Van Duzee, E. P., *1916*, 659
Van Steenis, C. G. G. J., *1957*, 137, 618
Van Tyne, J. & Berger, A. J., *1959*, 65
Vanzolini, P. E. & Guimaraes, L. R.,
1955, 627
Vaurie, P., *1962*, 253
Vavilov, N. I., *1940*, 22, 609
Vetter, R. C., *1964*, 247
Vladykov, V. D., *1934*, 621, 648
Voss, E. G., *1952*, 299

Wagstaffe, R. & Fidler, J. H., 614
Wainstein, B. A., *1960*, 643
Wallace, A. R., 15
Wallace, G. J., *1955*, 611
Ward, H. B. & Whipple, G. C., *1918*,
71
Wardle, R. A., *1954*, 635
Warren, B. C. S., *1958*, 637
Waterhouse, C. O., 15
1902, 243
1912, 243
Weaver, C. E., *1955*, 610
Weller, J. M., *1955*, 652

Wetmore, A., 98
Wharton, G. W., *1952*, 241
White, F. N., *1956*, 661
White, M. J. D., *1957*, 623, 630
Whittaker, R. H., *1959*, 641
Whitten, J. M., *1959*, 622
Wiegmann, A. F. A., 15
Wilimovsky, N. J., *1955*, 614
Wille, A., *1958*, 623
Williams, C. B., *1940*, 661
Wilson, E. O. & Brown, W. L., Jr.,
1953, 639
Wolfe, H. P., *1939*, 626
Womble, W. H., *1951*, 631
Wood, C. A., *1931*, 251
Woodger, J. H., *1937*, 347, 620, 652
1952, 347, 653
Woods, R. S., *1944*, 62, 238, 318
1947, 62
Woolley, E. C. et al., *1944*, 319

Yamashina, Y., *1952*, 623
Young, D. A., Jr., *1958*, 661
Young, F. N., *1953*, 608

Zangerl, R., *1948*, 631
Zimmer, C., *1928*, 251
Zimmermann, K. et al., 645
Zweifel, F. W., *1961*, 320
Zweig, G. & Crenshaw, J. W., Jr., *1957*,
626

Index to Organisms

Omitted are names used merely as examples of spellings, sources, or gender, as well as the synonyms in the sample synonymies.

Acanthocephala, 114, 445, 446
Acanthonia gigantea, 546
Acarina, nomenclator, 241
Acnidaria, 218
Agamodistomum, 496
Agamofilaria, 496
Agamomermis, 496
Aimophila, 342
Amiskwia, 124
Ammonoids, 177
Amoeba, 226, 441
Amoebozoa, 441
Amphibia, glossaries, 64
Amphibians, common name list, 213
Amphimerycoidea, 448
Amphistomulum, 496
Animalia, 444
Anobiidae, 283
Anopheles gambiae, 32, 226, 227
Arachne, 441
Arachnida, 441
Araneida, nomenclator, 241
Archaeocyatha, glossaries, 63
Artemia, 342
Arthropoda, 114, 116, 124, 128, 129, 217, 220, 441, 445
glossaries, 63
Articulata, 219, 441, 442
Aschelminthes, 440
Asterozoans, 176
Aurelia aurita, 58, 189
Aves, 5
glossaries, 65

Basseroceratida, 441

Bilateria, 440
Bilharzia mansoni, 230
Birds, 5, 114, 123
common name list, 213
taxonomic history, 11
Bivalvia, 125, 440, 441
Blastoids, 176
Bolitobius, 529
Bonellia, 114
Borboropora, 522
Brachiopoda, 126, 129
Branchiopoda, 342
Bryozoa, 62, 114, 126, 129, 219, 440, 441
glossaries, 63
nomenclator, 241
Burgessiida, 441
Butterflies, common names, 213

Cactoblastis cactorum, 28, 30
Calophaena, 520
Calyssozoa, 114
Camponotus, 414
Campoporus, 414
Canis latrans lestes, synonymy, 273
Cephalopoda, 117, 176
Cercaria, 496
Cestoidea, 441, 442
Chaetognatha, 124
Charadriiformes, 342
Chelicerata, nomenclator, 241
Chloroperla, synonymy, 276
Chordata, 440
Cnidaria, 218, 219, 440, 442
Cnidospora, 446

675

Coccoliths, 176
Coelenterata, 61, 114, 125, 129, 189,
     218, 219, 440, 441, 442
     nomenclator, 241
Coelomata, 440
Coleoptera, 125, 220, 441, 447
     sample key, 298
Colpodota, 506, 523
Colpodota negligens, 546
Conidiophryidae, 456
Conidiophrys, 456
Conodonts, 123, 176, 177, 178, 179
Conularida, 441
Conulariida, 441
Conulata, 441
Coproporus, 519
Crataegus, 342
Crematogaster, synonymy, 275
Crematogaster laeviuscula, synonymy,
     272
Creophilus maxillosus, 228, 229, 231
Criconarids, 123
Crinoids, 176
Crustacea, 117, 118
Ctenophora, 128, 218, 219, 442
Curculionidae, 125
Cyclostomata, 219, 440
Cynips, 342
Cysticercus, 496
Cystoids, 176

Dalotia, 524
Daphnia, 342
Decapoda, 218, 440
Deroderus, 523
Deuterostomia, 440
Diatoms, 25, 43
Diplopoda, nomenclator, 241
Diplostomulum, 496
Diptera, 441
Drosophila, 21, 144, 358
Drosophila melanogaster, 52, 53, 54
Dubium, 496
Dugesia, 196

Echinodermata, 128, 129, 155, 445
     nomenclator, 241
Echinoidea, 53, 117, 122, 128, 176
Ectoprocta, 219, 440
Edentata, 441

Endoprocta, 441
Entoprocta, 441
Eutheria, 447

Felis leo, 211
Fishes, common name list, 213
Flagellata, 18
Fleas, 32
Flies, 130
Foraminifera, 27, 28, 43, 129
     nomenclator, 241

Gambusia, 31
Gasteropoda, 441
Gastropoda, 117, 124, 125, 176, 441
Gastrotricha, 114
Glaucothoe, 496
Gnathostomata, 442
Goerius, 95, 508
Graphognathus, 342
Graptolites, 126
Graptolithina, 441
Graptolithus, 441
Graptoloidea, 441
Graptozoa, 441
Gymnostomata, 447
Gymnostomatida, 447

Hemichordata, glossaries, 64
Hirudinea, 117
Holothuria, 441
Holothurioidea, 128, 176, 441
Homoeochara, 531
Hydra, 116, 225, 441
Hydrariae, 441
Hydrida, 441
Hydroida, 441
Hydrozoa, 125, 126, 441
Hydrozoaria, 441
Hyolithids, 123
Hyracoidea, 441
Hyrax, 441

Ignotus aenigmaticus, 537
Inopeplus, 92
Insecta, 113, 114, 117, 124, 129, 159
     glossaries, 64
     nomenclator, 241
Insects, common name list, 213
     number of, 13

Insects, taxonomic history, 12
Invertebrates, glossaries, 63
Ischnopoda, sample synonymy, 417

Jellyfish, 131

Laverna, 524
Lecanium, 114
Lepidoptera, 441
Leptinotarsa, 27
Leucochloridium, 496
Leucopaederus, 518
Ligula, 496
Lycaenidae, 175, 200
Lycopteridae, 448
Lycopteroidei, 448

Macropterum, 525
Mammalia, 122, 131
  glossaries, 65
  nomenclator, 241
Mammals, common name list, 213
Mastigophora, 18
Matthevia, 124
Mesaxonia, 447
Metatheria, 447
Metaxya, 528
Metazoa, 444, 445
Millepores, 126
Mites, 5, 131
Mollusca, 58, 124, 129
  glossaries, 63
  nomenclator, 241
Monoblastozoa, 445
Monoplacophora, 441
Monotremata, 124
Muscidae, 342
Myxosporidia, 441
Myzostomida, 126, 218

Nemathelminthes, 440
Nematoda, 114, 116, 126, 128
Nematodes, 5
  taxonomic history, 13
Nemertinea, 128
Neopilina, 77
Notemigonus, synonymy, 275

Octocorals, 176
Ocypus olens, 94

Oecidiophilus, 524
Oedodactylus, 531
Oligochaeta, 126, 446
Onthostygnus, 518
Opuntia, 29
Osorius, 507
Ostracoda, 43
  nomenclator, 241

Paederus, synonymy, 417
Pantotheria, 447
Parapleurolophocerca, 496
Pauropoda, 124
Pelecypoda, 125, 440
Perissodactyla, 447
Peromyscus, 342
Philonthus discoideus, synonymy, 269, 272
Philorinum, 529
Phloeonomus, 94
Phoronida, 129
Phoronis, 58
Phytamastigophorea, 446
Phytomastigina, 446
Pigmentata, 496
Pilisuctoridae, 456
Pisces, glossaries, 64
  nomenclator, 241
Planaria tigrina, 196, 197
Platypsyllidae, 125
Plerocercoides, 496
Pogonophora, 129, 297
  nomenclator, 241
Pogonophorans, taxonomic history, 14
Polychaeta, 58, 126, 446
Polyzoa, 441, 442
Porifera, 129
  glossaries, 63
  nomenclator, 241
Priapuloidea, 442
Proboscidea, 441
Protozoa, 114, 115, 129, 300, 301, 444, 445
  glossaries, 63
  nomenclators, 241
  sample classifications, 300, 301
Pterobranchia, 114
Pycnogonida, 122

Quedius, synonymy, 274

Reptiles, common name list, 213
Reptilia, glossaries, 64
Rodentia, 124
Rotifera, 114, 128
Rubus, 342

Sableta, 521
Salinella, 445
Salix, 342
Sarcodina, 446
Schistosoma mansoni, 230
Schizocoela, 440
Scolecodonts, 176
Scyphozoa, 125, 189
Sipunculid worms, 35
Sparganum, 496
Sponges, 176
Sporozoa, 131, 445
Staphylinus maxillosus, synonymy, 303
Stenostoma, 496
Stenus, 522
Strepsiptera, 114, 447
Stromatoporoids, 126

Symphyla, 124

Taenia pisiformis, 20
Taraxacum, 342
Tragulina, 448
Trematoda, nomenclator, 241
Trilobita, 445
Trilobitoidea, 445
Trilobitomorpha, 445
    nomenclator, 241
Trochelminthes, 440
Tunicata, 114
Turbellaria, 342

Vertebrata, 129, 155, 220, 440, 441
    glossaries, 64
    nomenclators, 241

Xiphosura, 446
Xiphosurida, 446

Zoa, 444
Zoraptera, 124

Subject Index

Abbreviations, of journal names, 322
  list of, 265
  in literature, 264
  in publication, 321
  in taxonomy, 265
Aberrations, as taxa, 175, 432
  names of, 227
Academies of science, in taxonomy, 46
Acarology courses, 41
Accent, in pronunciation, 233
Accent marks, in names, 456, 479
Acceptances, 350, 364
Acknowledgements, in publications, 313
Acronyms, as epithets, 583
Adjectives, as epithets, 537
Agamospecies, 170
Allotypes, in descriptive taxonomy, 293
  in nomenclature, 591, 600
Alpha-taxonomy, 4, 14
Amateur taxonomists, 38, 43
American Men of Science, 246
Anagrams, as epithets, 582
Analogy, 141
Anatomical data, 144
Anatomy, 337
Anonymous publication of names, 428,
  543
Apostrophes, in names, 225, 390, 441,
  456, 479, 480, 540, 570, 584, 588
Applied Systematics, 24–37
Archetype, 4
Archiv für Naturgeschichte, 12, 15, 16,
  243, 250
Aristotelian logic, 10
Asexual individuals in classification,
  370
Association, method of classifying, 207
Atlases, in publication, 98
Australia, prickly pear control, 28, 29

Autecology, 152
Authority, double, 271
  of the code, 384
  of the Commission, 385
Authors, directories of, 245, 246
  of books, 259
  of names, see category involved
  responsibilities, 306
Author's extras, 326
Authorship, multiple, 311
  of papers, 311
  of species, 311
Autogenotype, 509
Availability, in 1961 Code, 394, 396,
  511
  criteria of, 424

Beta-taxonomy, 4, 14
Bibliographia Zoologia, 250
Bibliographic references, in literature,
  276
Bibliographies, by author, 249
  of books, 254
  in cataloging, 92, 93
  on collections, 252
  in descriptive taxonomy, 306
  on expeditions, 252
  on institutions, 253
  journal on, 249
  on localities, 253
  on methods & equipment, 251
  on nomenclature, 255
  on periodicals, 253
  personal, 248
  publication of, 316
  as publications, 100
  by subject, 251
Binomen, specific name, 229, 391, 396,
  414, 536, 540, 547

679

Binomen, subspecific name as, 569, 571
Binominal nomenclature, 422
Biochemical data, 152, 153, 154
Biochemical diversity, 127
Biochemistry, 20, 337
    taxonomic, 23
Biologia Centrali-Americana, 13
Biological Abstracts, 16, 80, 99, 243
Biological species concept, 4, 8, 331
Biological supply houses, list, 82
Biometry, 159
Biospecies, 4, 170
Biostatistics, 159
Biosystematics, 4, 302, 331
Blanchard Code, 7
Blastovariations, as taxa, 175
Book reviews, as publications, 100
Books, authorship, 259
    author's intent, 261
    bibliographies, 254
    issuance in parts, 257
    new editions, 258
    reprintings, 258
    title page, 256
Borings, work of an animal, 535, 565, 566
Botanical names, 214, 458, 478, 490, 491, 492
Branch, category, 220
Branches, names of, 448
British Association for the Advancement of Science, 15
British Museum (Natural History), 12, 28, 29
Bulletin of Zoological Nomenclature, 44, 255, 380, 384, 387, 389, 406, 598
Burrows, fossil, 535

Card catalogs, 90, 94
Cases, work of an animal, 535, 566
Castes, 113, 115
Castings, work of an animal, 535, 566
Casts, fossils, 535, 566
Cataloging, in curating, 76, 89, 279
    of literature, 89, 91
    in The New Systematics, 342, 345
    of species, 89, 90
    of specimens, 89, 90

Catalogs, card, 90, 94
    as nomenclators, 245
    as publications, 99
Catalogue of the Library of the British Museum (Natural History), 16
Categories, basic, 58
    changing, 217
    in classification, 57, 203, 351, 353
    coordinate, 424, 425
    definition, 351, 353, 354
    vs groups (taxa), 351, 355
    "higher," 435, 439
    infrasubspecific, 431, 432, 575, 576
    interpolated, 448
    membership in, 358
    names of, 439
    of names, 395
    naturalness, 367
    number needed, 219, 220
    objectivity, 358
    obligatory, 216, 219
    reality, 358
    species-group, 423
Category, in classification, 54, 55, 182
    definition, 439
    misuse of term, 367, 401
    subgenus, optional, 415
    subspecies, 172
    variety, 431
    see also under each level
Cedilla, in names, 456
Cell diversity, 127
Century Dictionary, 1889–1891, 238, 240
Chance similarity, 140
Changes, of names, 230, see also category involved
    of names, caused by Code, 402
    in rules, 403
Character, expression, 148, 203
    variant, 148
Characters, availability, 286
    in common, 149
    comparability, 286
    definition, 148
    diagnostic, 149, 285
    distinctiveness, 286
    generic, 203
    good and bad, 203, 286, 287
    key, 149

Characters, measurements, 287
  numerical, 151
  phylogenetic, 286
  qualitative, 151
  quantitative, 151
  selection of, 202, 286
  stability, 156, 286
  statistical, 287, 288
  taxonomic, 143, 144, 145, 148, 149, 285, 338
  variability, 156
Checklists, as publications, 99
Citation, of authors, *see* category involved
  of dates, *see* category involved
  of names, *see* category involved
Class, category, 56, 57, 219, 220, 436
  (group) defined, 137
Classes, names of, 216, 446
Class-group, categories, 220
  taxa, 220
Classification, 4, 20, 143
  archetypal, 184, 188
  by association, 207
  based on previous work, 182
  basis of, 191, 200
  in this book, 4
  data of, 138, 200
  definition, 3, 51
  evolutionary, 184, 186
  of genus-group names, 409
  Greek systems, 9
  horizontal, 184, 188
  identification, 183
  methods of, 206
  natural, 361, 364
  nature of, 351–364
  neo-Adansonian, 188
  omnispective, 184, 190, 191
  phenetic, 184, 187, 190
  phylogenetic, 184, 186, 187, 360, 364
  in phylogeny, 183
  vs phylogeny, 187, 191, 193, 339
  populations in, 204
  vs population studies, 339
  practice of, 182–209
  prediction, 183
  producing data, 183

Classification, purpose of, 182, 200, 299, 360
  reclassification, 209
  recording data of, 183
  as re-evaluation, 183
  of species-group names, 426
  stability of, 195, 196
  by subdivision, 208
  taxonomic, 190
  types in, 196
  typological, 184, 188
  units of, 204, 205, 344
  unnatural, 361
  use of, 51–59
  vertical, 184, 188
  what is classified, 193
Classifications, adult, 184, 188
  artificial, 184, 186, 361, 364
  basis in phylogeny, 301
  in descriptive taxonomy, 299
  general, 184, 189
  kinds of, 184
  larval, 184, 188
  natural, 184, 186
  "practical," 361
  publication of, 315
  as publications, 98
  special, 184, 189, 190
Classifying, in curating, 76
  groups, 351
  species, 351
Clines, stepped, 173
  in taxonomy, 173
Co-authorship, of papers, 312
Cocoons, as work of animals, 566
Code of Zoological Nomenclature, International, 1961, 377, 381
  completeness, 398
  complexity, 402
  new features, 390
  new problems, 401
  vs older rules, 405
  retroactive rulings, 398
  unsatisfactory aspects, 398
Code . . . 1964, 377
  differs from 1961, 403
  new problems, 403
Cohorts, names of, 448
Collecting, 76, 279
  specialization in, 76

Collections, bibliographies of, 252
  permanence, 82
Collective groups, 495
Colloquiums, on nomenclature, 215,
  216, 381, 382, 383, 384
Colonies, diversity in, 110
  as taxa, 175
Color forms, as taxa, 175
Color phases, as taxa, 175
Combining, families, 466
  genera, 494
  species, 558
  subfamilies, 470
  subgenera, 500
  subspecies, 575
  tribes, 472
Commission, see International Commission on Zoological Nomenclature
Common names, lists of, 213
  standardized, 212, 213
Comparative, anatomy, 20, 152, 153
  anatomy courses, 42
  biochemistry, 152, 153, 154
  cytology, 153
  data, 137–161
  ecology, 152, 153
  embryology, 152, 153
  endocrinology, 154
  ethology, 152, 153
  genetics, 152
  histology, 153
  morphology, 146, 337
  osteology, 153
  parasitology, 153
  physiology, 152, 153, 154
  psychology, 130, 152
  structure, 146
  zoology, 67, 151, 153, 338
Comparative studies, in publication, 97
Comparing methods, graphic, 158
  statistical, 159
Comparison, 137, 156, 161
  in identification, 68
  methods of, 155
Computers, 158, 160
Concepts, definition, 351, 354
  vs definitions, 351, 352
  "validity" of, 352
Concilium Bibliographicum, 243, 250

Conferences, The Needs of Systematics,
  22
  Taxonomic Biochemistry, Physiology,
  and Serology, 23
Conflicts among names, see category
  involved
Congress of Entomology, International,
  6th, 382
Congresses, membership, 386
Congress of Zoology, International, 384
  1st, 7
  5th, 215, 380, 385
  9th, 515
  13th, 215, 380, 382, 386, 436
  14th, 216, 217, 381, 382, 436
  15th, 180, 214, 216, 381, 437
Convergence, 139, 141
Co-ordinate, categories, 419, 420, 422
  taxa, 424, 425
Copenhagen Decisions on Zoological
  Nomenclature, 7, 216, 381, 384,
  392, 406, 436, 437, 438, 449, 561,
  599
Coprolites, work of an animal, 535,
  538, 566
Cores, fossils, 535, 566
Correlation, of attributes, 362
Correspondence, 137, 141, 156, 161
  analogy, 141
  homology, 141, 142, 143
Cotypes, definition, 600
  in nomenclature, 591, 600
  in taxonomy, 293
Courses in taxonomy, 40, 41
Courtesy in publication, 312, 313
Credit, giving, in publication, 312, 313
Curating, 76–83
Cyclomorphosis, 110
Cytogenetics, 337
Cytological data, 153
Cytology, 20, 337

Data, anatomical, 144, 146
  behavioral, 146, 336
  biochemical, 145
  of classification, 148, 200
  comparative, 137–161, 337
  from conchology, 146
  cytological, 146, 336
  cytotaxonomic, 146, 147

Data, distributional, 145
  ecological, 145, 147, 335, 336, 349
  from endocrinology, 146
  ethological, 146
  from fossils, 201
  genetical, 147, 336, 349
  geographical, 145, 147, 349
  from haematology, 146
  histological, 146
  kinds of, 145
  mensural, 287
  morphological, 144, 146
  from neurology, 146
  numerical, 287
  from osteology, 146
  outbreeding, 369
  from paleontology, 146, 147
  parasitological, 147
  physiological, 145, 146, 336
  publication of, 308–328
  recovery from literature, 52, 239
  from serology, 146, 147
  stratigraphic, 147
  structural, 144, 146
  of taxonomy, 144, 338
Data sheets, 85, 87, 88
Date of names, see category involved
Date of publication of names, 399
Dealers in secondhand books, 327
Declarations, of I.C.Z.N., 386, 387
Definitions, 3
Deme, 190, 206
Derogatory labels, 341
Describing, genera, 279
  kinds, 4
  species, 279
  two aspects, 289, 314
Description, 279, 289
  graphic, 291
  of new species, 284
  series or sample, 284
  species vs specimens, 288
  types in, 292
Descriptions, in identification, 72
  length, 288
  publication of, 281, 314
Descriptive publications, 97
Descriptive taxonomy, 143, 207, 279–307

Designation of genotypes, 511, 512
Developmental data, 152, 153
Developmental stages, in identification, 67
Diacritic marks, listed, 263, 399
  in literature, 262
  in names, 225, 390, 441, 456, 479, 480, 540, 570, 584, 588
Diaeresis, in names, 456, 479, 480, 584, 588
Diagnosis, in description, 296
Diagnostic characters, 149
Differences, 137, 140, 156
  between groups, 126, 195
  in classification, 195
Dimorphism, sexual, 114
Diphthongs, in pronunciation, 232
  see also Ligatures
Directories, 74, 246
Directory of Zoological Taxonomists of the World, 43, 246
Distinguishing, 182
Distribution, diversity in, 130
  of publications, 326
Distributional data (geographic), 152, 153
Diversity, 131
  in behavior, 130
  biochemical, 127
  causes, 110
  in cells, 127
  constant/sporadic, 110
  continuous/discontinuous, 110
  in development, 130
  in distribution, 130
  extent of, 123
  forms of, listed, 107
  in group breadth, 125
  in group size, 124
  of groups (taxa), 106, 121–132
  of individuals, 105–120
  inherited/non-inherited, 110
  intrinsic/extrinsic, 110
  kinds of, 105
  of kinds, 121–132
  limitation to, 122
  in organs, 128
  in organ systems, 128
  in reproduction, 128
  skeletal, 128

Diversity, taxonomic, 4
  vs taxonomic features, 143
  within a species, 109
Dividing, a family, 465
  a genus, 493
  a higher taxon, 218
  a species, 558
  a subfamily, 469
  a subgenus, 500
  a subspecies, 575
  a tribe, 472
Divisions, names of, 448
Documentation in publication, 317

Ecological data, 152, 153
Ecology, 7, 20, 26, 159, 183, 337
Ecotype, 190
Editors, instructions to, 392
Electrophoretic data, 147
Embryology, 20, 183, 337
  data from, 152, 153
  in taxonomy, 18
Emendations, *see* category involved
  defined, 587
  of epithets, 586, 587
  generic, 409, 412, 482
  specific, 426, 430
Encyclopaedia Britannica, 11th ed.,
  1911–1912, 240, 241
Endocrinological data, 154
Entomology courses, 41
Epistemology, 350
Epithets, 577–588
  acronyms, 583
  adjectives, 579, 580
  anagrams, 582
  arbitrary combinations of letters,
    582, 583
  descriptive, 582
  ecological, 582
  emendation, 586, 587
  formation, 583
  gender, 580
  geographical, 583
  gerunds, 579, 581
  grammar, 579
  meaning, 582
  mythological names, 582
  non-classical, 583
  nouns, 579, 581

Epithets, original spellings, 583, 585,
  586
  orthography, 583
  participles, 579, 580
  patronyms, 582, 583
  sources, 579, 581
  specific, 230, 422, 579
  subsequent spellings, 586
  subspecific, 230, 579
Equipment, bibliographies on, 251
  in curating, 82
Errors, *see* Spelling, under each category
  copyists', 409, 414
  printers', 409, 414
  in synonymy, 304
Essays, as publications, 100
Ethics, in identification, 73
  in publication, 312
Ethological data, 146
Evolution, 20, 27, 122, 159, 186, 191,
  192, 194, 302, 331, 350
  in classification, 184, 191, 193, 361
  effect on classification, 7, 361
  part of systematics, 3, 15
Evolutionary species, 164, 170
Expeditions, bibliographies, 252

Families, changing rank, 466
  combining, 466, 470
  nominate subordinate taxa, 468
  rules for names, 220, 454
  taxa, 453
  types, 199, 454
Family category, 56, 57, 219, 221, 436,
  453
Family-group categories, 220
Family-group names, changing level,
  466
Family-group taxa, 220, 449
  names of, 220
Family names, 224, 454
  author, 459
  changes of, 467
  citation, 460
  conflicts, 458
  date, 460
  emendations, 458, 465
  endings, 221, 224, 454, 455, 456
  errors, 465

Family names, first reviser, 462, 463
 formation, 455
 history, 449
 homonymy, 458, 462, 463
 multiple original spellings, 458
 priority, 462, 463
 proposal, 455
 publication, 457
 rejection, 464
 replacement, 464
 sources, 458
 spelling, 456
 stability, 467
 subsequent spellings, 465
 synonymy, 223, 391, 458, 461
 type genus, 454
Family, type genus, 221
Fauna of British India, 13
Faunal studies, 98
Feces, as work of an animal, 538, 566
Field data, recording, 80
Figures, in identification, 72
First reviser, *see* category involved
Footprints, work of an animal, 535, 566
Formation of names, *see* category involved
Formenkreise, as taxa, 175
Forms, names of, 227, 576
 as taxa, 175
Fossils, casts, 535, 566
 in classification, 201
 data from, 201
 evidence of, 565
 fragmentary, 176
 impressions, 565, 566
 molds, 535, 565, 566
 in nomenclature, 535
 replacements, 535, 565, 566
 in taxonomy, 28, 41
 see also names in Index to Organisms
Fragments of animals, 176
Freedom of individual, 387

Galls, work of an animal, 535, 538, 565, 566
Gamma-taxonomy, 4, 14
Gender of names, 391, 479, 498
Genera, changing rank, 494

Genera, characters of, 203
 combining, 494
 dividing, 494
 names of, *see* Generic names
 types of, *see* Genotypes
 types of families, 199
 uninominal names, 224
Generic character, 203
Generic names, 414, 415, 419, 474, 577, 578, 579
 author, 482
 changes, 494
 citation, 486
 conflicts, 486
 date, 483
 emendations, 409, 480, 482, 492
 errors, 480, 481, 482
 first reviser, 489
 formation, 478
 gender, 479
 homonymy, 409, 411, 412, 486, 490
 junior homonyms, 409, 411, 412
 junior synonyms, 409, 411, 412
 manuscript, 409, 410
 misapplication, 486, 492
 multiple publication, 486, 491
 nature, 419
 original spellings, 478
 priority, 488, 489
 proposal, 474
 publication, 478
 rejection, 492
 replacement, 492
 rules, 474
 sources, 478
 subsequent spellings, 480
 synonymy, 197, 225, 416, 417, 486
 used alone, 322
 uses of, 225
Generitype, 196, 505
Generotype, 196, 505
Genetic data, 152
Genetic recombination, 110, 117
Genetics, 20, 26, 159, 183, 331, 337
 in taxonomy, 18
Genolectotype, 509
Genosyntypes, 509
Genotype, 397, 398, 399, 502, 505, 509
 autogenotype, 509
 automatic fixation, 515

Genotype, citation, 508, 512
in classification, 196, 198
conflicts, 507
designation, 397, 511, 512, 515
determination, 512
by elimination, 513, 528, 531
fixation, 397, 511, 512, 514, 522, 528
genolectotype, 509
genosyntypes, 509
indication, 397, 511
by isogenotypy, 514, 515
by Linnaean tautonymy, 513, 515, 526
misidentified, 529
by monotypy, 513
n.g., n.sp. rule, 512, 515
by objective synonymy, 513, 514, 515, 516
orthotype, 509
permanence, 506
selection, 511
by special systems, 513, 524, 530
subgeneric elimination, 521, 528, 532
by subgeneric monotypy, 521
by subsequent monotypy, 514, 521
by tautonymy, 513, 515, 524, 525
term proscribed, 399, 505, 509
typical species, 524, 531
typicus or *typus*, 513, 515, 524
by virtual monotypy, 513, 520
Genotypy, 198, 502–534
generic names, 225, 503
subgeneric names, 225, 503
Genus, category, 56, 57, 219, 436, 474
in classification, 54, 55
concept of, 226
description of, 290
nominal, 503
subordinate taxa, 495
taxa, 474
Genus-group categories, 474
coordinate, 419, 420, 422
Genus-group names, change of rank, 486
Genus-group taxa, 224, 408, 474
Geographical distribution, 152, 153, 337
Gerunds, as epithets, 537
Glossaries of terms, 62–65
Grade, category, 220

Grouping, 182, 351
see also Classification
Groups, in classification, 54, 56, 351, 353
definition, 351, 353, 354
difficult, 342
diversity of, 121
existence, 194
more-inclusive, 55, 56
natural occurrence, 138
objectivity, 359
origin of, 194
still-more-inclusive, 55, 56
Gynandromorphs, 110

Handbook of Biological Data, 127
Handbooks, as publications, 100
Hard parts, variety of, 128
Hectographing, in publication, 393
Hermaphroditism, 107, 116, 119
Herpetology courses, 41
Heteromorphosis, 116, 119
Hierarchy, of categories, 351
in classification, 53, 54, 55, 182
definition, 439
of groups, 53
of name categories, 395
Higher-category taxa, nomenclators, 240
Histological data, 153
History, of higher-category names, 436
of nomenclature, 380
of taxonomy, 6, 11
Holomorph, 4
Holotype, 589, 591, 592, 593, 596, 599
definition, 593
in descriptive taxonomy, 293
specified, 592, 599
unique, 592, 599
Homoeotypes, 295, 591
Homogeny, 143
Homology, 141, 142, 143
Homonyms, generic, 412, 490
in families, 462
in higher categories, 218
junior, 490
senior, 490
specific, 430
subgeneric, 415, 430

Homonymy, diagrams, 551
 gender terminations, 553
 Law of, 180, 496
 of names, *see* category involved
 one-letter rule, 552
 primary & secondary, 550, 553
 replacement, 554
 in same species, 392
 secondary, 390, 400, 553
 species vs subspecies, 553
 spelling differences, 553
 in synonymy, 305
Homoplasy, 143
Homotypes, 591
Honeycomb, work of an animal, 535, 566
Humility, in publication, 312, 314
Hybridization, 117
Hybrids, names based on, 409, 410
 names of specific, 426, 567
 names of subspecific, 576
Hyphens, in names, 390, 399, 441, 456, 479, 540, 570, 584, 588
Hypodigm, 4, 167, 168
Hypotypes, 591
 in descriptive taxonomy, 295

Ichthyology courses, 41
Identification, 60–75, 279
 vs classification, 194
 classification permits, 194
 by comparison, 68
 in curating, 76
 definition, 60
 from descriptions, 72
 difficulties, 60
 ethics in, 73
 keys, 69
 list of books, 70
 literature, 65
 from literature, 69
 methods of, 61, 65, 68
 necessary experience, 61
 from pictures, 72
 by specialists, 73
 special terms, 62
Illustrations, of species, 291
 preparation of, 319
Impressions, as work of an animal, 565, 566

Indication, of genotypes, 397, 511
 vs description, 475, 538
Individuals, in classification, 194
 diversity of, 105, 119, 120
 in taxonomy, 288
Information, recovery of, 20
Infraclass, category, 220, 436
Infraclasses, names of, 447
Infraorder, category, 57, 220, 436
Infraorders, names of, 448
Infrasubspecific categories, 575, 576
Institutions, bibliographies, 253
Integrity in publication, 312
International Catalogue of Scientific Literature, 15, 243
International Code, 1964, 380, 387, 389
International Code of Zoological Nomenclature, 1961, 7, 149, 216, 311, 315, 316, 380, 381, 385, 387, 389
International Commission on Zoological Nomenclature, 7, 215, 380, 381, 382, 385, 386, 388, 391, 505, 510
 duties of, 386
 powers of, 386
International Congress, *see* Congress
International Rules of Zoological Nomenclature (1905), 292, 380, 385
 see also Règles
International Trust for Zoological Nomenclature, 44, 387
Intersexes, 110
Isogenotypy, 514, 515
-ites, -ytes, -ithes, 463, 478, 491

Journal, notebook, 88
Journals, abbreviating, 322
Jurists, 382, 384

Key characters, 149
Keys, books about, 298
 in descriptive taxonomy, 297, 299
 faults, 70
 in identification, 69
 publication of, 315
 as publications, 99
 purposes, 70
 requirements, 297
 sample, 298
Killing, 79

Kinds, 163, 365
  in classification, 54, 56
  diversity of, 105
  in taxonomy, 3
  see also Species
Kingdom, category, 57, 219, 220, 435
Kingdoms, names of, 444

Language forms, in literature, 261
Languages, foreign, quoting, 323
  quoting foreign, 323
  in taxonomy, 262, 317
Language, use of, 347
Lapsus calami, family names, 456
  generic names, 480
  specific names, 430, 586
  in synonymy, 304, 409, 414
Latin alphabet, 262, 478
Latin language, in names, 214
Latin names, 214, 227
Latin phrases, in literature, 266
  listed, 267
Law of Homonymy, 180, 496
Law of Prescription, 383, 390
Law of Priority, 177, 383, 390, 546,
  548, 565
Law of Recency, 383, 390
Lectotypes, 591, 592, 593, 596, 599
  definition, 594
  in descriptive taxonomy, 293
Legions, names of, 448
Levels, in hierarchy, 217
  see also Categories
Life cycles, in identification, 67
Ligatures, in literature, 263
  in names, 456, 478, 585
Lineages, 171, 192
  in vertical classification, 188
Linnaean binominal system, 212, 214
Literature, cataloging of, 89, 91
  in identification, 65
  interpretation of, 256
  "primary zoological," 390, 391
  recovery of data, 239
  in taxonomy, 281
  use of, 239–278
Localities, bibliographies of, 253
Logic, Aristotelian, 10
  Major, 6
  symbolic, 151

M', in names, 540, 570, 585, 587
Mac, in names, 540, 570, 585, 587
Mc, in names, 540, 570, 585, 587
Majority views, 384, 401
Malaria and taxonomy, 31, 32
Mammalogy courses, 41
Manuscript names, generic, 409, 410
  specific, 426, 427
Measurements, 281
Metataxonomy, 4
Metatypes, 591
  in descriptive taxonomy, 295
Methodological taxonomy, 4
Methods, bibliographies of, 251
Methods of study, as publications, 99
Microform, publication in, 428, 538,
  539
Mimeographing, in publication, 393
Mimicry, 141
Misidentifications, 409, 411
Misidentified genotypes, 529
Misspellings, in epithets, 546, 552, 585,
  586
  generic, 409, 414
Molds, fossils, 535, 565, 566
Monographs, 97
  in descriptive taxonomy, 297
Monotypy, generic, 512, 513, 514, 517
  subgeneric, 514, 521
  subsequent, 521
  virtual, 514, 520
Morphae, as taxa, 175
Morphological data, in identification, 68
Morphology, definition, 146
Morphospecies, 171
Morphotype, 4
Multiple publication, 486, 491, 545
Museum labels, as names, 409, 410,
  426, 427
  as publication, 538, 539
Museums, 47
Mutations, 110, 116
  as taxa, 175
Mythical animals, 537
Mythological names, 582
Myths, of nomenclature, 210

Names, for age forms, 211
  agreement in gender, 229
  anonymous, 409, 410, 428, 543

Names, apostrophe, 225, 390, 441, 456, 479, 480, 540, 570, 584, 588
for assemblages, 211
author, 214, 229
available, 396
back-dating of, 399
barbarous, 451, 583
binominal, 211, 227, 407
of branches, 448
capitalization, 214
of classes, 216, 446
of cohorts, 448
of collective groups, 409, 412, 495
common, 210, 213, 426, 427
conditional, 409
correct, 396, 401
in descriptive taxonomy, 305, 306
diacritic marks, 225, 390, 441, 456, 479, 480, 540, 570, 584, 588
of divisions, 448
of families, 454
of family-group taxa, 407, 408, 449–473
of forms, 409, 426, 428
of genera, 407, 408, 409, 411, 412, 414, 419, 474, 577, 578, 579
genus-group coordinate, 420, 421, 422
of higher-category taxa, 435–448
of hybrids, 409, 410, 426, 427, 428, 491, 567
hyphenated, 224, 225
hyphens in, 390, 399, 441, 456, 479, 540, 570, 584, 588
hypothetical, 491
of hypothetical taxa, 409, 410
inappropriateness of, 492
of infraclasses, 447
of infraorders, 448
infrasubspecific, 428, 432
of kingdoms, 444
Latin, 214, 217
of legions, 448
of major groups, 407
misapplied, 409, 414, 426, 431
misspellings, 426, 431
museum labels, 409, 410, 426, 427
nature of, 407–434
nomina nuda, 398, 409, 410
non-taxonomic, 409, 410

Names, numerals in, 483, 584
objectionable, 394
occupied, 396
of orders, 216, 447
of parataxa, 409, 410, 428
of phalanx, 448
of phyla, 216, 445
of plants, 214, 458, 478, 490, 491, 492
preoccupied, 395, 396
pronunciation of, 231
publication of, 316
published, 396
published in synonymy, 428, 487
purpose of, 281
quadrinominal, 211, 407
quinquenominal, 211, 407
rejected, 396
scientific, 214, 217, 407–434
of sections, 409, 412, 501
for sexes, 211
signs as part of, 584
of species, 227, 407, 422, 426, 429, 535, 536, 577, 578, 579
of species-group taxa, 227, 424, 425, 426
specific, 422, 535, 536
of specific parataxa, 428
spelling of, 214, 229
stability of, 210, 403
of subclasses, 446
of subfamilies, 469
of subgenera, 226, 408, 409, 411, 412, 414, 415, 419
of subkingdoms, 444
of suborders, 448
of subphyla, 445
of subspecies, 227, 407, 423, 426, 429, 432, 434, 559, 568–576, 577, 578, 579
of superclasses, 445
of superfamilies, 223, 452
of superorders, 447
of superphyla, 444
of taxa, 407
for terata, 409, 410
of tribes, 471
trinominal, 211, 407
trivial, 400, 423, 450, 577, 578
uninominal, 211, 216, 407, 408

Names, use of, 210–238
valid, 396, 401
of varieties, 407, 409, 426, 428
vernacular, 211
what is named, 424
Naming kinds, 4, 182
National Research Council, 22
Natios, as taxa, 175
Natural history, 19
in identification, 66
Neo-Adansonian, derogatory label, 4
Neo-Latin alphabet, 262, 390
Neotypes, 390, 591, 594, 596, 597, 599, 602
definition, 594, 599
in descriptive taxonomy, 293
Neo-typology, 169
Nests, work of an animal, 535, 538, 566
New species, proposal of, 279
New Systematics, The (1940 book), 7, 9, 21, 334, 335, 336, 340, 365
derogatory label, 4
(idea), 8, 331–348
phylogeny, 302
New systematists, 341, 344, 349
Nomenclatorial, 212
Nomenclators, 212, 240
for families, 241
for genera, 242
for higher taxa, 240
listed, 243, 244
as publications, 100
for species, 244
Nomenclator Zoologicus (Neave), 241
Nomenclatural, 212
publications, 98, 99
Nomenclature, binominal, 211, 212, 422
history, 7, 215
history of rules, 380
literature on, 255
in The New Systematics, 343, 344
rules of, 379
type-founded, 346
zoological, 210, 212, 373–603
Nomen dubium, in 1961 Code, 396, 398, 425
Nomen novum, in 1961 Code, 398
Nomen nudum, 409, 410
in 1961 Code, 396, 398, 426

Nomen oblitum, in 1961 Code, 398, 426, 464, 489, 492
Nominal genus, 503, 510
Nominal species, 503, 517, 590
Nominal taxa, 394, 405, 425, 503, 510
Nominate subgenus, type of, 508
Nominate subordinate taxa, family-group, 391, 398, 470, 472
Nominate subspecies, 426, 568, 572
Nominotypical subspecies, 568
Nominotypical taxa, 391, 398, 426
Non-dimensional, derogatory label, 4
Non-taxonomists, instructions to, 392
Notebooks, field, 84
journal, 88
Notes, field, 84, 86, 87
research, 88, 89
Nouns, as epithets, 537
Numerals, in names, 391, 483, 540, 570, 584
Numerical, characters, 151
methods, 158
taxonomy, 4, 158, 160

Objective synonymy, generic names, 512
specific names, 548
Objectivity, of categories, 358
Observation, of features, 281, 282
new methods, 283
Official Indexes, 386, 390
Official Lists, 386, 389, 598
Old Systematics, derogatory label, 340, 342
Omnispective classification, 184, 190, 191
Onomatophore, 4
Opinions, of I.C.Z.N., 381, 382, 386, 387, 388
Order, category, 56, 57, 219, 220, 436
Order/Class group of taxa, 436
Order-group, categories, 220
taxa, 220
Orders, names of, 216, 447
Organizations of taxonomy, 43
Ornithology courses, 41
Orthotype, 509
Osteological data, 153
Outbreeding animals, 369

Paleontology, courses, 41
  vertebrate, 153
Paleospecies, 171
Paralectotypes, 591, 592, 599, 600
  definition, 600
  in descriptive taxonomy, 293
Parallelism, 139, 141
Parasitology, 152
  courses, 41
Paraspecies, 171
Parataxa, 171, 176, 390, 496
  definition, 176
  names of, 227, 409, 410
Parataxonomy, 180
Paratypes, 589, 591, 599, 600
  definition, 600
  in descriptive taxonomy, 293, 295
Parentheses, use of, 229, 418, 545
Participles, as epithets, 537
Patronyms, 582, 583
Periodicals, bibliographies of, 253
Phalanx, names of, 448
Phases, as taxa, 175
Phenetic classification, 187, 190
Phosphoarginine, 155
Phosphocreatine, 155
Photographs, as description, 292
  publication as, 428, 538, 539
Phyla, diversity among, 123
  names of, 216, 445
  number of, 124
  number of classes, 126
Phylogenetic trees, as publications, 301
Phylogeny, 8, 331, 350
  basis of classification, 187, 191, 192, 301, 339
  in classification, 184, 185, 192, 360, 363
  postulated, 363
  reflection of, 361
Phylum, category, 56, 57, 219, 220, 435
Phylum-group, categories, 220
  taxa, 220
Physiological data, 152, 153, 154
Physiology, 183, 337
  taxonomic, 23
Plant kingdom, names from, 458
Plastotypes, 591, 601
  in descriptive taxonomy, 295

Plenary Powers, of I.C.Z.N., 7, 385, 389, 428, 514, 515, 548
Plesiotypes, 591
Ploidy, 110, 117
Polymorphism, 108, 110, 112
  alternating generations, 111
  body forms, 111
  castes, 111, 113
  colonial, 113
  developmental, 110, 116
  functional, 113
  genetic, 111
  sexual, 110, 114
Polyphasy, 113
Populations, dynamics, 159, 331
  in classification, 79, 204, 206
  interbreeding of, 170
  in The New Systematics, 339
  sampling, 79
  species as, 159, 160, 167, 168
  subspecies as, 174
  in taxonomy, 159, 160, 284
  vs taxonomy, 339
Preoccupied names, 395
Preparation, of illustrations, 319
  of papers for publication, 318
  of specimens, 283
Preserving, in curating, 76
  in field, 80
Principle of Conservation, 383, 390
Priority, among names, *see* category involved
  Law of, 177, 383, 390, 546, 548, 565
  limitation on, 399
Pronunciation, accent, 233
  diphthongs, 232
  general rules, 237
  letter sounds, 233
  of names, 231
  syllabification, 232
  vowel length, 232
Proof sheets, publication as, 428, 538, 539
Proposal of names, *see* category involved
Protozoology courses, 41
Pseudotaxa, 174
Psychological data, 152
Publication, comparative studies, 97
  criteria of, 424

Publication, of data, 308–328
　of descriptions, 281
　descriptive, 97, 281
　ethics in, 312
　general problems, 308
　by hectographing, 393
　historical documentation in, 317
　library deposit, 428
　by mimeographing, 393
　multiple, 258, 259
　of names, *see* category involved
　of names simultaneously, *see* category involved
　nomenclatural aspects, 316
　simultaneous, 487, 489, 546
　of taxonomic data, 84, 96, 306, 310
　technical problems, 314
　what to publish, 306, 310
　where to publish, 306, 310
Publications, atlases, 98
　author vs authors, 311
　bibliographies, 100
　book reviews, 100
　catalogs, 99
　checklists, 99
　classifications, 98
　descriptive, 296
　distribution of, 326
　errors in, 317, 320
　essays, 100
　faunal studies, 98
　handbooks, 100
　keys, 297
　large vs small, 310
　monographs, 97, 297
　nomenclators, 100
　nomenclature studies, 98, 99
　preparation of, 318
　revisions, 297
　spelling of names in, 317
　types of, 96
Publishing, in curating, 76

Quadrinominal names, 211
Quinquenominal names, 211
Quotation, foreign language, 323

Races, names of, 227
　as taxa, 175, 432
Rassenkreise, as taxa, 175

Reality, of categories, 358
　of genera, 359
　of groups, 360
　of species, 360
Recommendations in Code, 393, 401
Recording, of data, 84–101
　by description, 281
　field data, 80
　kinds, 4
Record of Zoological Literature, 15
Recovery, of data, 52
　of information, 52, 53, 58
Reference books, history, 15
Reference publications, 99
Règles Internationales de la Nomenclature Zoologique, 7, 215, 380, 381, 437, 438
Rejection of names, *see* category involved
Replacement of names, *see* category involved
Replacements, fossils, 535, 565, 566
Reprints, 326
Resemblance, 137, 139, 140
　causes of, 141
　chance, 140
　convergence, 139
　environmental, 140
　hereditary, 139
　parallelism, 139
Retroactive rulings in Code, 399
Revisionary studies, 282
Revisions, in descriptive taxonomy, 297
Rockefeller Foundation, 31, 32
Royal Society Catalog of Scientific Papers, 16
Rules, complexity of, 388
Rules, *see* Règles or Code
Rules of nomenclature, 215

Samples, bias in, 202
　in classification, 202
　statistical, 78
　in taxonomy, 284
Sampling, 78, 167
Science, definition, 331
　publication in, 307
Second Law of Thermodynamics, 122
Section on Nomenclature, 382, 383

Sections, names of, 501
  vs subgenera, 501
Segregating kinds, 4, 279
Selection, of characters, 202, 286
  of genotypes, 511
Semantics, 350
Separata, 326
Separates, 326
Series, as a category, 220, 448
  or sample, 284
Serological data, 147
Serology, taxonomic, 23
Shelters, work of an animal, 535
Shipment, of specimens, 81
Similarities, in classification, 195
  universal, 122
Similarity, 139, 156
Societies in taxonomy, 43–48
Society of Systematic Zoology, 22, 43,
  74, 346
Sources of names, see category involved
Specialists, in identification, 73
  lists of, 246
Speciation, 7, 20, 331, 350
Species, 162–181, 364–372
  agamospecies, 170
  allopatric, 165
  biological, 170
  biological concept, 369, 371
  biological definition, 369
  cataloging of, 89, 90
  category, 219, 366, 535
  changing rank, 558
  in classification, 54, 204, 206, 351,
    353
  combining, 558
  continental, 165
  controversy, 365
  cosmopolitan, 165
  cryptic, 165
  defined nomenclaturally, 425
  definitions, 163, 164, 354
  description of, 289
  diverse usage, 366
  dividing, 558
  ecological, 165
  evolutionary, 164, 170
  evolutionary importance, 371
  fictitious, 537
  form-species, 171

Species, genetic, 170
  group of individuals, 366
  incipient, and subspecies, 171
  inquirenda, 425
  insular, 165
  monotypic, 170
  montane, 165
  morpho-geographical, 165, 170
  morphological, 342, 343
  morphospecies, 171
  mythical, 537
  names of, 227, 422, 535–567
  naturalness, 358
  nature of, 364–372
  nominal, 503
  non-dimensional, 171
  non-taxonomic, 169
  not representative, 155
  objectivity, 358, 367
  paleontological, 171
  panmictic, 171
  pantropical, 165
  phenotypic, 347
  philopatric, 171
  physiological, 165
  polymorphic, 170
  polytypic, 170
  reality of, 366, 367
  sibling, 165, 170
  "The Species Problem," 366, 371
  subdivisions of, 175
  subjectivity, 366
  subordinate taxa, 559
  successional, 171
  sympatric, 165
  taxa, 535
  taxonomic, 162, 164
  taxonomic importance, 372
  transient, 171
  tropicopolitan, 165
  types, 165, 589
  typification, 165
  typological, 369
  typological concept, 369
  units of classification, 204
Species-group, categories, 423
  change of rank, 546
  definition, 369
  names, 578
  taxa, 227, 423

Specific epithet, 422, 536, 540, 579
Specific names, 535–567, 577, 578, 579
  author, 539
  causes of changes, 560
  citation, 229, 231, 541
  conflicts, 545
  date, 541
  errors, 426, 431
  first reviser, 547, 549
  formation, 539
  homonymy, 426, 546, 548, 549
  manuscript, 426, 427
  meaning of, 227, 229, 394, 400, 422
  misapplied, 557
  misidentifications, 546
  objective synonymy, 548
  priority, 546, 548
  proposal, 536
  publication, 536
  rejection, 554
  replacement, 554, 557
  sources, 539
  spelling, 546
  subspecific names co-ordinate, 559
  synonymy, 426, 545, 546, 547
  tautonymy, 546
  unacceptable, 427
Specific trivial name, 577, 578
Specimens, number available, 343, 349
Spelling, of authors' names, 324
  of names, *see* category involved
Stability of names, *see* category involved
Statement of concept, 352
Statistical analysis, 281
Statistical characters, 287, 288
Statistical comparisons, 159
Stem, of Latin words, 450
Storage, in curating, 76, 82
  of data, 51
Stratigraphy, 152
  data of, 147, 202
Strickland Code, 6
Style manuals, 318
Subclass, category, 57, 220, 436
Subclasses, names of, 446
Subdivision, method of classification.
  207
Subfamilies, changing level, 470
  dividing, 469
  endings of names, 224

Subfamilies, nominate subordinate taxa,
  470
  taxa, 469
  type genera, 221, 469
Subfamily, category, 57, 221, 436, 468
Subfamily names, 224, 469
  author, 469
  changes, 470
  citation, 469
  conflicts, 469
  dates, 469
  first reviser, 469
  formation, 469
  homonymy, 469
  priority, 469
  proposal, 469
  publication, 469
  rejection, 469
  replacement, 469
  spellings, 469
  stability, 470
  synonyms, 469
Subgenera, changing the category, 500
  combining, 500
  dividing, 500
  genotypes, 225, 497
  names of, 226, 497
  subordinate taxa, 501
  types of, 497
  uninominal names, 224
Subgeneric monotypy, 514, 521
Subgeneric names, 414, 415, 419
  author, 499
  changes, 501
  citation, 227, 231, 499
  conflicts, 499
  dates, 499
  emendation, 499, 500
  employment, 415
  first reviser, 500
  formation, 224, 499
  gender, 499
  homonymy, 225, 415, 499, 500, 553
  misapplication, 500
  nature of, 419
  priority, 499
  proposal, 498
  publication, 499
  rejection, 500
  replacement, 500

Subgeneric names, sources, 499
  spelling, 224, 499
  synonymy, 225, 415, 416, 499, 500
Subgenus, category, 57, 436, 497
  nominate, 415
  taxa, 497
  "typical," 415
Subkingdom, category, 57, 220, 435
Subkingdoms, names of, 444
Suborder, category, 57, 220, 436
Suborders, names of, 448
Subordinate taxa, see category involved
Subphyla, names of, 445
Subphylum, category, 57, 220, 435
Subsequent spellings of names, see
  category involved
Subspecies, 162–181
  category, 568
  changing rank, 575
  combining, 575
  definitions, 172
  dividing, 575
  in genotypy, 519
  names of, 227, 229, 407, 423, 426,
    429, 432, 434, 559, 568–576, 577,
    578, 579
  in The New Systematics, 342, 343
  nominate, 568, 572
  nominotypical, 568
  as pseudotaxa, 174
  single, 323
  subordinate taxa, 575
  taxa, 568
  taxonomic nature, 172
  types, 165, 589
  typical, 568
  typification, 165, 568
Subspecific epithet, 579
Subspecific names, 227, 229, 407, 423,
    426, 429, 432, 434, 559, 568–576,
    577, 578, 579
  author, 569
  as binomina, 571
  causes of changes, 575
  citation, 571
  conflicts, 571
  date, 570
  first reviser, 574
  formation, 569
  homonymy, 574

Subspecific names, misapplication, 574
  priority, 573
  proposal, 569
  publication, 569
  rejection, 574
  replacement, 574
  sources, 569
  spelling, 569
  synonymy, 571
Subspecific trivial name, 577, 578
Subspecific types, 602
Subtribe, category, 57, 221, 436, 473
Subtribes, names, 224, 473
  taxa, 473
Superclass, category, 57, 220, 435
Superclasses, names of, 445
Superfamilies, changing rank, 453
  combining, 453
  dividing, 452
  nominate subordinate taxa, 453
  taxa, 451
  types, 451
Superfamily, type genus, 221
Superfamily category, 57, 221, 436, 451
Superfamily names, 452
  author, 452
  changes, 453
  citation, 452
  conflicts, 452
  date, 452
  endings, 223, 393
  first reviser, 452
  formation, 452
  homonymy, 452
  priority, 452
  proposal, 452
  publication, 452
  rejection, 452
  replacement, 452
  spelling, 452
  stability, 453
  synonymy, 452
  uniform endings, 452
Superorder, category, 57, 220, 436
Superorders, names of, 447
Superphyla, names of, 444
Superphylum, category, 220, 435
Supertribe, category, 221, 436, 470
Supertribes, endings, 224
  names, 224, 470

Supertribes, taxa, 470
Suspension of the Rules, 385, 514, 515, 597
Syllabification, in pronunciation, 232
Synonymies, as publications, 302
  publication of, 316
Synonyms, absolute, 197, 412
  citation of, 231
  conditional, 197
  generic, 197, 412
  in genotypy, 519
  in higher-category taxa, 218, 440
  isogenotypic, 412, 417
  junior, 197, 412, 488, 492
  nomenclatural, 197, 412
  objective, 197, 304, 412, 417
  partial, 420
  in publication, 302
  rejected, 269, 302, 488
  senior, 197, 412, 488
  specific, 426, 429
  stillborn, 197, 304, 512
  subjective, 197, 304, 305, 412, 417
  zoological, 197, 412
Synonymy, absolute, 488
  conditional, 488
  examples of, 95, 231, 269, 302–305
  interpretation of, 268
  nomenclatural, 488
  objective, 488, 499, 548
  purpose, 305
  in revisionary work, 302
  sample, 95, 197, 198, 231, 269, 302–305
  subjective, 488, 499, 548
  subjective-objective, 488
  what is included, 302
  zoological, 488
  see also category involved
Syntypes, 293, 591, 592, 596, 599, 600
  definition, 600
Systematics, applied, 24–37
  "biological," 340
  in this book, 4
  definition, 3, 4
  in everyday life, 24
  "modern," 340
  "multidimensional," 340
  new vs old, 340
  non-dimensional, 340

Systematics, "objective," 340
  typological, 340
Systematics Association, 43
Systematic Zoology (journal), 43
Systematist, tasks of, 183

Tabulation of data, 281
Tautonymy, genotype fixation, 515
  Linnaean, 525, 526
Taxa, in classification, 55, 351, 355
  coordinate, 425
  definition, 439
  in higher categories, 435, 437, 439
  nature of, 173, 174
Taxon (journal), 48
Taxonomic biochemistry, 23
Taxonomic characters, 143, 144, 148, 149, 285
  in identification, 61
Taxonomic classifications, 190
Taxonomic courses, 40, 41
Taxonomic data, 138, 200
Taxonomic judgment, 163
Taxonomic physiology, 23
Taxonomic serology, 23
Taxonomic sets, 359
Taxonomic species, 162, 164, 170
Taxonomic thinking, 8
Taxonomists, 38, 279
  activities of, 4
  amateurs, 38, 40
  professional, 38
  status, 39
  training of, 40
  what they do, 16, 18
Taxonomy, 1
  academic, 40, 42
  applied, 24–37
  in biological control, 28
  and the Black Death, 32
  in this book, 4, 5
  the broader, 333, 334, 341
  in chronology, 19
  classical, 339
  data of, 144, 338
  definition, 3, 4, 279, 332
  descriptive, 207, 279–307
  and ecology, 26
  and engineering, 26
  in epidemiology, 31

Taxonomy, in everyday life, 24
  in evolution studies, 21
  in fisheries biology, 35
  in genetics studies, 26
  goals of, 5, 6, 16
  historical aspects, 239
  history of, 6, 11
  identification, 60
  importance of, 17, 22
  in insect quarantine, 30
  metataxonomy, 4
  methodological, 4
  numerical, 4, 158, 160
  old vs new, 341
  organizations of, 43
  orthodox, 336
  paleontological, 22, 23
  phylogeny in, 360
  vs phylogeny, 331
  practical, 49–101
  problems of, 350
  professional, 42
  purpose of, 350
  as a science, 158, 331
  in species sanitation, 32
  theoretical, 329–372
  of today, 348
  typological, 345, 346
  use of, 19, 20, 22
  in wartime, 33
Teachers in taxonomy, 42
Terata, names based on, 409
  types of, 119
Teratology, 119
Terms, in identification, 62
  technical, books on, 324
  technical, over-use of, 324
  use in 1961 Code, 394
Tilde, in names, 456
Titles, common errors, 320, 325
Topotypes, 591
  in descriptive taxonomy, 295
Traces of animals, 566
Tracks, work of an animal, 535, 538,
  565, 566
Trails, work of an animal, 535, 566
Transition forms, names of, 227
  as taxa, 175, 432
Translation, of literature, 266
Transliteration, in literature, 266

Treatise on Invertebrate Paleontology,
  241
Tribe, category, 57, 221, 436, 471
Tribe names, 224, 471
  author, 472
  changes, 472
  citation, 472
  conflicts, 472
  date, 472
  endings, 393, 472
  first reviser, 472
  formation, 471
  homonymy, 472
  priority, 472
  proposal, 471
  publication, 472
  rejection, 472
  replacement, 472
  spellings, 471, 472
  stability, 472
  synonymy, 472
Tribes, changing rank, 472
  combining, 472
  dividing, 472
  as higher-category taxa, 448
  nominate subordinate taxa, 472
  taxa, 471
  type genera, 221, 471
Trinomen, 569
Trinominal names, 211
Trivial names, 400, 423, 540, 577, 578
Tubes, work of an animal, 535, 538,
  565
Type genus, of family, 221, 454
Type locality, 295, 601
  definition, 601
  vs published locality, 601
Type series, 392, 592, 595
Type species, of genus, 196, 502, 505
Types, archetype, 4
  in classification, 196
  conflicts, 595
  in descriptions, 292
  of families, 199, 454
  function of, 166, 589
  of genera, 397, 398, 502, 505
  of higher-category taxa, 436
  morphotype, 4
  in The New Systematics, 345

Types, in objective synonymy, 597
ownership, 602
permanence, 599
preservation, 602
primary, 591, 592
replacement, 598
secondary, 591, 600
selection, 293, 294, 592
of species, 165, 589
of subfamilies, 469
of subspecies, 165, 568, 589
in taxonomy, 345, 346
tertiary, 591, 600
of tribes, 471
Type-species, 196
Typical, in genotypy, 524, 531
subspecies, 426, 568
taxa, 391, 398
see also Genotype
Typicus or typus, in genotypy, 515, 524
Typological, derogatory label, 4, 345
Typological taxonomy, 345, 346
Typology, 168, 169, 345, 589

Umlaut, in names, 441, 456, 479, 480,
540, 570, 588
non-German, 463, 479, 480
Uniform endings, family-group names,
221
"higher taxa" names, 438
subfamily names, 469
superfamily names, 393
tribe names, 393
Union List of Serials, 322
U.S. Department of Agriculture, 14, 30,
32
U.S. National Museum, 27, 28
Units of classification, 204, 205, 344
Universities, in taxonomy, 47
taxonomy courses, 40

Validity, in 1961 Code, 394, 396, 401,
511
Variability, of characters, 156
Variants, as taxa, 175
Variation, accidental, 118
autotomy, 118

Variation, in colonies, 110, 113
cyclomorphosis, 110
developmental, 110, 116
environmental, 110, 117
genetic recombination, 110, 116
individual, 109, 119, 144
mutations, 110, 116
ploidy, 110
sexual, 110, 116
teratology, 119
willful, 118
Varieties, in genotypy, 519
names of, 227, 576
in synonymy, 305
Vernacular names, 211
Vowel length, in pronunciation, 232

Waiting periods, 392
Webs, as work of an animal, 535, 566
Weighting, 149, 150, 187
correlation, 150
a posteriori, 150
a priori, 150
Weighting of characters, definition, 149
rejection, 150
residual, 150
Who Knows and What, 246
Who's Who, 246
Word-books, 319
Words, use of, 347, 348
Work of an animal, 535, 538, 565, 566,
576
definition, 565
Writing manuals, 319

Zeitschrift für wissenschaftliche Zool-
ogie, 243
Zoogeography, 152, 153, 183
data from, 152, 153
Zoological nomenclature, 212, 373–603
Zoological Record, 12, 15, 16, 80, 99,
243, 244, 250, 254, 311
Zoological Record Association, 15
Zoological Society of London, 15
Zoologischer Anzeiger, 250